17th International Conference on Stability, Handling and Use of Liquid Fuels (IASH 2022)

Stability, Handling and Use of Liquid Fuels

Dresden, Germany
11-15 September 2022

Editor:

Robert E. Morris

ISBN: 978-1-7138-6334-2

Printed from e-media with permission by:

Curran Associates, Inc.
57 Morehouse Lane
Red Hook, NY 12571

Some format issues inherent in the e-media version may also appear in this print version.

Printed with permission by Curran Associates, Inc. (2023)

International Association for Stability, Handling and Use of Liquid Fuels (IASH)
3525 Piedmont Road
5 Piedmont Center
Suite 300
Atlanta, GA 30305
USA

Phone: +1 404.760.2829

www.iash.net

Additional copies of this publication are available from:

Curran Associates, Inc.
57 Morehouse Lane
Red Hook, NY 12571 USA
Phone: 845-758-0400
Fax: 845-758-2633
Email: curran@proceedings.com
Web: www.proceedings.com

TABLE OF CONTENTS

SESSION 3: IMPROVED FUEL QUALITY CONTROL

Mickael Clarinard, Session Chair

SESSION 4: GROUND FUELS - GASOLINE, DIESEL, BIODIESEL AND GAS OIL

Gareth Williams, Session Chair

SESSION 5: FUEL TESTING AND TEST METHODS
Marcella Frauscher and Mark Romanczyk, Session Co-Chairs

SESSION 7: IMPACT OF IMO 2020 ON MARINE FUELS

Dave Evans, Session Chair

SESSION 8: FUEL CHEMISTRY RESEARCH
PART I: FUEL PROPERTY PREDICTION AND MEASUREMENT

Krege Christison, Session Chair

SESSION 9: FUEL CHEMISTRY RESEARCH
PART II: THERMAL-OXIDATIVE STABILITY R&D

Steve Zabarnick and Simon Blakey, Session Co-Chairs

SESSION 10: AVIATION FUEL AND AVGAS APPLICATIONS

JP Belieres, Session Chair

SESSION 11: POSTER SESSION

Mickaël Sicard, Session Chair

IASH ®, The International Association for

Stability, Handling and Use of Liquid Fuels

www.iash.net

Proceedings of the 17th International Conference on Stability Handling and Use of Liquid Fuels

Edited by

Dr. Robert E. Morris
IASH Secretary
Nova Research, Inc.
1900 Elkin Street, Suite 230
Alexandria, VA 22308

Dresden, Germany
September 11-15, 2022

PREFACE

Theme: **Liquid Fuels: A Changing World**

Location: **Dresden, Germany**
Dates: **September 11-15, 2022**

The Hilton Dresden in Dresden, Germany was the venue for the 17th International Conference organized by the International Association for the Stability, Handling and Use of Liquid Fuels. This Conference was planned to be held in 2021, but was rescheduled to 2022 due to Covid-19 related health, travel and economic concerns. Despite continuing global health and economic issues, the Conference was well-attended with a technical level commensurate with previous Conferences.

There were 125 delegates and 17 spouses/guests in attendance from 19 countries. The poster session was comprised of 18 technical presentations

The Keynote address was presented by Dr.-Ing. Manfred Aigner, DLR Institute for Combustion. The title of his presentation was *"DLR Research for Sustainable Aviation – Sustainable Fuels & Clean Combustion"*. The presentation discussed current challenges the aviation industry is facing related to emissions/pollutants, and increasing production capacity for Sustainable Aviation Fuels, followed by overviews of recent efforts and research initiatives for effective near- and long-term implementation and transition.

There were six (6) paper submissions for the Chevron Research Award of Excellence in Honor of John Bacha from the following individuals:

- Charlie Adams, University of Sheffield, UK
- Soraya Aminane, ONERA, France
- Ryma Benrabah, University of Lorraine, France
- Clemons Hall, German Aerospace Center, DLR, Germany
- Mark Romancyzk, Naval Research Laboratory, USA
- Judith Ugbeh-Johnson, University of Cranfield UK

All of the submissions were of high technical quality and merit. The award was made to Clemens Hall from the German Aerospace Center, DLR. His presentation and paper were entitled: "Comparison of Probabilistic Jet Fuel Property Models for the Fuel Screening and Design". Four of the other paper submissions were accepted for presentation in the General Sessions.

A total of forty-eight (48) technical presentations were given in ten sessions that included the following topics:

- Session 1: Sustainable Aviation Fuels Implementation
- Session 2: Processing Technology Development for Sustainable Aviation Fuels
- Session 3: Improved Fuel Quality Control

- Session 4: Ground Fuels - Gasoline, Diesel, Biodiesel and Gas Oil
- Session 5: Fuel testing and Test Methods
- Session 6: Microbial Contamination/Monitoring
- Session 7: Impact of IMO 2020 on Marine Fuels
- Session 8: Fuel Chemistry Research.
  Part I: Fuel Property Prediction and Measurement
- Session 9: Fuel Chemistry Research.
  Part II: Thermal-Oxidative Stability R&D
- Session 10: Aviation Fuel and AvGas Applications

Conference Proceedings are available to members on-line through the IASH Library

During the technical program, the first-ever Panel Discussion was convened at an IASH Conference. The topic of the discussion was *"Relevant Issues Related to Aviation Fuel Quality Assurance Throughout the Supply Chain and End-Use"*. The panel was moderated by Alisdair Clark (BP International Limited) and comprised of Joanna Bauldreay (Bauldreay Jet Fuel Consulting Limited), Moshe Rabaev (Israeli Air Force) and Patrick Bosmans (CEPS – Central Europe Pipeline System). The moderator and panelists discussed various issues related to fuel quality assurance and surveillance throughout the supply chain, potential challenges related to the introduction and use of alternatively-derived (i.e., non-petroleum) fuels in these systems, and impact on producers, distributers and users. There was extensive audience participation and discussion during the Panel Discussion, bringing forth various perspectives from across the supply chain, use and testing spectrum.

The IASH Awards Dinner was held on Wednesday evening in the courtyard of the historic Residenz Castle. There were four awards given at this Conference. The Nahum Por Award was presented to Matthew Fielder, for his leadership, support, and promotion of the Association over the past 10+ years. IASH Lifetime Achievement Awards were given to the following individuals for their multifaceted and dedicated efforts in Research, Development and Use of hydrocarbon-based fuels and related technologies throughout their careers:

- David Abdallah, ExxonMobil Research and Engineering
- Frederick Passman, Biodeterioration Control Associates, Inc.
- Simon Blakey, University of Birmingham

The Conference Chairman and the IASH Board of Directors are greatly appreciative of the presenters for their time and effort in producing excellent technical presentations and the delegates for supporting the conference by their attendance and active participation.

Special thanks are given to the 9 sponsoring organizations and exhibitors without whose financial support we would not have been able to convene the 17th International Conference:

Diamond Sponsor:

- ECHA Microbiology

Premium Sponsors:

- Borsig
- The Energy Institute
- University of Dayton Research Institute
- University of Sheffield – Transitional Energy Research Centre

Classic Sponsors:

- LuminUltra
- Southwest Research Institute

Sponsors:

- Aviation Fuels Projects Consulting
- Baere Aerospace Consulting, Inc.

Media Partner:

- F+L Asia

Respectfully submitted,

Matthew J. DeWitt
Conference Chair / 1st Vice-Chair IASH

PREVIOUS CONFERENCES

1st Conference: Tel Aviv, Israel, July 11-14, 1983.

2nd Conference: San Antonio, Texas, USA, July 29-August 1, 1986.

3rd Conference: London, England, September 13-16, 1988.

4th Conference: Orlando, Florida, USA, November 19-22, 1991.

5th Conference: Rotterdam, the Netherlands, October 3-7, 1994.

6th Conference: Vancouver, BC, Canada, October 13-17, 1997.

7th Conference: Graz, Austria, September 24-29, 2000.

8th Conference: Steamboat Springs, Colorado, USA, September 14-19, 2003.

9th Conference: Sitges, Spain, September 18-22, 2005.

10th Conference: Tucson, Arizona, USA, October 5 – 11, 2007.

11th Conference: Prague, Czech Republic, October 18 – 22, 2009.

12th Conference: Sarasota, Florida, USA. October 16 – 20, 2011.

13th Conference: Rhodes, Greece, October 6 – 10, 2013.

14th Symposium: Charleston, South Carolina, USA, October 4 – 8, 2013.

15th Conference: Rome, Italy, September 10-14, 2017.

16th Conference: Long Beach, California, USA, September 8-12, 2019.

17th Conference: Dresden, Germany, September 11 – 15, 2022.

CONFERENCE SPONSORS

Aviation Fuel Projects Consulting

Baere Aerospace

BORSIG Membrane Technology

ECHA Microbiology

Energy Institute

F&L Asia

LuminUltra

Southwest Research Institute

University of Dayton Research Institute

University of Sheffield - Translational Energy Research Centre

CONFERENCE EXHIBITORS

BORSIG Membrane Technology

ECHA Microbiology

Energy Institute

LuminUltra

Southwest Research Institute

University of Dayton Research Institute

University of Sheffield - Translational Energy Research Centre

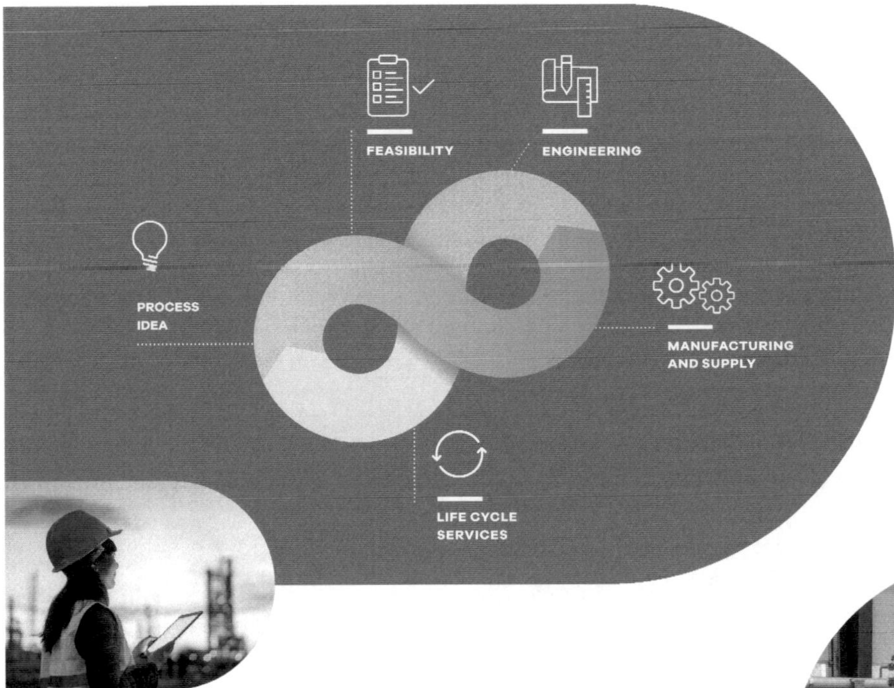

FEASIBILITY
ENGINEERING
PROCESS IDEA
MANUFACTURING AND SUPPLY
LIFE CYCLE SERVICES

BORSIG

The University of Dayton Research Institute (UDRI) is the research arm of the University of Dayton, located in Dayton, Ohio USA. UDRI is an international leader in scientific and engineering research, serving government, industry, and not-for-profit customers. Our full-time professional staff of 700 engineers and scientists conducts research and provides support in a wide variety of technical areas.

UDRI
UNIVERSITY *of* DAYTON
RESEARCH
INSTITUTE

The Research Institute's efforts in the area of advanced fuels include studies in:

- Development and evaluation of alternative fuels and fuel technologies
- Improved high temperature thermal stability fuels and additives
- Improved low temperature flowable fuels and additives
- Development of advanced fuels and additives for reduced combustion particulate and gaseous emissions
- Detection, identification, and mitigation of fuel biocontamination
- Advanced chemical analyses of fuels and additives
- Chemical kinetics model development and fluid dynamics modeling of fuels and combustion
- Development and testing of advanced combustion concepts

UDRI is a pioneer in **testing and evaluation of alternative jet fuels**. We coordinate the FAA D4054 Clearinghouse for ASTM certification and qualification of alternative jet fuels. Our researchers conducted many of the detailed laboratory evaluations and supported field trials of Fischer-Tropsch, HEFA, HC-HEFA, and ATJ fuels and fuel blends which resulted in ASTM and military certification.

UDRI has significant experience in **identification and mitigation of biocontaminants** for ground and aircraft fuel systems. We pioneered DNA sequencing methods for identification and quantification of microbial contamination. We have studied the role of fuel composition (e.g., alternative fuel species) on the growth of fuel microbes. We continue to study the development of non-toxic advanced fuel antimicrobial molecules and the molecular mechanisms of fuel species metabolism.

UDRI is a world leader in the **study of jet fuel thermal stability** in both oxidative and pyrolytic environments. Our researchers have many years expertise in thermal stability additive testing, evaluation, and formulation. We have pioneered the development and interpretation of thermal stability tests such as:

- Quartz crystal microbalance (QCM)
- Extended duration thermal stability test (EDTST)
- Isothermal corrosion oxidation test (ICOT)
- Near isothermal flowing test rig (NIFTR)
- Phoenix & ECAT thermal stability rigs
- AFRL fuel system simulator (FSS)

UDRI is a leader in the development of **unique analysis techniques** for the identification and quantification of important fuel species. We specialize in difficult analyses which require custom preseparation methods, as well as quantification of fuel species classes. Preseparation techniques pioneered by UDRI researchers for use for fuel include solid-phase extraction and high performance liquid chromatography. These techniques are then combined with gas chromatography with various detectors or with multi-dimensional gas chromatography (GCxGC). Our researchers also employ unique derivatization techniques which greatly improve

von Ohain
Fuels & Combustion
Center

difficult analyses, such as silylation for analysis of metal deactivator additive and reaction with triphenylphosphine for analysis of hydroperoxides.

UDRI is the primary U.S. Air Force technical contractor responsible for the development, testing, and evaluation of the advanced **high temperature thermal stability additive package** JP-8+100 and the improved low temperature flowable fuel JP-8+100LT. These improved fuels have saved the Air Force millions of dollars per year in maintenance and logistics costs.

UDRI has expertise in improving the **low temperature flowability of fuels** via development of flow improving additive packages. We have developed numerous techniques to study fuel crystallization and flowability including:

- Scanning Brookfield viscometry
- Cold-stage optical microscopy
- Differential scanning calorimetry
- Custom fuel system simulators
- Custom additive screening tests

In addition we have developed numerous **computational fluid dynamic models of test devices and aircraft fuel system components** to better understand fuel tank temperature profiles and the effects of viscosity, heat transfer, and solidification. These models have supported the conversion of the U.S. Air Force from JP-8 to Jet A fuels at CONUS locations.

UDRI researchers are actively involved in performing measurement of **combustion particulates and gaseous emissions** for test rigs as well as actual gas turbine engines. We employ numerous methods for characterization of emissions including particle number density, particle size distribution, particle mass concentration, and particle count using instruments such as:

- Condensation nuclei counter
- Scanning mobility particle sizer
- Differential mobility analyzer
- Tapered element oscillating microbalance
- Laser particle counter

In addition, we have pioneered the evaluation and development of emissions reducing additives.

UDRI researchers are pioneering techniques for **prediction of thermal-oxidative fuel deposition**. A methodology has been developed, which employs a combination of unique chemical kinetic mechanisms with computational fluid dynamics techniques which will allow the prediction of fuel deposition and oxidation for various fuels as a function of temperature and time in complex fuel system geometries. Such models can be used to improve fuel system design, make better use of the fuel heat sink, predict nozzle lifetime, and improve reliability and time between maintenance.

UDRI has a Cooperative Research and Development Agreement (**CRADA**) with the U.S. Air Force Research Laboratory Fuels & Energy Branch at Wright-Patterson Air Force Base which allows use of their one-of-a-kind facilities for third party research and testing. Your organization gains the dual benefits of research and testing using these unique facilities along with analysis and interpretation of results by UDRI's fuel experts.

For additional information please contact:

Steven Zabarnick, Ph.D., Division Head, Fuels & Combustion
University of Dayton Research Institute
300 College Park, Dayton, OH USA 45469
Telephone: (937) 229-3961
Email: Steven.Zabarnick@udri.udayton.edu

Please visit our web page: https://www.udri.udayton.edu

The University Of Sheffield. Energy Institute.

Sustainable Aviation Fuels Innovation Centre.

Sustainable Aviation Fuels Innovation Centre (SAF-IC)

The Sustainable Aviation Fuels Innovation Centre (SAF-IC) is a bespoke research centre for the development and testing of sustainable aviation fuels, based at the University of Sheffield. SAF-IC is the first centre of its kind in the UK to produce, develop, test, assess and characterise new SAFs in one location.

The facility works in combination with the neighbouring Translational Energy Research Centre to support state-of-the-art research and provide much needed testing capabilities to help ready sustainable aviation fuels for commercial use.

SAF-IC helps producers and users of sustainable aviation fuels to work towards approval for their products for application in the aviation sector, and offers consultations with world-leading fuels experts, laboratory and testing capabilities and coordination and networking facilities.

To significantly reduce the costs, fuel usage and time it takes to test SAFs, SAF-IC offers facilities for:

- Characterising the chemical composition of aviation fuels using analytical methods
- Determining a wide range of fuel chemical, physical, and electrical properties
- Testing the performance of fuels and fuel system components/architecture
- Assessing combustion gaseous emissions and particulate matter

State-of-the-art equipment

- Full testing capabilities for fuel samples to ensure they meet the required ASTM standards
- Aviation fuel characterisation lab covering ASTM D4054 Tiers 1 and 2 and more
- Small and medium aviation fuel thermal degradation testing facilities (ASTM D3241 and D6811).
- Large scale SAF thermal oxidative stability & component test rig covering ASTM D4054 Tier 3.
- Auxiliary Power Unit (APU) testing for SAF (ASTM D4054 Tier 3), plus APU with hydrogen capabilities
- TRL-5 SAF production facility: Fischer-Tropsch plant with Reverse Water Gas Shift (Power-to-Liquid)

Superior expertise and full-spectrum capabilities

- Detailed quantitative analysis of the chemical composition of aviation fuel
- Pre-screening methods that can provide feedback and facilitate chemical process modelling
- Valuable insight into fuel chemistry and fuel systems performance
- Numerical modelling (CFD) of deposition in complex geometries, as well as numerical simulation of complete fuel systems for assessment of fuel thermal stressing.

Work with us

- Support to scale up production and formally test fuels and fuel systems
- Access to multidisciplinary technical advice and consultancy in high-priority research and innovation topics from world-leading experts
- Reduced costs and timescales compared to full-scale engine testing
- Next-generation facilities and collaboration space

SAF-IC is jointly funded by the European Regional Development Fund and the University of Sheffield.

European Union
European Regional
Development Fund

NORTHERN POWERHOUSE
HM Government

DLR Research for Sustainable Aviation
- Sustainable Fuels & Clean Combustion-

Prof. Dr. Manfred Aigner, Dr. B. Rauch, Dr. C. Arndt, Dr. P. Le Clercq
IASH 2022 Keynote Lecture

DLR

Research Center + Space Agency + Project Management Agency

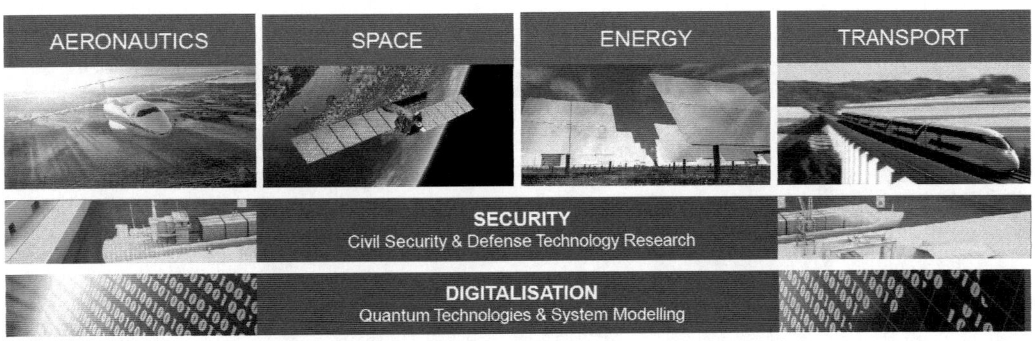

- Europe's largest research center for aeronautics and space
- Close cooperation with academia, businesses and industry
- Funding by BMWK, additional institutional funding by BMVg, project funding by BMI, BMU, BMZ and others

DLR

DLR: 10.000 Employees, 55 Institutes, 30 Sites

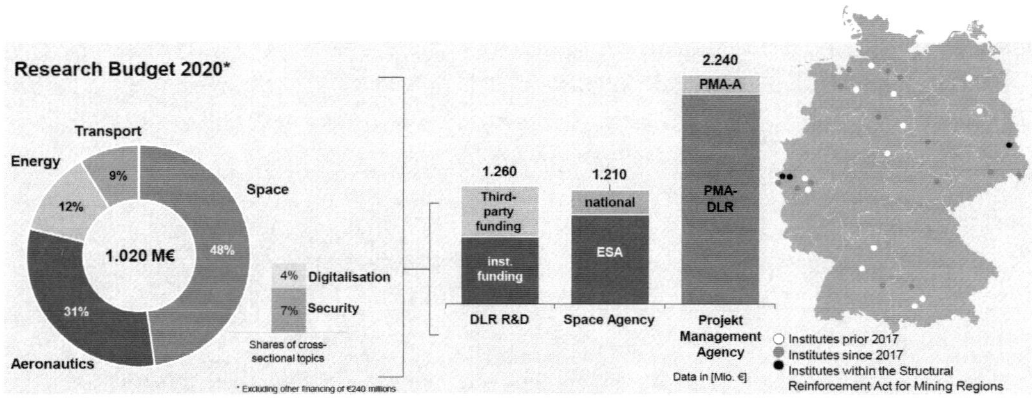

- Research and innovation for economy and society
- Technology transfer from research
- Science-based consulting for decision makers

DLR

DLR Research Infrastructure with Focus on Aviation

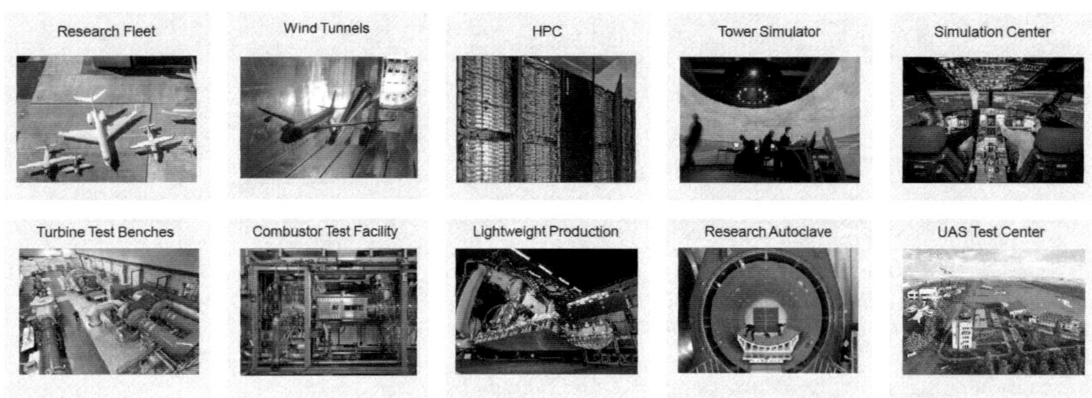

DLR

DLR Institute for Combustion Technology
Competences and Philosophy

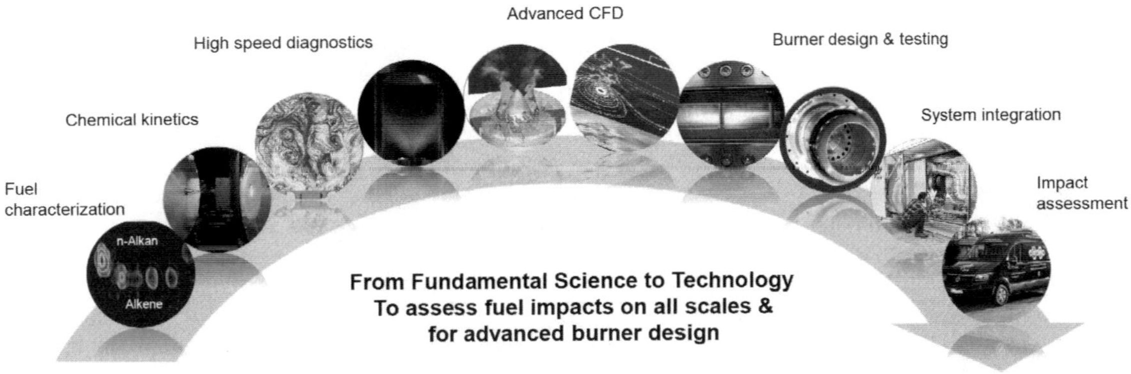

DLR

Challenges Aviation Industry is Facing

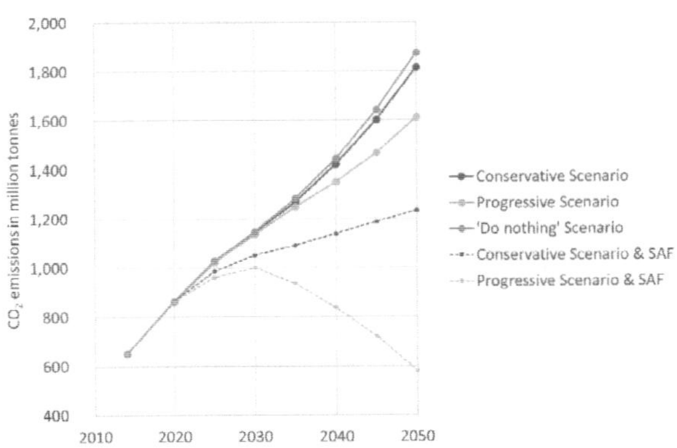

Leipold, A. et al. „DEPA 2050 Development Pathways for Aviation up to 2050", DLR project DEPA 2050, final report, 2021 https://www.dlr.de/fw/Portaldata/42/Resources/dokumente/aktuelles/DEPA2050_StudyReport.pdf

Brazzola, Nicoletta; Patt, Anthony; Wohland, Jan (2022): Definitions and implications of climate-neutral aviation. In *Nat. Clim. Chang.* 12 (8), pp. 761–767. DOI: 10.1038/s41558-022-01404-7.

"We demonstrate that simply neutralizing aviation's CO2 emissions, if nothing is done to reduce non-CO2 forcing, causes up to 0.4 °C additional warming, thus compromising the 1.5 °C target."

Major challenges:
Impact on Climate: CO2- and non-CO2-effects
Impact on Environment by Pollutants

Chance:
SAF has the potential to reduce nearly all impacts on climate and environment

DLR

7

Challenges in Upscaling SAF

In Mission Possible Partnership's „prudent scenario" (high biomass availability, moderate renewable electricity costs) to achieve net zero by 2050, it requires:

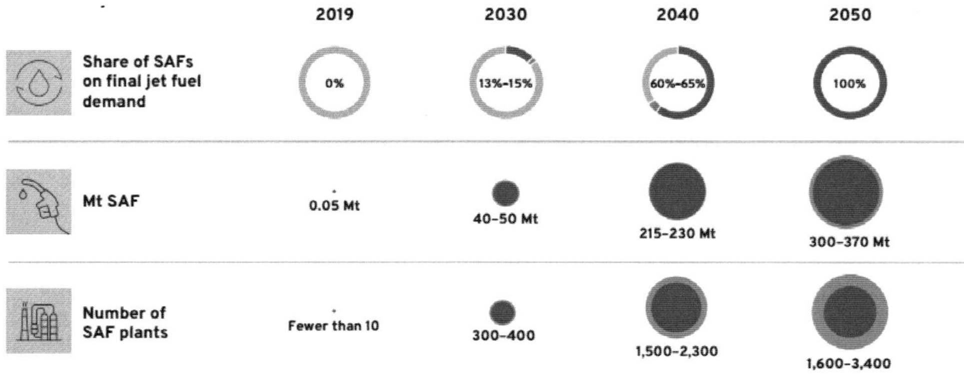

Mission Possible Partnership (2022): Making Net Zero Aviation Possible. Hg. v. Mission Possible Partnership.
Online: https://www.energy-transitions.org/publications/making-net-zero-aviation-possible/

DLR

8

Challenges in Upscaling SAF

In Mission Possible Partnership's „prudent scenario" (high biomass availability, moderate renewable electricity costs) to achieve net zero by 2050, it requires:

2019 2030 2040 2050

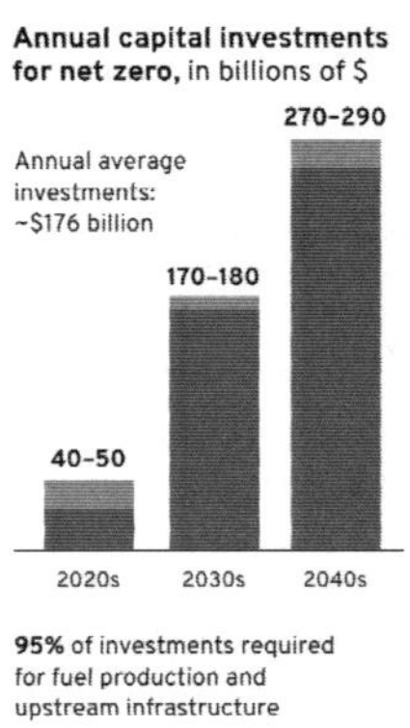

That requires massive investments in building SAF capacities (as well as renewable electricity).

Results agree very well with EC-funded SWAFEA study (2012), which concluded that ~ 250b€ until 2050 are required.

Not much did happen yet!

Mission Possible Partnership (2022): Making Net Zero Aviation Possible. Hg. v. Mission Possible Partnership.
Online: https://www.energy-transitions.org/publications/making-net-zero-aviation-possible/

DLR

9

Challenges in Upscaling SAF
Focus on Realizing Required Investments

- Investors: **revenue certainty** is essential, main barriers are

  I. High feedstock cost and product price **volatility**
  II. First of A Kind Plant challenges:
     a) High degree of **technology risk**
     b) High **capital cost** compared to further plants
  III. Market mechanism uncertainty

Potential inclusion & mechanism for Non-CO2 climate impacts still very unclear

- Intensified by competition with other sectors, esp. road transportation with higher revenue certainty

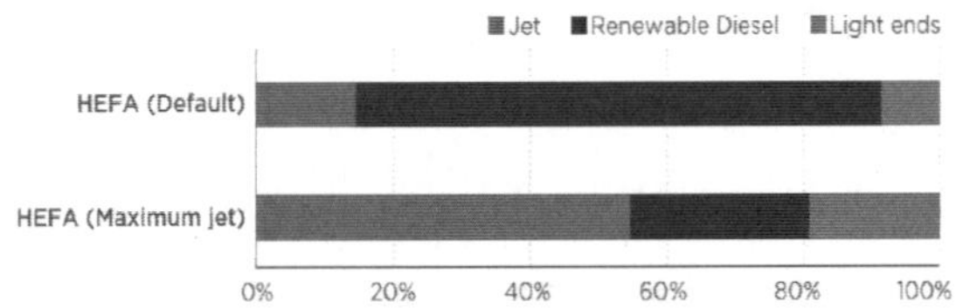

Pavlenko, Nikita; Searle, Stephanie; Christensen, Adam (2019): The cost of supporting alternative jet fuels in the European Union. INTERNATIONAL COUNCIL ON CLEAN TRANSPORTATION.

DLR

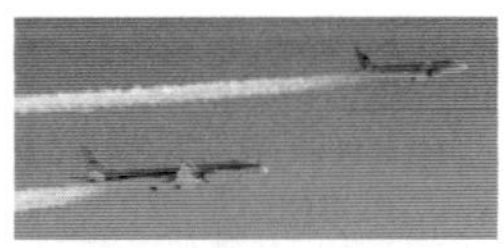

Enablers for an effective transition
1. Collaborative Climate Impact Assessments

2014 ACCESS-II
NASA/DLR/NRC/FAA; Falcon/Falcon/DC8
50% SAF (HEFA) → ~50% reduction in soot emissions

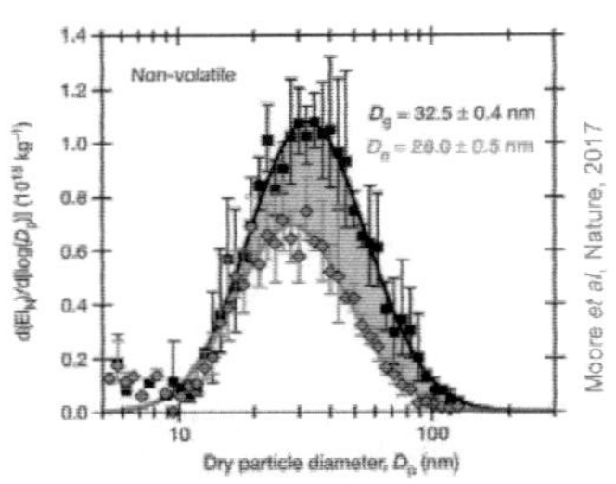

1. SAF (HEFA) reduces not only CO_2 but also nvPM emissions

2015 ECLIF-1
DLR/NASA; Falcon/ATRA-A320
Systematic investigation → SAF optimization possible

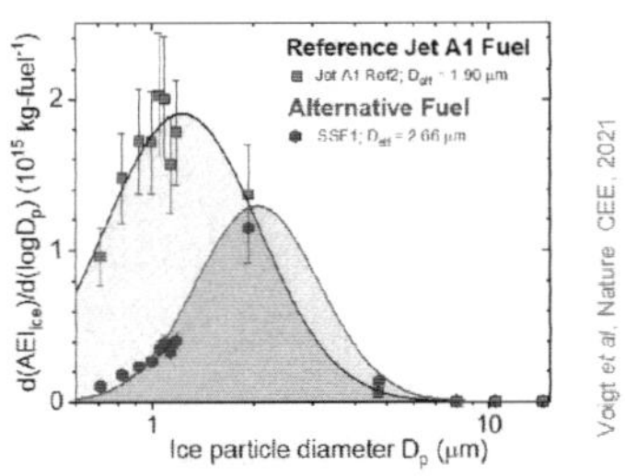

2. Aromatics & hydrogen content impact on nvPM and contrail formation

2018 ECLIF-2/ND-MAX
DLR/NASA/NRC/FAA; DC8/ATRA-A320
50% SAF (HEFA) → ~50% reduction in nvPM and of ice particles in contrails

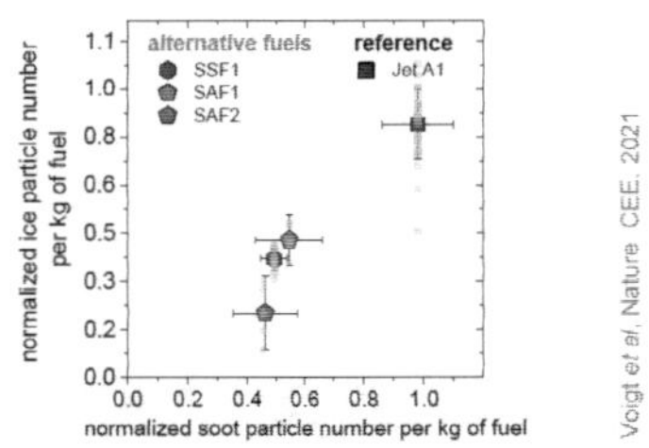

3. Fuel Design: 30% HEFA yields stronger nvPM and ice crystal reduction w/r 50%

DLR

Enablers for an effective transition
1. Collaborative Climate Impact Assessments

2021 ECLIF-3	2021 Volcan (F)
DLR/AIRBUS/Rolls-Royce/Neste; Falcon/A350	DGAC/AIRBUS/ONERA/SAFRAN/Dassault; Falcon/A320 neo

World's first emissions measurements of 100% SAF (HEFA) behind large passenger aircraft: A350, equipped with RR Trent XWB Engines

Emissions measurements of 100% SAF (HEFA) & HT Jet A-1 behind single aisle aircraft: A320 with CFM Leap

4. Demonstrate use and benefits of 100% SAF (SPK) in RQL Comb. Syst.

5. Use of 100% SAF (SPK) in Lean Burn Comb. Syst. Analysis still in progress

Future: Creating evidence for climate mitigation effectiveness of 100% Drop-In vs 100% Non-Drop-In SAF

Enablers for an Effective Transition
DLR Proposal for a PtL Research, Technology and Demonstration Plattform

- Demonstration of **PtL production** on a s semi-industrial scale (10.000 t/a PtL)
- Combination with a research module (100 t/a PtL)
- Integration and scale-up of all process compoents
- Optimization of operation and reduction of cost
- Optimization and evaluation of PtL,
- Use of the optimised PtL in flight campaigns for research and in daily operation

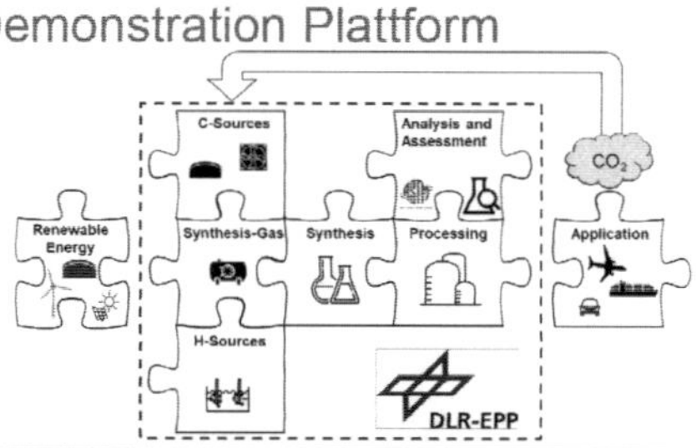

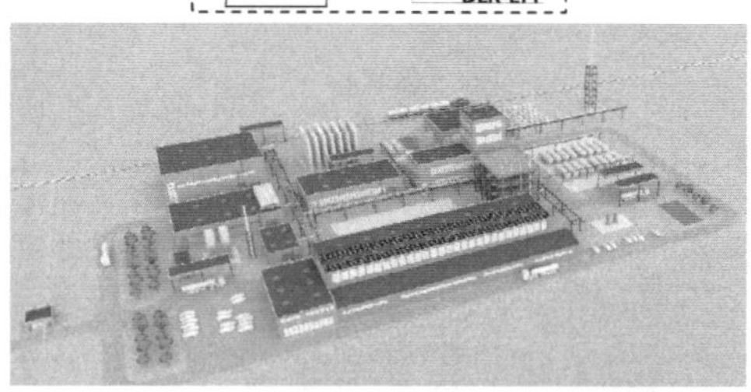

DLR

Enablers for an effective transition
Systematic Technical Fuel Assessment

SAF prescreening process

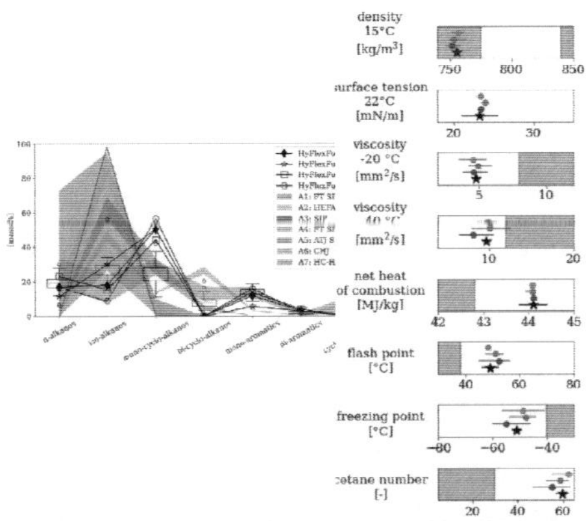

Joshua Heyne, Bastian Rauch, Patrick Le Clercq, Meredith Colket, Sustainable aviation fuel prescreening tools and procedures, Fuel, 2021.

Data-assissted tools to assess and optimize fuels

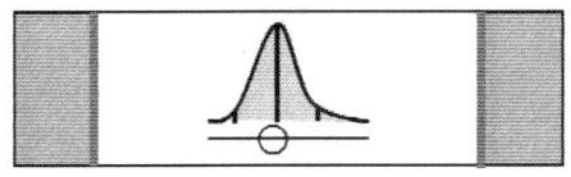

Probabilistic machine learning models
(see talks of Clemens Hall)

Reinforcement learning
(see talk of Florian Pütz)

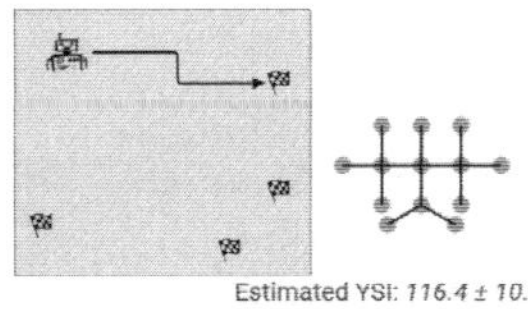

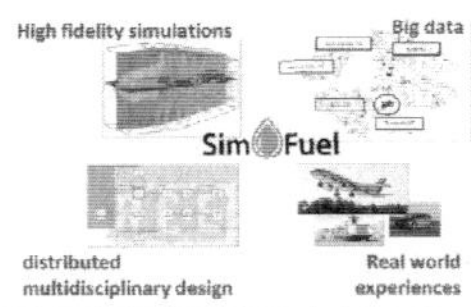

Simulation and Assessment Plattform for Fuels
(see poster of Bauder & use online)

DLR

Enablers for an effective transition
Feedback loop supported by Digital Twin/Thread

- Intention: Maximize positive impacts of SAF
- Method: Support the optimization of the whole value chain

Support development of efficient fuel production processes

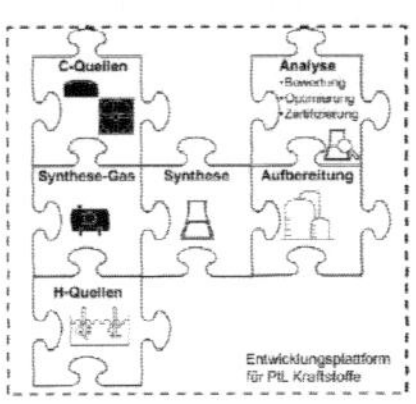

Technical Fuel Assessment

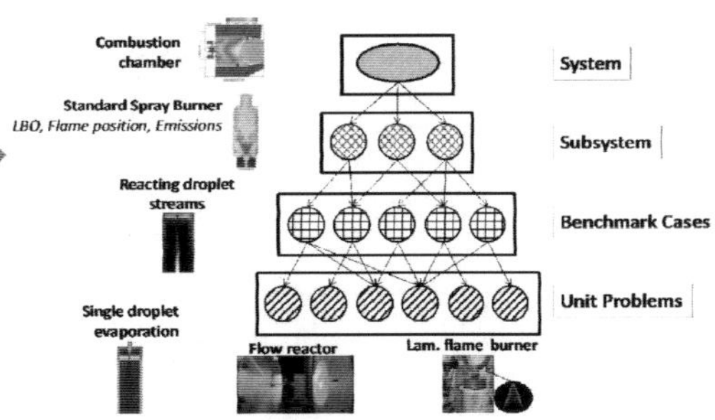

Fuel impact assessment

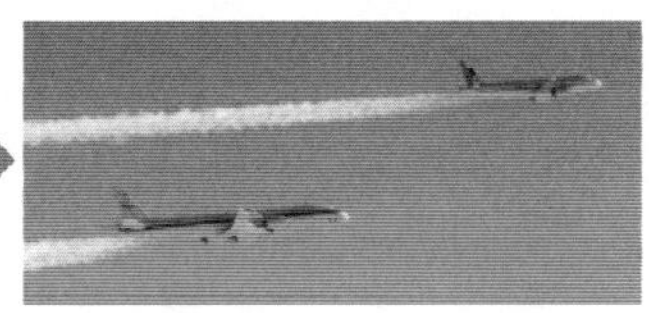

Climate and local air quality

DLR

Research supporting Long Term Perspectives
Co-Optimization of Sustainable Aviation Fuel and Combustor Design

What do we mean by Co-Optimization?

Fuel-Design ⇔ Combustion chamber design
→ Design loop to maximize performance and minimize emissions

Utilize SAF added values

- Increased thermal stability
- Reduced aromatics content
- Reduced no of species

+ SAF production process optimization

- New fuel specification probably needed

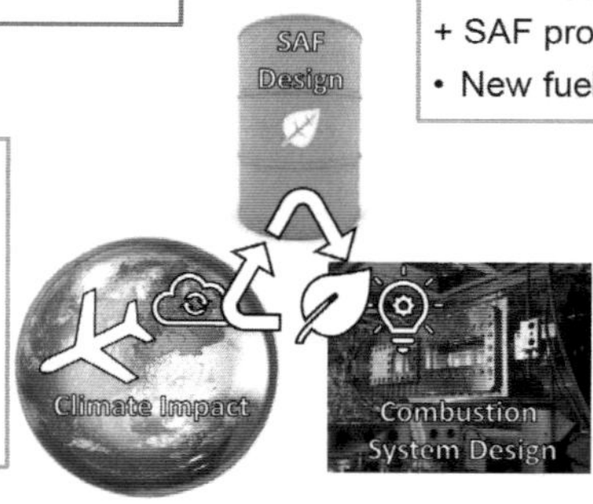

Combustor Concept

- Innovative optimization for combustion system possible due to new properties of the fuel
- Synthetic fuel as potential enabler for systems with specific limitations

Utilize fuel sensitive tool chain

- Detailed analysis of performance, emissions and operability

SimFuel

DLR

Research supporting Long Term Perspectives

- Pollutant generating molecules such as (di-)aromatics or sulfur are not needed for safe and efficient operation.
- An important corner stone will be the co-optimization of a synthetic fuel and the combustion system to enhance pre-vaporization, pre-mixing, and homogeneous burning. This is the basis for reduction of NOx by a factor of up to 100 and of soot and UFP to zero emission.
- Most SAF production pathways offer a wide compositional space to explore the sweet spot for best performance and best pollutant emissions reduction.

The Vision „Zero Emission Aviation" is feasible!

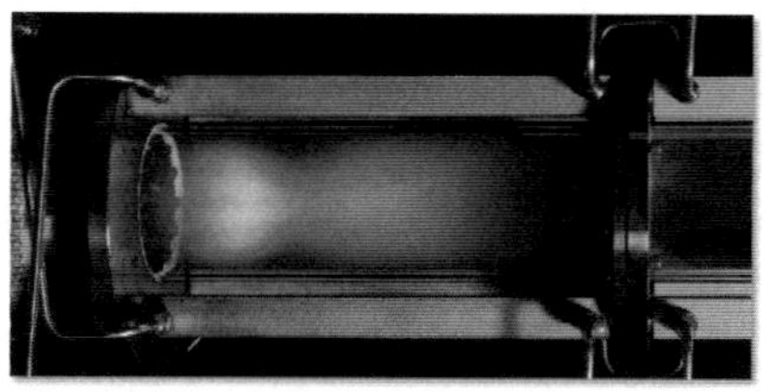

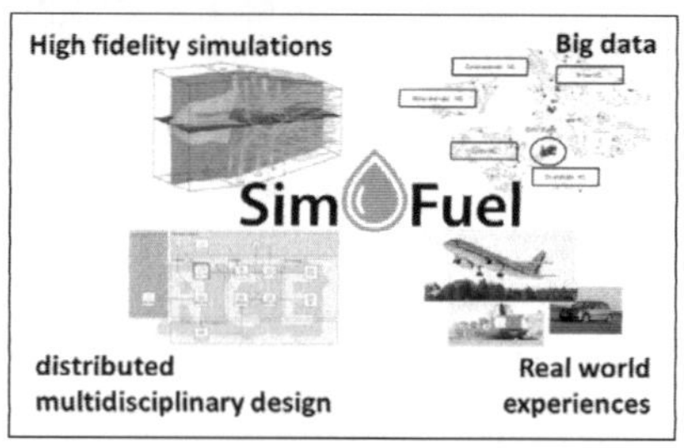

DLR

Research supporting Long Term Perspectives

Co-Optimization of Sustainable Aviation Fuel and Combustor Design

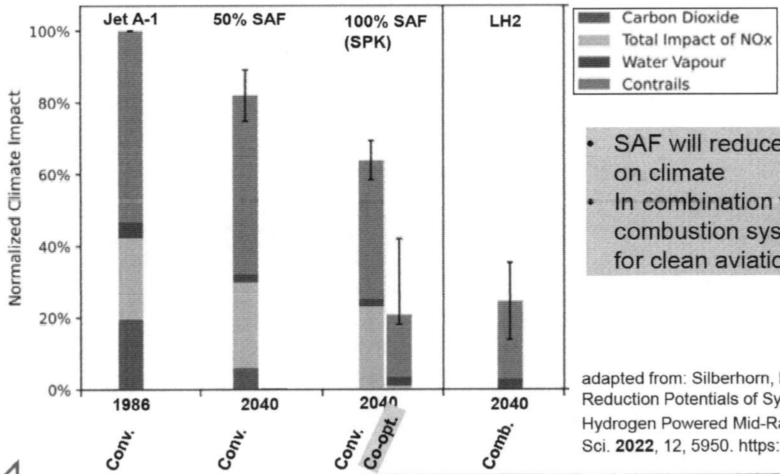

- SAF will reduce CO2- and non-CO2 impact on climate
- In combination with a co-optimization of the combustion system a major break-through for clean aviation is possible

adapted from: Silberhorn, D. et.al. Climate Impact Reduction Potentials of Synthetic Kerosene and Green Hydrogen Powered Mid-Range Aircraft Concepts. Appl. Sci. **2022**, 12, 5950. https://doi.org/10.3390/app12125950

DLR

How can a research organization like the DLR support the transition in the most effective way?

DLR's Roadmap for SAF

Vision: (Near) Zero Emission Aviation

Co-Optimisation of SAF and combustion

Support certification process

Market introduction and scale up of PtL-SAF

Fuel evaluation and prescreening

PtL Research, Technology and Demonstration Platform

Ramp-up production/ use of Bio-SAF

Digital thread for fuels

Fuel Design to optimize performance

Best use of available SAF

Collaborative research regarding climate impacts

DLR

Institute of Combustion Technology

DLR Stuttgart

Knowledge for Tomorrow

DLR

Research supporting Long Term Perspectives

Co-Optimization of Fuel Specifications and Fuel Production

- **Where do specification limits restrict the production?**
- **Can these limits safely be optimized to maximize SAF yields and minimize costs?**

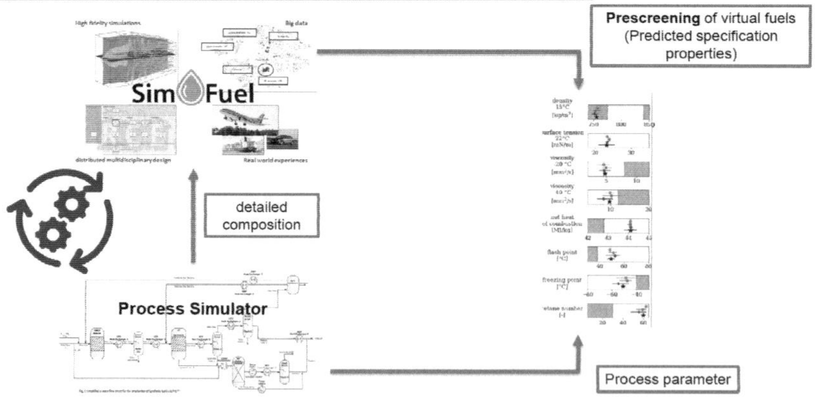

DLR

Research supporting Long Term Perspectives
Co-Optimization of Fuel Specifications and Fuel Production

- Maximize SAF yields & minimize production costs

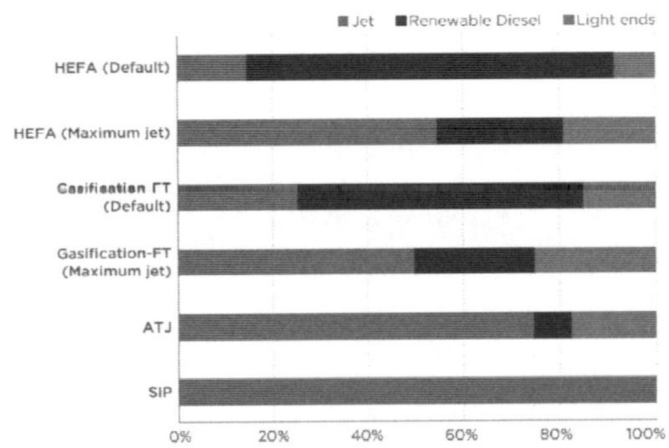

Pavlenko, Nikita; Searle, Stephanie; Christensen, Adam (2019): The cost of supporting alternative jet fuels in the European Union. INTERNATIONAL COUNCIL ON CLEAN TRANSPORTATION.

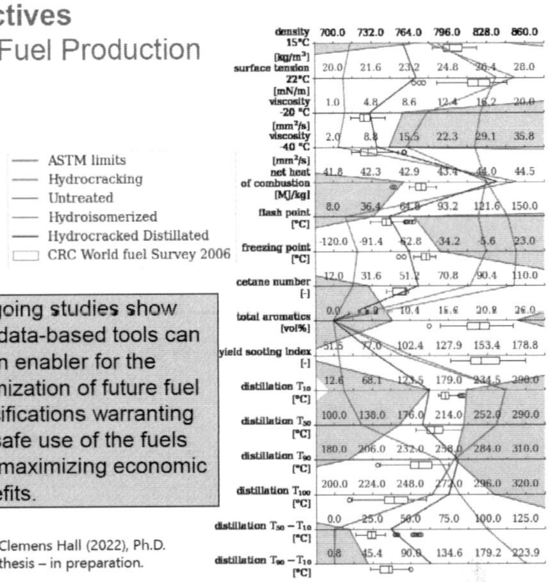

On-going studies show that data-based tools can be an enabler for the optimization of future fuel specifications warranting the safe use of the fuels and maximizing economic benefits.

Clemens Hall (2022), Ph.D. thesis – in preparation.

DLR

IASH 2022, the 17TH INTERNATIONAL CONFERENCE ON STABILITY, HANDLING AND USE OF LIQUID FUELS
Dresden, Germany
11 – 15 September 2022

COMPARISON OF PROBABILISTIC JET FUEL PROPERTY MODELS FOR THE FUEL SCREENING AND DESIGN

Clemens Hall, Bastian Rauch, Uwe Bauder, Manfred Aigner

DLR, German Aerospace Center, Institute of Combustion Technology, MAT, 70569 Stuttgart, Germany, clemens.hall@dlr.de

ABSTRACT

Model-based property prediction is considered a key enabler in the screening and design of new jet fuel candidates. Combined with two-dimensional Gas Chromatography (GCxGC) for the compositional analytics, they allow the prediction of fuel properties critical for the assessment of fuel candidates already from small volumes (5ml). As safety relevant use case the assessment of new aviation fuel candidates makes the consideration of uncertainties due to unidentifiable isomers, measurement and model uncertainty necessary. Recent developments yield several probabilistic modelling approaches that can estimate uncertainties as part of the property prediction. In this work we present three of those approaches: Direct correlation (DC), Mean Quantitative Structure-Property Relationship Modelling (M-QSPR) and Quantitative Structure-Property Relationship Modelling (QSPR) with sampling. The approaches use Monte-Carlo Neural Networks (MCNN), a Machine Learning algorithm for the underling regression of the properties from the GCxGC measurements. We assess their predictive capabilities for the modelling of eight safety-relevant properties like the freezing point and the cetane number for conventional and synthetic jet fuels. The predictive capability assessment is based on metrics that quantify accuracy, reliability and precision of the models. Identifying strengths and limitations of the different approaches, we provide recommendations for the practical application of the different models and the use of model-based property prediction for the support of jet fuel assessments in general.

KEYWORDS: Alternative Fuels, Specification Properties, Modelling, Probabilistic Modelling, Machine Learning, Fuel Assessment, Fuel Prescreening

INTRODUCTION

High safety and performance regulations make the development and market penetration of novel sustainable aviation fuel (SAF) candidates arduous. In the past the ASTM D4054 fuel evaluation process for new SAF production pathway required a tremendous commitment of hundreds to thousands of liters of fuel and a financial budget of three to five millions of dollars for the fuel property and engine tests[1]. Especially in early stages of the jet fuel development, the volumes required for extensive property testing and reformulation iterations often exceed the capabilities of the production process. Screening methods, as developed by the Jet Fuel Screening and Optimization (JETSCREEN) EU project[2] and the National Jet Fuels Combustion Project [3] are considered key enablers in this stage of the jet fuel development. They allow the assessment of fuel candidates on low volumes, enabling fast feedback and iterations while minimizing the time and cost for extensive testing. Heyne and Rauch published a fuel prescreening procedure that allows the screening of SAF candidates at volumes below 5ml [4]. Combined with modern analytical methods, model-based prediction methods play a central role in this prescreening procedure. In the Tier α stage of the prescreening procedure critical fuel properties are predicted of compositional measurements with two-dimensional Gas-Chromatography (GCxGC). Of this measurement property models can predict physicochemical properties considered critical for the early stage of jet fuel development[2,3]: density, kinematic viscosity, surface tension, net heat of combustion, flash point, freezing point, cetane number and the distillation line.

Jet fuels are highly complex mixtures, composed of hundreds of possible hydrocarbon molecules. Presently the exact identification of each molecule in the fuel is not possible with state-of-the-art GCxGC methods (using both mass spectroscopy and flame ionization detector)[5,6]. The GCxGC measurements classify detected species in the fuel in bins of their chemical family and number of contained carbon atoms e.g. iso-alkanes with 10 carbon atoms. Isomers are not identified. This induces uncertainty for the property modelling since their physicochemical properties can vary significantly. E.g. for iso-alkanes with 10 carbon atoms a value range of -110.15 °C to 12.6 °C is recorded for the freezing point. To account for the uncertainty, Heyne and Rauch use probabilistic modelling approaches for the estimation of the fuel properties. These models do not return one deterministic value but rather a distribution of values as part of the prediction. Depending on the modelling method the uncertainty can stem from unidentifiable isomers in the compositional analysis, as well as uncertainty in the property measurements and model uncertainty.

There exist two fundamental approaches of modelling a fuel: 1) modelling a fuel as mixture of constituents and 2) modelling a fuel as entity. Recent research yield probabilistic methods from both of the described approaches. The first approach models fuels as mixtures of compounds. Properties are estimated for a set of components, which are assumed to be present in the mixture. The bulk property is subsequently calculated using an appropriate mixing rule. Yang et al. presented a modelling method that uses sampling of property measurements tabulated in databases[7]. This method estimates uncertainty from the compositional analysis by sampling measurements from different isomers and uncertainties from property measurements by sampling measurements of the same isomers from different laboratories. The approach strongly relies on the availability of measurement of the components assumed to be present in the fuel. The second approach directly correlates a compositional measurement or a chemometrical measurement signal of the fuel with the physicochemical properties using a probabilistic regression algorithm. Rocha and Sheen utilized bootstrapped Partial Least Square and Support Vector Machine regression algorithms to correlate the measurement signals of one-dimensional Gas-Chromatography measurements with the density and kinematic viscosity of fuel at 15, 40 and 60 °C[8]. We published a model that uses GCxGC measurements of fuels for the direct correlation with the properties, modelling the density over a temperature range from -40 to 140 °C with a Gaussian Process Regressor as probabilistic regressor[9]. Models using the direct correlation approach can only be trained on measurements of fuels not on data from pure components. The method can therefore estimate uncertainty due to isomers and the property measurement only if these exist as variance in the training data. This variance can be expressed e.g. by similar GCxGC measurements with different isomers and property measurements from different laboratories. If this is not the case, the model only estimates the intrinsic uncertainty of the training data, e.g. the distribution and distance of the individual datapoints. Since GCxGC measurements are only available in limited numbers presently, direct correlation models can only be trained on a limited number of fuel compositions. To overcome this limitation we developed the Mean Quantitative Structure-Property Relationship (M-QSPR) modelling method [10]. The M-QSPR approach extends the approach of Ajmani et al. of a mole averaged quantitative pseudo structure of a mixture [11] to jet fuels described by GCxGC measurements. This method converts the GCxGC measurement of a fuel into a pseudo structure representation similar to the quantitative structure representation of Quantitative Structure-Property Relationship (QSPR) models for pure compounds. This pseudo structure representation can be directly used for the correlation with the physicochemical property together with the data of pure compounds. A M-QSPR model can therefore be trained on both pure compound data like a QSPR model and fuel data like a direct correlation model. This increases the size of available training data and allows for a direct learning of the mixing behavior, resolving the need for an appropriate mixing rule.

The use case of fuel prescreening requires accurate and highly reliable models that correctly estimate predictions and uncertainties over the possible range of jet fuel compositions, since false predictions might significantly lower the trust in the model-based fuel prescreening and increase cost due to additional measurements. Hence, the predictive capabilities and limitations of the different methods have to be known in order to choose the most appropriate model for a property prediction or clearly indicate the need for further information and model improvement. So far, no systematic comparison of the different probabilistic modelling approaches has been done in the published literature. In this work we compare three models from both approaches for the modelling of the described critical fuel properties: 1) Direct correlation (DC), directly correlating GCxGC measurements of fuels with properties, 2) QSPR predictions of fuel component properties with sampling and 3) M-QSPR, correlating the mean pseudo structure of fuels with properties. The three methods considered in this work model the fuels in fundamentally different ways. We intentionally choose these different approaches to compare their individual advantages, disadvantages and limitations. We assess the predictive capability of the models on 75 conventional crude oil-based jet fuels and 56 synthetic fuels composed of SAF, SAF blends and research fuels. The predictive capability is calculated using a recent introduced concept, quantifying the accuracy of the predictions and the reliability and precision of the estimated uncertainty[9]. We furthermore illustrate the applicability of the models on the use case of Tier α jet fuel prescreening and review the possibility of competitive modelling, considering not one but all model predictions. For the Tier α prescreening we assess 3 jet fuels; one conventional fuel and two SAFs, one produced by the Alcohol-To-Jet process, the other produced by a Fischer-Tropsch process. By comparing the different modelling approaches on the modelling of a wide range of possible fuel compositions we identify strengths and limitations of the different approaches and provide recommendations for the practical application of model-based property prediction for the support jet fuel assessments in general.

BACKGROUND

The possible composition range of jet fuels is vast and the composition of SAF fuels can drastically differ from conventional crude oil-based jet fuels. An exemplary illustration of the possible composition range is given in Figure 1, showing a side by side comparison of an exemplary conventional Jet A-1 fuel on the left and a SAF fuel produced by the Alcohol-to-Jet process on the right. The GCxGC composition measurement of the fuels are shown as barplots, with mass fractions of the detected species for the different chemical families and their respective number of contained carbon atoms.

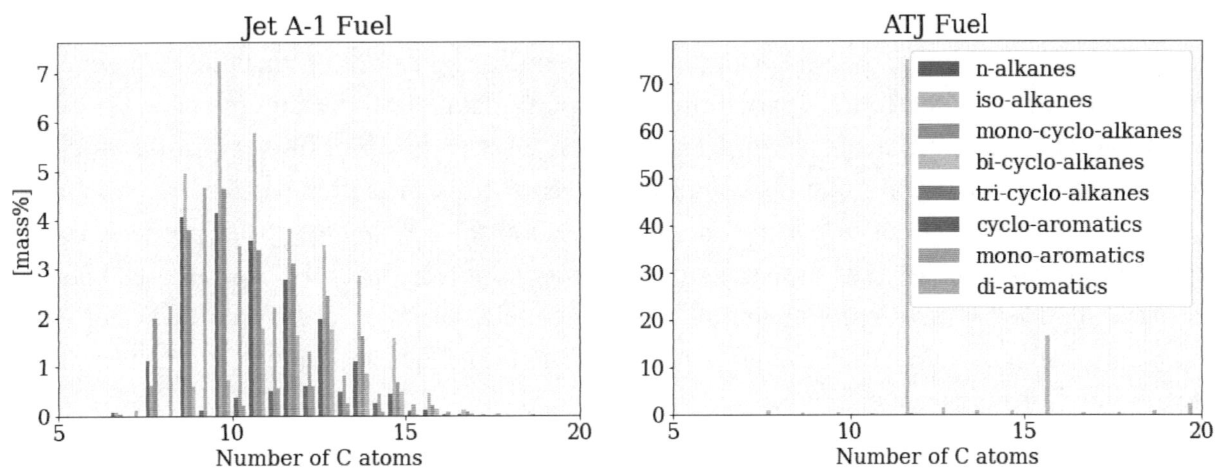

Figure 1: Barplot of GCxGC measurement of an exemplary conventional oil-based jet fuel (left) and an Alcohol-to-Jet fuel (right)

While the composition of the conventional fuel shows distributions of all considered chemical families over a broad number of contained carbon atoms, the composition of the ATJ SAF is dominated by a distribution with distinct peaks of only iso-alkane family with a specific number of carbon atoms. The jet fuel specifications ASTM D4054 and ASTM D1655 do not directly specify the range of possible compositions for jet fuel but clearly state minimum and maximum limits of selected physicochemical properties and fractions of certain chemical families. This allows a vast range of possible fuel compositions, with strong possible variations, that pose a challenge for the model-based predictions. Properties need to be correctly predicted for completely unrecorded compositions with sufficient accuracy to classify the candidate fuel as potential SAF candidate. In addition, the predicted uncertainties have to be valid and correct cover the possible value range. In this work we investigate the ability of the models from the three outlined methods to comply with those requirements modelling conventional fuels, SAF fuels and research fuels.

Modelling Methods

Sampling with QSPR method

Sampling methods model a fuel as a mixture of components. Since GCxGC can presently not identify all isomers contained in the fuel, the composition has to be approximated by representative components for each GCxGC bin that are assumed to be part of the fuel. Up to this point a proven list of possible representative molecules, that describe the variability of unidentified isomers relevant for jet fuels, does not exit. Yang et al. therefore approximated the fuel composition by all molecules tabulated in physicochemical property databases[7]. The method samples measured property values from databases like the NIST Standard Reference Database 103a[12] or DIPPR 801 database[13]. The method has great extensibility, since new hydrocarbon families due to changing fuel compositions can easily be added, if property measurements of the respective hydrocarbon family are available. Yang et al. illustrated the predictive capability of this measurement sampling method for the modelling of 20 SAF using measurements of up to 1223 molecules. However, the method is strongly limited to the availability of property data. If no measurements are available for a GCxGC bin, measurements of similar components with a lower or higher number of carbon atoms have to be chosen. Furthermore, outlier and erroneous measurement have to be filtered before the sampling. This can induce a modelling error. For this reason, we utilize property models that predict the physicochemical properties of all components assumed to be in the fuel, based on the structural information of individual molecules. The method can therefore predict properties for components, as long as the structural information of the molecules are known, it is not limited to the availability of measurement data.

There exist various approaches for the modelling of physicochemical properties based on the structural information of molecules. Saldana et al. demonstrated the applicability of Quantitative Structure-Property Relationship (QSPR) models for the modelling of most of the desired properties[14–16]. The QSPR modelling method postulates a relationship between the structure and physicochemical property and assumes similar property values for molecules with similar structures and substructures. In this approach quantitative structures like the number of contained carbon atoms or the number of methyl groups are quantified for every molecule. This approximates the molecular structure by a quantitative description, that can be correlated with the property. The selection of the molecular features depends on the respective modeling task. From conclusions drawn in previous work by Saldana et al.[14–16] we solely consider functional group count descriptors. These descriptors quantify relevant functional chemical substructures that characterize the molecules of the considered chemical families. The substructures are quantified based on the simplified molecular input line entry system (SMILES) of the respective molecule, using SMILES arbitrary target specifications (SMARTS)[17]. Figure 2 shows a visual example of the quantification of the functional count descriptors for 2,3-hydro-2-methly-1h-idene, with colored substructures and the count of the occurrence of the quantitative structure behind its SMARTS key in the legend: e.g. for the substructure [CX4H3], the count of 1 indicates that the molecule contains one methyl-group, while 9 [R] show that the molecule has 9 ring atoms etc. A full table of the utilized 49 substructures, the corresponding SMART key and a description is listed in the Supplementary Material Table S1. The SMART keys are quantified using the RDKit python package [18].

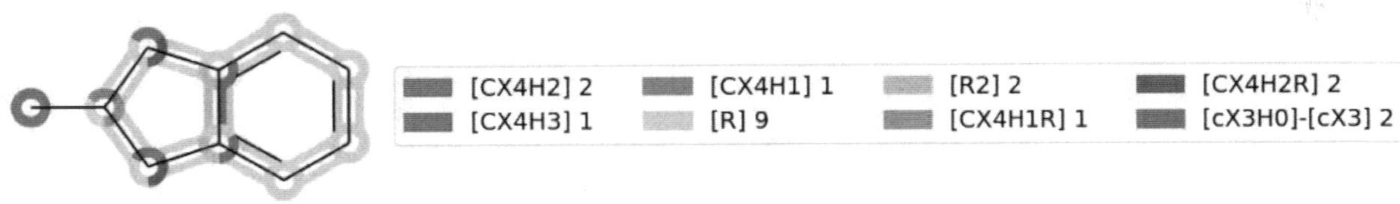

Figure 2: Quantified molecular features of 2,3-hydro-2-methly-1h-idene, number behind SMART key shows count of molecular feature (For reasons of clarity only features that do appear in the properties are shown in the legend) [10].

A schematic representation of the whole process from the quantification of the molecular structure of the components to the prediction of the property and the estimation of the bulk property with the random sampling is shown in Figure 3. We illustrate the process for two exemplary isomers of the GCxGC bin of iso-alkanes with 10 C-atoms with four exemplary substructures.

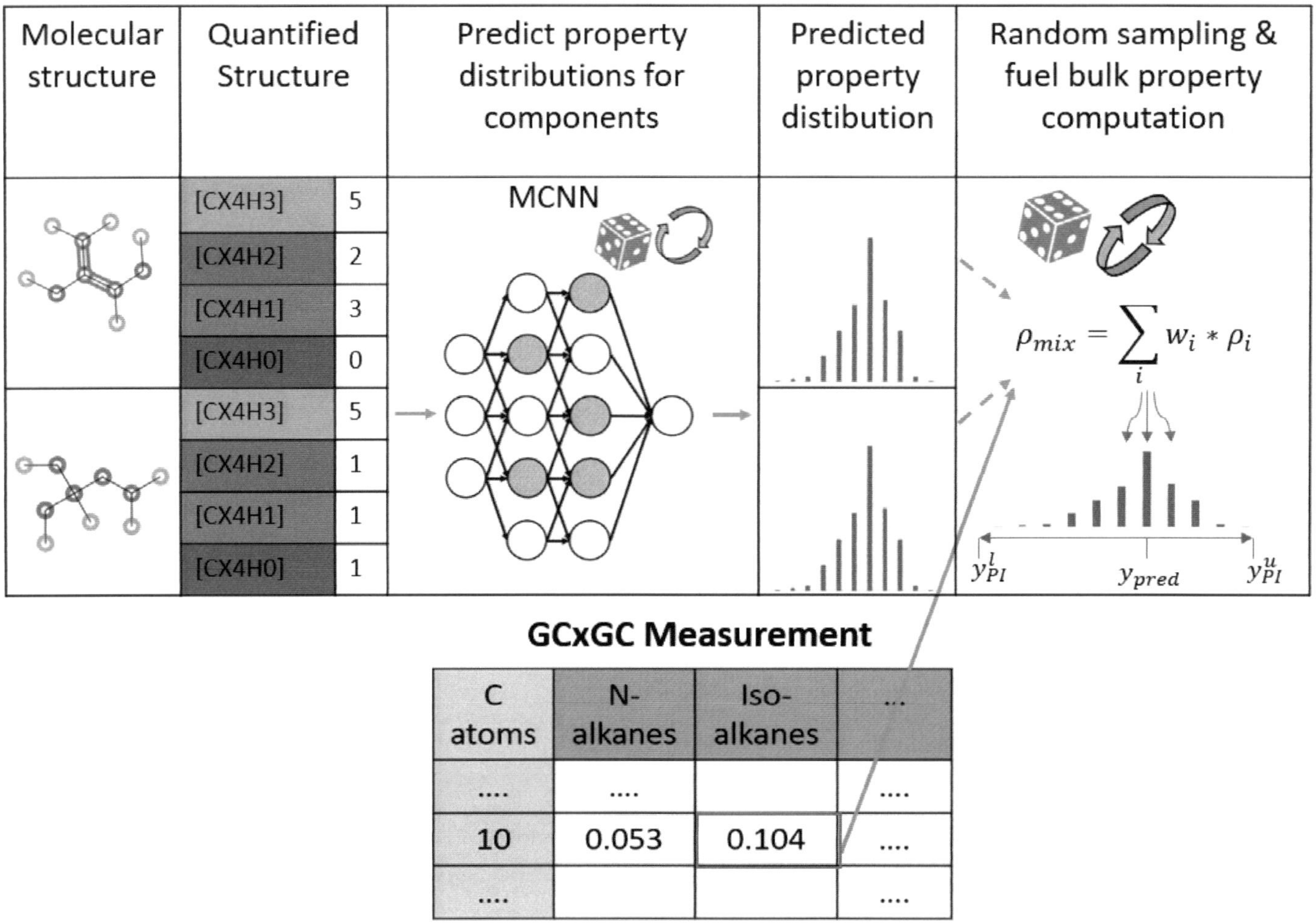

Figure 3: Schematic illustration of QSPR sampling modelling method from quantification of the molecular structure of components to the property estimation of the fuel.

In the first step representative molecules that are assumed to be present in the fuel, based on the GCxGC, are selected and converted into the quantified structure representation by the functional count descriptors. We consider up to 1870 molecules as described in section Data and assume that all recorded isomers of a GCxGC bin are representative and potentially present in the fuel. In the second step property values of all components are predicted with a trained probabilistic QSPR property model, computing a distribution of property values for each molecule. The pure compound property models are trained on the data of pure compounds described in section Data. Unity plots of the cross-validation results of the training and testing for fuels and pure compound of all models can be found in the Supplementary Material in Figures S2-S42. In a last step random samples are drawn from the predicted value distributions of randomized isomers, indicated by the rolling dice in Figure 3. The property samples are multiplied with the fraction given by the GCxGC measurements and summed up using a suitable mixing rule to compute the bulk property of the fuel. By sampling random values from the predicted value distributions of random isomers both uncertainties from the predicted property values of the individual component and uncertainty due to unidentifiable isomers are considered. The sampling process is repeated 100 times to create a value distribution of the bulk property of the fuel. The distribution is described by the median value y_{pred} and the lower and upper prediciton intervals y_{PI}^{l} and y_{PI}^{u} with 95 % confidence intervals.

The modelling of different properties requires suitable mixing rules for each property. For the modelling of the density, surface tension, net heat of combustion, flash point, freezing point and cetane number a linear mixing rule after Equation 1 is used, where the bulk property value y_{mix} is calculated by summing over the predicted property value y_i of the individual components weighted by their fraction in the fuel w_i. All isomers are assumed to be present with the same probability, and are not weighted individually. For the density, surface tension and net heat of combustion this is based on the mass fractions of the components, for the cetane number the calculated mole fractions of the components are used and for the flash point and freezing point the volume fractions of the components yield the best results. For the calculation of the volume fraction we use the predicted densities at 20 °C predicted with the QSPR density property model.

$$y_{mix} = \sum_i w_i * y_i \qquad \textit{Equation 1}$$

For the kinematic viscosity the Grunberg-Nissan mixing rule without binary interaction coefficient is utilized, see Equation 2 [19].

$$ln(v_{mix}) = \sum_i w_i * ln(v_i) \qquad \textit{Equation 2}$$

All mixing rules were chosen based on experience and the best accordance with measurements.

To predict the viscosity at low temperatures (< -10 °C) we utilize a formula proposed by the ASTM D341 see Equation 3 and Equation 4. We therefore predict viscosity values at -10, 0, 20, 60 °C as base points using the QSPR property model. Afterwards we use a linear Bayesian regression model to fit the linearized formula of Equation 3 and draw random samples from the fitted Bayesian regression model to estimate the property value distribution.

$$\log(log(Z)) = A - B \log(T) \qquad \textit{Equation 3}$$

$$Z = v + 0.7 + \exp(-1.47 - 1.84v - 0.51v^2) \qquad \textit{Equation 4}$$

For the calculation of the distillation line the simulated distillation approach after ASTM D2887 is chosen. The distillation is thereby based on predicted boiling points of the components. In this approach components are assumed to evaporate immediately when the temperature reaches the predicted boiling point. The predicted line is corrected using the correction formula defined by ASTM D2887. To calculate the evaporated volume, the corresponding densities of the components are calculated at 15 °C.

M-QSPR method

The M-QSPR approach extends the approximation of the structure of molecules by the quantified structure representation to mixtures and fuels. In contrast to a QSPR modelling method the functional group count descriptors are not calculated for the individual components but directly for the fuel. The fuel is represented as a mean quantitative pseudo structure, described by the same 49 structural features used for the QSPR sampling method. To calculate this M-QSPR representation, the molar fractions of the GCxGC measurement are multiplied with the mean occurrence of the structural features and then summed up over all GCxGC bins. The mean occurrence of a quantitative structure is calculated by averaging the quantitative structure representation of all isomers of a GCxGC bin for all GCxGC bins, creating a mean occurrences matrix for all structural features. Likewise to the QSPR sampling method we assume all isomers to be representative and potentially present in the fuel, both for conventional and synthetic fuels. However, if further information about the possible components of a certain fuel or a certain production pathway exist, an individual mean occurrence matrix can be calculated for that fuel or production pathway. A schematic illustration of the calculation of the quantitative pseudo structure representation is given in Equation 5 on the example of iso-alkanes with 8 carbon atoms and the methyl group descriptor [CX4H3].

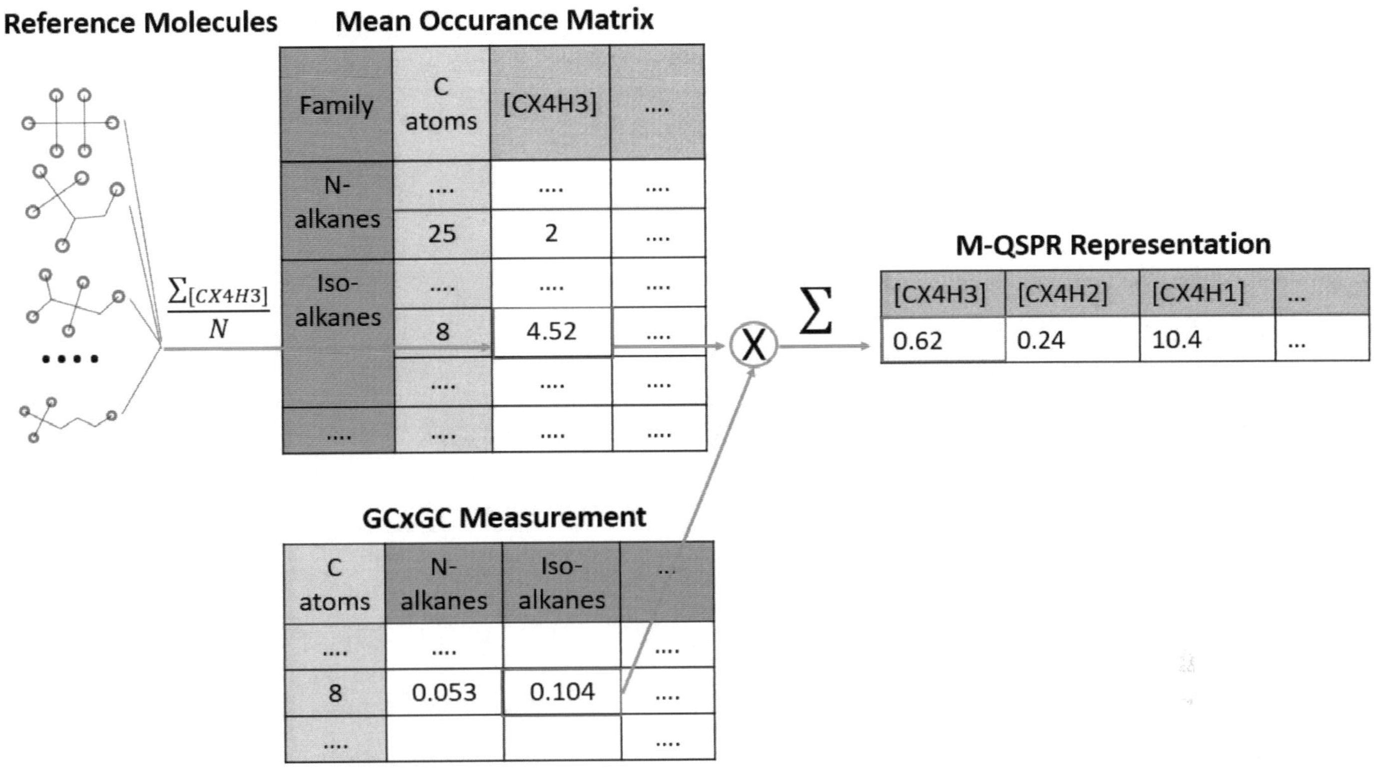

Figure 4: Schematic illustration of pseudo mean quantitative structure estimation of a jet fuel using GCxGC measurement and a mean occurrence matrix calculated from averaging structural features of representative components.

The M-QSPR representation can be directly used for the training and validation of the property, together with the quantitative structure representation of the pure compounds. The distribution of possible property values is directly computed by the used probabilistic regressor, see section Probabilistic Machine Learning model -Monte-Carlo Dropout Neural Network. The distribution is again described by the median value y_{pred} and the lower and upper prediciton intervals y_{PI}^{l} and y_{PI}^{u} with 95 % confidence. The use of data from both fuels and pure compounds increases the training data of the models and makes the need for appropriate mixing rules, needed for the QSPR sampling method, obsolete.

Direct correlation method

In this method, the GCxGC measurement, a two-dimensional table, is reshaped into a one-dimensional vector that is directly used for the correlation with the measured properties. Assumptions about representative molecules, data about pure compounds and appropriate mixing rules are not necessary. The property model is directly trained and validated on the fuel data (and only on the fuel data). The probability distribution of property values is calculated directly by the probabilistic regressor, likewise to the M-QSPR method. To describe the distribution, we use the median value y_{pred} and the lower and upper prediciton intervals y_{PI}^{l} and y_{PI}^{u} with 95 % confidence, likewise to the other methods. A schematic representation of the reshaping of a GCxGC fuel measurement and the property prediction is shown in Figure 5.

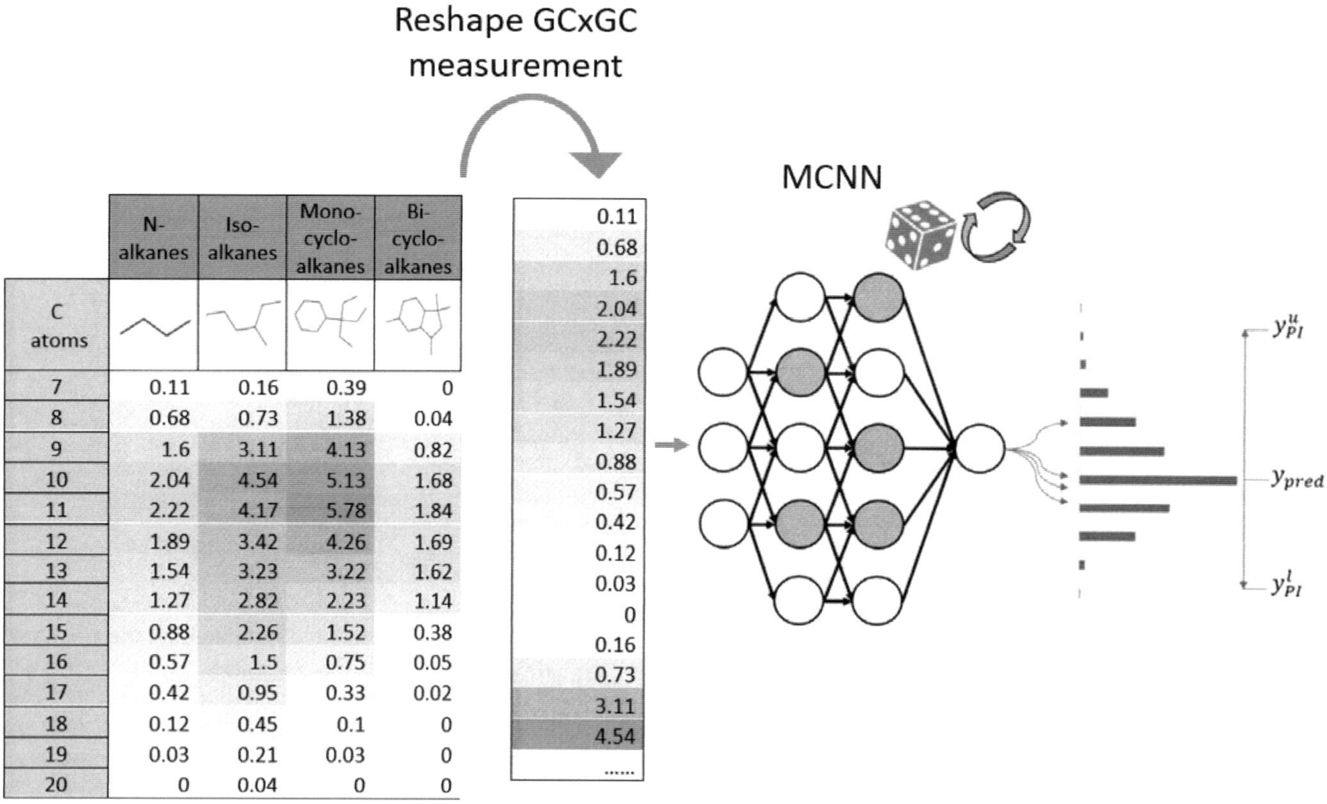

C atoms	N-alkanes	Iso-alkanes	Mono-cyclo-alkanes	Bi-cyclo-alkanes
7	0.11	0.16	0.39	0
8	0.68	0.73	1.38	0.04
9	1.6	3.11	4.13	0.82
10	2.04	4.54	5.13	1.68
11	2.22	4.17	5.78	1.84
12	1.89	3.42	4.26	1.69
13	1.54	3.23	3.22	1.62
14	1.27	2.82	2.23	1.14
15	0.88	2.26	1.52	0.38
16	0.57	1.5	0.75	0.05
17	0.42	0.95	0.33	0.02
18	0.12	0.45	0.1	0
19	0.03	0.21	0.03	0
20	0	0.04	0	0

0.11
0.68
1.6
2.04
2.22
1.89
1.54
1.27
0.88
0.57
0.42
0.12
0.03
0
0.16
0.73
3.11
4.54
......

Figure 5: Schematic illustration of direct correlation modelling method.

The limitation of the DC method to only train on fuel data, restricts flexibility and extensibility of the GCxGC method. Up to this point the number of available GCxGC fuel measurements and corresponding property measurements is low (in the few hundreds up to now) compared to the data available for the other two methods. The potential composition and value range for training the property model is therefore very limited. Combined with the high number of input features (8 families * 25 carbon atoms = 200) this can lead to potential overfitting of the model and the limitation of a high predictive capability of the property model to the training and validation domain. Predictions outside of the training and validation domain can consequently deviate significantly from the true values. Since the direct correlation method estimates uncertainty based on the intrinsic noise of the training data, uncertainties due to isomers and measurements are only estimated if this is reflected in the data. Multiple GCxGC measurements with different isomers and property measurements from different laboratories have to be available to estimate these uncertainties as part of the predicted uncertainty of the model.

Probabilistic Machine Learning model -Monte-Carlo Dropout Neural Network

For the correlation of the three composition descriptions of the fuels (reshaped GCxGC for the DC method, mean quantitative pseudo structure for the M-QSPR method and quantitative structure of components for the QSPR sampling method), we use a deep neural network algorithm with the Monte-Carlo dropout technique (MCNN). The MCNN algorithm was developed by Gal and Ghahramani[20] and uses an artificial neural network for the correlation of the input features with the desired output. Artificial neural networks mimic biological neural systems. They are composed of connected layers of individual neurons that propagate an input signal if a threshold of an activation function is exceeded. The connections are weighted to regulate the importance of inputs. The algorithm is trained by adjusting the weights of the individual neurons. The Monte-Carlo dropout technique uses the popular regularization technique of randomly deactivating neurons of the neural network, originally developed to prevent overfitting of the model, not only during the training but also during the prediction. The prediction of a fuel property value is thereby repeated multiple times, each time deactivating random neuros. This produces varying property values, that resemble a probability distribution. A schematic illustration of prediction process of the MCNN is given in Figure 6.

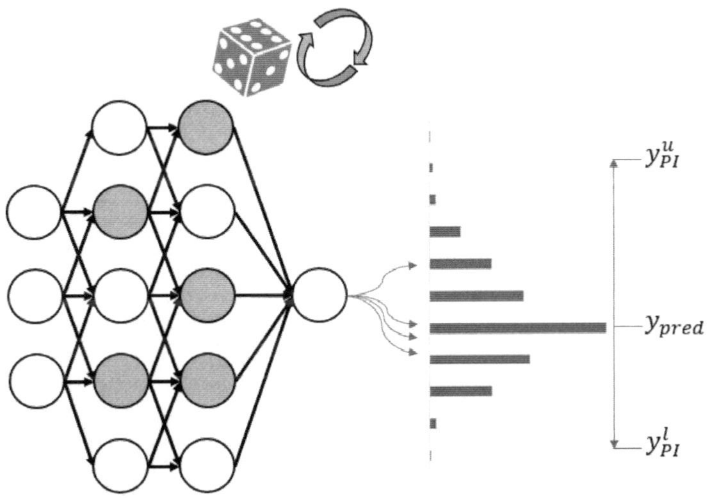

Figure 6: Schematic representation of Monte-Carlo dropout neural network dropout functionality during prediction. Network neurons are deactivated randomly (gray) to generate a distribution of prediction values.

The predicted distribution is interpreted as a Bayesian approximation that describers the prediction uncertainty due to noise in the data (conflicting measurements) as well as dissimilarity of the training and the test data. The predicted distribution therefore reflects the certainty of the prediction, with a narrow distribution indicating a certain prediction and a wider predicted distribution indicating an uncertain one[20]. The distribution itself is described by the median value y_{pred} and the lower and upper prediction interval y^l_{PI} and y^u_{PI}. To assess the validity of the predicted uncertainty intervals as prediction intervals (PI), we review the validity and precision of the prediction intervals as part of the predictive capability assessment, see Predictive Capability Assessment. The MCNN is written in Python using the PyTorch library[21].

Before the actual training of the MCNN for the property prediction a hyperparameter optimization of the topology as well as the learning parameters of the neural network is executed. For this process we used a Bayesian optimization with a Gaussian Process, to find an optimal configuration of the neural networks before the actual training. A more detailed description of the hyperparameter optimization is given in our recent publication[10], a complete list of the optimized hyperparameters is given in Supplementary Material Table S35.

Predictive Capability Assessment

To determine the predictive capability of the models we use a concept based on three metrics that was especially developed for the assessment of probabilistic models in safety relevant use cases[9,10]. By quantifying both the accuracy of the mean prediction and the validity and precision of the computed uncertainty distribution we review if the predicted uncertainty, e.g. due to unidentified isomers, correctly describes the possible value range.

As metric of the accuracy we calculate the mean absolute error (MAE) of the median of the predicted distribution $y_{pred,i}$ and the measurement value $y_{test,i}$, see Equation 5. As a metric for the validity of the predicted distribution, described by the PI, we calculate the Prediction Interval Coverage Probability (PICP), see Equation 6. The PICP calculates the average probability, that a measured value lies inside the predicted upper and lower PI ${y_{PI}}^l$ and ${y_{PI}}^u$. In Equation 6, c_i is a boolean value; it is 1 if $y_{pred,i}$ lies inside the interval and therefore ${y_{PI}}^l < y_{test} < {y_{PI}}^u$ and 0 otherwise. If the PICP and the set confidence level of PI of the prediction are comparable, predictions do on average lie inside the PI and the PI can be considered valid. If this is true for training and testing, the PI is considered reliable. As metric for the precision of the distribution we calculate the Normalized Mean Prediction Interval Width (NMPIW). The NMPIW calculates the mean width of the PI relative to a reference Δ_{ref}, see Equation 7. Note, that different references are possible. For the scope of this work we chose the reproducibility of the property measurement methods. This allows a comparison of the predicted uncertainty with the uncertainty of the measurements.

$$MAE = \frac{1}{n_{Test}} \sum_{i=1}^{n_{Test}} y_{pred,i} - y_{test,i}$$ *Equation 5*

$$PICP = \frac{1}{n_{Test}} \sum_{i=1}^{n_{Test}} c_i * 100\ \%$$ *Equation 6*

$$\text{NMPIW} = \frac{1}{n_{Test}} \sum_{i=1}^{n_{Test}} \frac{{y_{PI,i}}^u - {y_{PI,i}}^l}{\Delta_{ref}} * 100\ \%$$ *Equation 7*

To handle cases where the predicted PI are not valid since the calculated PICP differs significantly from the set confidence level, we use an additional correction factor. The Mean Absolute Error of Outliers (MAOE) quantifies the mean deviation of outliers to the next upper or lower PI, see Equation 4. The size of invalid PI can therefore be increased by the MAOE to achieve valid PI on average, with the price of an increased NMPIW.

$$MAOE = \frac{1}{n_{test}} \sum_{i=1}^{n_{test}} c_i * min\left(\left|y_{PI,i}{}^{u} - y_{test,i}\right|, \left|y_{test,i} - y_{PI,i}{}^{l}\right|\right)$$

Equation 8

Data

The QSPR sampling method needs property data of jet fuels and the M-QSPR method is able to train on both data from pure compounds and fuels. Hence, composition and property data for both fuels and pure compounds are needed. To train the QSPR and the M-QSPR model pure compound data is used from the NIST Standard Reference Database 103a[12], DIPPIR 801 database[13] as well as the public databases Pubchem [22] and Chemspider[23]. Data from Pubchem and Chemspider are especially used for the properties flash point and freezing point. Data for the cetane number is taken from publications of the National Renewable Energy Laboratory [24] and Li et al. [25]. All in all, data of 1870 pure compounds are used for the training and validation of the QSPR and the M-QSPR models. As fuel data we used 75 conventional crude oil-based jet fuels and 56 synthetic fuels containing SAF fuels, blends of SAF as well as research fuels. Data for the conventional fuels mainly originates from the world fuel survey of 2006 by the Coordinating Research Council [26]. The data of the synthetic fuels was systematically gathered in the DLR jet fuel database from research project like JETSCREEN, NJFCP and Emission and Climate Impact of Alternative Fuels (ECLIF) [27] . Table 1 gives a detailed summery of the number of data points utilized for the training and validation of the models for each property. Column #FL thereby lists the number of unique fuels and column #DP lists the total number of individual measurements. The summary is given for fuel components (Comp.), conventional fuels (Conv.), and non-conventional synthetic fuels and blends (Syn.) as well as the sum of all. For the jet fuel screening property values are needed at 15 °C for the density, 22 °C for the surface tension and -20 °C and -40 °C for the kinematic viscosity at 1 atm. Flash point, freezing point, net heat of combustion, distillation line and cetane number are measured under laboratory standard conditions. In the scope of this work, measurements for the density, surface tension and kinematic viscosity are used from an extended temperature range of -40 to 140 °C, due the availability of data. For the kinematic viscosity, dynamic viscosity measurements are taken and divided by density measurements at the same temperature, since most of the viscosity measurements of pure compounds are only available as dynamic viscosity. For dynamic viscosity measurements without a corresponding density measurement, the density is predicted using the trained QSPR model.

	Comp.		Conv.		Syn.		All	
Properties	#FL	#DP	#FL	#DP	#FL	#DP	#FL	#DP
Density	1865	27347	75	425	56	260	1996	28032
Surface tension	294	4455	62	157	22	79	377	4691
Kinematic viscosity	1795	20046	74	197	60	166	1929	20409
Net heat of combustion	1503	1875	74	85	45	62	1622	2022
Flash point	268	640	75	80	62	54	397	791
Freezing point	402	1053	74	80	50	58	526	1201
Cetane number	159	211	7	22	29	48	195	281
Distillation / boiling point	1825	4333	75	772	63	524	1963	5629

Table 1: Number of unique fuels and pure compounds #FL and corresponding datapoints #DP used for the training and validation of the models.

Fuel data

In this work GCxGC measurements are used that describe fuels as matrix of eight hydrocarbon families: n-alkanes, iso-alkanes, mono-cyclo-alkanes, bi-cyclo-alkanes, tri-cyclo-alkanes, cyclo-aromatics, mono-aromatics and di-aromatics, with species in a range of 1-25 carbon atoms. To visualize the variety of fuel compositions Figure 7 shows a scatter plot for the mass fractions of each considered hydrocarbon family for conventional fuels (Conv.) with the fuel types Jet A, Jet A-1, JP-5, JP-8 and TS-1 in blue and for synthetic fuels (Syn.) in green. The synthetic fuels contain fuels produced by the Alcohol-To-Jet process (ATJ), the Fischer-Tropsch process (FT), from Hydroprocessed Esters and Fatty Acids, the Integrated Hydropyrolysis process of Shell with a high cyclo-alkanes content Cyclo Paraffinic Kerosene, the Catalytic Hydrothermal Conversion Jet fuel process as well as fuel blends.

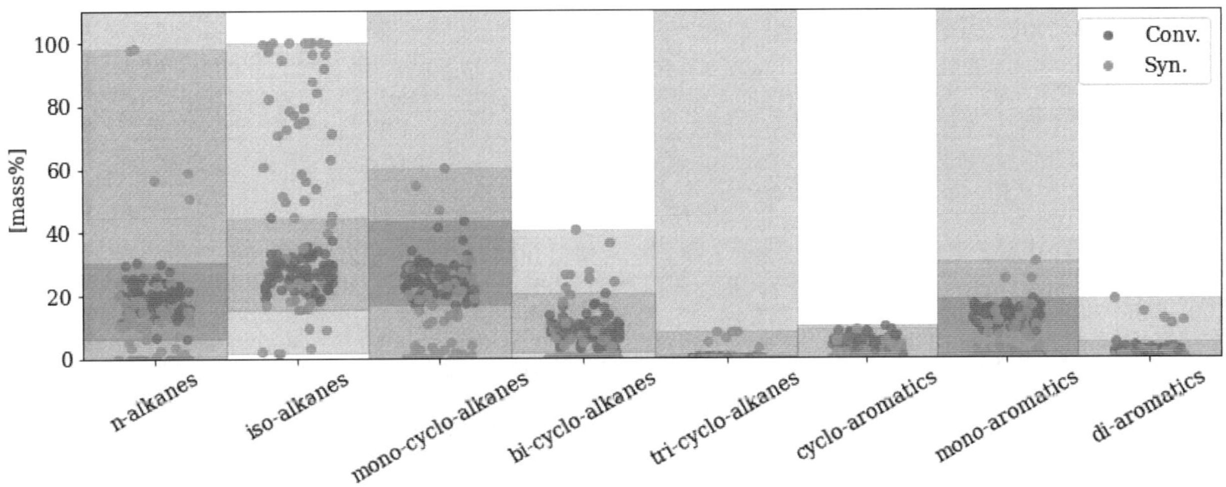

Figure 7: Scatter plot of GCxGC measurements with summed hydrocarbon families. Blue: conventional fuels, green: synthetic fuels and blends. Blue and green shaded areas indicate the observed range.

For all families except the cyclo-aromatics synthetic fuels cover a broader and more distributed composition range. The broader composition range of the synthetic fuels can be explained by the large variety of different feedstock and production pathways and the lower number of hydrocarbon components compared to conventional ones.

Pure compound data

As explained in the introduction, the exact molecular composition can presently not be identified with state-of-the-art GCxGC measurements. To estimate the uncertainty, due to unidentified isomers, we assume that all pure compounds, that belong to the considered hydrocarbon families, can potentially be present as fuel components. This is a worst-case approximation since the range of possible components that could actually exist in conventional and synthetic fuels might be significantly smaller. The pure compound databases used for this work list 1870 relevant compounds. However, these are only a small fraction of all theoretically possible molecules. Figure 8 shows a comparison of the number of theoretically possible compounds and the number of compounds listed in the databases and considered in this work. The comparison is taken from our recent publication [10]. The theoretically possible compounds were systematically generated using the molecule generator MOLGEN v. 5[28]. Their number is listed in the column MG, the number of molecules listed in databases and used for this work is shown in the column DB. The compounds are classified into the eight considered hydrocarbon families using characteristic molecular substructures outlined in the Supplementary Material, see Table S36. Due to the exponential increase of possible isomers and computational limitations for the classification, only calculations up to molecules with 12 C atoms were executed. For iso-alkanes the numbers above C 12 were taken from theoretical calculations [29].

C number	n-alkanes		iso-alkanes		mono-cyclo-alkanes		bi-cyclo-alkanes		cyclo-aromatics		mono-aromatics		di-aromatics	
	DB	MG	DB	MG	DB	MG	DB	MG	DB	MG	DB	MG	DB	MG
1	1	1												
2	1	1												
3	1	1			1	1								
4	1	1	1	1	2	2	0	1						
5	1	1	2	2	5	5	2	4						
6	1	1	4	4	12	12	5	14	0	4	1	1		
7	1	1	8	8	27	29	5	46	0	9	1	1		
8	1	1	17	17	43	73	8	150	0	31	4	4		
9	1	1	32	34	52	185	18	477	2	75	8	8		
10	1	1	49	74	58	475	18	1503	8	218	22	22	0	5
11	1	1	37	158	33	1231	12	4680	16	588	42	51	2	21
12	1	1	45	354	29	3232	18	14461	27	1657	60	136	9	103
13	1	1	34	801	14	>3232	15	>14461	32	>1657	51	>136	24	>103
14	1	1	29	1857	17	>3232	16	>14461	31	>1657	60	>136	38	>103
15	1	1	27	4346	18	>3232	14	>14461	29	>1657	34	>136	17	>103
16	1	1	32	10358	8	>3232	15	>14461	16	>1657	22	>136	9	>103
17	1	1	11	24893	6	>3232	8	>14461	7	>1657	15	>136	3	>103
18	1	1	12	60522	8	>3232	17	>14461	6	>1657	22	>136	13	>103
19	1	1	14	147283	13	>3232	6	>14461	12	>1657	11	>136	4	>103
20	1	1	20	366318	4	>3232	10	>14461	9	>1657	13	>136	9	>103
21	1	1	9	>366317	6	>3232	2	>14461	0	>1657	3	>136	1	>103
22	1	1	14	>366317	5	>3232	4	>14461	6	>1657	14	>136	4	>103
23	1	1	5	>366317	3	>3232	0	>14461	0	>1657	4	>136	0	>103
24	1	1	17	>366317	4	>3232	3	>14461	1	>1657	9	>136	0	>103
25	1	1	2	36797587	4	>3232	5	>14461	1	>1657	7	>136	1	>103

Figure 8: Comparison of the number of representative molecules available in the database (DB) and the number of theoretically possible molecules, calculated by MOLGEN (MG) [10]

Fuel screening data

To test the applicability of the property models for the use case of Tier α fuel prescreening, we selected three fuels that are not part of the training and validation process of the models. Predictions are therefore performed as blind test to simulate a real fuel prescreening process. One conventional oil-based fuel and two SAF fuels, one produced by the Alcohol-To-Jet process (ATJ), one produced by a Fischer-Tropsch process (FT), are selected. The conventional fuel and the ATJ fuels were composed and measured as part of the JETSCREEN project, whereas the FT fuel with the POSF 5018 originated from the NJFCP project. Figure 9 shows the GCxGC composition of the fuels as bar plots.

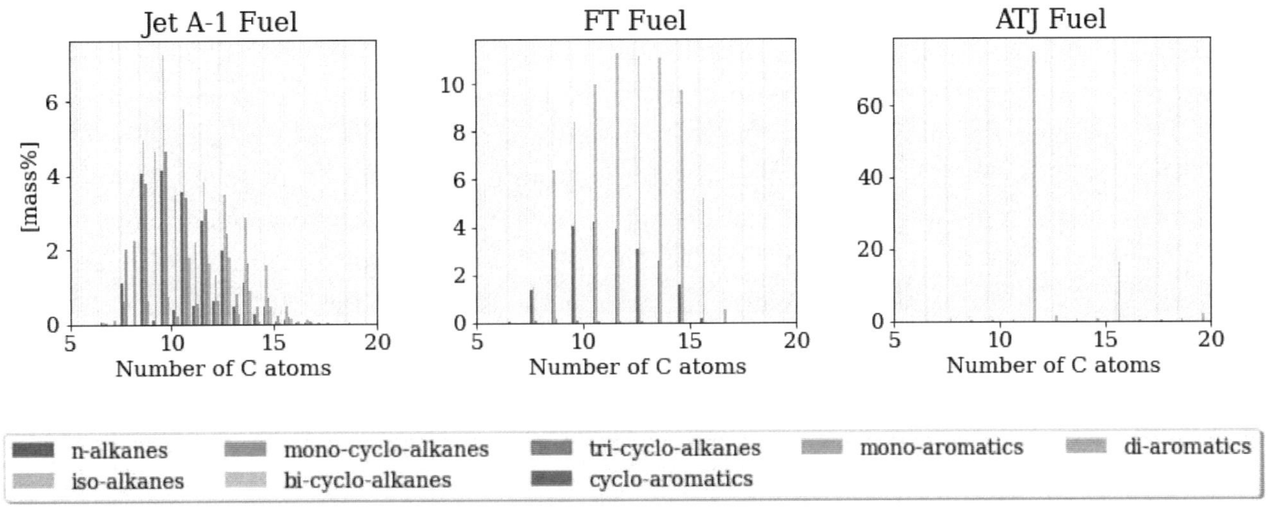

Figure 9: Bar plot of jet fuels used for Tier α prescreening: conventional oil-based fuel (left), SAF produced by the Fischer-Tropsch process (middle) and SAF produced by Alcohol-to-Jet process (right).

The three fuel compositions differ significantly. The fuels are chosen intentionally to investigate the predictive capability of the models over a broad range of possible fuel compositions: a conventional fuel with a composition composed of broad distributions of the mass fractions for all hydrocarbon families, a FT fuel with a composition dominated by two hydrocarbon families with a broad distribution and an ATJ fuel with a composition dominated by just two distinct GCxGC bins of one hydrocarbon family.

RESULTS AND DISCUSSION

The results of this work are presented in two parts. In the first part we present the results of the validation of the property prediction for both the conventional (Conv.) and synthetic (Syn.) fuels. Since the DC model and the M-QSPR model are directly trained on fuel data, validation results are taken from the testing of the four-fold cross-validation of the models, that were performed with a 15 % fraction. The cross-validation results of the training and testing for the property prediction of the fuels and pure compounds are given in the Supplementary Material Figure S2 to Figure S34 for all models. To furthermore investigate the importance of synthetic fuels for the predictive capability of the DC and M-QSPR models, we present validation results of a DC and a M-QSPR model trained solely on data from conventional fuels.

Property prediction validation

In the following we outline the general schema of the presentation of the validation results. The schema is used for all properties, while the results and discussion of the properties are presented individually. The validation results are presented as unity plots, the predicted values are thereby plotted against the corresponding measurements. Markers for conventional fuels (Conv.) are shown in blue and synthetic fuels (Syn.) in green. If the prediction and the measurement are in perfect accordance, the markers lie on the unity line, displayed in black. The predicted prediction intervals (PI) of the models are indicated as error bars for a 95 % confidence level. As reference for the accuracy and precision of the predictions we indicate the reproducibilities of the measurements taken from the CRC Report No. AV-23-15/17[30] as grey dashed lines. The reproducibilities of the corresponding measurement methods are listed in Table 2.

Property	ASTM Method	Reproducibility
Density, kg/m3	D4052	0.52
Flash point, °C	IP 170	3.2
Freezing point, °C	D5972	0.8
Net heat of combustion, MJ/kg	D4809	0.324
Surface Tension, mN/m	D971	0.1*X
Viscosity, mm2/s	D445	0.019*X
Cetane number,	D6890	0.0385*(X+18)
Distillation 10 vol%	D86	3.15
Distillation 50 vol%	D86	3.46
Distillation 90 vol%	D86	383

Table 2: Reproducibilities of ASTM property measurement methods.

The calculated predictive capability metrics are provided in separate tables given in the Supplementary Material Table S37 to Table S46. The PICP metric that quantifies the validity of the predicted PI, is thereby calculated with respect to the measurement reproducibilities. If a measurement lies within the predicted PI +/- the measurement reproducibilities, the PI are considered valid.

Density

For the density the predictions of all models, except the ones of the DC -no Syn. model, lie close to the unity line, as visible in Figure 10. The density is therefore modelled accurately with all modelling methods, if synthetic fuels are part of the training data. The deviations of the DC -no Syn. model predictions correspond to synthetic fuels with a low fraction of aromatics. Synthetic fuels with a high fraction of aromatics are still predicted correctly. The striking outlier at 726 kg/m^3 is visible for almost all models. Since this is the only consistent outlier observed for different models, an erroneous GCxGC measurement is possible. The measurement was not made in house, a verification of this observation is not possible. The estimated validity of the PI quantified by the PICP, is above or is close to the desired 95 % for the conventional fuels. This means that on average 95 % of the measurements are enclosed by the PI. For the synthetic fuels the PICP threshold is not reached by any model. Therefore, the use of the estimated MAOE for the predictions is recommended for a safety relevant use case, to statistically enclose the measurements, as described in chapter Predictive Capability Assessment. The NMPIW exceeds the reproducibilities of the measurements multiple times. Since the utilized composition and property measurements originate from different laboratories the higher uncertainty is comprehensible. For the application of the density models for the use case of fuel screening, all methods can be utilized. With a MAE of 0.96 kg/m^3 to 3.59 kg/m^3 for conventional and 2.39 kg/m^3 to 4.69 kg/m^3 for synthetic fuels highly accurate results with errors below 0.5 % are achieved.

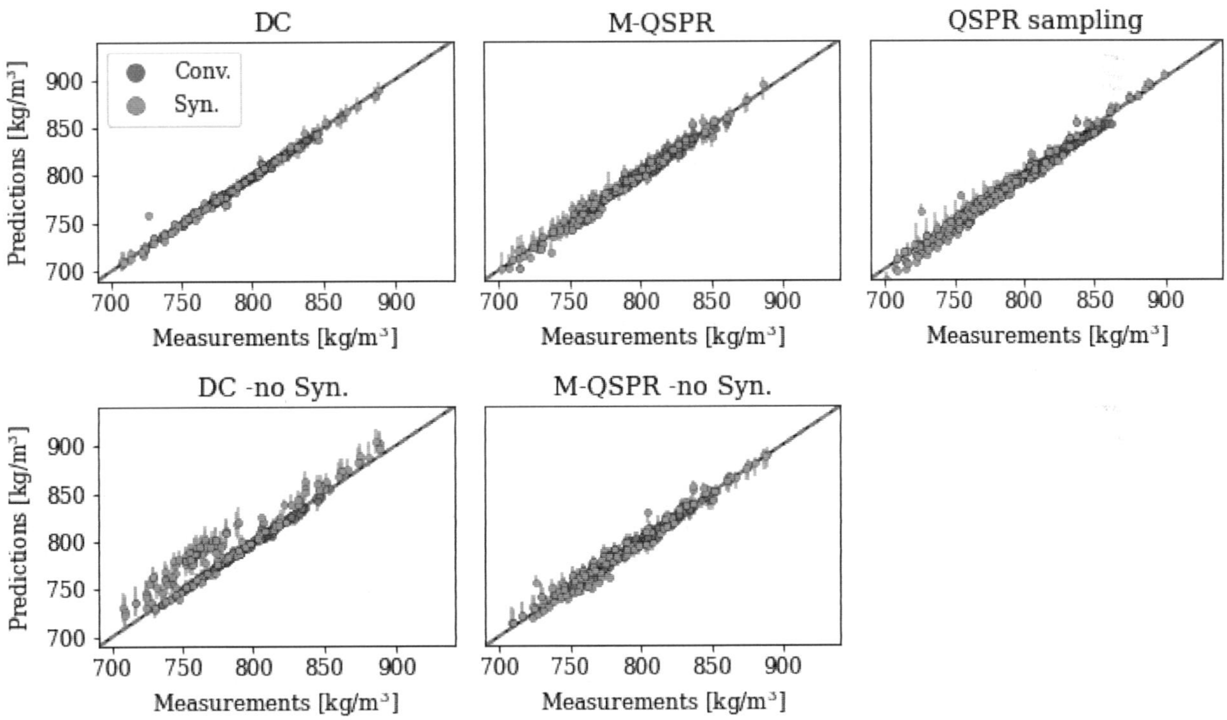

Figure 10: Validation results of the density prediction. Results for conventional fuels are displayed in blue, results of synthetic fuels in green.

Surface tension

Similarly, to the density, the predictions for the surface tension closely follow the unity line and most of the predictions are located inside the reproducibility region, with the exception of the DC -no Syn. model, see Figure 11. Striking is the systematic deviation of predictions of the QSPR sampling model for a few synthetic fuels with values above 30 mN/m. These predictions correspond to research fuels from the JETSCREEN project with an unusual high fraction of aromatics (up to 30 mass%). This could be due to a false prediction of the pure compounds by the QSPR model or a false selection of isomers for the corresponding GCxGC bins. The metrics show a similar predictive capability for all models, with a PICP over 95 % and a NMPIW which is comparable to the reproducibility of the measurements. With respect to the accuracy, MAE of 0.36 mN/m to 0.54 mN/m are achieved for conventional fuels and 0.35 mN/m to 0.95 mN/m for synthetic fuels. The surface tension can therefore be accurately predicted with methods and all methods can be utilized for the use case of fuel screening.

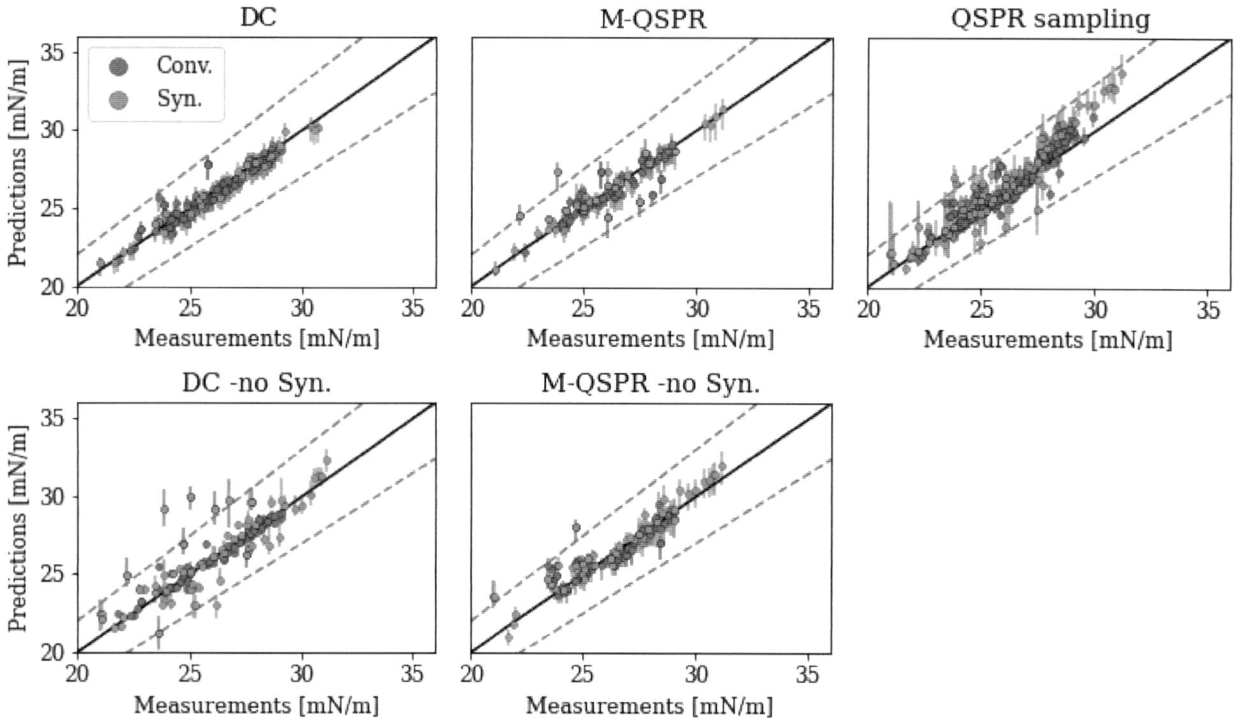

Figure 11: Validation results of the surface tension prediction. Results for conventional fuels are displayed in blue, results of synthetic fuels in green.

Net heat of combustion

For the net heat of combustion most of the predictions for both conventional and synthetic fuels lie inside the reproducibility region of the measurement for all models. The predictions follow the unity line, while a systematic positive offset of the predictions for conventional fuels is visible for the QSPR sampling model. This positive offset was also observed for a M-QSPR model trained only on pure compounds, that basically utilizes a molar fraction based mixing rule for the prediction of the net heat of combustion[10]. The offset is therefore not due to the inapplicability of the mass fraction based mixing rule, but probably rather due to differences in the measurement methods of pure compounds and fuels. For the M-QSPR models trained on fuels and pure compounds this offset is not observed, the model is therefore able to adapt the offset with the presence of fuels in the dataset. Since the offset is smaller than the reported reproducibility the offset is not significant, but further investigations might be necessary to understand the observation. A further striking systematic error is observed for the M-QSPR -no Syn. model and the QSPR sampling model for a group of synthetic fuels in the value range 43.8 to 44.3 MJ/kg. The models predict very similar values for different fuels with different measurement values. These systematic errors were also observed by Yang et al[7]. The corresponding synthetic fuels all contain a large fraction of iso-alkanes. Yang et al. explained the systematic errors with errors in the net heat of combustion measurements of the fuels and the high uncertainties due to the low reproducibility of the measurement method. Considering the reproducibility of 0.324 MJ/kg of the measurement method, all models are able to predict the net heat of combustion accurately, with MAE ranging from 0.05 MJ/kg to 0.18 MJ/kg for conventional fuels to 0.09 MJ/kg to 0.19 MJ/kg for synthetic fuels. All models can therefore be utilized for the use case of fuels screening.

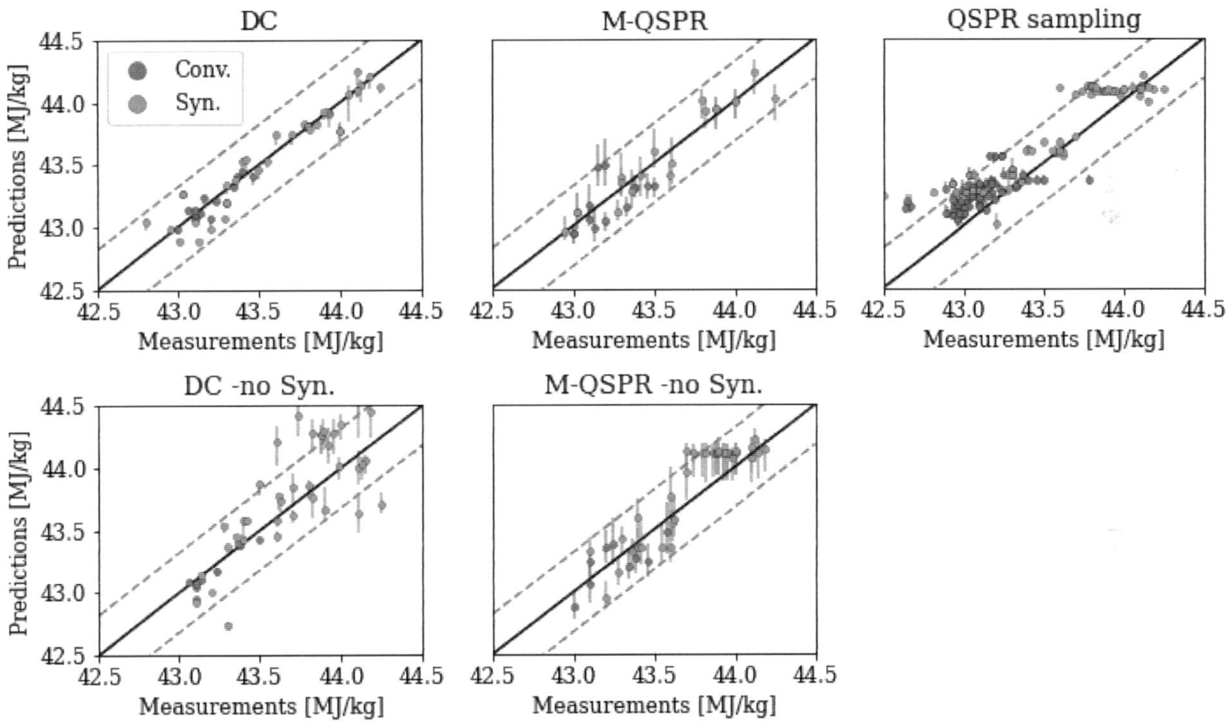

Figure 12: Validation results of the net heat prediction. Results for conventional fuels are displayed in blue, results of synthetic fuels in green.

Kinematic viscosity

For the kinematic viscosity the predictions of all models except the DC -no Syn. model closely follow the unity line for measurements up to 12 mm^2/s, as shown in Figure 13. Above this range significant deviations for a group of synthetic fuels are visible, especially for the QSPR sampling model and M-QSPR -no Syn model. The erroneous predictions again belong to research fuels from the JETSCREEN project with an unusual high fraction of aromatics. The highest accuracies are observed with the DC model and the M-QSPR model with MAE of 0.24 mm^2/s and 0.33 mm^2/s for conventional fuels and 0.61 mm^2/s and 0.58 mm^2/s for synthetic fuels. Models that directly learn on fuel data therefore show the highest accuracies. The scattering and deviations of the DC -no Syn. model and the M-QSPR -no Syn. model illustrate the need for synthetic fuel data as part of the training data for those models. The deviations of the predictions from the QSPR sampling model above 12 mm^2/s can probably be explained with the low number of available measurements at low temperature ranges, below -10 °C. For the pure compounds only a small number of viscosity measurements are available for aromatic components. The underlying QSPR model might therefore underestimate the true kinematic viscosity value of the sampled aromatic components, which subsequently leads to an underestimation of the kinematic viscosity of the fuel. Intermolecular interactions of the fuel components at the lower temperature ranges of -20 °C to -40 °C and the wrong selection of the isomers could also play a role.

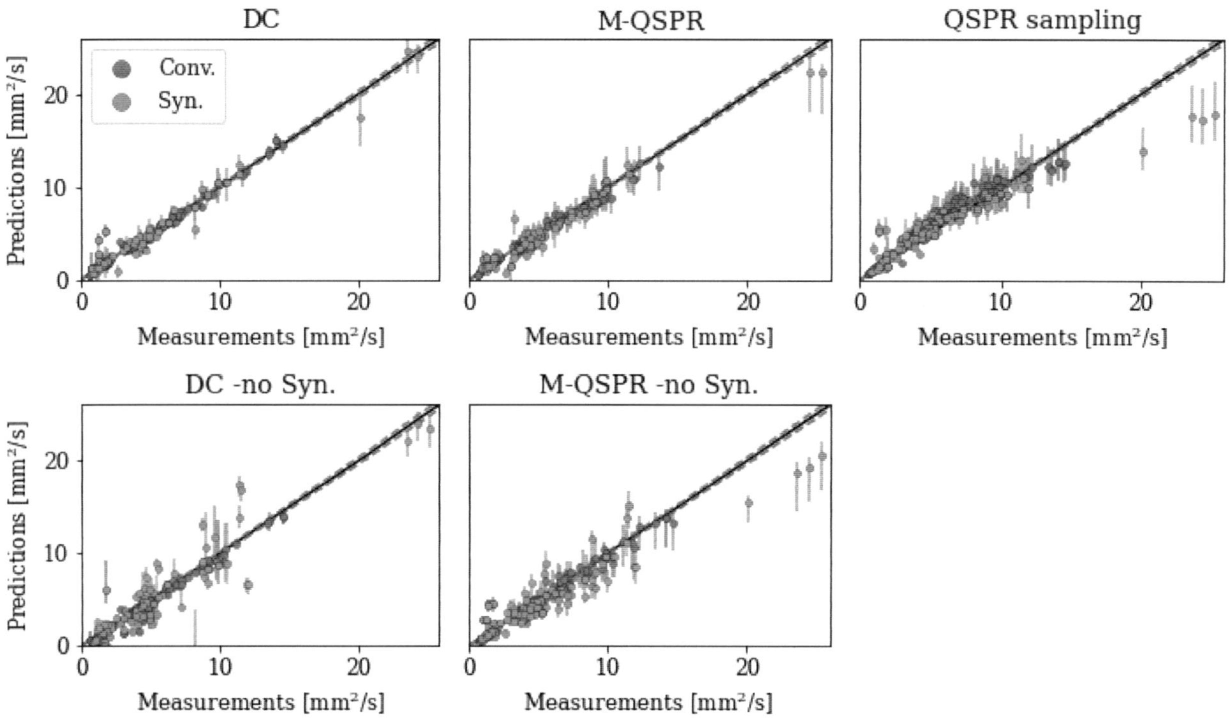

Figure 13: Validation results of the kinematic viscosity prediction. Results for conventional fuels are displayed in blue, results of synthetic fuels in green.

All in all, the DC model and the M-QSPR model show the highest predictive capability and should therefore be utilized for the use case of fuel screening. For the QSPR sampling methods significant deviations could occur for higher predicted viscosity ranges, above 12 mm^2/s. Furthermore, the predicted PI should be increased by the estimated MAOE since the calculated PICP for synthetic fuel predictions of the models lie significantly below the set confidence level of 95% with a PICP of 78.75 % for the DC model and 86.8 % for the M-QSPR model.

Flash point

The results of the flash point are presented in Figure 14. With the exception of the DC -no Syn. model most predictions follow the unity line, however stronger scattering and significant deviations are observed especially for synthetic fuels. Striking is the erroneous prediction at 49 °C, which is observed for all models. It corresponds to a synthetic fuel produced by the Alcohol-To-Jet process, which is solely composed of iso-alkanes. Other outliers also correspond to synthetic fuels composed of only one or two hydrocarbon families with GCxGC measurements dominated by only a few bins.

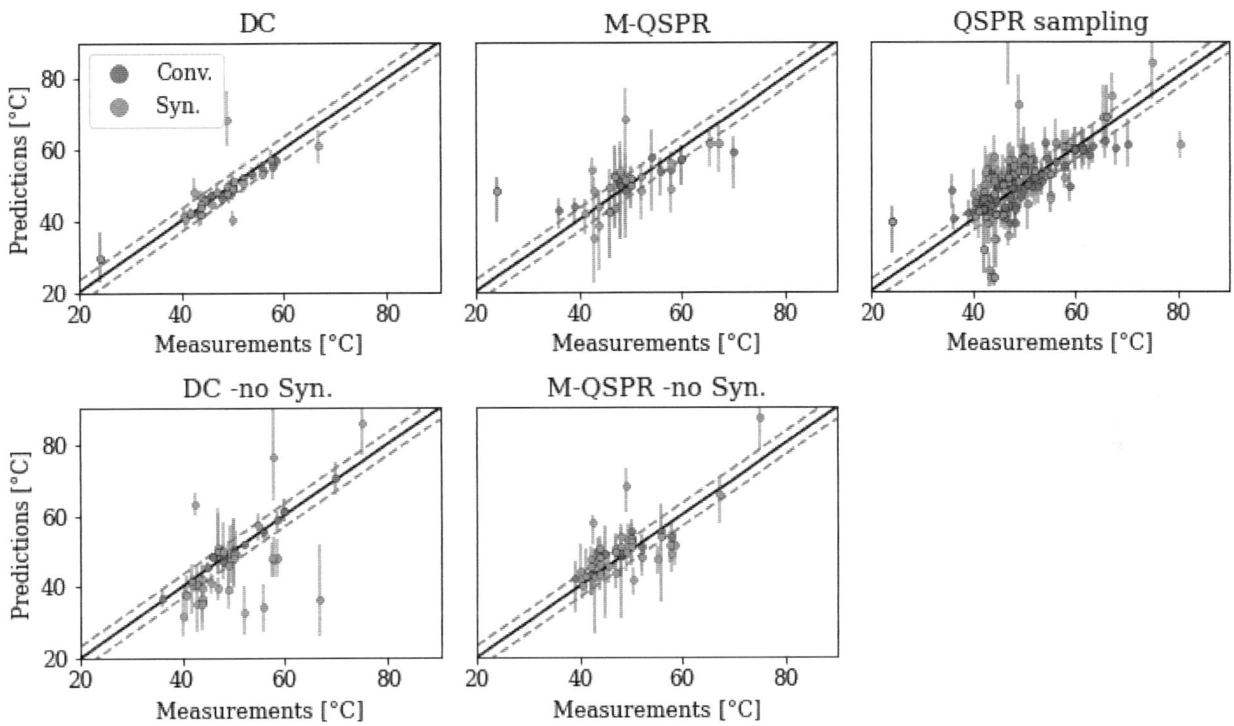

Figure 14: Validation results of the flash point prediction. Results for conventional fuels are displayed in blue, results of synthetic fuels in green.

In general, the higher scattering can be explained with the smaller dataset and larger variance of the property values of the flash point for the different fuels and isomers. Only data for 397 unique components and fuels is available for the flash point, compared to 1622 unique components and fuels for the net heat of combustion. Simultaneously the variance of the property values increases for different isomers. This has a significant influence on the QSPR sampling model, where strong deviations and uncertainties for synthetic fuels can be explained by those variances. E.g. for iso-alkanes with 10 carbon atoms the values in the database range between 15.85 °C to 51.7 °C. Since the QSPR sampling model assumes all isomers of the considered hydrocarbon families to be potentially present in the fuels, this can induce significant deviations and uncertainties. For conventional fuels the influence of isomers is smaller due to more distributed compositions. For the QSPR sampling method more detailed compositional information down to the molecular level or a restriction of the possible isomers is probably needed, to accurately model the described synthetic fuels. In comparison the DC model shows the highest predictive capability with a MAE of 0.99 °C for conventional and 2.43 °C for synthetic fuels. The higher accuracy and lower uncertainty can be explained with the training and validation dataset of the DC model. The training and validation dataset of the DC model does not contain variances of the flash point due to possible isomers. For the use case of fuel screening the utilization of the DC model is therefore recommended.

Freezing point

For the freezing point only the predictions for conventional fuels accurately follow the unity line over the full value range, see Figure 15. For the DC model and M-QSPR strong deviations are observed for synthetic fuels below -60 °C.

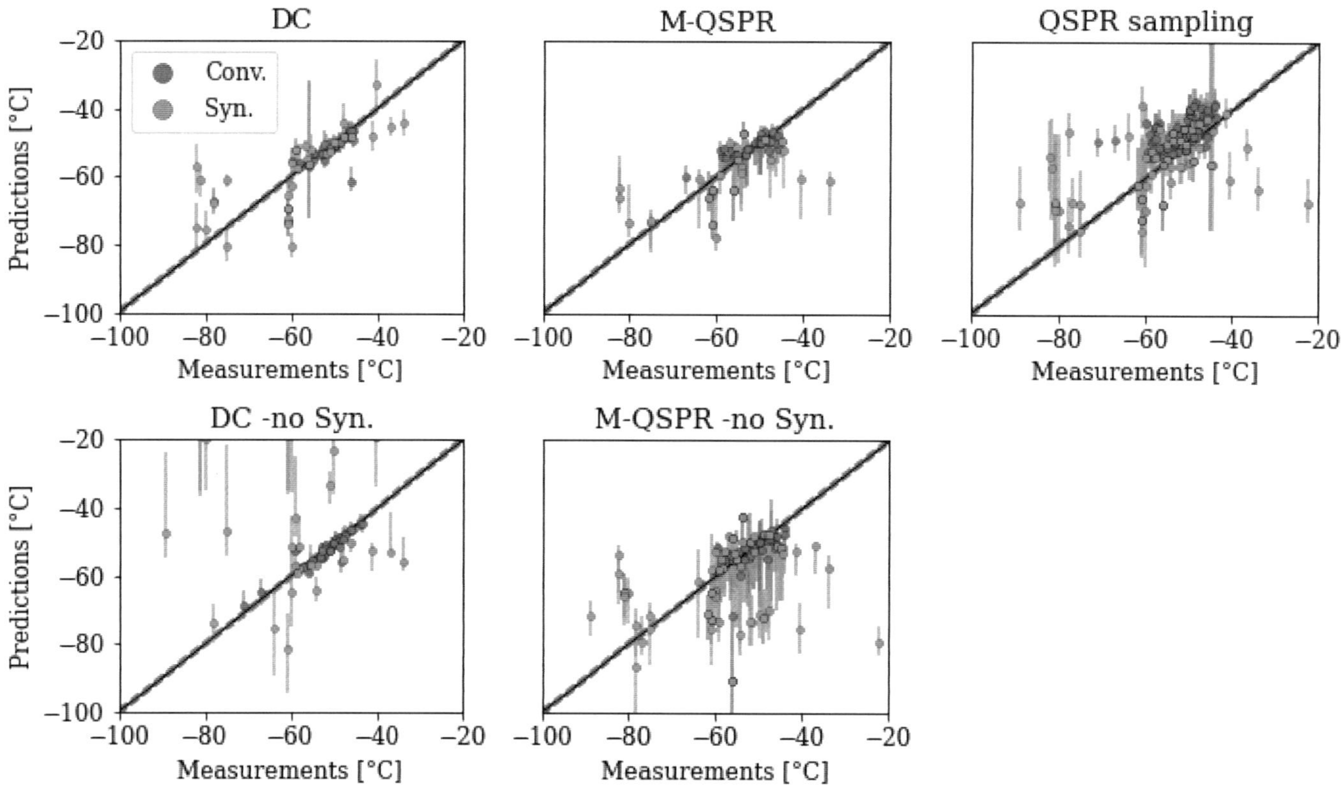

Figure 15: Validation results of the freezing point prediction. Results for conventional fuels are displayed in blue, results of synthetic fuels in green.

For the QSPR sampling model large uncertainties are visible, while the M-QSPR no-Syn. model underestimates and the DC -no Syn. model strongly overpredicts the freezing point of synthetic fuels. Likewise, to the flash point, the large uncertainties and significant deviations of the M-QSPR model and the QSPR sampling model can be explained with the smaller dataset and strong influence of the different isomers on the freezing point. For iso-alkanes with 10 carbon atoms value differences of -110.15 °C to -12 °C are listed in the pure compound databases. This explains the strong deviations for the QSPR sampling model, and shows the need for more detailed compositional information down to the molecular level, or a restriction of the sampled isomers. For the application of the screening of fuels the DC model and the M-QSPR model show the highest accuracies with MAE of 1.51 °C and 2.24 C for conventional fuels and 7.14 °C and 7.3 °C for synthetic fuels. The strong deviations observed for fuels with freezing point below -60 °C would however limit the application domain to a value range above -60 °C. Furthermore, an increase of the PI is recommended since the calculated PICP are significantly lower compared to the confidence level of 95 %, with 89.7 % and 79 % for conventional fuels and 48.65 % and 50 % for synthetic fuels for DC model and M-QSPR model respectively.

Cetane number

Since the number of available fuel data for the cetane number with only 36 datapoints is severely limited, see Table 1, and significantly smaller than the number of input features for the DC correlation, significant overfitting of the algorithm was observed. Therefore, the mass fractions of the range 1-12 and 12-25 carbon atoms had to be summed up, creating the features n-alkanes 1-12, n-alkanes 12-25, iso-alkanes 1-12 etc. for the DC correlation which reduced the number of features from 200 to 16 and successfully prevented overfitting. The results for this modified DC model and all other models are shown in Figure 16. The unity plots show that accurate prediction results for both conventional and synthetic fuels could only be achieved for the DC model and the M-QSPR model. Accuracies with a MAE of 1.02 and 1.4 for the conventional fuels and 3.74 and 3.7 for the synthetic fuels are calculated for the DC model and the M-QSPR model respectively. For the other models only the predictions of conventional fuels lie close to the unity line. For the QSPR sampling method the predictions for all synthetic fuels are in a similar value range of 35 to 45, with large uncertainties especially for measurements below 20. This can be explained with the strong variance of the cetane number for different isomers in a GCxGC bin. The cetane number is strongly influenced by the branching of a component, low branched components similar to cetane (hexadecane) have a high cetane number, while highly branched isomers show lower cetane numbers. Since the QSPR sampling method assumes all isomers of the considered hydrocarbon families to be equally present, significant deviations and large uncertainties are possible for synthetic fuels composed of only a few specific isomers. For the QSPR sampling model a more detailed composition information down to the molecular component level of the fuel or a restriction of possible isomers might be necessary to improve the accuracy for those synthetic fuels. For the use case of fuel screening only the utilization of the DC model and the M-QSPR model is therefore recommended.

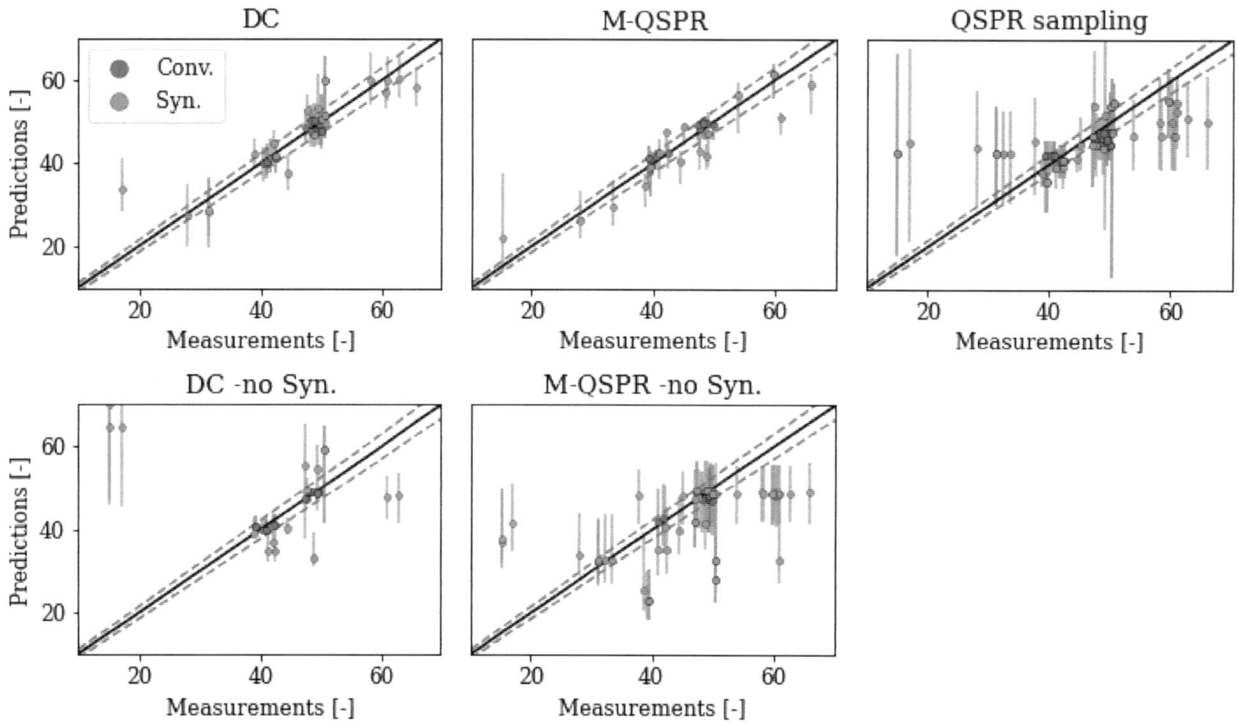

Figure 16: Validation results of the cetane number prediction. Results for conventional fuels are displayed in blue, results of synthetic fuels in green.

Distillation line

ASTM D4054 lists thresholds for the distillation of jet fuels at 10, 50 and 90 vol% recovery. The results are shown in Figure 17, Figure 18 and Figure 19 respectively, for the DC model and the QSPR sampling model. A modelling with the M-QSPR method does not bring any benefit compared to the DC method, since only fuel data can be used for the correlation. For both the DC model and the QSPR model highly accurate results are achieved for conventional fuels with a MAE below 5 °C for both models. For synthetic fuels significant deviations are observed for the DC model trained solely on conventional fuels and partially also for the QSPR sampling model. For the DC model this again shows the need for fuels synthetic fuels as part of the training data. For the QSPR sampling model the deviations correspond to fuels with compositions dominated by only a few GCxGC bins like fuels produced by the ATJ process and fuel surrogates. These fuels are composed of only a small number of distinct components, which results in an unsteady distillation line. It was observed that the utilized correlation of the ASTM D2887, which is used to convert the simulated distillation line calculated on the basis of the boiling points to ASTM D86 values, returns erroneous and unphysical values for these synthetic fuels: Increasing temperatures are calculated partially for decreasing evaporated volumes. In those cases, we utilized the temperature at the higher evaporated value. An update of the ASTM D2887 correction for synthetic fuels like the ATJ fuels might therefore be necessary in the future.

The large uncertainties of the QSPR sampling model are in large part due to the reproducibilities of the ASTM D2887 correction of 10.73 % for 10 vol%, 6.96 % for 50 vol% and 8.85 % for

90 vol%, which are add to the predicted PI of the QSPR sampling model. For the conventional fuels the reproducibilities clearly overstate the observed uncertainty and value range with PICP values close to 100 %, whereas for synthetic fuels the uncertainties are on average sufficient with PICP close to 95 %. This underlines the necessity of a more detailed composition information for synthetic fuels composed of only a few molecules. This would probably not only increase the accuracy of the predictions, but also allow a reduction of the reproducibilities of the ASTM D2887 correction model. As for the use case of fuel screening both models can be utilized. However, the large uncertainties of the QSPR sampling model, due to the reproducibilities of the ASTM D2887 method and some sporadic outliers for fuels with compositions like ATJ fuels show that significant deviations are possible.

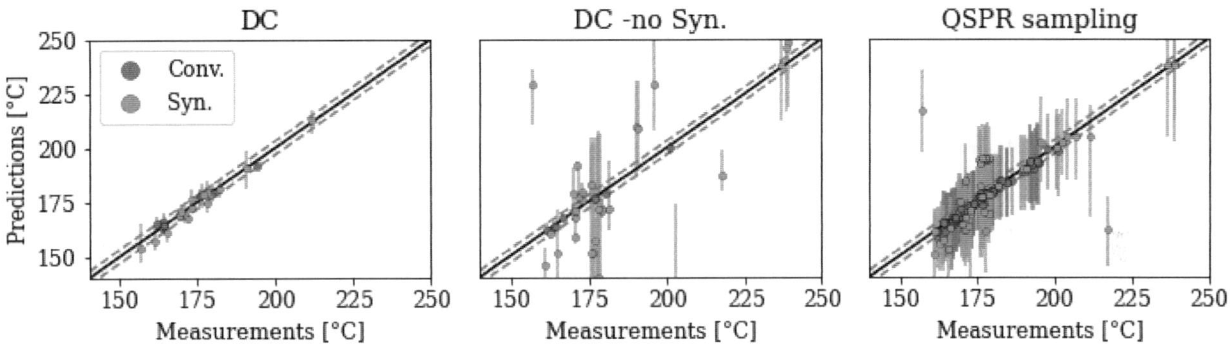

Figure 17: Validation results of the distillation with 10 vol% evaporated volume prediction. Results for conventional fuels are displayed in blue, results of synthetic fuels in green.

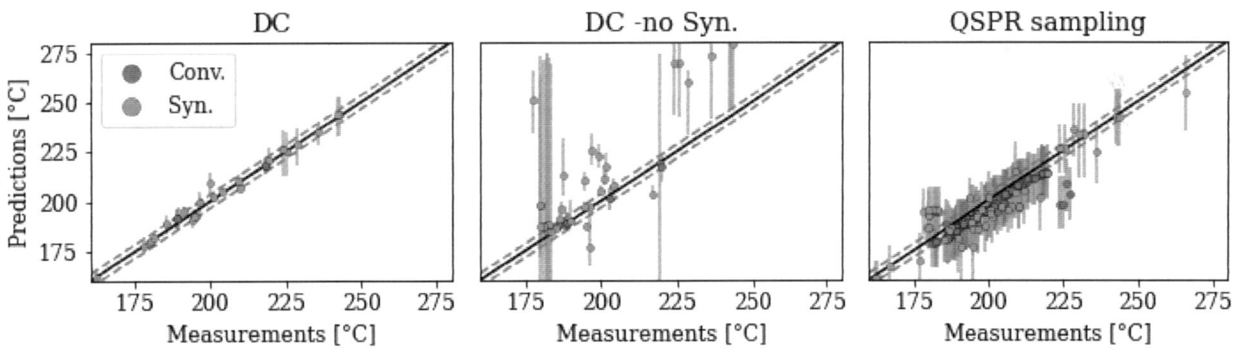

Figure 18: Validation results of the distillation with 50 vol% evaporated volume prediction. Results for conventional fuels are displayed in blue, results of synthetic fuels in green.

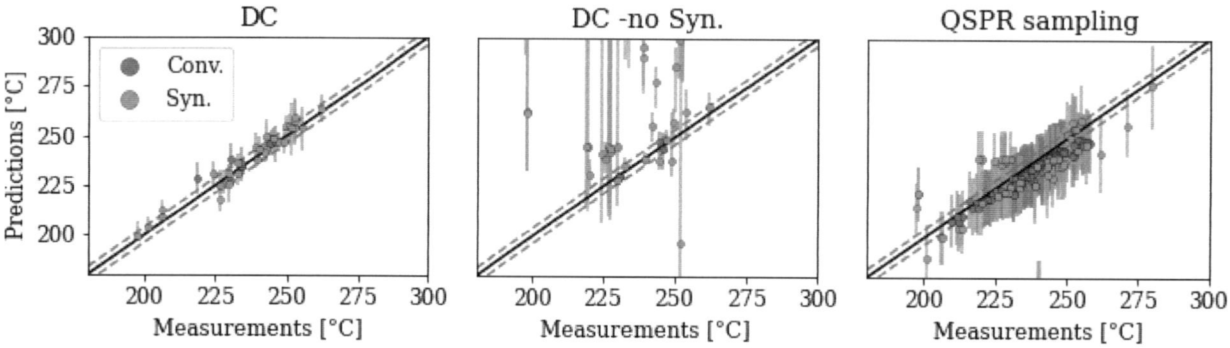

Figure 19: Validation results of the distillation with 90 vol% evaporated volume prediction. Results for conventional fuels are displayed in blue, results of synthetic fuels in green.

Tier α jet fuel candidate prescreening

To test the applicability of the three models we simulate a blind test according to the Tier α jet fuel prescreening protocol developed by Heyne and Rauch [4]. The models are therefore trained on the full datasets listed in Table 1. In the prescreening protocol properties are predicted relative to the specification limits of ASTM D4054 in order to assess their conformity as drop-in jet fuels that could potentially be utilized without an additional blending with conventional fuel. The prediction results for the prescreening are shown in Figure 20 with separate columns for each screened fuel. For this blind test a high accuracy as well as the correct classification of the fuel property inside or outside of the specification limits under consideration of the predicted uncertainties is of importance. The results of the prescreening are displayed in Figure 20. Predictions of the DC model are shown in blue, predictions of the M-QSPR model in green and predictions of the QSPR model in purple. The property measurements of the fuels are displayed in black, with error bars indicating the reproducibility of the measurement method. The specification limits of the respective properties are shown as red lines, value ranges out of these specifications are shaded in red.

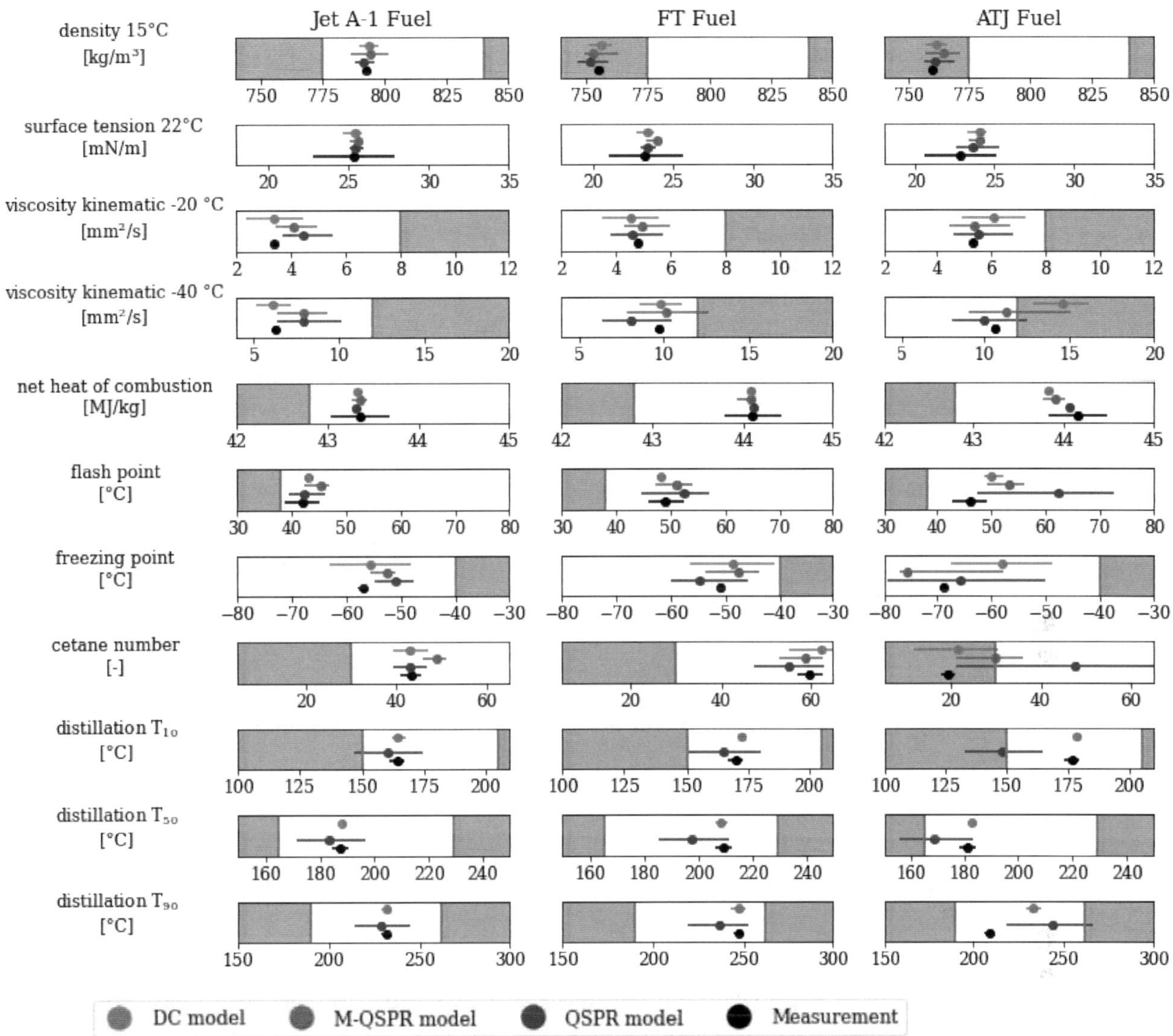

Figure 20: Results of Tier α prescreening: Jet A-1 fuel (left row), FT SAF (middle) and ATJ SAF (right). DC model predictions are displayed in blue, of the M-QSPR model in green and the QSPR sampling model in purple.

For the Jet A-1 fuel and the FT fuel the predictions of all models are highly accurate and the PI sufficiently precise. All properties are estimated correctly either inside or outside the specification limits, no predicted PI crosses the specification lines with the exception of the distillation prediction of the QSPR sampling method. As explained in the distillation validation section, the high uncertainty can be contributed to the conversion correlation of the ASTM D2887. For the DC model the high precision and high accuracy of the Jet A-1 and FT fuel prediction can be explained by the presence of similar fuels in the training data. For the M-QSPR model and the QSPR sampling model the high accuracy is can be explained by the composition of the fuels; broad distributions over the GCxGC bins of families and neglectable effects of isomers. This corresponds to the assumptions of the M-QSPR method of an averaged pseudo structure and the assumption of the QSPR sampling method of sampling of all possible isomers. For the ATJ fuel a good accordance of all the model predictions is only observed for the density, surface tension and kinematic viscosity at -20 °C. For the other properties the predictions contradict each other and the predictions of the QSPR sampling model predictions show large uncertainties. The PI cross the specification limits in multiple cases. In a real prescreening scenario, property measurement would be recommended since the predictions are highly conflicting and uncertain. The conflicting predictions and the uncertainty can again be explained with the significant effect of unidentified isomers. The assumptions of the QSPR model are not valid for the ATJ fuel composition dominated by the two GCxGC bins iso-alkanes 12 and iso-alkanes 16. Detailed information beyond the GCxGC measurement might be necessary for the QSPR sampling method.

Since the QSPR sampling method is a very flexible modelling approach that allows the definition of the considered components down to the individual molecule, we investigated the influence of a constrained component selection of the QSPR sampling method. We repeated the prediction with the QSPR model, considering only components with high amounts of branching, since it is known that ATJ fuels are produced from branched alcohols. We therefore constrained the sampled components to the 33% with the most methyl groups, described by the [CX4H3] descriptor. A high number of methyl groups corresponds to molecules with multiple branches and branch endings. The results of the predictions can be seen in Figure 21. The predictions of the original QSPR sampling model are again shown in purple, the predictions of the QSPR sampling model with constrained molecules is shown in pink. The results show improvements for the accuracy of the flash point, freezing point, flash and cetane number prediction, while the accuracy of the density is reduced. This illustrates, that the restriction of the isomers has strong potential to improve the overall accuracy of the QSPR modelling method. However, the decrease of the previously highly accurate density shows that information about the exact molecules is necessary to correctly model fuels like the ATJ fuel.

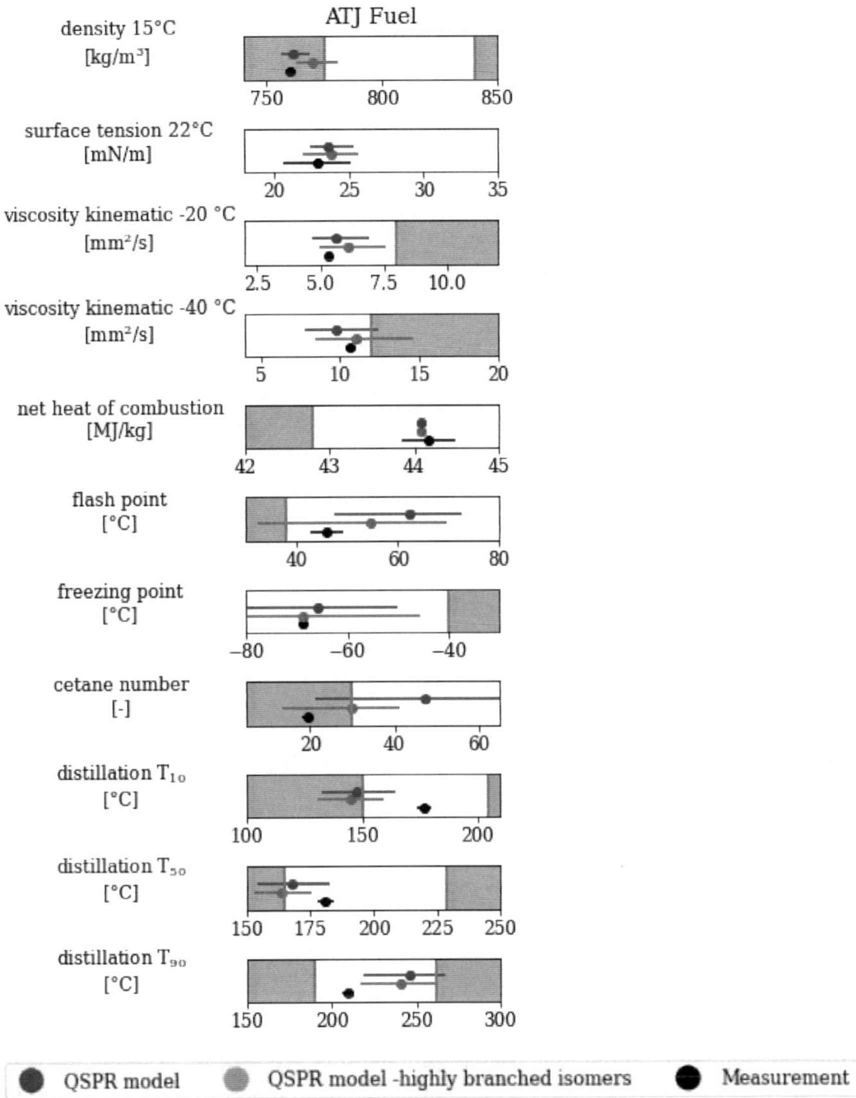

Figure 21: Results of Tier α prescreening for ATJ SAF, predictions of the QSPR sampling model with all isomers displayed in purple and predictions of the QSPR sampling model sampling only highly branched iso-alkanes in pink.

CONCLUSION

Probabilistic modelling methods play a central role in fuel screening methods, that enable the assessment of critical jet fuel candidates on the bases of low volume composition measurements. Their ability to predict critical physicochemical properties and corresponding uncertainties, due to e.g. unidentified isomers, accurately and reliably, allows risk informed decision making in the early development of fuel candidates.

In this work, we compared three different probabilistic property modelling methods that exist in the current scientific literature: Direct correlation (DC), Mean Quantitative Structure-Property Relationship Modelling (M-QSPR) and Quantitative Structure-Property Relationship Modelling (QSPR) with sampling. We compared their predictive capability, explored their individual strengths and limitations and investigated the benefit of considering predictions of multiple models simultaneously. Using a Monte-Carlo Dropout Neural Network as probabilistic regression algorithm, we trained and validated models from those methods and assessed their predictive capability for the prediction of 8 critical physicochemical properties. The assessments were conducted on the property prediction of 75 conventional oil-based jet fuels and 56 synthetic fuels produced by Alcohol-To-Jet process, the Fischer-Tropsch process, Hydroprocessed Esters and Fatty Acids, the Integrated Hydropyrolysis process of Shell with a high cyclo-alkanes content Cyclo Paraffinic Kerosene, the Catalytic Hydrothermal Conversion Jet fuel process as well as fuel blends.

In the validation the highest predictive capability was achieved on average by models of the DC method. With the exception of the freezing point highly accurate and precise results were achieved for fuels of all considered fuel types and production paths. The comparison with DC models trained solely on conventional fuels however, revealed the need of this method for fuels similar to the predicted fuels as part of the training data. If the DC model is trained solely on conventional fuels, accuracy and precision for the prediction of synthetic fuels are significantly reduced. The M-QSPR model showed less dependence on the synthetic fuel data, and the second-best performance for most of the properties. The possibility to learn from both data of pure compounds and fuels proved to be beneficial, especially for cases with a low number of available fuel measurements, like for the modelling of the cetane number. For the QSPR sampling model high predictive capability is only achieved for properties with an insignificant influence of different isomers on the properties: density, surface tension and net heat of combustion or fuels with a broad distribution over the different hydrocarbon families. This became especially visible for the modelling of the flash point, freezing point and cetane number, for which large uncertainties and significant deviations were observed for synthetic fuels with distinct compositions, dominated by just a few GCxGC bins. For these properties the unknown isomers significantly affect the property value. The underlying assumptions of the M-QSPR and the QSPR sampling model, that all known isomers are equally present, is therefore not valid for some synthetic fuels. Since the QSPR sampling model is the most flexible modelling method, that allows the definition of the fuel down to the molecular level, higher accuracies could probably be achieved, if more detailed fuel composition is known, or if the range of possible isomers is constrained.

To review the applicability of the individual models for the practical use case of fuel screening, we also conducted a blind test for one conventional Jet A-1 fuel and two synthetic fuels produced by the Fischer-Tropsch process (FT) and the Alcohol-to-Jet process (ATJ). As for the model validation, the highest predictive capability was on average observed for the DC model, followed by the M-QSPR model and the QSPR sampling model. For the Jet A-1 fuel all models accurately and precisely predicted all properties. None of the predicted uncertainty range crossed the specification lines with the exception of the distillation prediction of the QSPR sampling method. The fuels were therefore correctly classified as drop-in or non-drop-in fuels. For the ATJ fuel however significant deviations and disagreements between the different models were observed especially for properties with a significant influence of the different isomers. Since the ATJ fuel is predominantly composed of a few distinct iso-alkanes, the underlying assumption of the QSPR sampling model is again not valid. Generally, the consideration of predictions from multiple models was found to be beneficial for both the accuracy and the trustworthiness of predicted properties in the fuel prescreening. If predictions from the different models were in high accordance with each other, the predicted values were always close to the true measurement. If the predictions differed and large uncertainties were observed, significant errors could occur and the use of additional property measurements would be recommended.

In sum, this study demonstrates the capabilities and limitations of model-based predictions for jet fuels up to now. Some of the properties can be predicted with high accuracy and precision for fuels that are either similar to the known fuels used for the training of the model or for fuels with compositions complying with the underlying assumptions. For fuels where this is not the case, a more detailed compositional information that either constrains the selection of possible molecules or directly identifies them is needed for the accurate modelling. We illustrated this by the modelling of the ATJ fuel with a composition constrained to molecules with a high amount of intramolecular branching. If this is not possible, property measurements are still recommended.

Abbreviations

ATJ	Alcohol-To-Jet
Comp.	Pure compounds
Conv.	Conventional fuels
DC	Direct correlation
GCxGC	Two-dimensional Gas-Chromatography
FT	Fischer-Tropsch
MAE	Mean absolute error
MAOE	Mean absolute outlier error
M-QSPR	Mean Quantitative Structure-Property Relationship Modelling
NMPIW	Normalized mean prediction interval width
PI	Prediction interval
PICP	Prediction interval coverage probability
QSPR	Quantitative Structure-Property Relationship Modelling
SAF	Sustainable aviation fuel
SMILES	Simplified Molecular Input Line Entry System
SMARTS	SMILES arbitrary target specifications
Syn.	Synthetic fuels

ACKNOWLEDEMENTS

The research presented in this paper has been performed in the framework of the JETSCREEN project (JET fuel SCREENING and optimization) and has received funding from the European Union Horizon 2020 Programme under grant agreement n◦ 723525.

REFERENCES

(1) Rumizen, M. ASTM D4054 Users Guide **2018**.

(2) B. Rauch et al. *JET fuel SCREENing and optimization (JETSCREEN) project:European Union's Horizon 2020 research and innovation programme under grant agreement No 723525.* https://cordis.europa.eu/project/id/723525, checked on 3/3/2022. (accessed 2022-03-03).

(3) Colket, M.; Heyne, J.; Rumizen, M.; Gupta, M.; Edwards, T.; Roquemore, W. M.; Andac, G.; Boehm, R.; Lovett, J.; Williams, R.; Condevaux, J.; Turner, D.; Rizk, N.; Tishkoff, J.; Li, C.; Moder, J.; Friend, D.; Sankaran, V. Overview of the National Jet Fuels Combustion Program. *AIAA Journal* **2017**, *55* (4), 1087–1104. DOI: 10.2514/1.J055361.

(4) Heyne, J.; Rauch, B.; Le Clercq, P.; Colket, M. Sustainable aviation fuel prescreening tools and procedures. *Fuel* **2021**, *290*, 120004. DOI: 10.1016/j.fuel.2020.120004.

(5) Vozka, P.; Modereger, B. A.; Park, A. C.; Zhang, W. T. J.; Trice, R. W.; Kenttämaa, H. I.; Kilaz, G. Jet fuel density via GC × GC-FID. *Fuel* **2019**, *235*, 1052–1060. DOI: 10.1016/J.FUEL.2018.08.110.

(6) Tim Nelson. Application of GC×GC-VUV and GC×GC-FID for the analysis of common gasoline samples, middle distillates and crude oil distillation cuts using Zebron® ZB-35HT and ZB-1PLUS GC Columns **2019**.

(7) Yang, Z.; Kosir, S.; Stachler, R.; Shafer, L.; Anderson, C.; Heyne, J. S. A GC × GC Tier α combustor operability prescreening method for sustainable aviation fuel candidates. *Fuel* **2021**, *292*, 120345. DOI: 10.1016/j.fuel.2021.120345.

(8) Rocha, W. F. d. C.; Sheen, D. A. Determination of physicochemical properties of petroleum derivatives and biodiesel using GC/MS and chemometric methods with uncertainty estimation. *Fuel* **2019**, *243*, 413–422. DOI: 10.1016/j.fuel.2018.12.126.

(9) Hall, C.; Rauch, B.; Bauder, U.; Le Clercq, P.; Aigner, M. Predictive Capability Assessment of Probabilistic Machine Learning Models for Density Prediction of Conventional and Synthetic Jet Fuels. *Energy Fuels* **2021**, *35* (3), 2520–2530. DOI: 10.1021/acs.energyfuels.0c03779.

(10) Hall, C.; Creton, B.; Rauch, B.; Bauder, U.; Aigner, M. Probabilistic Mean Quantitative Structure–Property Relationship Modeling of Jet Fuel Properties. *Energy Fuels* **2022**, *36* (1), 463–479. DOI: 10.1021/acs.energyfuels.1c03334.

(11) Ajmani, S.; Rogers, S. C.; Barley, M. H.; Livingstone, D. J. Application of QSPR to mixtures. *Journal of chemical information and modeling* **2006**, *46* (5), 2043–2055. DOI: 10.1021/ci050559o.

(12) NIST. *NIST Standard Reference Database 103a*. https://www.nist.gov/mml/acmd/trc/thermodata-engine/srd-nist-tde-103a (accessed 2022-01-17).

(13) *DIPPR 801 Database*. https://www.aiche.org/dippr/events-products/801-database (accessed 2022-01-18).

(14) Saldana, D. A.; Starck, L.; Mougin, P.; Rousseau, B.; Pidol, L.; Jeuland, N.; Creton, B. Flash Point and Cetane Number Predictions for Fuel Compounds Using Quantitative Structure Property Relationship (QSPR) Methods. *Energy Fuels* **2011**, *25* (9), 3900–3908. DOI: 10.1021/ef200795j.

(15) Saldana, D. A.; Starck, L.; Mougin, P.; Rousseau, B.; Ferrando, N.; Creton, B. Prediction of Density and Viscosity of Biofuel Compounds Using Machine Learning Methods. *Energy Fuels* **2012**, *26* (4), 2416–2426. DOI: 10.1021/ef3001339.

(16) Saldana, D. A.; Starck, L.; Mougin, P.; Rousseau, B.; Creton, B. On the rational formulation of alternative fuels: melting point and net heat of combustion predictions for fuel compounds using machine learning methods. *SAR and QSAR in environmental research* **2013**, *24* (4), 259–277. DOI: 10.1080/1062936X.2013.766634. Published Online: Apr. 10, 2013.

(17) *Daylight Theory: SMILES*. https://www.daylight.com/ (accessed 2022-03-04).

(18) *RDKit*. https://www.rdkit.org/ (accessed 2022-03-04).

(19) GRUNBERG, L.; NISSAN, A. H. Mixture law for viscosity. *Nature* **1949**, *164* (4175), 799. DOI: 10.1038/164799b0.

(20) Gal, Y.; Ghahramani, Z. *Dropout as a Bayesian Approximation: Representing Model Uncertainty in Deep Learning*. http://arxiv.org/pdf/1506.02142v6.

(21) *PyTorch*. https://pytorch.org/ (accessed 2022-03-09).

(22) PubChem. *PubChem*. https://pubchem.ncbi.nlm.nih.gov/ (accessed 2022-03-04).

(23) *ChemSpider*. https://www.chemspider.com/ (accessed 2022-03-04).

(24) J. Yanowitz; M.A. Ratcliff, R.L. McCormick, and J.D. Taylor; M.J. Murphy. Compendium of Experimental Cetane Numbers **2017**.

(25) Li, R.; Herreros, J. M.; Tsolakis, A.; Yang, W. Machine learning regression based group contribution method for cetane and octane numbers prediction of pure fuel compounds and mixtures. *Fuel* **2020**, *280*, 118589. DOI: 10.1016/j.fuel.2020.118589.

(26) CRC. World Fuel Sampling Program Final Report with CRC Cover.12Oct06.doc **2006**.

(27) Voigt, C.; Kleine, J.; Sauer, D.; Moore, R. H.; Bräuer, T.; Le Clercq, P.; Kaufmann, S.; Scheibe, M.; Jurkat-Witschas, T.; Aigner, M.; Bauder, U.; Boose, Y.; Borrmann, S.; Crosbie, E.; Diskin, G. S.; DiGangi, J.; Hahn, V.; Heckl, C.; Huber, F.; Nowak, J. B.; Rapp, M.; Rauch, B.; Robinson, C.; Schripp, T.; Shook, M.; Winstead, E.; Ziemba, L.; Schlager, H.; Anderson, B. E. Cleaner burning aviation fuels can reduce contrail cloudiness. *Commun Earth Environ* **2021**, *2* (1). DOI: 10.1038/s43247-021-00174-y.

(28) molgen group. *MOLGEN*. https://www.molgen.de/ (accessed 2022-03-10).

(29) Henze, H. R.; Blair, C. M. THE NUMBER OF ISOMERIC HYDROCARBONS OF THE METHANE SERIES. *J. Am. Chem. Soc.* **1931**, *53* (8), 3077–3085. DOI: 10.1021/ja01359a034.

(30) Thom MA. *CRC Project No. AV-23-15/17:Review of Existing Test Methods Used for Aviation Jet Fuel and Additive Property Evaluations with Respect to Alternative Fuel Compositions*; CRC, 2018.

COMPARISON OF PROBABILISTIC JET FUEL PROPERTY MODELS FOR THE FUEL SCREENING AND DESIGN

Clemens Hall, Bastian Rauch, Uwe Bauder, Manfred Aigner

DLR, German Aerospace Center, Institute of Combustion Technology, MAT, 70569 Stuttgart, Germany, clemens.hall@dlr.de

Table S1: Description of utilized structural molecular features

Feature	Description
[CX4H3]	Aliphatic C with 1 further total connection with 3 further hydrogen
[CX4H2]	Aliphatic C with 2 further total connections with 2 further hydrogen
[CX4H1]	Aliphatic C with 3 further total connections with 1 further hydrogen
[CX4H0]	Aliphatic C with 4 further total connections with 0 further hydrogen
[CX4H1]	Aliphatic C with 1 further total connection with 2 further hydrogen
[CX3H1]	Aliphatic C with 2 further total connections with 1 further hydrogen
[CX3H0]	Aliphatic C with 3 further total connections with 0 further hydrogen
[CX2H1]	Aliphatic C with 1 further total connection with 1 further hydrogen
[CX2H0]	Aliphatic C with 2 further total connections with 0 further hydrogen
[CX4H2R]	Aliphatic C with 2 further total connections 2 further hydrogen, in a ring
[CX4H1R]	Aliphatic C with 3 further total connections 1 further hydrogen, in a ring
[CX4H0R]	Aliphatic C with 4 further total connections 0 further hydrogen, in a ring
[CX3H1R]	Aliphatic C with 2 further total connections 1 further hydrogen, in a ring

[CX3H0R]	Aliphatic C with 3 further total connections 0 further hydrogen, in a ring
[cX3H1](:*):*	Aromatic C with 0 further total connections with 1 further hydrogen
[cX3H0](:*)(:*)*	Aromatic C with 0 further total connections with 0 further hydrogen
[cX3H0](:*)(:*):*	Aromatic C with 0 further total connections with 0 further hydrogen
[cX3H0]-[cX3]	Aromatic C with 2 further total connections with 0 further hydrogen
[cX3H0](:*)(:*)(:*)	Aromatic C with 0 further total connections with 0 further hydrogen
[cX3H0](:*)(:*)(-[CX4H2R])	Aromatic C with 0 further total connections with 0 further hydrogen with connection to an aliphatic C with 1 further total connection with 2 further hydrogen, in a ring
[CX4H2]-[CX4H1]-[CX4H2]	Bound of aliphatic C with 1 further total connection with 1 further hydrogen
[C][C]([!CX1])([!CX1])[!CX1]	Bound of aliphatic C with one further aliphatic C and 3 non C atoms with 0 further total connections
[!C][C]([C])([C])[C]	Bound of aliphatic C with two 3 further aliphatic C and one non C atom
[C][C]([C])([C])[C]	Bound of aliphatic C with 4 further aliphatic C
[C][CR]([!C])([!C])[C]	Bound of aliphatic C in a ring with two further aliphatic C and two non C atoms
[C][CR]([!C])([C])[C]	Bound of aliphatic C in a ring with three further aliphatic C and one non C atoms
[C][CR]([C])([C])[C]	Aliphatic C in a ring with 4 further aliphatic C
[cH1]	Aromatic C with 1 further connection
[cH0]	Aromatic C with 0 further connections
[C]=[C]([!CX1])[!CX1]	Bound of 2 aliphatic C with branched double bound and 2 non C atoms with 0 further connections
[C]=[C]([C])[!C]	Bound of 3 aliphatic C and one non C atom, with branched double bound between first and second C

[C]=[C]([C])[C]	Bound of 4 aliphatic C, with branched double bound between second and third C
[C]=[C]=[C]	3 aliphatic C with double bounds
[!CX1][C]#[C]	Bound of aliphatic C with triple bound to aliphatic C and C with 0 further total connections
[!X1][C]#[C]	Bound of aliphatic C with triple bound to aliphatic C and one atom with 0 further connections
[!#1][CH2][!#1]	Bound of aliphatic C with two further hydrogen and two atoms that are not hydrogen atoms
[CX3H1]=[CX3H0]	Bound of aliphatic C with 1 further total connection with 1 further hydrogen and aliphatic C with 2 further total connections with 0 further hydrogen
[CX3H0]=[CX3H0]	Bound of two aliphatic C with 2 further total connections with 0 further hydrogen
[CX3H2]=[CX3H1]	Bound of aliphatic C with 0 further total connections with 2 further hydrogen with aliphatic C with 1 further total connection with 1 further hydrogen
[CX3H1]=[CX3H1]	Bound of two aliphatic C with 1 further total connection with 1 hydrogen
[CX3H2]=[CX3H0]	Bound of aliphatic C with 0 further total connections with 2 further hydrogen with aliphatic C with 2 further total connections with 0 further hydrogen
[c][CX4H3]	Bound of aromatic C with aliphatic C with 0 further total connections with 3 further hydrogen
[c][CX4H2]	Bound of aromatic C with aliphatic C with 1 further total connection with 2 further hydrogen
[c][CX4H1]	Bound of aromatic C with aliphatic C with 2 further total connections with 1 further hydrogen
[CX2H1]#[CX2H0]	Triple bound of aliphatic C with 0 further total connections with 1 further hydrogen and aliphatic C with 1 further total connection with 0 further hydrogen
[CX2H0]#[CX2H0]	Aliphatic C with triple bond and 1 further total connection 0 further hydrogen
[R]	Number of atoms in ring

Figure S2: Cross-validation results training and testing for density prediction of the DC model

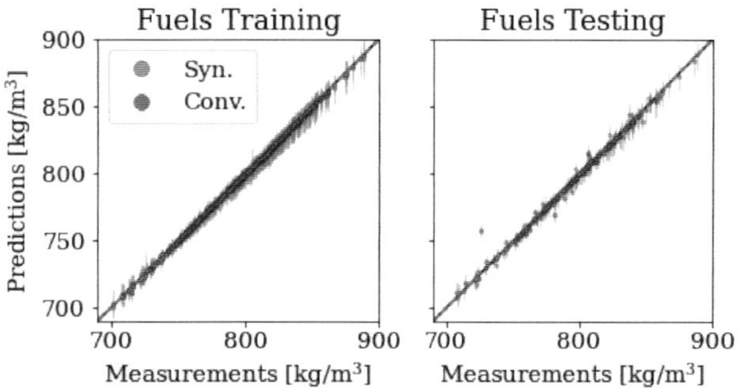

Fuels Training:

	MAE	PICP	NMPIW	MAOE
Syn.	0.72	99.27	8.83	0.85
Conv.	0.64	100.0	8.85	0.51

Fuels Testing:

MAE	PICP	NMPIW	MAOE
2.5	84.17	9.5	4.04
0.9	99.25	8.58	1.45

Figure S3: Cross-validation results training and testing for density prediction of the DC -no Syn. model

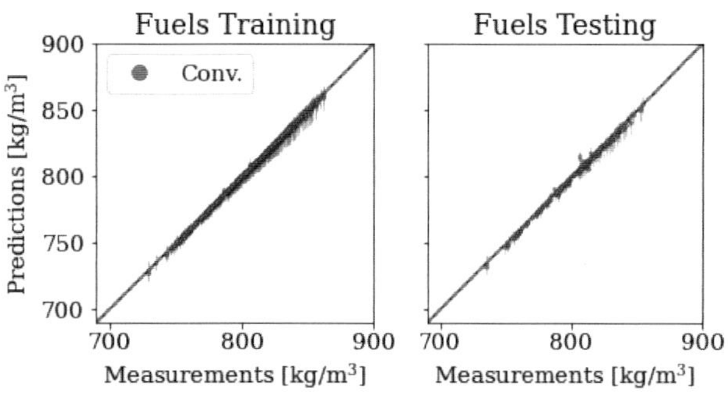

Fuels Training:

	MAE	PICP	NMPIW	MAOE
Conv.	0.89	99.86	6.58	0.48

Fuels Testing:

MAE	PICP	NMPIW	MAOE
1.22	96.09	6.46	1.14

Figure S4: Cross-validation results training and testing for density prediction of the M-QSPR model

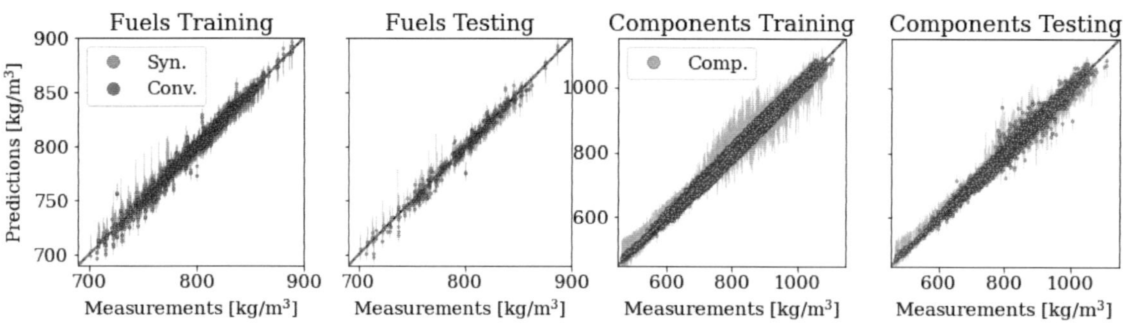

Fuels Training:

	MAE	PICP	NMPIW	MAOE
Syn.	4.69	84.96	16.96	3.84
Conv.	2.59	98.41	15.51	2.4

Fuels Testing:

MAE	PICP	NMPIW	MAOE
4.61	86.63	16.82	3.13
2.81	97.25	15.44	4.28

Components Training:

MAE	PICP	NMPIW	MAOE
5.84	85.29	24.79	5.01

Components Testing:

MAE	PICP	NMPIW	MAOE
6.56	83.41	24.92	6.54

Figure S5: Cross-validation results training and testing for density prediction of the M-QSPR -no Syn. model

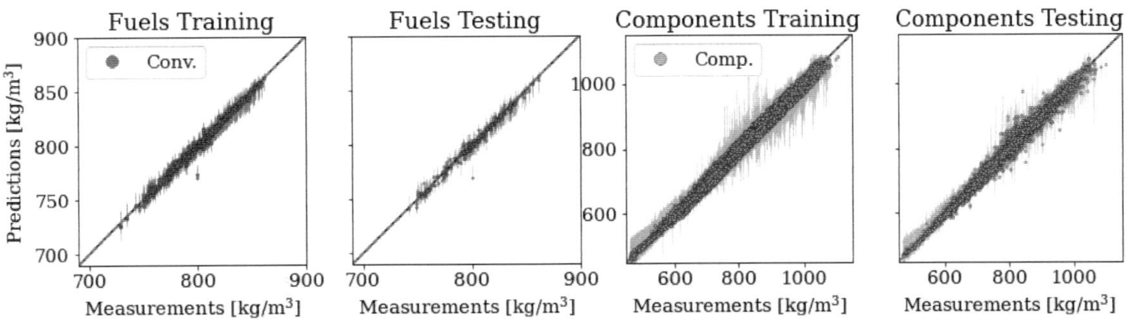

Fuels Training:

	MAE	PICP	NMPIW	MAOE
Conv.	1.92	98.55	15.02	3.52

Fuels Testing:

MAE	PICP	NMPIW	MAOE
2.12	98.43	15.21	3.89

Components Training:

MAE	PICP	NMPIW	MAOE
5.53	85.3	24.95	5.08

Components Testing:

MAE	PICP	NMPIW	MAOE
6.15	84.06	25.26	6.41

Figure S6: Cross-validation results training and testing for density prediction of the QSPR model

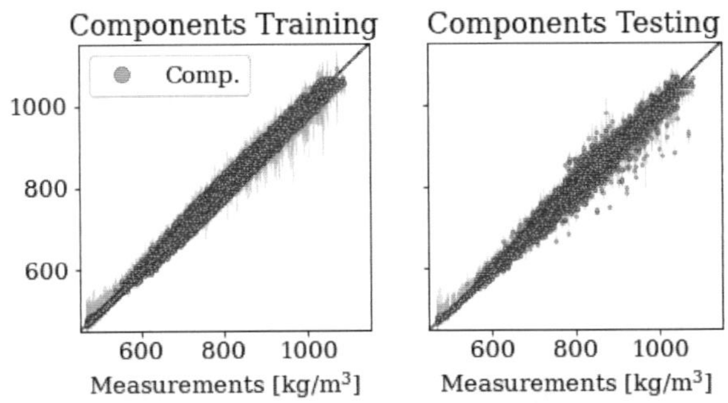

Components Training:

MAE	PICP	NMPIW	MAOE
5.97	78.34	19.41	5.63

Components Testing:

MAE	PICP	NMPIW	MAOE
6.78	76.58	19.62	7.52

Figure S7: Cross-validation results training and testing for surface tension prediction of the DC model

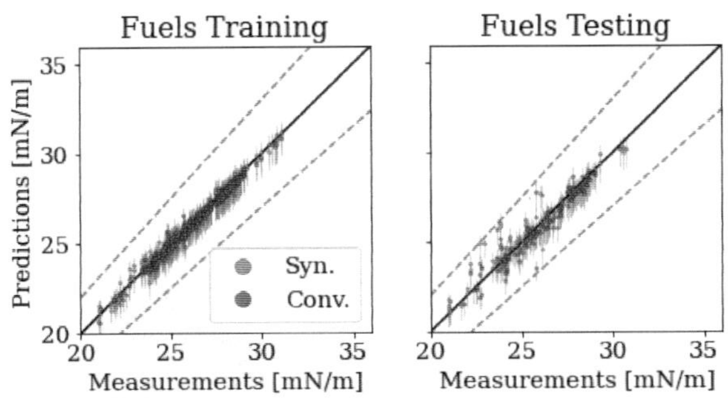

Fuels Training:

	MAE	PICP	NMPIW	MAOE
Syn.	0.22	100.0	1.39	0.16
Conv.	0.2	100.0	1.34	0.11

Fuels Testing:

MAE	PICP	NMPIW	MAOE
0.54	100.0	1.38	0.64
0.36	100.0	1.32	0.36

Figure S8: Cross-validation results training and testing for surface tension prediction of the DC -no Syn. model

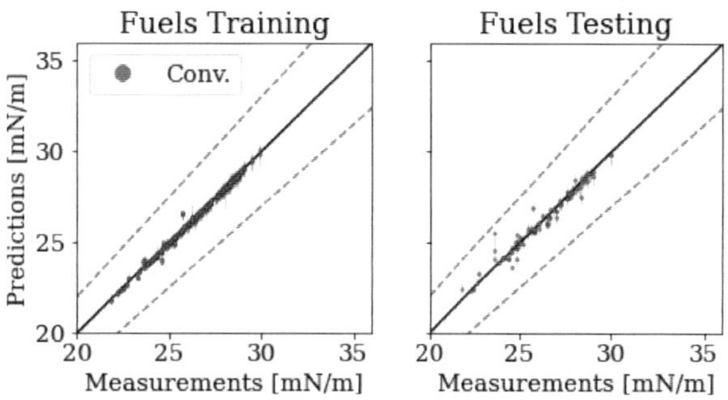

	MAE	PICP	NMPIW	MAOE
Conv.	0.07	100.0	0.43	0.24

MAE	PICP	NMPIW	MAOE
0.22	100.0	0.45	0.27

Figure S9: Cross-validation results training and testing for surface tension prediction of the M-QSPR model

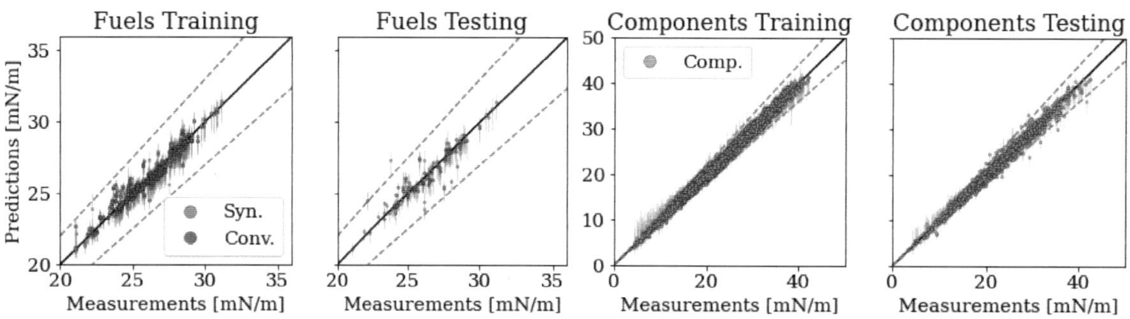

	MAE	PICP	NMPIW	MAOE
Syn.	0.37	100.0	1.33	0.4
Conv.	0.31	100.0	1.14	0.32

MAE	PICP	NMPIW	MAOE
0.66	100.0	1.4	1.15
0.32	100.0	1.19	0.42

MAE	PICP	NMPIW	MAOE
0.43	99.97	1.67	0.42

MAE	PICP	NMPIW	MAOE
0.53	99.73	1.7	0.53

Figure S10: Cross-validation results training and testing for surface tension prediction of the M-QSPR -no Syn. model

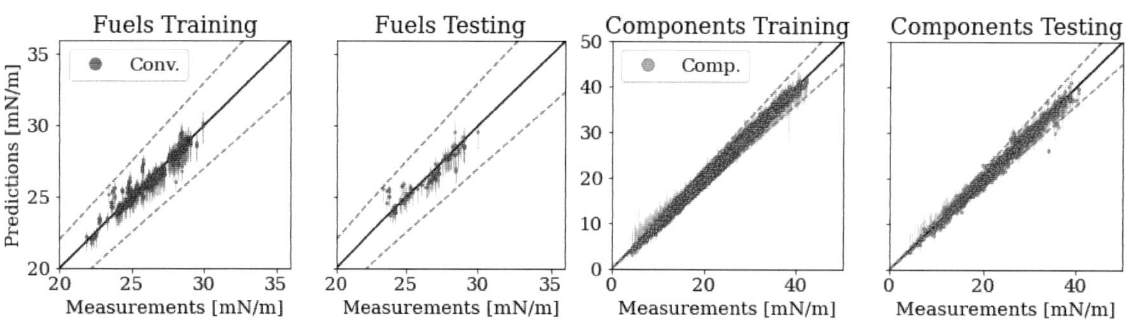

	MAE	PICP	NMPIW	MAOE
Conv.	0.32	100.0	1.23	0.34

MAE	PICP	NMPIW	MAOE
0.38	100.0	1.27	0.45

MAE	PICP	NMPIW	MAOE
0.45	99.94	1.76	0.42

MAE	PICP	NMPIW	MAOE
0.54	99.66	1.77	0.52

Figure S11: Cross-validation results training and testing for surface tension prediction of the QSPR model

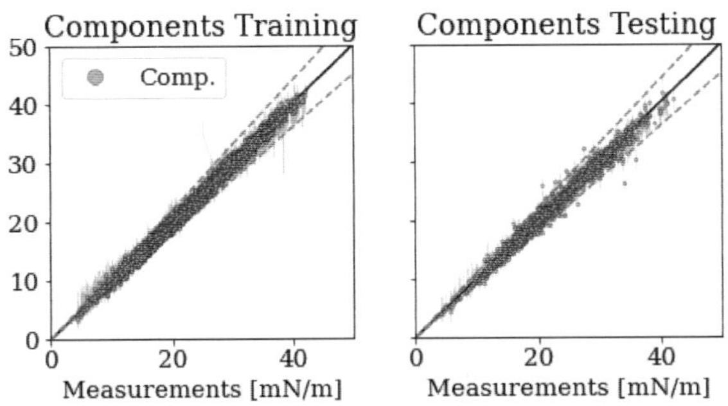

Components Training:

MAE	PICP	NMPIW	MAOE
0.43	99.97	1.58	0.4

Components Testing:

MAE	PICP	NMPIW	MAOE
0.53	99.66	1.57	0.52

Figure S12: Cross-validation results training and testing for kinematic viscosity prediction of the DC model

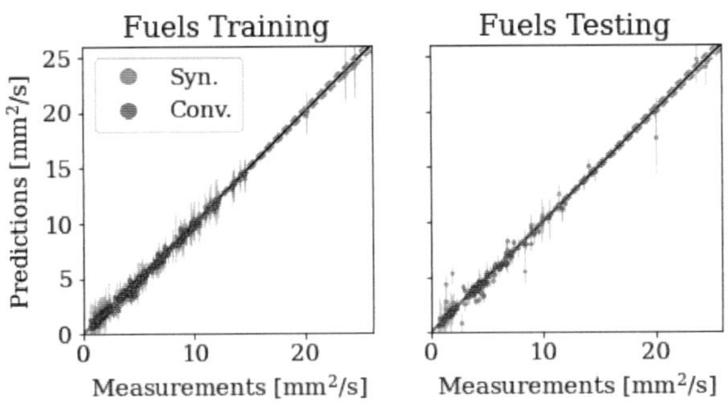

Fuels Training:

	MAE	PICP	NMPIW	MAOE
Syn.	0.18	96.78	1.36	0.36
Conv.	0.16	97.12	0.99	0.13

Fuels Testing:

MAE	PICP	NMPIW	MAOE
0.47	81.0	1.46	0.69
0.23	91.67	0.97	0.19

Figure S13: Cross-validation results training and testing for kinematic viscosity prediction of the DC - no Syn. model

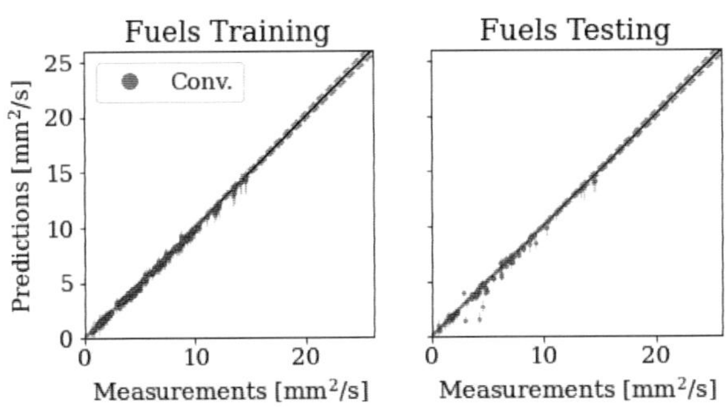

Fuels Training:

	MAE	PICP	NMPIW	MAOE
Conv.	0.11	97.88	0.67	0.1

Fuels Testing:

MAE	PICP	NMPIW	MAOE
0.24	86.67	0.68	0.37

Figure S14: Cross-validation results training and testing for kinematic viscosity prediction of the M-QSPR model

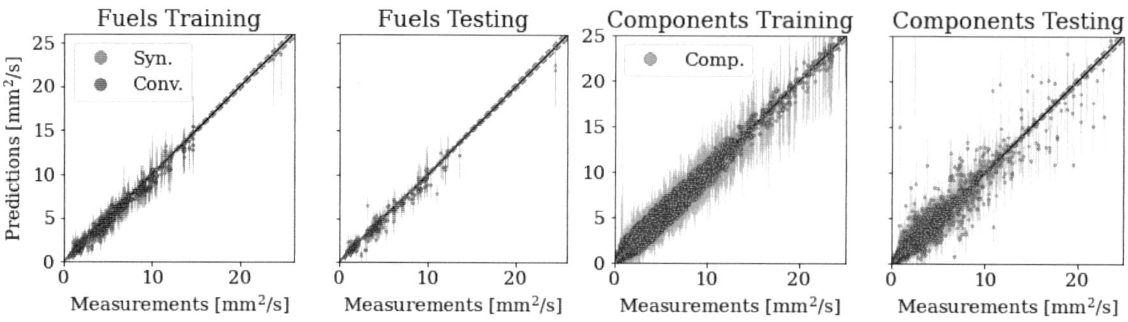

Fuels Training:

	MAE	PICP	NMPIW	MAOE
Syn.	0.39	89.83	1.85	0.27
Conv.	0.31	97.87	1.91	0.26

Fuels Testing:

MAE	PICP	NMPIW	MAOE
0.55	84.4	2.02	0.66
0.3	98.4	1.79	0.41

Components Training:

MAE	PICP	NMPIW	MAOE
0.21	95.32	1.26	0.27

Components Testing:

MAE	PICP	NMPIW	MAOE
0.32	92.07	1.28	0.64

Figure S15: Cross-validation results training and testing for kinematic viscosity prediction of the M-QSPR -no Syn. model

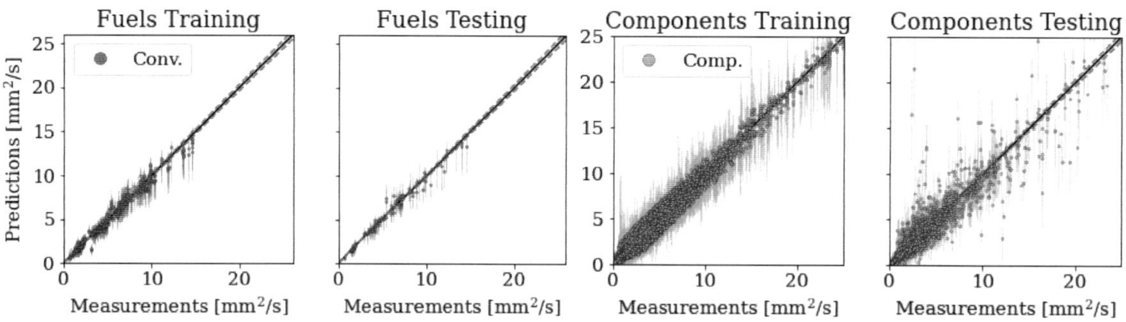

Fuels Training:

	MAE	PICP	NMPIW	MAOE
Conv.	0.3	96.59	1.82	0.28

Fuels Testing:

MAE	PICP	NMPIW	MAOE
0.31	97.22	1.99	0.28

Components Training:

MAE	PICP	NMPIW	MAOE
0.21	95.3	1.3	0.31

Components Testing:

MAE	PICP	NMPIW	MAOE
0.3	92.38	1.32	0.73

Figure S16: Cross-validation results training and testing for kinematic viscosity prediction of the QSPR model

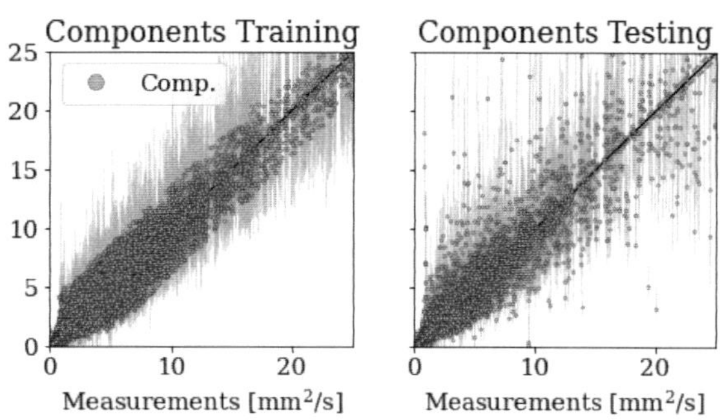

Components Training:

MAE	PICP	NMPIW	MAOE
0.34	80.36	1.29	0.31

Components Testing:

MAE	PICP	NMPIW	MAOE
0.45	78.53	1.34	0.51

Figure S17: Cross-validation results training and testing for net heat of combustion prediction of the DC model

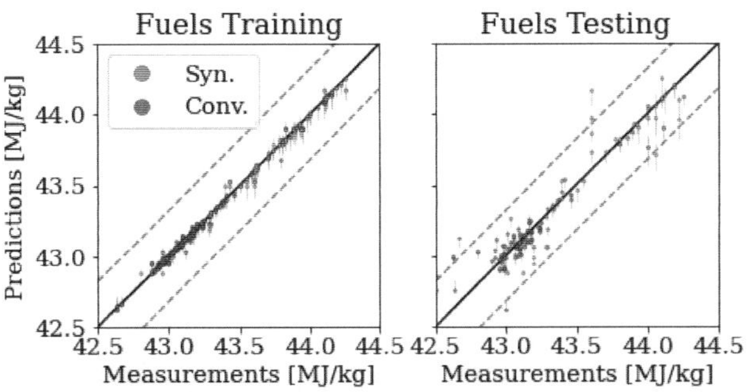

	MAE	PICP	NMPIW	MAOE
Syn.	0.02	100.0	0.08	0.03
Conv.	0.01	100.0	0.04	0.03

MAE	PICP	NMPIW	MAOE
0.12	96.55	0.09	0.13
0.06	97.78	0.04	0.05

Figure S18: Cross-validation results training and testing for net heat of combustion prediction of the DC -no Syn. model

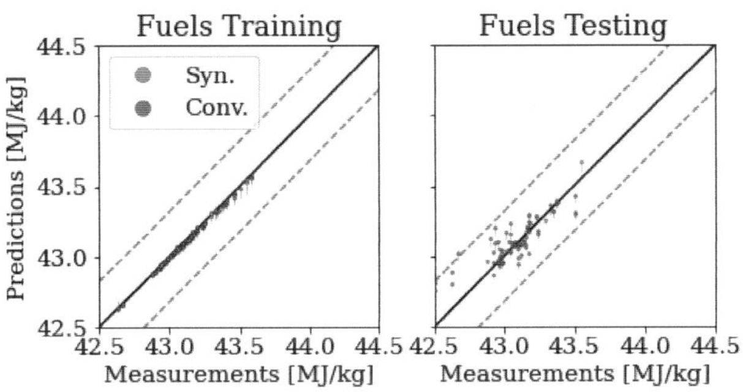

	MAE	PICP	NMPIW	MAOE
Syn.	0.01	100.0	0.07	0.01
Conv.	0.01	100.0	0.04	0.01

MAE	PICP	NMPIW	MAOE
0.18	100.0	0.05	0.16
0.07	100.0	0.04	0.06

Figure S19: Cross-validation results training and testing for net heat of combustion prediction of the M-QSPR model

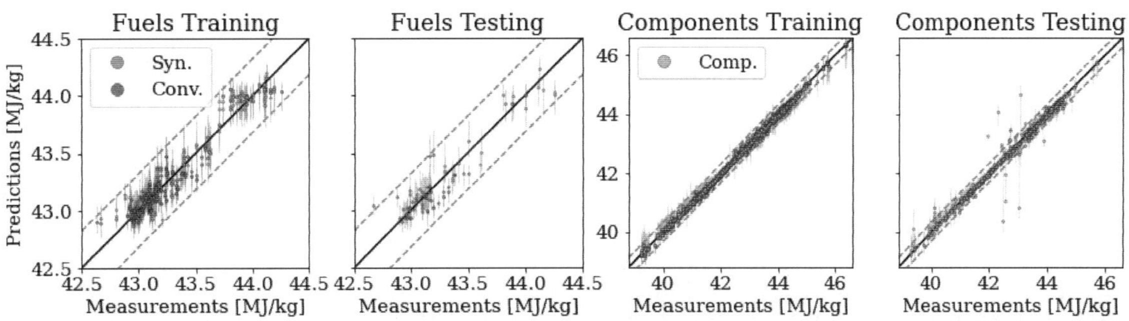

	MAE	PICP	NMPIW	MAOE
Syn.	0.09	100.0	0.27	0.05
Conv.	0.07	100.0	0.21	0.05

MAE	PICP	NMPIW	MAOE
0.09	100.0	0.26	0.08
0.09	100.0	0.21	0.1

MAE	PICP	NMPIW	MAOE
0.04	100.0	0.3	0.05

MAE	PICP	NMPIW	MAOE
0.07	99.11	0.31	0.12

Figure S20: Cross-validation results training and testing for net heat of combustion prediction of the M-QSPR -no Syn. model

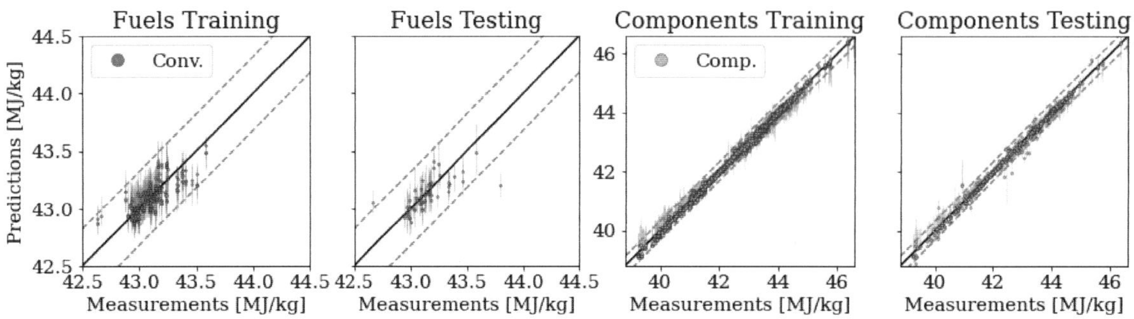

	MAE	PICP	NMPIW	MAOE
Conv.	0.08	100.0	0.27	0.06

MAE	PICP	NMPIW	MAOE
0.08	98.21	0.27	0.1

MAE	PICP	NMPIW	MAOE
0.04	100.0	0.31	0.05

MAE	PICP	NMPIW	MAOE
0.05	99.64	0.32	0.08

Figure S21: Cross-validation results training and testing for flash point prediction of the QSPR model

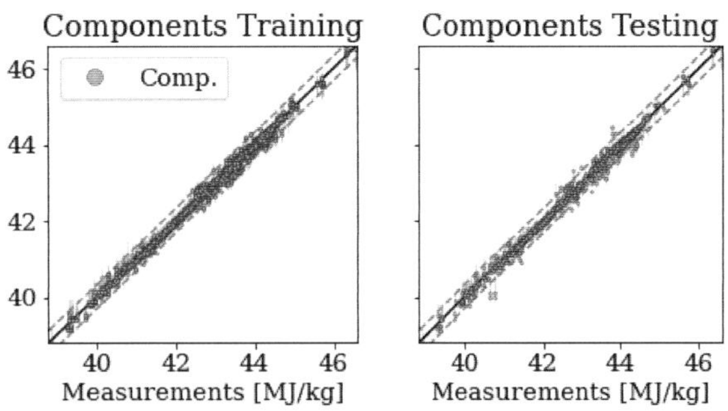

MAE	PICP	NMPIW	MAOE
0.05	100.0	0.24	0.08

MAE	PICP	NMPIW	MAOE
0.06	99.34	0.24	0.09

Figure S22: Cross-validation results training and testing for flash point prediction of the DC model

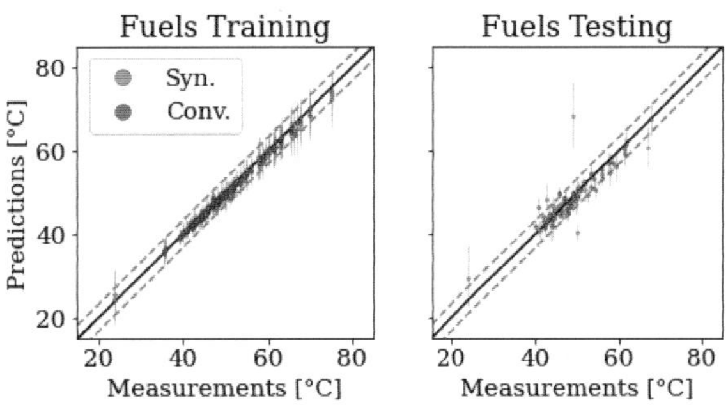

	MAE	PICP	NMPIW	MAOE
Syn.	0.49	100.0	3.63	0.86
Conv.	0.41	100.0	3.33	0.2

MAE	PICP	NMPIW	MAOE
2.14	92.11	4.32	2.22
1.4	98.15	3.23	1.29

Figure S23: Cross-validation results training and testing for flash point prediction of the DC -no Syn. model

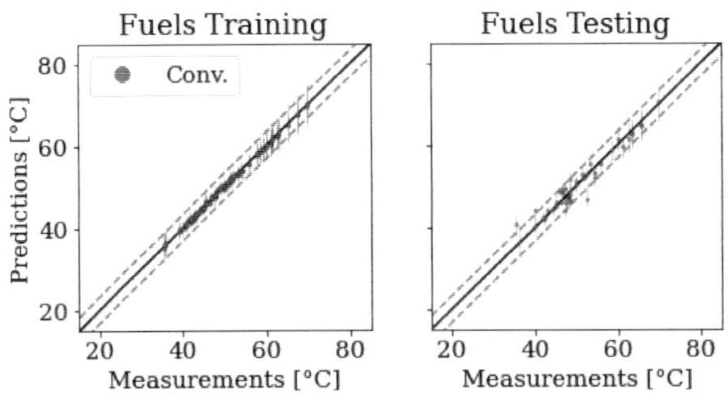

Fuels Training:

	MAE	PICP	NMPIW	MAOE
Conv.	0.26	100.0	2.95	0.0

Fuels Testing:

MAE	PICP	NMPIW	MAOE
1.27	100.0	3.69	1.22

Figure S24: Cross-validation results training and testing for flash point prediction of the M-QSPR model

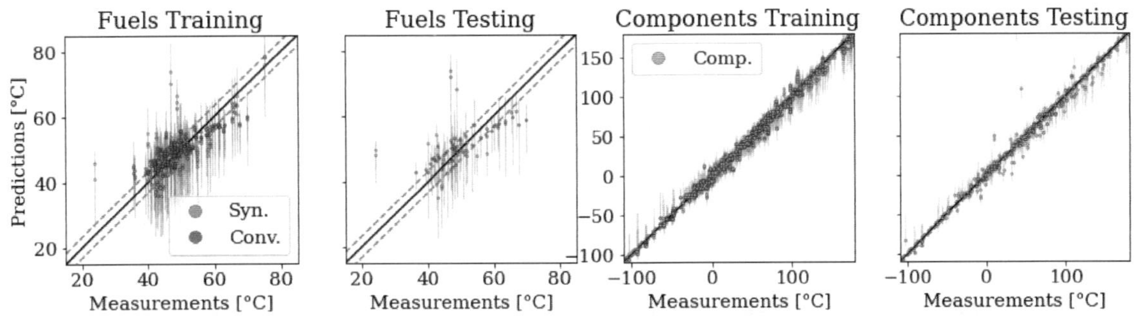

Fuels Training:

	MAE	PICP	NMPIW	MAOE
Syn.	4.0	93.62	13.51	2.77
Conv.	3.16	95.24	13.05	2.15

Fuels Testing:

MAE	PICP	NMPIW	MAOE
7.5	75.76	14.5	6.96
3.55	97.87	13.54	2.04

Components Training:

MAE	PICP	NMPIW	MAOE
4.01	94.84	19.01	4.0

Components Testing:

MAE	PICP	NMPIW	MAOE
5.72	89.97	19.92	7.44

Figure S25: Cross-validation results training and testing for flash point prediction of the M-QSPR -no Syn. model

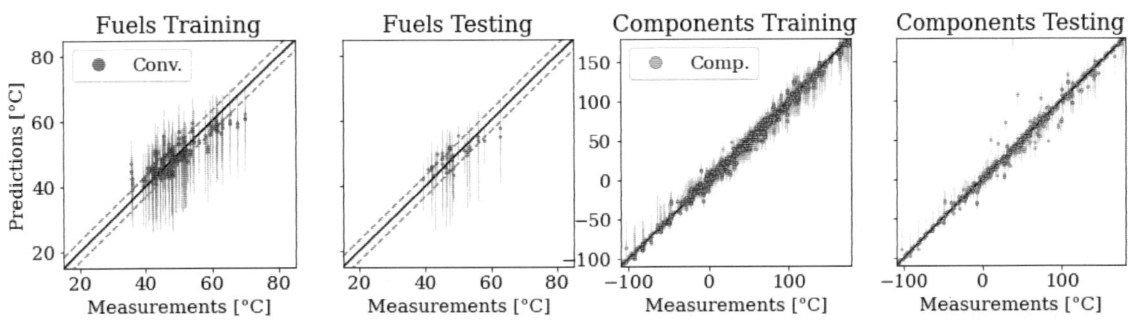

Fuels Training:

	MAE	PICP	NMPIW	MAOE
Conv.	3.06	97.1	14.69	1.99

Fuels Testing:

MAE	PICP	NMPIW	MAOE
3.1	97.5	15.85	1.5

Components Training:

MAE	PICP	NMPIW	MAOE
4.04	94.44	19.92	3.98

Components Testing:

MAE	PICP	NMPIW	MAOE
7.17	87.13	20.41	12.5

Figure S26: Cross-validation results training and testing for flash point prediction of the QSPR model

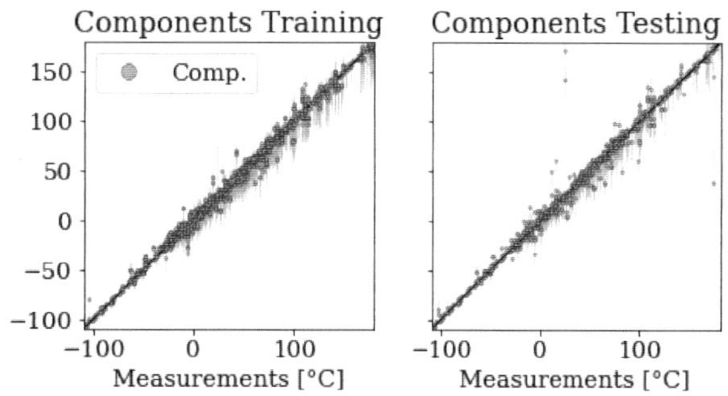

Components Training:

MAE	PICP	NMPIW	MAOE
4.1	90.61	17.0	4.31

Components Testing:

MAE	PICP	NMPIW	MAOE
6.19	85.6	16.93	6.97

Figure S27: Cross-validation results training and testing for freezing point prediction of the DC model

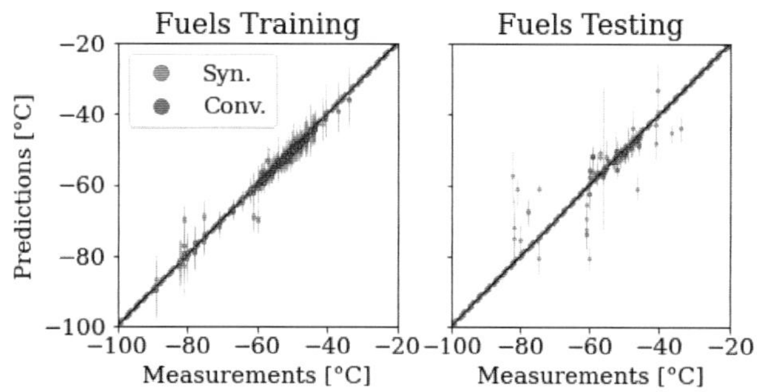

Fuels Training:

	MAE	PICP	NMPIW	MAOE
Syn.	1.09	94.82	6.88	3.44
Conv.	0.45	99.64	4.68	1.0

Fuels Testing:

MAE	PICP	NMPIW	MAOE
6.18	55.81	8.52	5.66
1.68	88.89	4.77	3.13

Figure S28: Cross-validation results training and testing for freezing point prediction of the DC -no Syn. model

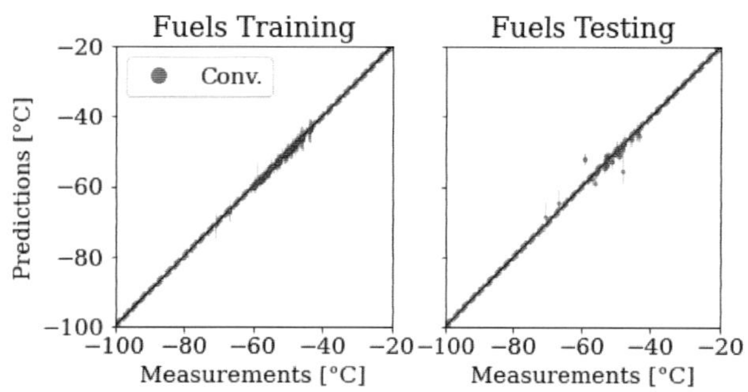

Fuels Training:

	MAE	PICP	NMPIW	MAOE
Conv.	0.23	100.0	2.95	0.11

Fuels Testing:

MAE	PICP	NMPIW	MAOE
1.41	89.58	3.59	1.62

Figure S29: Cross-validation results training and testing for freezing point prediction of the M-QSPR model

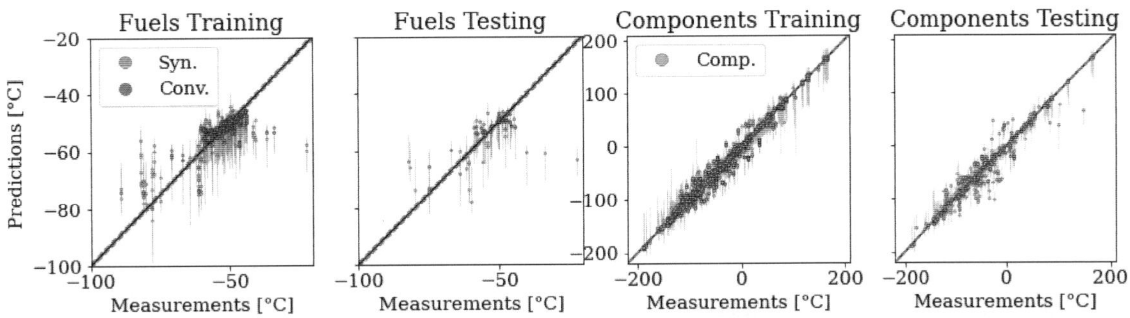

Fuels Training:

	MAE	PICP	NMPIW	MAOE
Syn.	5.94	59.13	11.6	5.77
Conv.	2.62	81.62	8.44	2.19

Fuels Testing:

MAE	PICP	NMPIW	MAOE
7.81	54.29	12.53	8.37
2.34	79.17	8.28	1.68

Components Training:

MAE	PICP	NMPIW	MAOE
6.19	83.56	25.05	9.9

Components Testing:

MAE	PICP	NMPIW	MAOE
9.78	77.23	25.22	13.66

Figure S30: Cross-validation results training and testing for freezing point prediction of the M-QSPR -no Syn. model

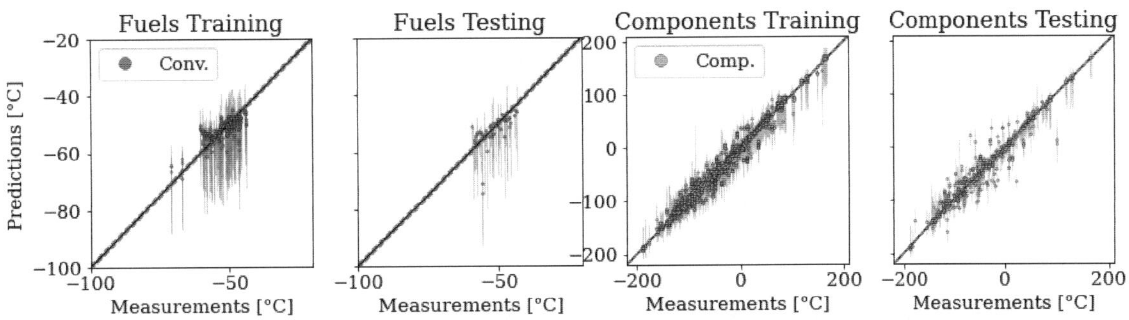

Fuels Training:

	MAE	PICP	NMPIW	MAOE
Conv.	2.57	87.13	15.91	2.32

Fuels Testing:

MAE	PICP	NMPIW	MAOE
2.79	85.42	16.2	2.54

Components Training:

MAE	PICP	NMPIW	MAOE
7.16	83.57	30.41	9.97

Components Testing:

MAE	PICP	NMPIW	MAOE
10.38	77.08	31.33	14.5

Figure S31: Cross-validation results training and testing for freezing point prediction of the QSPR model

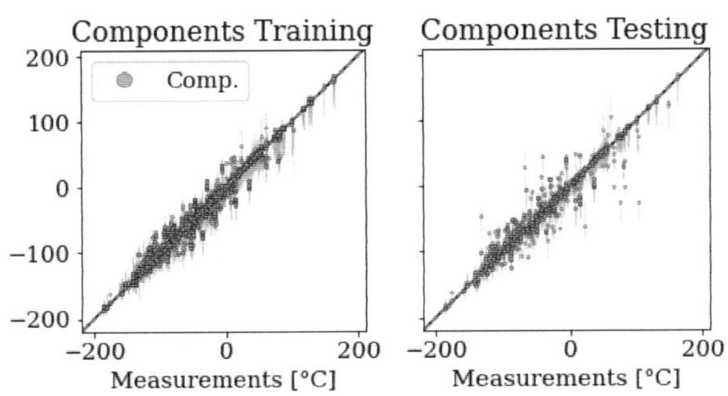

Components Training:

MAE	PICP	NMPIW	MAOE
7.06	79.59	22.19	10.64

Components Testing:

MAE	PICP	NMPIW	MAOE
9.81	74.71	22.09	14.24

Figure S32: Cross-validation results training and testing for cetane number prediction of the DC model

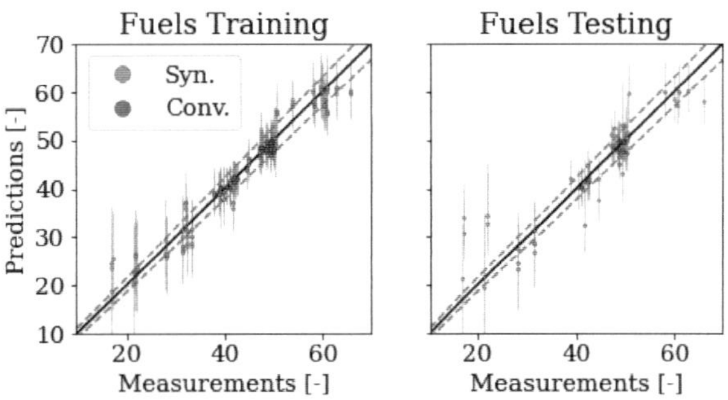

Fuels Training

	MAE	PICP	NMPIW	MAOE
Syn.	2.05	100.0	11.86	0.72
Conv.	0.85	100.0	7.24	0.0

Fuels Testing

MAE	PICP	NMPIW	MAOE
4.01	83.72	12.45	4.16
0.91	100.0	7.64	0.0

Figure S33: Cross-validation results training and testing for cetane number prediction of the DC -no Syn. model

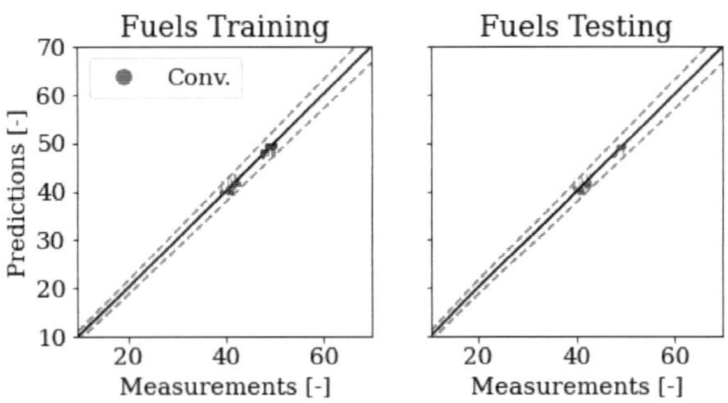

Fuels Training

	MAE	PICP	NMPIW	MAOE
Conv.	0.48	100.0	2.0	0.22

Fuels Testing

MAE	PICP	NMPIW	MAOE
0.63	100.0	2.48	1.06

Figure S34: Cross-validation results training and testing for cetane number prediction of the M-QSPR model

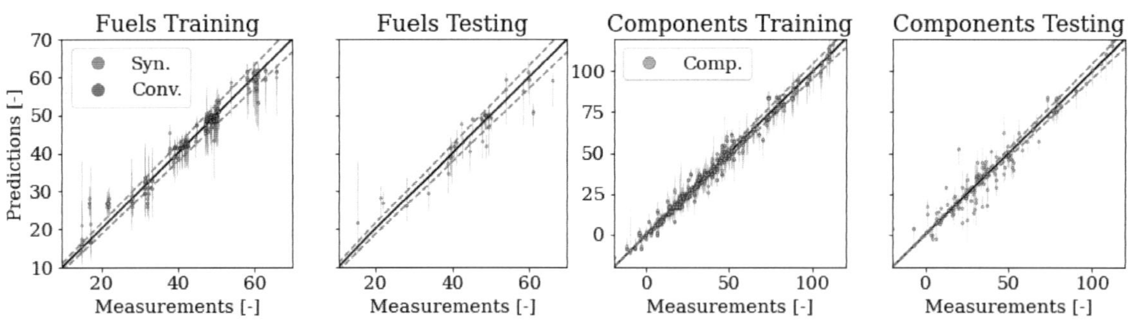

Fuels Training

	MAE	PICP	NMPIW	MAOE
Syn.	2.41	89.31	9.17	2.63
Conv.	0.9	100.0	6.27	0.16

Fuels Testing

MAE	PICP	NMPIW	MAOE
4.08	76.0	9.05	3.29
1.21	100.0	5.35	0.0

Components Training

MAE	PICP	NMPIW	MAOE
2.17	90.67	9.33	2.29

Components Testing

MAE	PICP	NMPIW	MAOE
5.59	70.0	10.76	4.84

Figure S35: Cross-validation results training and testing for cetane number prediction of the M-QSPR -no Syn. model

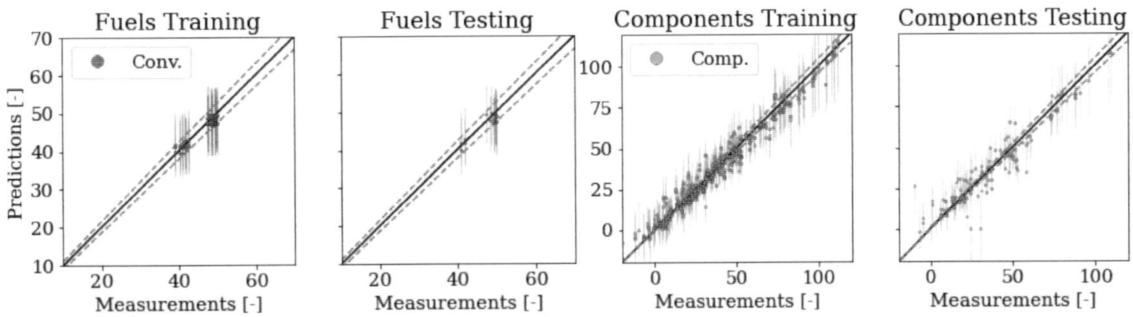

	MAE	PICP	NMPIW	MAOE
Conv.	1.0	100.0	14.42	0.0

MAE	PICP	NMPIW	MAOE
1.22	100.0	14.54	0.0

MAE	PICP	NMPIW	MAOE
3.67	93.8	19.93	4.17

MAE	PICP	NMPIW	MAOE
6.38	82.79	20.94	4.83

Figure S36: Cross-validation results training and testing for cetane number prediction of the QSPR model

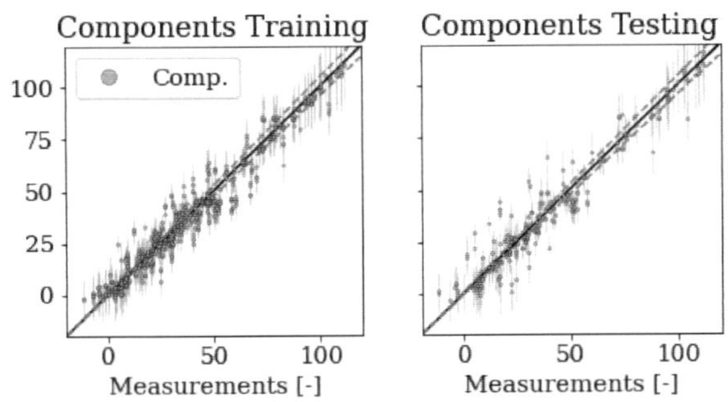

MAE	PICP	NMPIW	MAOE
4.55	81.53	14.48	4.66

MAE	PICP	NMPIW	MAOE
5.77	75.68	15.55	5.69

Figure S32: Cross-validation results training and testing for distillation prediction of the DC model

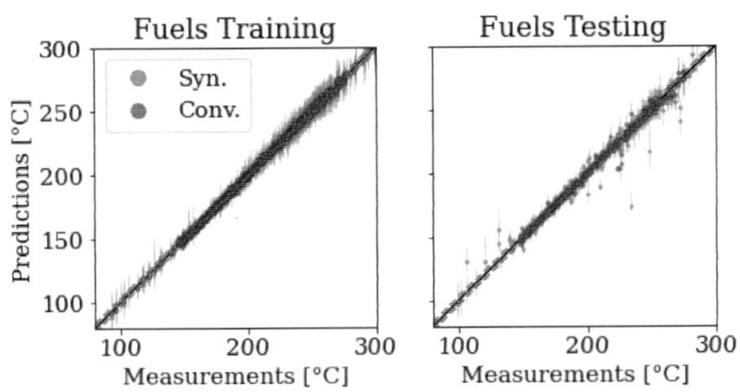

	MAE	PICP	NMPIW	MAOE
Syn.	1.31	100.0	10.93	0.79
Conv.	1.19	99.96	8.2	0.91

MAE	PICP	NMPIW	MAOE
3.08	94.38	11.33	5.51
1.71	97.81	8.1	3.63

Figure S33: Cross-validation results training and testing for distillation prediction of the DC -no Syn. model

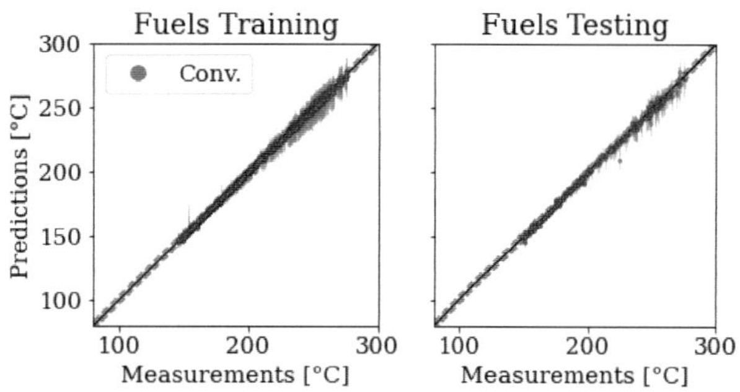

	MAE	PICP	NMPIW	MAOE
Conv.	0.97	100.0	8.07	0.47

MAE	PICP	NMPIW	MAOE
1.38	99.57	8.28	1.46

Figure S34: Cross-validation results training and testing for distillation / boiling point prediction of the QSPR model

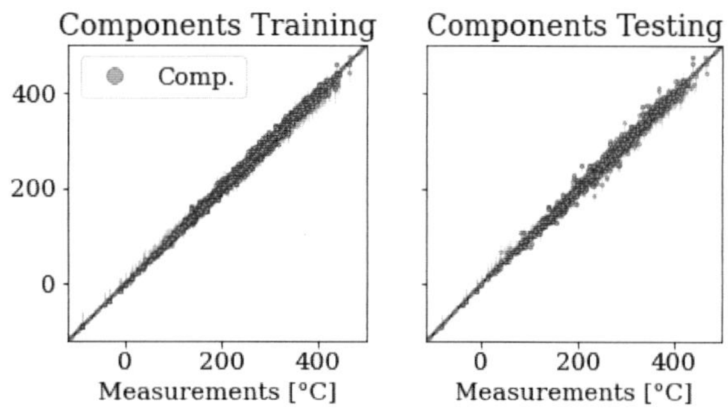

MAE	PICP	NMPIW	MAOE
3.47	92.49	14.25	3.41

MAE	PICP	NMPIW	MAOE
4.44	88.9	14.34	4.63

Tables S35: Utilized parameters for hyperparameter optimization of Monte-Carlo Neural Networks

Parameter	*Optimization range*
(Neurons per Layer)	(num_{feat}), (num_{feat}, num_{feat}), $(num_{feat}, num_{feat}, num_{feat})$, $(2*num_{feat})$, $(2*num_{feat}, 2*num_{feat})$, $(2*num_{feat}, 2*num_{feat}, 2*num_{feat})$,
Droprate	*0.001, 0.01, 0.05, 0.1, 0.2, 0.3, 0.4, 0.5*
Learning rate	*1e-4, 1e-3, 1e-2*
Learning weight decay	*0, 1e-8, 1e-6, 1e-4, 1e-2*
Number of training epochs	*500, 1000, 1500, 2000*
Activation function	*ReLU, ELU, Sigmoid*
X feature scaler	*None, MinMaxScaler, RobustScaler*
y feature scaler	*None, MinMaxScaler, RobustScaler*

Utilized Scalers:

$$MinMaxScaler = \frac{x_i - \overline{x_i}}{x_{i,max} - x_{i,min}}$$

$$RobustScaler = \frac{x_i - Q_{50}(x_i)}{Q_{75}(x_i) - Q_{25}(x_i)}$$

Table S36: Considered molecular families for GCxGC measurements and classification criteria

Family	Formula	Structural criteria	Example
N-alkanes	C_nH_{2n+2}	No atoms in rings No branched carbon bonds	
Iso-alkanes	C_nH_{2n+2}	- No atoms in rings - At least one branched carbon bond	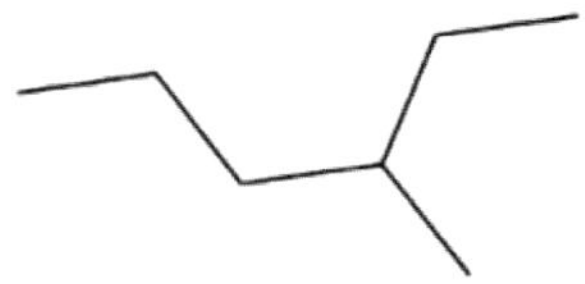
Mono-cyclo-alkanes	C_nH_{2n}	- One aliphatic ring - No aromatic rings	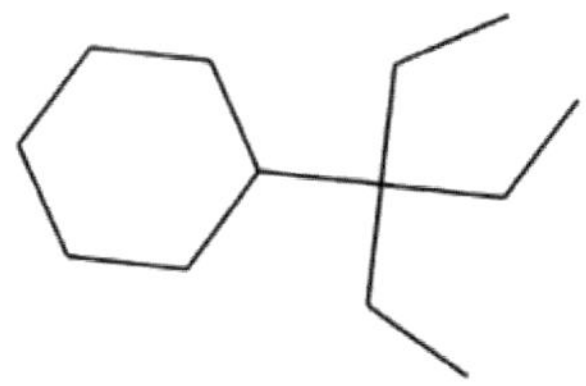
Bi-cyclo-alkanes	C_nH_{2n-2}	- Two aliphatic rings - No aromatic rings	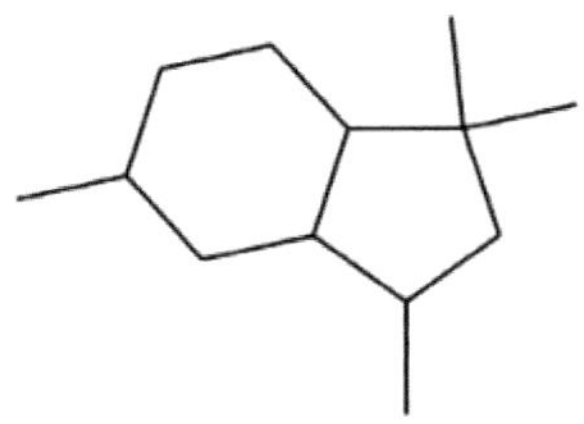
Mono-aromatics	C_nH_{2n-6}	- No aliphatic rings - One aromatic ring	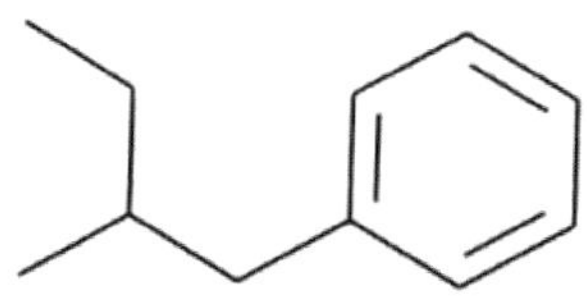
Cyclo-aromatics	C_nH_{2n-8}	- One aliphatic ring - One aromatic ring	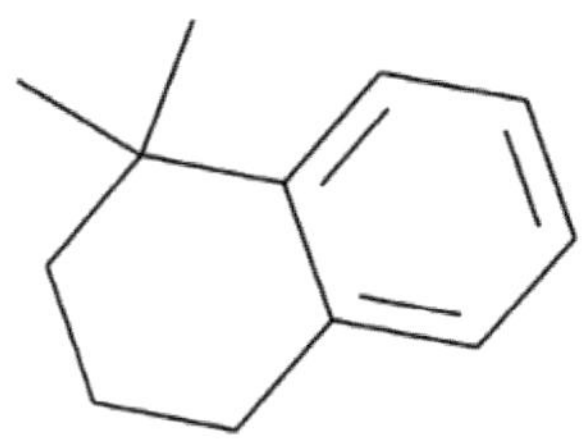
Di-aromatics	C_nH_{2n-12}	- No aliphatic rings - Two aromatic rings	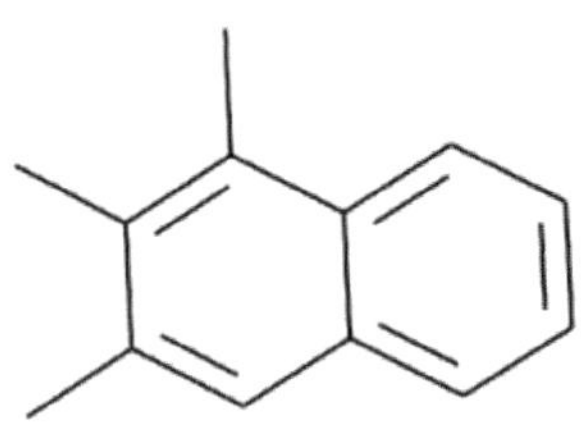

Table S37: Predictive capabilities density

	Conv. Fuels				Syn. Fuels			
	MAE	PICP	NMPIW	MAOE	MAE	PICP	NMPIW	MAOE
DC	0.96	99.44	867.3	0.42	2.39	81.82	960.73	2.86
M-QSPR	2.59	98.08	1522.31	0.52	4.59	88.24	1589.83	3.37
QSPR sampling	3.59	90.71	975.93	0.84	4.69	69.85	1245.04	3.82
DC -no Syn.	1.23	96.55	665.77	1.11	14.1	43.98	1817.93	11.97
M-QSPR -no Syn.	2.16	98.82	1502.64	1.52	5.55	74.7	1530.5	4.14

Table S38: Predictive capabilities surface tension

	Conv. Fuels				Syn. Fuels			
	MAE	PICP	NMPIW	MAOE	MAE	PICP	NMPIW	MAOE
DC	0.36	100.0	134.01	0.37	0.35	100.0	142.13	0.08
M-QSPR	0.38	100.0	122.79	0.55	0.78	100.0	149.76	1.19
QSPR sampling	0.54	100.0	117.97	0.5	0.95	100.0	206.28	0.79
DC -no Syn.	0.22	100.0	46.44	0.26	1.49	93.55	134.33	1.2
M-QSPR -no Syn.	0.42	100.0	130.57	0.53	0.82	98.33	130.1	0.8

Table S39: Predictive capabilities kinematic viscosity

	Conv. Fuels				Syn. Fuels			
	MAE	PICP	NMPIW	MAOE	MAE	PICP	NMPIW	MAOE
DC	0.24	92.45	105.12	0.24	0.61	75.9	161.97	0.86
M-QSPR	0.33	97.09	184.71	0.46	0.58	84.95	216.13	0.58
QSPR sampling	0.89	30.33	153.53	0.46	0.8	56.9	161.0	0.66
DC -no Syn.	0.26	86.41	73.8	0.45	1.44	59.46	251.99	1.07
M-QSPR -no Syn.	0.32	97.85	203.62	0.21	0.94	63.47	152.69	0.93

Table S40: Predictive capabilities net heat of combustion

	Conv. Fuels				Syn. Fuels			
	MAE	PICP	NMPIW	MAOE	MAE	PICP	NMPIW	MAOE
DC	0.05	100.0	5.67	0.03	0.1	100.0	9.31	0.1
M-QSPR	0.14	100.0	21.73	0.1	0.09	100.0	27.15	0.09
QSPR sampling	0.18	91.86	9.29	0.16	0.19	83.33	5.79	0.19
DC -no Syn.	0.05	100.0	4.89	0.04	0.26	87.18	18.82	0.19
M-QSPR -no Syn.	0.13	100.0	29.54	0.03	0.17	100.0	28.69	0.07

Table S41: Predictive capabilities flash point

	Conv. Fuels				Syn. Fuels			
	MAE	PICP	NMPIW	MAOE	MAE	PICP	NMPIW	MAOE
DC	0.99	100.0	318.51	0.08	2.43	91.89	518.81	2.59
M-QSPR	3.77	91.67	1461.15	2.78	6.53	83.33	1512.95	6.85
QSPR sampling	3.67	89.29	838.51	2.31	7.68	77.22	1120.82	4.94
DC -no Syn.	0.81	100.0	378.28	1.59	6.26	73.17	1309.13	6.14
M-QSPR -no Syn.	2.13	100.0	1439.91	0.24	4.37	85.37	958.19	3.71

Table S42: Predictive capabilities freezing point

	Conv. Fuels				Syn. Fuels			
	MAE	PICP	NMPIW	MAOE	MAE	PICP	NMPIW	MAOE
DC	1.51	89.74	506.65	3.66	7.14	48.65	1027.0	6.29
M-QSPR	2.24	78.95	918.39	1.45	7.3	50.0	1246.17	5.74
QSPR sampling	5.76	40.0	764.23	3.91	9.04	65.71	1886.49	7.1
DC -no Syn.	1.14	92.86	370.4	0.96	30.57	21.05	2476.04	23.8
M-QSPR -no Syn.	2.33	88.89	1649.29	2.24	11.64	42.03	1424.05	10.65

Table S43: Predictive capabilities cetane number

	Conv. Fuels				Syn. Fuels			
	MAE	PICP	NMPIW	MAOE	MAE	PICP	NMPIW	MAOE
DC	1.02	100.0	789.43	0.0	3.74	82.61	1078.53	3.96
M-QSPR	1.4	100.0	577.99	0.0	3.7	84.21	979.12	3.23
QSPR sampling	2.53	100.0	790.0	1.01	7.53	82.22	2051.77	3.1
DC -no Syn.	0.71	100.0	245.21	1.06	15.79	62.5	1543.38	13.39
M-QSPR -no Syn.	1.21	100.0	1451.26	0.0	9.56	57.78	1548.5	7.2

Table S44: Predictive capabilities distillation 10

	Conv. Fuels				Syn. Fuels			
	MAE	PICP	NMPIW	MAOE	MAE	PICP	NMPIW	MAOE
DC	1.27	100.0	649.98	0.78	1.8	100.0	906.11	1.45
DC -no Syn.	0.65	100.0	494.92	0.0	38.5	62.16	4591.25	44.22
QSPR sampling	1.29	100.0	3120.12	0.0	9.2	94.55	3424.96	20.93

Table S45: Predictive capabilities distillation 50

	Conv. Fuels				Syn. Fuels			
	MAE	PICP	NMPIW	MAOE	MAE	PICP	NMPIW	MAOE
DC	1.6	100.0	922.22	0.0	1.98	100.0	978.34	1.55
DC -no Syn.	0.92	100.0	844.46	0.0	30.13	52.63	5339.07	24.46
QSPR sampling	4.71	96.75	2582.34	10.35	9.08	93.22	2690.4	7.82

Table S46: Predictive capabilities distillation 90

	Conv. Fuels				Syn. Fuels			
	MAE	PICP	NMPIW	MAOE	MAE	PICP	NMPIW	MAOE
DC	3.67	100.0	1267.79	2.35	3.22	100.0	1357.8	1.71
DC -no Syn.	1.54	100.0	1100.6	0.0	35.62	52.78	6464.84	32.75
QSPR sampling	4.48	100.0	2997.17	0.0	10.7	87.72	3158.49	19.43

IASH 2022, the 17th INTERNATIONAL SYMPOSIUM ON STABILITY, HANDLING AND USE OF LIQUID FUELS
Dresden, Germany
September 11 - 15, 2022

100 % DROP IN SAF FROM VARIOUS D7566 PATHWAYS: A FIT FOR PURPOSE PROPERTIES EVALUATION.

David Evans*, Tony Marasigan, Nathan Matheson, Paul M. Rawson

Defence Science and Technology Group, 506 Lorimer Street, Fishermans Bend, 3207 Australia

ABSTRACT

With the aviation industry looking at net zero carbon emissions by 2050, the use of Sustainable Aviation Fuels (SAF) is vital to success to meeting this challenge. Currently there are seven pathways approved in D7566 but rely on the being blended with fossil fuel based jet fuel. The current availability and scalability of production of SAF is varied, leaving options of combining multiple pathways to produce a 100 % SAF without the requirement of blending with fossil fuels. Using fuel produced via the catalytic hydrothermal conversion route as the base fuel due to the aromatic levels, blends using ATJ-SPK, FT-SPK, HEFA-SPK, and SIP-SPK at the maximum blend ratios were prepared. The fuels were tested against the D7566 requirements for a finished fuel as well as thermal conductivity, enhanced thermal stability, chemical and physical ignition delay properties.

*Corresponding Author: DR David Evans; David.Evans@dst.defence.gov.au; 0438653371

IASH 2022, the 17th INTERNATIONAL SYMPOSIUM ON STABILITY, HANDLING AND USE OF LIQUID FUELS
Dresden, Germany
September 11 - 15, 2022

THE ROLE OF SYNTHETIC FUELS IN A SUSTAINABLE ENERGY SYSTEM

Achim Schaadt*, Rober Szolak

Head of Department Thermochemical Processes, Division Hydrogen Technologies, Fraunhofer ISE, Heidenhofstr. 2, 79110 Freiburg, Germany

ABSTRACT

The manufactures of internal combustion engines and gas turbines face currently a lot of challenges. Among these are the required drastic emission reductions of carbon dioxide (CO2) as well as harmful local emissions the most prominent. Therefore, the state of the art of synthetic fuels based on hydrogen as an essential part of the energy system is discussed. The utilization of hydrogen and hydrogen-based fuels is a promising option to reduce greenhouse gas emissions of heavy-duty vehicles, ships and air crafts. Hydrogen can be further converted together with CO2 via Power to X (PtX) concepts into synthetic fuels such as methanol and Fisch-Tropsch products offering both CO2 reductions and reduction of pollutants. Methanol can be used both as a fuel and as feedstock for further upgrading to fuels like dimethyl ether (DME), oxymethylene ether (OME), methanol-to-gasoline (MtG) and methanol-to-jet fuel. Methanol based jet fuel synthesis opens new opportunities in the sustainable aviation fuel (SAF) production field regarding process efficiencies, production costs and sustainability. It has been shown in process simulations that the target jet fuel product selectivity and the H2 efficiency (e.g. amount of hydrogen needed per liter or kg of the end product) are the main advantages of this value chain relative to other SAF production pathways e.g. Fischer Tropsch processes (FT).

*Corresponding Author: Dr.-Ing. Achim Schaadt; achim.schaadt@ise.fraunhofer.de; +4976145885428

IASH 2022, the 17th INTERNATIONAL SYMPOSIUM ON STABILITY, HANDLING AND USE OF LIQUID FUELS
Dresden, Germany
September 11 - 15, 2022

A NEW MANUFACTURING PROCESS FOR THE PRODUCTION OF SUSTAINABLE AVIATION FUEL FROM RENEWABLE FEEDSTOCKS

Dan Kadlecek*

ExxonMobil Research & Engineering, 1545 US Highway 22 East, Annandale, NJ 08801

ABSTRACT

Sustainable Aviation Fuels (SAF) are establishing themselves as a viable segment within the aviation fuel industry. Many aviation stakeholders have committed to long term reduction targets regarding greenhouse gas emissions, and this places significant pressure on the industry to develop technologies to meet the growing SAF demand. SAFs produced from primarily biological feedstocks are commercially available today and production capacity is anticipated to grow substantially over the coming decades. However, the socio-environmental expectations set forth by international bodies such as ICAO, as well as the corporate aspirations of fuel suppliers, OEMs and airlines means that the necessary volumes of sustainable, biologically-derived SAF feedstocks may be limited. One possible solution to the potential limitations of biologically-sourced SAFs is the development of eFuels, which are derived from carbon dioxide and hydrogen produced from renewable sources. Fischer-Tropsch technology is today the only pathway that provides access to eFuels for the aviation space. In this paper, a new eFuel pathway will be discussed that provides optionality in the production of SAF from advantaged feedstocks.

*Corresponding Author: Principal Aviation Fuel Scientist Dan Kadlecek;
daniel.e.kadlecek@exxonmobil.com; +1 215 756 3235

IASH 2022, the 17th INTERNATIONAL CONFERENCE ON STABILITY, HANDLING AND USE OF LIQUID FUELS
Dresden, Germany
11-15 September 2022

CO-HYDROPROCESSING – A ROUTE TO SUSTAINABLE AVIATION FUEL

Ulrich Balfanz[1], Drew Braden[2], Robert Brown[2], Alisdair Clark[3], Eric Doskocil[2], Vincent Kispersky[2], Tegan Roberts[4], Michaela Seebold[1], John Shabaker[2], Erik Vierling[1], Zhuoran Xu[2], Michael Zahnhausen[1], Dongting Zhao[2]

[1] *BP Europa SE, Querenburger Strasse 46, D-44789, Bochum, Germany;*
[2] *BP Naperville Complex, 150 West Warrenville Road, Naperville, Illinois, 60563, USA;*
[3] *BP International Ltd, Centre for Business & Technology, Sunbury-on-Thames, Middlesex, TW16 7LN, United Kingdom;*
[4] *BP International Ltd, Saltend Lane, Hull, East Yorkshire, HU12 8DS, United Kingdom.*

ABSTRACT

Aviation represents a unique and critical part of the global transportation system, carrying 1.8 billion passengers in 2020 to link the countries of the world across oceans and continents. The industry is built upon scientific understanding, with attention to detail for aircraft/engine design and fuel selection to deliver high levels of reliability, efficiency and performance. With environmental focus now on transport CO_2 emissions, the industry have sought to apply these principles to meet a new challenge, to achieve[1]

- Carbon neutral growth from 2021
- Net zero CO_2 emissions by 2050

Numerous options are being progressed to meet these goals, including engine/aircraft design, flight profile optimisation, carbon management[2] and fuel technology. The focus of the present paper is on fuel technology, where the industry have approved a number of aviation kerosenes derived from alternative sources to mitigate CO_2 emissions. Technology options include stand-alone manufacture or utilization of existing refinery units for co-processing[4,5]. Studies suggest the latter may reduce capital expenditure requirement by $0.33/gallon[6], a significant benefit to meet targets of 2% and 5% sustainable aviation fuel (SAF) production by 2025 and 2030 respectively. Currently co-processing of sustainable feedstocks is limited to 5% v/v in mixture with crude oil derived distillates. In this paper, the impact on aviation turbine fuel quality is examined through pilot plant studies where this limit has been increased to 30% v/v for bio-oils and Fisher Tropsch wax, both independently and in mixture.

1. INTRODUCTION

Aviation represents a unique and critical part of the global transportation system, carrying 1.8 billion passengers in 2020 to link the countries of the world across oceans and continents. The industry is built upon scientific understanding, with attention to detail for aircraft/engine design and fuel selection to deliver high levels of reliability, efficiency and performance. With environmental emphasis now on transport CO_2 emissions, the industry have sought to apply these principles to meet a new challenge, to achieve[1]

- Carbon neutral growth from 2021
- Net zero CO_2 emissions by 2050

Numerous options are being progressed to meet these goals, including engine/aircraft design, flight profile optimisation, carbon management[2] and fuel technology. The focus of the present paper is on fuel technology where scientific research has been applied to process engineering to manufacture sustainable, hydrocarbon aviation kerosene blending components or finished fuels, which can be used by current turbine powered aircraft. This initiative started in 2006 with the formation of the Commercial Alternative Aviation Fuels Initiative (CAAFI)[7] to promote and progress non-petroleum manufacturing paths. To ensure technical rigour, the ASTM specification D4054 Standard Practice for Evaluation of New Aviation Turbine Fuels and Fuel Additives[8] was developed and revised, including knowledge gained from earlier UK Defence Strategic Fuels Authority[9] assessment / approval of Fischer Tropsch (FT) jet fuel manufacture. With an appropriate assessment structure in place a new ASTM specification to contain stand-alone synthetic jet manufacturing process was developed, ASTM D7566 Standard Specification for Aviation Turbine Fuel Containing Synthesized Hydrocarbons[10], the first publication in 2009 featuring the FT process. To date, seven Annexes have been added to D7566 covering a range of synthetic processes which, with appropriate selection of feedstock, offer sustainable aviation fuel manufacturing processes to reduce CO_2 emissions, Table 1.

Table 1
ASTM D7566 Synthetic Kerosene Blend Component Production Processes

ASTM D7566 Annex	Process	Maximum Blending % v/v
1	Fischer Tropsch (FT)	50
2	Hydroprocessed Esters and Fatty Acids	50
3	Iso-Paraffins from Hydroprocessed Fermented Sugars	10
4	FT Synthesized Paraffinic Kerosine plus Aromatics	50
5	Alcohol-to-Jet Synthetic Paraffinic Kerosene	50
6	Hydrothermal Conversion of Fatty Acid Esters and Fatty Acids	50
7	Hydroprocessed Algae Esters and Fatty Acids	10

Maximum blending limits for the synthetics have been set based on the following criteria. To:

- ensure a managed transition of the synthetic blend component to the aviation fuel pool / allow aircraft maintenance and overhaul rates to be monitored and compared with historical data.
- align with D4054 assessment protocol allowing up to 50 or 10% v/v for the above depending on standard or Fast Track approaches respectively.
- resolve any technical issues.

To meet the IATA 2050 carbon neutral target the industry are currently considering if fully synthetic jet fuel production could be approved in the near to mid-term. Two ASTM Task Forces are active in this area covering:

- 100% Drop-in Sustainable Aviation Fuel
- 100% Non-Drop in Sustainable Aviation Fuel

The former would provide a product for the current aviation fleet and builds on earlier industry approvals. The latter is potentially for the future fleet with reduced aromatic content / lower density among other changes to possibly maximise environmental performance.

Preceding, and now in parallel to ASTM D7566, synthetic blend component manufacturing routes have also been included directly in jet fuel specifications Defence Standard 91-091 and ASTM D1655, for example:

- Def Stan 91-091 Issue 14 Annex B Additional Requirements Applicable to Fuels from Non-Conventional Sources
- ASTM D1655-22 Annex A1. Fuels from Non-Conventional Sources

This has primarily been the result of historical development where synthetic blend component production has been integrated with refinery operation manufacturing Jet and/or diesel fuels together with crude oil feedstocks. Four manufacturing processes are currently featured, Table 2.

Table 2
Def Stan 91-091 and ASTM D1655 Synthetic Kerosene Production Processes

Annex	Processes	Maximum Blending % v/v
Def Stan 91-091 Annex B	Sasol Fischer Tropsch Semi-Synthetic	50
	Sasol Fully Synthetic (up to 5 blended streams)	100
	Co-processed mono-, di-, and triglycerides, free fatty acids and fatty acid esters	5
	Co-processed hydrocarbons derived from synthesis gas via the Fischer Tropsch process using Iron or Cobalt catalyst	5
ASTM D1655 Annex A1	Sasol Fischer Tropsch Semi-Synthetic	50
	Sasol Fully Synthetic (up to 5 blended streams)	100
	Co-processed mono-, di-, and triglycerides, free fatty acids and fatty acid esters	5
	Co-processed hydrocarbons derived from synthesis gas via the Fischer Tropsch process using Iron or Cobalt catalyst	5

Integration of sustainable aviation fuel manufacturing processes with traditional refining is technically and commercially attractive:

- Hydroprocessing, a key requirement for current synthetics, is readily available at refineries with integrated utilities / hydrogen supply / engineering support.
- Experienced processing / laboratory personnel are available 24 / 7 to ensure consistent unit operation / product quality.
- Product tankage and distribution networks to airport are in place / optimized.
- During the period of energy transition from crude oil to sustainable product, airport supply can be maintained by balancing production volumes of the available feedstocks.

International Civil Aviation Organisation (ICAO) Carbon Offsetting and Reduction Scheme for International Aviation (CORSIA) calculations highlight that both ASTM D7566 and refinery co-hydroprocessing offer similar benefits for aviation CO_2 reduction per sustainable fuel molecule MJ, Table 3[11] . Co-processing may therefore offer a rapid route for aviation carbon footprint reduction. In addition, refinery units might fully transition to optimised co-processing operation over time, rather than crude oil, to gain equivalent efficiency to purpose built D7566 units.

Table 3
CORSIA Default Life Cycle Emissions Values for Eligible Fuels Manufactured by the Hydroprocessed Esters and Fatty Acids (HEFA) Fuel Conversion Process – Life Cycle Per Sustainable Fuel Molecule

Region	**Feedstock**	**Pathway**	**Life Cycle Emissions Factor (gCO_2e/MJ)**	**Life Cycle Emissions versus Crude Oil Jet, %**
BaseLine	Crude Oil	Conventional	89.0	0
Global	Tallow	D7566 Annex 2 Unit	22.5	-75
	Tallow	Refinery Co-Processing	27.2	-69
Global	Used Cooking Oil	D7566 Annex 2 Unit	13.9	-84
	Used Cooking Oil	Refinery Co-Processing	16.7	-81

Given these significant benefits, work is in progress to determine if increasing the permissible co-processing concentration of:

- hydrocarbons derived from Fischer Tropsch synthesis gas
- mono-, di-, and triglycerides, free fatty acids and fatty acid esters

from 5 to 30% v/v is technically viable. The current status of this initiative is reported below.

2. PROCESS TECHNOLOGY

2.1 Fischer Tropsch (FT)

FT technology utilizes synthesis gas as a feedstock. Synthesis gas can be produced from a wide range of raw materials, for example biomass, municipal solid waste, coal or natural gas. Production broadly seeks to balance a range of exothermic and endothermic reactions[12] :

Exothermic

$C + ½ O_2 \rightarrow CO$ (-111 MJ/kmol)

$CO + ½ O_2 \rightarrow CO_2$ (-283 MJ/kmol)

$H_2 + ½ O_2 \rightarrow H_2O$ (-242 MJ/kmol)

$C + 2H_2 \leftrightarrow CH_4$ (-75 MJ/kmol)

Endothermic

$C + H_2O \leftrightarrow CO + H_2$ (+131 MJ/kmol)

$C + CO_2 \leftrightarrow 2CO$ (+172 MJ/kmol)

The resultant product is a mixture of carbon monoxide and hydrogen which is then converted by the FT process into a hydrocarbon product of wide carbon number distribution (SynCrude) and water. The process is carried out in the presence of an iron or cobalt catalyst at moderate temperature and pressure for example 200–350 °C and 20–50 bar respectively. The reaction is summarized by the chemical equation:

$$nCO + 2nH_2 + \text{heat} \rightarrow n(\text{-}CH_2\text{-}) + nH_2O$$

The carbon number distribution of the FT SynCrude is generally too heavy to offer a useful kerosene yield. However, the product may be hydrocracked and fractionated, in a similar manner to heavy crude oil distillates, to give a kerosene of suitable quality as a component for semi-synthetic or fully synthetic jet fuel production. Extensive aviation industry assessments of FT derived kerosene has been undertaken[13,14,15,16] with the product included in both ASTM D7566, D1655 and Defence Standard 91-091 approvals. Additional work has also demonstrated co-hydroprocessing of the FT SynCrude at 10% v/v in mixture with standard petroleum distillates as satisfactory[17].

Given the versatility of FT synthesis gas feedstocks and positive industry assessment of FT kerosene above, the present study seeks to examine co-processing of SynCrude in mixture with petroleum crude oil feedstock at up to 30% v/v.

2.2 Hydroprocessing of Fats and Oils

Fats and oils from vegetable or animal sources (Bio-Oils) are constructed of triglyceride building blocks resulting from the combination of one unit of glycerol and three units of fatty acids, Figure 1. Such products are classified as "lipids" which is a category that embraces a broad variety of chemical substances for example mono- and diglycerides, phosphatides, cerebrosides, sterols, terpenes, fatty alcohols, fatty acids, and fat-soluble vitamins. Bio-oils are insoluble in water but soluble in most organic solvents. Densities are lower than water and consistency may be solid, semisolid, or liquid at ambient temperature. Bio-oils which are solid at room temperature are commonly referred to as "fats" while liquid bio-oils are called "oils." For simplification purposes, the terms "fat" and "oils" are used interchangeably in this report and termed 'bio-oil'.

Figure 1
Natural Oils – Triglycerides

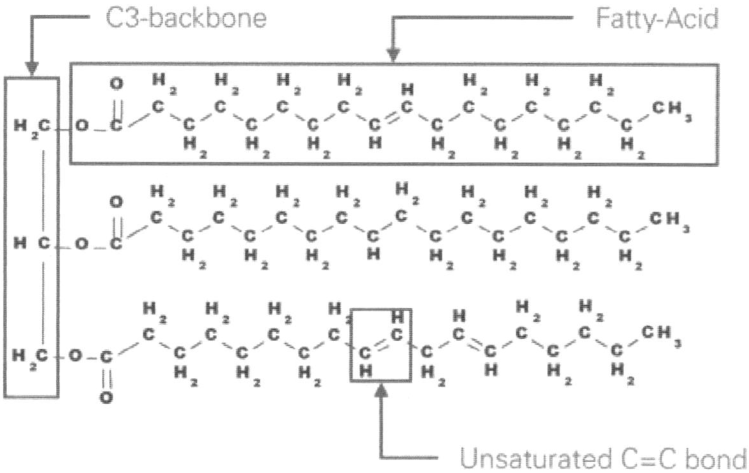

The detailed composition of bio-oils may vary, for example dependent on plant species, soil conditions, moisture content in the seeds and oil extraction method. Figure 2 shows the composition[18] of various oil types. A key difference is the carbon number range. Coconut oil contains largely C12-fatty acids while Rapeseed, Jatropha and Soybean oil contain predominantly C18-fatty acids.

Figure 2
Triglycerides – fatty acid distribution

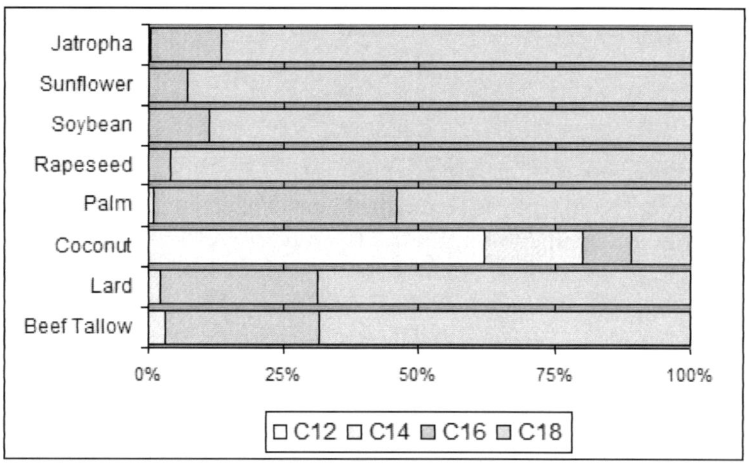

Hydroprocessing of such feedstocks results in hydrodeoxygenation, and other direct mechanisms such as hydrodecarbonylation and hydrodecarboxylation, to yield a hydrocarbon product together with a range of bi-products such as water, propane and

CO_2. With appropriate process severity any alkene bonds within the bio-oil are also eliminated to give a fully saturated, alkane product, Figure 3

Figure 3
Hydroprocessing Scheme

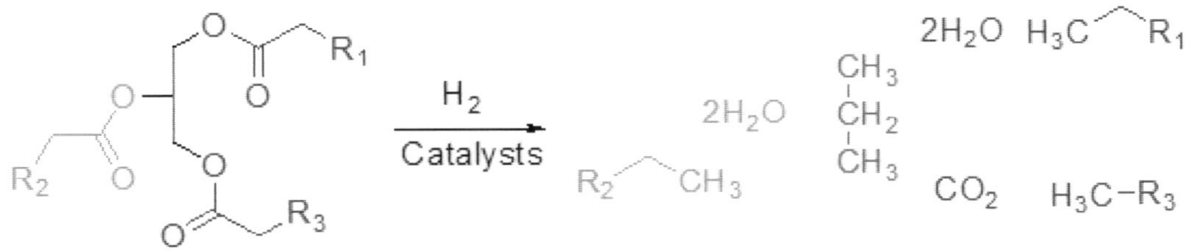

The exact composition of product depends on the bio-oil and process conditions. However, as a general rule the fatty acid carbon number may be retained or reduced by 1. For example, a C18 fatty acid could be converted into:

- C18 paraffins through hydrodeoxygenation with water as by-product.
- C17 paraffins through hydrodecarboxylation with CO_2 as by-product.

Similarly, to FT syncrude, hydroprocessing of Bio-Oils has been extensively evaluated for the production of synthetic kerosene with positive results[19,20]. Again, product has been included in ASTM D7566, and in ASTM D1655 and Defence Standard 91-091 via co-processing.

At the present time D7566 hydroprocessed Bio-Oils represent the major source of sustainable aviation fuel blending component with over 75 million gallons featured in a recent ASTM survey. Refinery co-processing of such feedstocks is also active for aviation industry supply at a number of locations[21, 22, 23].

The present project assesses the possibility of increasing the permissible concentration of co-processing Bio-Oils from 5 to up to 30% v/v to further promote supply and aviation transition to a low carbon future. Focus is on hydrocracking as, similar to the FT case, the carbon chain length of the Bio-Oils must be changed / isomerised to provide a viable yield of sustainable kerosene.

2.3 Hydrocracking

Hydrocracking represents a severe form of hydrotreatment where the carbon molecules in heavy crude oil fractions are broken and saturated with hydrogen to form lighter, more marketable products. The feed-stock is transformed by:

- Contacting with high pressure hydrogen, for example 80 to 160 bar in the temperature range 400 – 460 °C in the presence of a dual function catalyst:
    - Function 1: cracking, typically effected using particles of clay, alumina or silica alumina treated with hydrogen fluoride.
    - Function 2: hydrogenation, typically effected by coating the particles with nickel, tungsten, platinum and palladium.
- The product is separated from the catalyst and sulphur extracted as H_2S.

Feedstocks are generally purified by hydrotreatment before entering the main reactor due to the sensitivity and high cost of the precious metal catalyst. The process also requires significant hydrogen availability to ensure the olefins formed are rapidly saturated to alkanes before further degradation can occur. Example stoichiometry is given as:

$$RCH_2CH_2R'SH \rightarrow R{=}CH_2 + CH_2{=}R'SH$$
$$R{=}CH_2 + CH_2{=}R'SH + 2H_2 \rightarrow R\text{-}CH_3 + CH_3\text{-}R'H + H_2S$$

The result is a high quality product which can be fractionated to yield naphtha, kerosene and diesel. The use of hydrogen may prove advantageous in the future, allowing green hydrogen to be safely incorporated into aviation fuel and further reduce carbon foot-print.

Hydrotreatment is featured in standard petroleum refining and in ASTM D7566 FT and Bio-Oils processing, Annexes 1 and 2 respectively. In all three cases industry data demonstrates the technology can be applied to upgrade 100% of the feedstock. On this basis, it is reasonable to assume that hydrocracking, the most severe form of hydrotreatment, may be able to upgrade any mixture of the respective feeds with appropriate unit management, Figure 4. An evaluation of mixtures of crude oil, FT and Bio-Oil feedstocks has therefore also been included in the present study for aviation turbine fuel quality assessment.

Figure 4
Hydrotreatment – A Spectrum of Operation

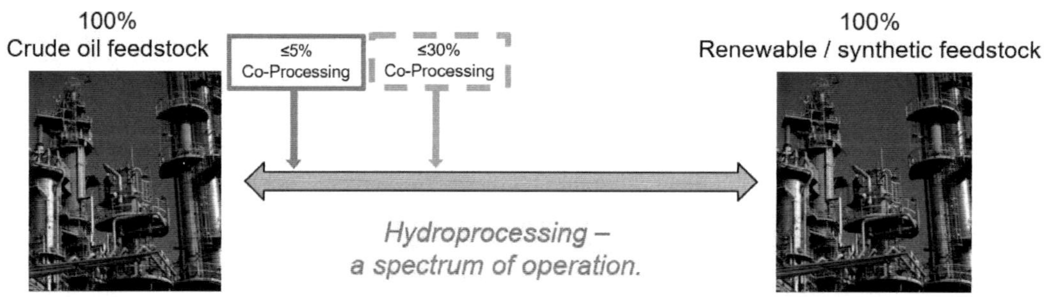

3. EXPERIMENTAL PROGRAMME

With the support of an aviation industry Task Force an experimental programme has been developed to evaluate the potential of hydrocracking for refinery co-processing. This features the following components in mixture with petroleum feedstock:

- Up to 30% v/v Fischer Tropsch SynCrude
- Up to 30% v/v Bio-Oils
- Up to 30% v/v mixtures of the above

The current status of the programme is presented in 6 sections:

- Pilot Plant
- Petroleum Feedstock
- Fischer Tropsch Feedstock
- Bio-Oil Feedstocks
- Test Matrix
- Analytical Analysis

3.1 Pilot Plant

Pilot plant options and operating conditions were assessed through expertise in bp Naperville US and bp Bochum Germany, the aim being to feature a commercial refinery catalyst system, petroleum feedstock and appropriate conditions for conversion. A pilot plant located in bp Bochum, Germany, was selected for the experimental work based on:

- Representative refinery catalyst / feed stock availability.
- Reactor scale.
- Operating system.
- On-site analytical laboratory.

A schematic of the pilot plant is given in Figures 5. Earlier use for aviation focused programmes had proved highly successful, for example in assessing Bio-Oil co-processing and jet fuel desulphurisation[20, 24, 25].

Figure 5
Pilot Plant Schematic

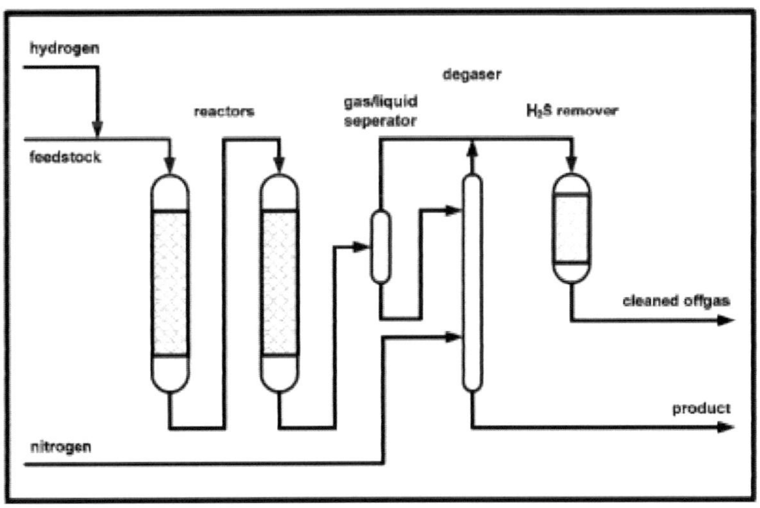

3.2 Petroleum Feedstock

A representative crude oil petroleum hydrocracking base-stock was sourced from a local refinery for co-processing. Inspection data are provided in Table 4.

Table 4
Crude Oil Base-Stock Analysis

Analysis	Units	Crude Oil Hydrocracking Base-Stock
Appearance	-	Black / Pasty
Density at 50°C	kg/m³	874.9
Density at 20°C calculated	kg/m³	895.6
Mono aromatics	% m/m	20.7
Di aromatics	% m/m	11.8
Total aromatics	% m/m	35.0
Sulphur	% m/m	1.279
Phosphorus	% m/m	<0.05
Zinc	% m/m	<0.05
Calcium	% m/m	<0.05

3.3 Fischer Tropsch Syn-Crude Feedstock

A sample of Fischer Tropsch wax was sourced from the open market. Properties and carbon range and given in Tables 5, 6 and Figure 7.

Table 5
Fischer Tropsch Sample Properties – Supplier Data

Property	Units	Result
Appearance	Visual	White, waxy, solid
Density	kg/m³	Ca. 764
Viscosity @100 °C	mm²/s	3 – 4
Flash Point	°C	≥250
Initial boiling point	°C	Ca. 280

Table 6
Fischer Tropsch Sample Carbon Number Distribution

n-Paraffin	%(m/m)	n-Paraffin	%(m/m)
n-C3	0.00	n-C28	8.68
n-C4	0.00	n-C29	7.47
n-C5	0.00	n-C30	6.22
n-C6	0.00	n-C31	4.93
n-C7	0.00	n-C32	3.60
n-C8	0.00	n-C33	2.44
n-C9	0.00	n-C34	1.52
n-C10	0.00	n-C35	0.89
n-C11	0.00	n-C36	0.50
n-C12	0.00	n-C37	0.28
n-C13	0.00	n-C38	0.16
n-C14	0.00	n-C39	0.09
n-C15	0.00	n-C40	0.05
n-C16	0.01	n-C41	0.03
n-C17	0.04	n-C42	0.02
n-C18	0.15	n-C43	0.012
n-C19	0.41	n-C44	0.007
n-C20	1.04	n-C45	0.004
n-C21	2.46	n-C46	0.003
n-C22	5.13	n-C47	0.002
n-C23	8.30	n-C48	0.001
n-C24	10.79	n-C49	0.001
n-C25	11.28	n-C50	0.001
n-C26	10.69	Trace / outside range	3.01
n-C27	9.77	Total	100.00

Figure 7
Fischer Tropsch Gas Chromatography Analysis

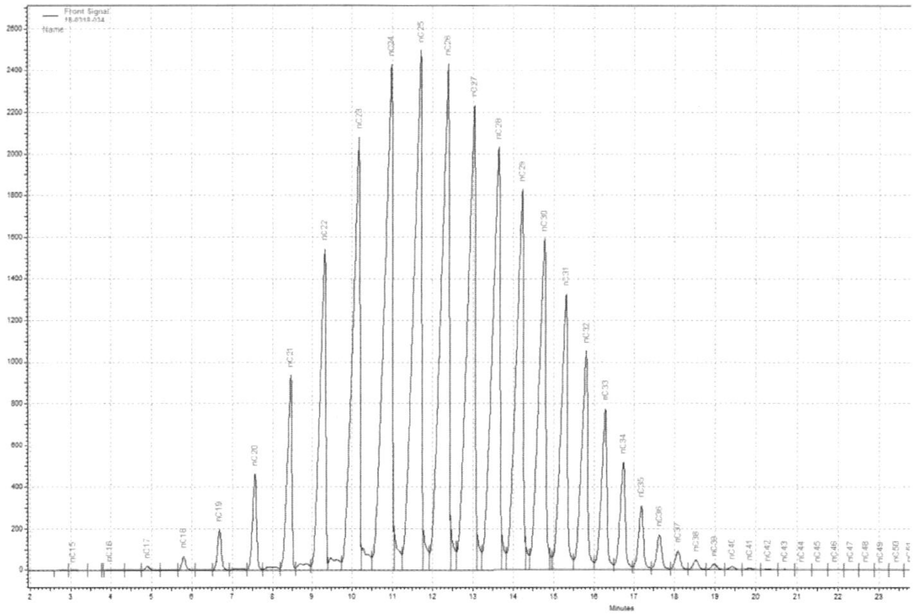

3.4 Bio-Oil Feedstocks

Given the variation in bio-oil carbon chain-lengths four oils were selected for the programme to give as wide a range as possible:

- Coconut oil, rich in C12 carbon molecules
- Rapeseed oil, rich in C18 carbon molecules
- Mustard seed oil, rich in C22 carbon molecules
- Used Cooking Oil (UCO), rich in C16 / C18 carbon molecules

The quality of each oil was determined by analysis of:

- Density
- Sulphur
- Water
- Composition of fatty acids

Results confirmed the anticipated composition of each feedstock and proposed variations for the study, Table 7.

Table 7
Coconut, Rapeseed, Mustard and Used Cooking Oil Analysis

Analysis	Units	Coconut Oil	Rapeseed Oil	Mustard Oil	UCO
Density @ 15 °C	kg/m³	924.1	919.9	914.2	896.9
Sulphur	mg/kg	<0.10	-	<0.10	<0.10
Water content	mg/kg	266	-	183	598
C06:0	% m/m	0.73	0.00	0.00	0.03
C08:0	% m/m	8.22	0.00	0.00	0.12
C10:0	% m/m	5.87	0.00	0.02	0.08
C12:0	% m/m	42.85	1.68	0.28	0.52
C14:0	% m/m	15.95	0.06	0.23	0.76
C16:0	% m/m	7.63	4.32	2.15	18.43
C16:1	% m/m	0.03	0.2	0.15	0.37
C18:0	% m/m	2.35	1.67	0.96	4.79
C18:1	% m/m	5.45	58.87	23.53	40.50
C18:2	% m/m	1.99	17.58	8.70	22.86
C18:3	% m/m	0.10	7.55	8.35	0.96
C20:0	% m/m	0.15	0.57	0.61	0.34
C20:1	% m/m	0.00	1.16	8.86	0.33
C22:0	% m/m	0.05	0.30	0.56	0.49
C22:1	% m/m	0.33	0.20	34.96	0.06
C24:0	% m/m	0.00	0.14	0.35	0.20
C24:1	% m/m	0.10	0.16	2.75	0.14

3.5 Test Matrix

To assess the impact of co-processing on kerosene quality a test matrix featuring the selected feedstocks at the maximum proposed concentration of 30% v/v in mixture with the crude oil base-stock was developed with 100% v/v base-stock representing the base-line run. In addition, a range of mixtures was also featured to evaluate combined feedstock operation, Table 8.

Table 8
Co-processing Test Matrix

Run	VGO % v/v	Fischer Tropsch % v/v	Coconut Oil % v/v	Rapeseed Oil % v/v	Mustard Seed Oil % v/v	Used Cooking Oil % v/v
1	100					
2	70	30				
3	70		30			
4	70				30	
5	70					30
6	70		10		10	10
7	70	20.1	3.3	3.3		3.3
8	70	10.2	6.6		6.6	6.6

3.6 Fractionation

Each pilot plant run provided a full boiling range liquid product covering:

- Naphtha
- Jet fuel
- Diesel fuel

The jet fuel fraction was recovered using a distillation column operated according to ASTM D2892 Standard Test Method for Distillation of Crude Petroleum (15-Theoretical Plate Column). Similar cut point to refinery operation were selected based on constraints of:

- Flash Point ≥ +38 °C
- Freeze Point ≤ -47 °C

The resultant kerosene was then analysed as detailed below.

3.7 Analytical Analysis

Co-processed Jet A / A-1 containing up to 5% v/v hydrotreated FT syn-crude / bio-oils must meet the analysis requirements detailed in ASTM D1655 or Defence Standard 91-091 Table 1 together with additional tests detailed in Tables A1.1 and Table 3 respectively. Where tests are duplicated, the more severe analysis applies. For example, thermal stability at 280 rather than 260 °C. With respect to 100% hydrotreated FT syn-crude / bio-oils, the requirements of ASTM D7566 Annexes 1 and 2, Tables A1.1, A1.2 and A2.1, A2.2 apply respectively. These analyses were combined to offer a comprehensive set to evaluate the quality of 30% v/v co-processed product, Table 9. In addition, GC x GC may be utilised to evaluate product composition and Carbon 14 to evaluate renewable content of the bio-oil products.

Table 9 – Part 1
30% v/v Co-Processing Analysis Strategy

Analysis	**Units**	**Method**
Acidity	mgKOH/g	D3242
Aromatics	% v/v	D1319
Mercaptane Sulphur	% m/m	D3227
Total Sulphur	% m/m	D1266
Distillation		D86
T10%	°C	
T50%	°C	
T90%	°C	
FBP	°C	
Residue	% v/v	
Loss	% v/v	
T50-T10%	°C	
T90-T10%	°C	
Simdis		D2887
10%	Report	
20%	Report	
50%	Report	
80%	Report	
90%	Report	
FBP	Report	
Flash Point	°C	D56
Density	kg/m3	D4052
Freeze Point	°C	D2386
Viscosity @-40°C	cSt	D445
Energy Content	MJ/kg	D3338
Naphthalenes	% v/v	D1840
Smoke Point	mm	D1322
Copper Corrosion	Rating	D130
JFTOT @ 280°C		D3241
JFTOT @ 325°C		"
Visual	Rating	
Thickness	ETR nm	
	ITR nm	
dP	mmHg	
Existent gum	mg/100ml	D381
MSEP	Rating	D3948
Lubricity	mm	D5001
Conductivity	pS/m	D2624
Trace FAME	mg/kg	D7797
Cycloparaffins	% m/m	D2425
Aromatics	% m/m	"
Paraffins	% m/m	"
Carbon + Hydrogen	% m/m	D5291

Table 9 – Part 2
30% v/v Co-Processing Analysis Strategy

Analysis	Units	Method
Nitrogen	mg/kg	D4629
Water	mg/kg	D6304
Sulphur	mg/kg	D5453
Metals		
Al	mg/kg	D7111
Ca	mg/kg	
Co	mg/kg	
Cr	mg/kg	
Cu	mg/kg	
Fe	mg/kg	
K	mg/kg	
Li	mg/kg	
Mg	mg/kg	
Mn	mg/kg	
Mo	mg/kg	
Na	mg/kg	
Ni	mg/kg	
P	mg/kg	
Pb	mg/kg	
Pd	mg/kg	
Sn	mg/kg	
Sr	mg/kg	
Ti	mg/kg	
V	mg/kg	
Zn	mg/kg	
Halogens	mg/kg	D7359

4. RESULTS AND DISCUSSION

At the time of this paper pilot plant testing is nearing completion and product analysis has commenced. A set of provisional, limited, data is available as follows:

Pilot Plant Full Boiling Range Product – Polycyclic Aromatics

The heavy crude oil base-stock selected for the project contained in excess of 12% m/m polycyclic aromatics. Aviation fuel specifications limit naphthalenes to 3.0% v/v maximum and it is important co-processing has no negative impact on removal of such species. Data generated by the present study indicates this is the case, with polycyclics falling at or below the limit of detection, 0.1% m/m, for all present runs, Table 10.

Table 10
Assessment of the Impact of Co-Processing on Di-Aromatic (Naphthalene) Reduction

Run	Blend	Full Boiling Range Material	Mono-Aromatics % m/m	Di-Aromatics % m/m	Total Aromatics % m/m
1	100% Crude Oil Base-stock	Total Feed Total Post Reactor	20.7 3.4	11.8 0.1	35.0 3.6
2	70% Crude Oil Base-stock 30% FT Wax	Total Feed Total Post Reactor	14.5 2.3	8.3 0.0	24.5 2.3
5	70% Crude Oil Base-stock 30% Used Cooking Oil	Total Feed Total Post Reactor	14.6 1.6	8.3 0.0	24.9 1.6
7	70% Crude Oil Base-stock 20.1% FT Wax 9.9% Mixed Vegetable Oils	Total Feed Total Post Reactor	14.8 2.3	8.4 0.0	24.7 2.3

Pilot Plant Full Boiling Range Product – Mono-Aromatics

The presence of some aromatics in jet can be advantageous for maintaining energy density by volume and elastomer seal-swell in older aircraft. As such an 8% v/v minimum limit has been set for this property in ASTM D7566. For the present study, the full range pilot plant product has highlighted the concentration of mono-aromatics is also reduced during co-processing. This is the result of two effects:

(i) Dilution of the aromatics present in the crude oil base-stock by the sustainable feedstock.

(ii) Hydrogenation of some of the aromatic rings during the hydrocracking process.

Current results indicate the full-range product contains between 1.6 and 3.4% m/m mono aromatics. Dependent on the boiling range distribution of the species, this may increase / decrease on recovery of the final jet fuel fractions following ASTM D2892 distillation.

ASTM D2892 Distillation

An ASTM D2892 distillation has been completed on the crude oil base-stock product. The initial kerosene fraction has been assessed for flash point and freeze point similar to refinery practice, Table 11. Data indicate a potentially viable jet fuel product. Distillation cut-points are now being optimised to maximise yield.

Table 11
Initial Analysis of Kerosene Fraction Derived from Crude Oil Base-stock Hydrocracking Run

Analysis	**Units**	**Result**
Density @ 15 °C	kg/m^3	806.3
Flash Point	°C	55.0
Freeze Point (D7153)	°C	-58.7
Mono Aromatics	% m/m	4.6
Naphthalenes	% m/m	<0.1

A further update regarding this project will be provided at the ASTM International D02.J Aviation Meeting, December 4-9, 2022, New Orleans USA.

5. CONCLUSIONS

- Refinery co-processing of sustainable feedstocks offers a potential path for the Aviation Industry to transition to carbon neutral operations by 2050.
- Dependent on feedstock / scenario, ICAO CORSIA calculations indicate a similar benefit per sustainable molecule for D7566 and co-processing, for example -84 versus -81% for used cooking oil respectively.
- To capture this benefit work is in progress to evaluate if current co-processing limits can be increased from ≤5 to ≤30% v/v for Fischer Tropsch and Bio-oil feedstocks individually and in mixture under hydrocracking conditions:

  - 6 feedstocks (crude oil, FT wax, coconut, rapeseed, mustard seed, used cooking oil).
  - A matrix of 8 pilot plant runs featuring both pure and mixed feedstocks to compare to base-line of crude oil processing.
  - ASTM D2892 distillation to recover the resultant jet fuel fractions.
  - Combined ASTM D1655 co-processing and D7566 Annex 1 & 2 analytical analysis.
- Initial results indicate:
  - No detrimental impact of co-processing on polycyclic aromatic / naphthalene reduction for hydrocracked product.
  - A reduction in mono aromatic content as a result of:
    - Sustainable feedstock incorporation.
    - Hydrogenation of aromatic rings during processing
  - Successful production of a kerosene fraction meeting aviation freeze point and flash point limits with additional optimisation / analysis in progress.
- A further update will be provided at the ASTM International D02.J Aviation Meeting, December 4-9, 2022, New Orleans USA.

6. ACKNOWLEDGEMENT

The expertise, enthusiasm and dedication of team members in bp Bochum and Naperville research centres and the bp refinery group is recognised for this technical study. 'Thank You' to all those involved.

7. REFERENCES

[1] International Air Transport Association, www.iata.org

[2] Carbon Offsetting & Reduction Scheme for International Aviation, www.icao.int

[4] D1655 Standard Specification for Aviation Turbine Fuels, ASTM International, 100 Barr Harbor Drive, PO Box C700, West Conshohocken, PA 19428-2959. USA.

[5] Defence Standard 91-091 Turbine Fuel, Kerosene Type, Jet A-1; NATO Code: F-35; Joint Service Designation: AVTUR, UK Defence Standardization, Kentigern House, 65 Brown Street, Glasgow, G2 8EX, UK.

[6] National Renewable Energy Laboratory, Commercial Aviation Alternative Fuels Initiative (CAAFI) January 19th, 2022.

[7] Commercial Alternative Aviation Fuels Initiative (CAAFI) www.caafi.org

[8] D4054 Standard Practice for Evaluation of New Aviation Turbine Fuels and Fuel Additives, ASTM International, 100 Barr Harbor Drive, PO Box C700, West Conshohocken, PA 19428-2959. USA.

[9] Defence Strategic Fuels Authority, Larch 3B, Ministry of Defence, Abbey Wood, Bristol, BS34 8JH.

[10] ASTM D7566 Standard Specification for Aviation Turbine Fuel Containing Synthesized Hydrocarbons, ASTM International, 100 Barr Harbor Drive, PO Box C700, West Conshohocken, PA 19428-2959. USA.

[11] International Civil Aviation Organisation (ICAO) Carbon Offsetting and Reduction Scheme for International Aviation (CORSIA) Default Life Cycle Emissions Values for CORSIA Eligible Fuels, June 2022.

[12] US Department of Energy National Energy Technology Laboratory: Detailed Gasification Chemistry.

[13] Southwest Research Institute (SwRI) report number 8531.

[14] Southwest Research Institute (SwRI) report number 08-04438.

[15] Southwest Research Institute (SwRI) report number 08-04438-2.

[16] Comparative Evaluation of Semi-Synthetic Jet Fuels, Final Report. CRC Project AV-2-04a. C Moses, 2008.

[17] Co-Processing of Fischer-Tropsch Feedstocks with Petroleum Hydrocarbons for Jet Production using Hydrotreating & Hydrocracking. B Miller, R Freerks, 2009. ASTM International Research Report RR:D02-1929.

[18] Food Fats and Oils, Institute of Shortening and Edible Oils 2006.

[19] Evaluation of Bio-Derived Synthetic Paraffinic Kerosenes (Bio-SPKs). J Kinder. ASTM International Report RR:D02-1739.

[20] Revision of D 1655, Specification for Aviation Turbine Fuels: Co-processing of HEFA Feedstocks with Petroleum Hydrocarbons for Jet Production using Hydrotreating and Hydrocracking. J Sorena et al, A Clark, U Balfanz, E Lodrigueza, G Heminghaus. ASTM International Research Report RR:D02-1886.

[21] Repsol Coprocesses Sustainable Aviation Fuel in Tarragona, Spain. Biobased Diesel Daily, 2021.

[22] Lingen refinery starts SAF production using waste and residues. Air bp, bp.com, 2022.

[23] Oil Majors Look to Co-processing as a Rapid Route to Producing Sustainable Aviation Fuels at Scale. GreenAir 2022.

[24] J Tucker, C Lewis, A Clark; Jet fuel Desulphurisation: An Investigation into the impact of hydrotreatment on product quality. IASH 2011 12th International Conference on Stability, Handling and Use of Liquid Fuels, Sarasota, Florida USA, 16th – 20th October 2011.

[25] A Clark et al; The impact of co-hydroprocessing on jet fuel quality. 13th International Symposium on the Stability, Handling and Use of Liquid Fuels, Rhodes, Greece. October 2013.

IASH 2022, the 17th INTERNATIONAL SYMPOSIUM ON STABILITY, HANDLING AND USE OF LIQUID FUELS
Dresden, Germany
September 11 - 15, 2022

CO-PROCESSING OF HYDROPROCESSED VEGETABLE OIL

Paul P. Wells*, Daniel E. Kadlecek

ExxonMobil Research & Engineering, 1545 US Highway 22 East, Annandale, NJ 08801

ABSTRACT

Refinery co-processing of hydroprocessed vegetable oil (HVO) can provide an effective means of increasing the bio-derived synthetic aviation fuel (SAF) content of jet fuel with minimal capital investment. Co-processing of bio-derived material is currently limited to 5 vol% of either triglycerides, free fatty acids, and fatty acid esters or Fischer-Tropsch liquids. Both feedstocks require hydroprocessing and fractionation at the refinery to produce a viable product. Spare hydroprocessing capacity at a refinery is rarely available and expansion of an existing hydroprocessing unit or installation or a new unit is quite costly. HVO, originally intended as a blending component for diesel fuel, is commercially available and (as the name implies) has been hydroprocessed to remove unwanted heteroatoms, primarily oxygen. The boiling range of HVO, dependent upon the feed source and degree of hydroprocessing, could exceed the 300°C endpoint of jet fuel and therefore fractionation may be required during refinery co-processing. This research will explore the impact of co-processing different types and amounts of HVO on key jet fuel properties as well as non-hydrocarbon composition.

*Corresponding Author: Advanced Research Associate Paul Wells;
paul.p.wells@exxonmobil.com; 1-832-624-5894

IASH 2022, the 17TH INTERNATIONAL CONFERENCE ON STABILITY, HANDLING AND USE OF LIQUID FUELS
Dresden, Germany
11-15 September 2022

NEW METHODS FOR ANALYSIS OF QUALITIES OF FUELS AND NOVEL FUEL COMPONENTS

Marcella Frauscher[1], Adam Agocs[1] and Martin Miedler[2].

[1] AC2T research GmbH, Viktor-Kaplan-Straße 2C, 2700 Wiener Neustadt, Austria. marcella.frauscher@ac2t.at

[2] OMV Downstream GmbH, Mannswörtherstraße 28, 2320 Schwechat, Austria

ABSTRACT

Changes in regulations and a growing awareness of the limit of fossil resources will lead to a rapid growth in terms of alternative future fuels due to a rising demand. In order to accelerate this development and make it more efficient design tools including stability studies and analytical methods may be used. Determination of the influence of promising fuel components in various blend level or certain additives can be achieved by tailor-made stability tests. These can either simulate fuel circulation in use, storage experiments or stability assessment tests. In combination with tools such as high-resolution mass spectrometry precursors of beginning degradation can be detected. Evaluation thereof is assisted with data evaluation techniques to ensure fast processing and easy interpretation. Hence, a determination of the stability performance depending on different blend components is possible. Moreover, ideal additive concentrations of stability additives can be elaborated. This approach can be applied to various fuels and fuel components, thus replacing and supporting the development of future fuels.

INTRODUCTION

The global transportation sector faces several challenges. Currently, sustainability of passenger cars as well as light commercial vehicles is in the focus of development work, where several different drivetrain technologies are competing. Nevertheless, the internal combustion engine (ICE) still dominates the market, as approx. 88 % of new vehicles utilized an ICE in their drivetrain in 2021 [1]. Hence, the development of sustainable fuels for internal combustion engines, based on biologically sourced or synthetic components is one of the most pressing issues of the decade. Novel base fuel components and additives offer several advantages, but also come with challenges for the manufacturers and the end users. One of the key aspects is oxidation stability of the fuels, as it influences storage stability and performance in an ICE to a great extent. Accordingly, studies about oxidation mechanisms of antioxidants using advanced analytical methods, such as mass spectrometry (MS) are increasingly common [2]. This study demonstrates a comprehensive methodology to assess the influence of bio components, cetane improver additives and stabilizers on the oxidation stability in various fuel blends.

MATERIALS AND METHODS

The presented Q.O.T. oxidation test is based on a modified version of ASTM D2272 – 14a Method B [3]. The original Quantum® oxidation tester was supplied by Tannas (Tannas CO. & King Refrigeration, INC, Midland, MI, USA). The oxidation cell was extensively modified (displayed in figure 1 a) to enable the measurements of fuels in a safe manner, also enabling the easy sampling of the liquid phase during oxidation tests. Alterations of various fuel blends were carried out in a closed, pressurized vessel at 130°C and 2 bar oxygen pressure. Pressure and temperature were monitored continuously throughout the whole alteration process. Oxidation initiation time (OIT) was defined as 10 % pressure loss from the maximal pressure in the oxidation cell, which corresponds to the definition given in DIN EN 16091 [4]. Accordingly, a good comparability between the two methods was achieved, where the OIT measured by the presented method and DIN EN 16091 shows a good correlation (see figure 1 b). Furthermore, the total oxidation time (TOT) was defined as a stable final pressure in the oxidation cell.

Subsequently, oxidation of the altered fuels was determined according to an in-house method via Fourier-transformed infrared spectrometry (FT-IR), described in detail in [5]. Furthermore, the neutralization number (NN) and the water content were also measured according to [6] and [7] respectively.

In this study, as an example to show the approach for accelerated R&D methods, a conventional, biofree diesel fuel, a diesel blend containing 20 % biocomponent as well as the influence of a cetane improver (cetane-number booster) and various stabilizers is presented.

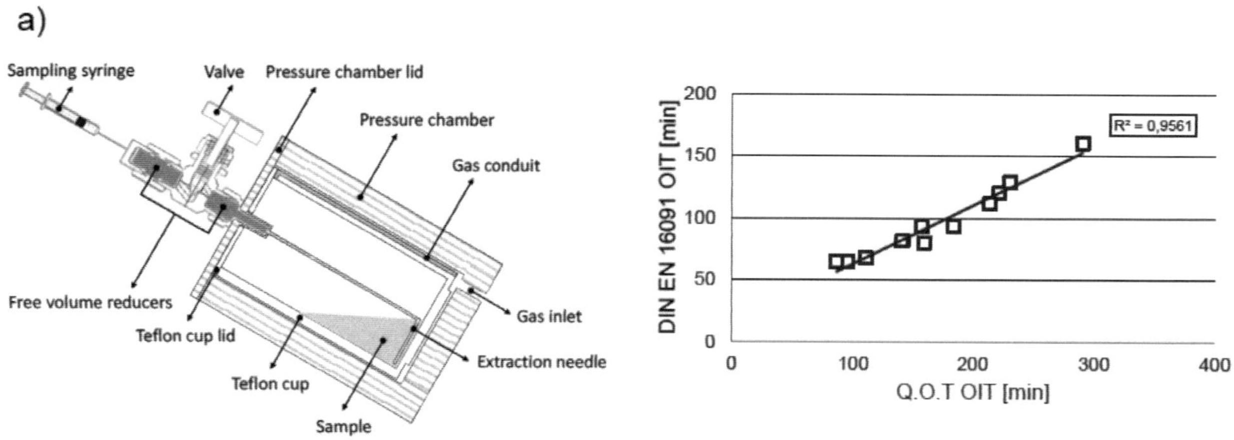

Figure 1 a): Schematics of the modified Quantum ® oxidation tester cell
b): Comparability of the Q.O.T. test and DIN EN 16091

RESULTS AND DISCUSSION

Figure 2 a) shows the pressure curves of the base fuels (biofree and 20 % bio) with and without added cetane improver. Figure 2 b) displays the calculated OIT and TOT. As shown, the base fuels showed very comparable oxidation behavior, as OIT and TOT as well as the overall pressure curve characteristics are very similar. The addition of the cetane improver influenced the oxidation stability in a negative manner, a significant decrease in both OIT and TOT was measured for both the biofree base fuel and the fuel with 20 % biocomponent.

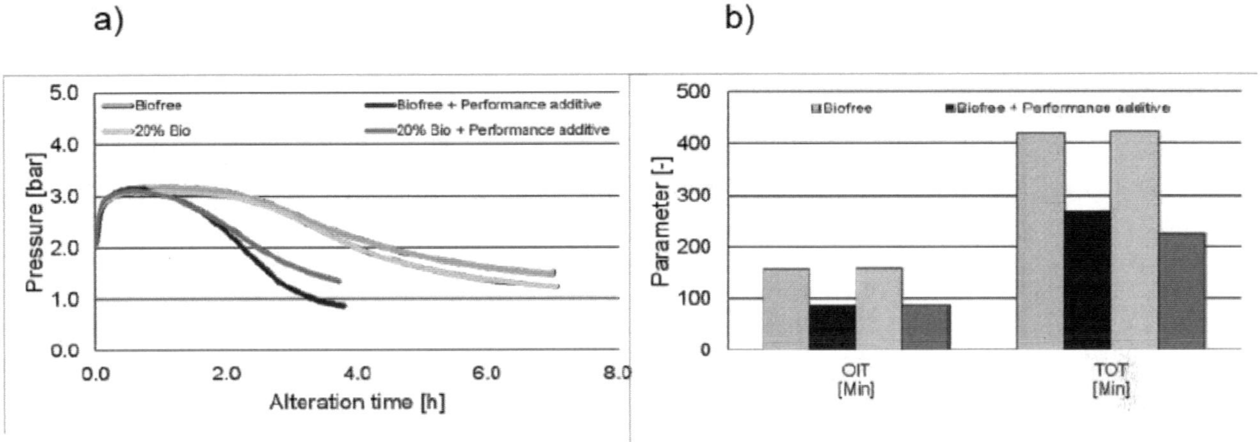

Figure 2 a): Pressure curves of the base fuels with and without cetane improver
b): OIT and TOT of the base fuels with and without cetane improver

Two commercially available stabilizer additives were tested to determine additive performance and the optimal additive concentration to counteract the negative effect of the cetane improver on the oxidation stability. Figure 3 shows the dependency of the OIT on the used stabilizer concentration for the biofree (a) and the 20% bio fuels (b) with added cetane improver. The dotted lines denote the OIT of the base fuels without cetane improver (reference value). OIT showed a non-linear dependency in case of both stabilizers, where the effect of the stabilizer is diminishing if the applied concentration is too high. This highlights the importance of selecting the optimal additive concentration, as it is the most efficient from an economical perspective and also the most sustainable, as only the required amount of additive is used. Furthermore, it is shown that differences in stabilizer performance can be determined, as stabilizer 2 showed a significantly higher OIT in both base fuels.

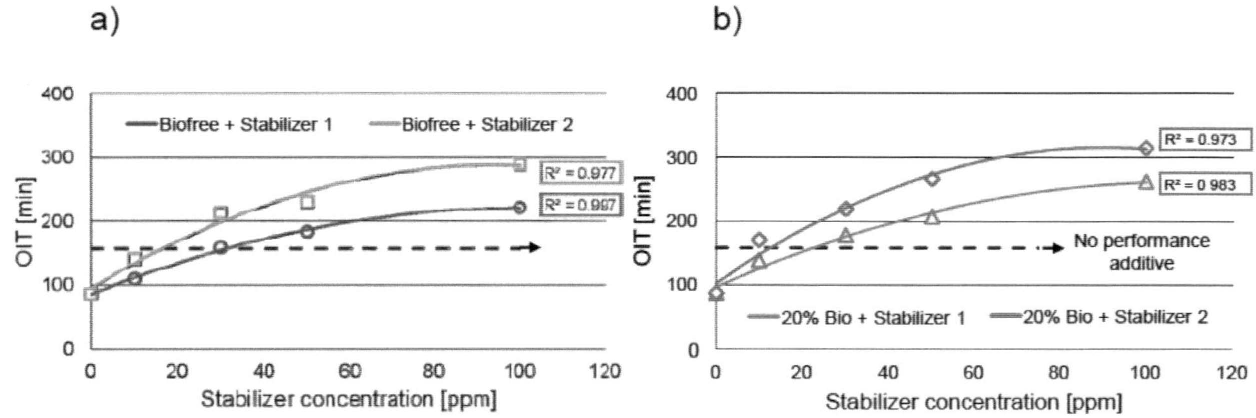

Figure 3: Dependency of the OIT on the stabilizer concentration. a): biofree base fuel with cetane improver. b): 20% bio base fuel with cetane improver. The dotted lines denote the OIT of the base fuels without cetane improver (reference OIT)

Figure 4 a) and b) shows the pressure curves of the fuel blends with cetane improver and stabilizer 2 for the biofree and 20% bio base fuels respectively. As shown, the applied stabilizer counteracts the oxidation (indicated by the pressure drop) in all concentrations. Furthermore, photo documentation of the altered samples is also shown. The amount of solid residue (sludge) decreases with the applied stabilizer concentration, and sludge formation is completely mitigated at higher stabilizer concentrations.

The selection of the optimal concentration was based on the analytical results of the altered fuels, where oxidation and NN are displayed exemplary in figure 4 c) and d). As shown, both oxidation and NN are decreasing with the stabilizer concentration and are very low (< 20%) compared to the not-stabilized blends at 30 ppm.

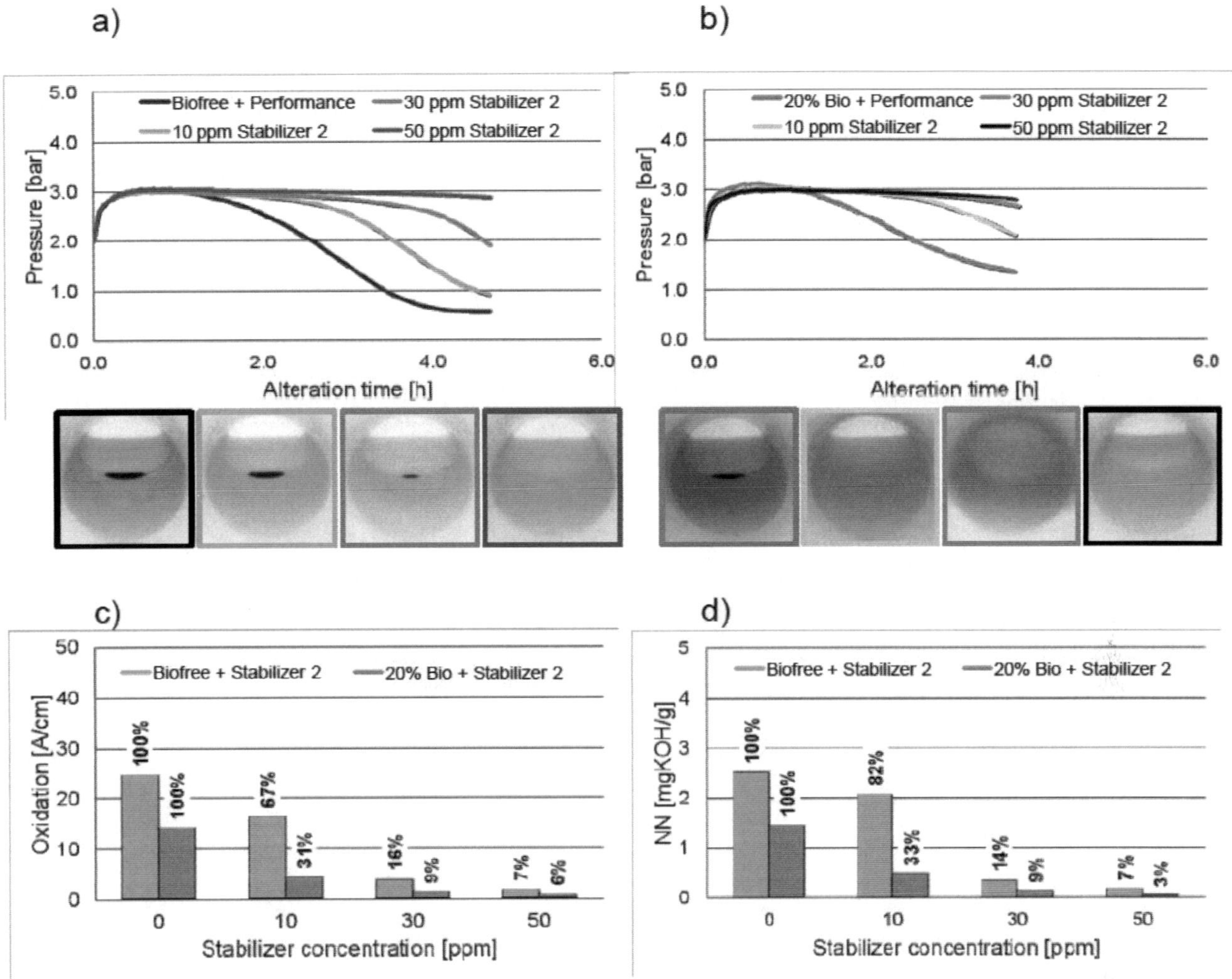

Figure 4 a-b): Pressure curves and photo documentation of stabilized fuel blends with cetane improver (biofree and 20% bio respectively).
c-d): Oxidation and NN of stabilized fuel blends with cetane improver (biofree and 20% bio respectively)

In order to support stability studies and application simulation of future fuel formulations analytical technique beyond standardized tests and conventional fuel analysis are applied. An example is presented where the behavior of future fuel components in a hybrid test rig is simulated over a period of 4 months. While conventional analytics such as determination of oxidation or NN do not show any change, a decrease in oxidation stability and was observed. By using high-resolution mass spectrometry, the degradation of the components and the beginning depletion of the additives was identified. Hence, a correlation of the component degradation with the performance was possible, although not yet measurable with conventional analytical methods.

CONCLUSIONS

The presented Q.O.T. oxidation test showed very good correlation with the commonly available DIN EN 16091. It was shown that a common cetane improver has a significant negative impact on the oxidation stability of both conventional as well as biologically sourced diesel fuel components.

Additionally, it was demonstrated, that comparison of commercially available stabilizer additives can be achieved with the Q.O.T oxidation test. OIT shows a non-linear dependency of the stabilizer additive concentration, where high stabilizer amounts have diminishing effects. The selection of the optimal additive concentration for fuel blends is possible by applying analytical methods on the altered fuel samples.

In combination with mass spectrometry the observed stability performance can be correlated with the fuel condition on the molecular level and provide knowledge crucial for optimized blend components and formulations. Due to the broad applicability to various fuel components the presented approach allows the selection of best-performer additives in the exactly right concentration, thus, results in economical, sustainable, high-performance fuels for internal combustion engines.

ACKNOWLEDGMENT

Presented results were realized in research projects with financial support from the participating project partners and the **Austrian COMET program** (Project InTribology, No. 872176).
The COMET program is funded by the **Austrian Federal Government** and concerning InTribology by the provinces of **Lower Austria** and **Vorarlberg**.

REFERENCES

[1] The European Automobile Manufacturers' Association (ACEA): Fuel types of new cars: battery electric 9.1%, hybrid 19.6% and petrol 40.0% market share full-year 2021 (https://www.acea.auto/fuel-pc/fuel-types-of-new-cars-battery-electric-9-1-hybrid-19-6-and-petrol-40-0-market-share-full-year-2021/) Accessed: 25.08.2022

[2] Marcella Frauscher, Adam Agocs, Charlotte Besser, Alexandra Rögner, Günter Allmaier, and Nicole Dörr: Time-Resolved Quantification of Phenolic Antioxidants and Oxidation Products in a Model Fuel by GC-EI-MS/MS Energy & Fuels 2020 34 (3), 2674-2682 DOI: 10.1021/acs.energyfuels.9b03483

[3] ASTM D 2272. Standard test method for oxidation stability of steam turbine oils by rotating pressure vessel. ASTM International; West Conshohocken, PA, USA; 2016

[4] DIN EN 16091. Liquid petroleum products - Middle distillates and fatty acid methyl ester (FAME) fuels and blends - Determination of oxidation stability by rapid small scale oxidation method: Deutsches Institut für Normung; 2012

[5] Agocs, A., Besser, C., Brenner, J. et al. Engine Oils in the Field: A Comprehensive Tribological Assessment of Engine Oil Degradation in a Passenger Car. Tribol Lett 70, 28 (2022). https://doi.org/10.1007/s11249-022-01566-7

[6] DIN ISO 6618:2015-07. Petroleum products and lubricants - Determination of acid or base number - Colour-indicator titration method. Berlin: Deutsches Institut für Normung; 2015

[7] DIN 51777:2020-04. Petroleum products - Determination of water content using titration according to Karl Fischer. Berlin: Deutsches Institut für Normung; 2020

IASH 2022, the 17TH INTERNATIONAL CONFERENCE ON STABILITY, HANDLING AND USE OF LIQUID FUELS
Dresden, Germany
11 - 15 September 2022

ADDITIVE TRAIL-BACK EFFECT IN MULTI-PRODUCT PIPELINES HANDLING LARGE VOLUMES

Hajar Essa, John Bennett, Asim Saleh, Khalid Butairi, Ayed Shammari, Abdulelah Tamimi.

Saudi Aramco Research & Development Center, GB208, Building 2296, Dhahran 31311, Saudi Arabia. hajar.essa@aramco.com

ABSTRACT

Conductivity improver additives are widely used in ultra-low sulfur diesel (ULSD). Where diesel is being transferred through pipelines also used for the movement of jet fuels, it is necessary that the additives are approved for such multi-product pipelines (MPP).

The potential for trail-back of additives from the ULSD into the jet fuel is investigated during the approval process by undertaking pipeline trials. A batch of diesel containing the additive is dispatched through a pipeline, then followed by a jet fuel batch that is monitored for the presence of the additive. These trial batches are usually small in size and of short duration due to the operational needs of the pipeline.

Following the introduction of ULSD into a network that had previously only transported higher sulfur diesel without additives, it was observed that conductivity in the interface between the ULSD and the jet was higher than either of the parent fuels. One possible explanation was greater than expected additive desorption following the prolonged exposure time of the pipeline to the ULSD during normal operation, where batches of diesel may take several days to pass through a pipeline compared to typical short duration trials.

A bench study was undertaken to look at the effect of a pipeline's ULSD exposure time on the conductivity of the following jet fuel batch. It was shown that increasing the exposure time would also increase the amount of additive that was present in the following jet fuel. It was also seen that the type of additive could have an effect, with some appearing to have a greater persistence than others. When the additive trail-back persists beyond the physical transition from diesel to jet fuel, then greater attention should be paid to this phenomenon, including during the selection of the additive.

INTRODUCTION

Saudi Arabia is currently is moving to the production and supply of ultra-low sulphur fuels (10 mg/kg maximum sulphur) for road vehicles. In the case of ultra-low sulphur diesel (ULSD), the fuel requires the use of additives to recover some properties that are adversely affected by the fuel production process, notably for conductivity and lubricity.

During the process of transitioning to ULSD, it was observed that the conductivity of jet fuel arriving at a remote location after being co-transported through a multi-product pipeline that was also handling ULSD was higher than expected. Although the fuel remained on specification, it was important to understand why this was happening and a program of work was undertaken to identify the root cause.

The fuels in this case are introduced into the pipeline at the refinery, with bulk plants (storage locations) at 70 km (referred to as Bulk plant A) and 490 km (referred to as Bulk plant B) from the source.

At Bulk Plant A, it was observed that the ULSD had a conductivity of 154 pS/m and when it arrived at Bulk Plant B, this had reduced to 78 pS/m. In contrast, the Jet fuel had increased from 91 pS/m at Bulk Plant A to 398 pS/m at Bulk Plant B. Note that all of these results are within the relevant conductivity specifications for such fuels, 50 pS/m – 600 pS/m.

Due to the large distance between the two locations, the time taken to move from Bulk Plant A to Bulk Plant B is significant, around 7 days, and due to the large volumes being handled, a single fuel type be present in the pipeline for the entire 7 days.

EXPERIMENTAL

Having identified the issue of conductivity increase, it was proposed that there were three potential mechanisms of why the conductivity value in jet fuel might have increased.

1/. Contamination during transit. External source of contamination introduced to the jet fuel during transit and increasing the conductivity value

2/. Direct interaction or reaction of different fuel types. ULSD passes through the MPP followed by a jet fuel product, with potential reaction between these two fuels and/or their additives, leading to increased conductivity value in the jet fuel.

3/. Trail-back of additives. Additives from ULSD adsorbs into the pipeline and later desorbs into the following batch of jet fuel

1- Contamination during transit

The obvious properties that could be related to external sources of contamination were tested. These were Water content using ASTM D6304 Karl Fisher titration technique, Salt (chloride) content using ASTM D3230 and Trace Element content ASTM D7111 ICP-AES.
The results are shown in Table 1.

Table 1: Jet Fuel - tests for externally sourced contamination

Test	Jet fuel location A	Jet fuel location B
Water, mg/kg (ASTM D6304)	83	90
Salt, mg/kg (ASTM D3230)	0	0
Trace elements, mg/kg (ASTM D7111)	ND / <0.5	ND / <0.5

All the results were within the normal range with no significant difference between the two locations, nor evidence of contamination being found in either of the Jet fuel samples.

2- Interaction between jet fuel and ULSD

To establish if a simple mixing mechanism between the two fuels was responsible for the increased conductivity seen in the Jet fuel at the remote location, samples of the Jet and ULSD taken at Bulk Plant A were mixed and the conductivity value of this mixture measured (Figure 1).

The 100% ULSD showed a conductivity value of 154 pS/m, whilst the 100% Jet fuel had a conductivity value of 91 pS/m. Two intermediate mixtures were made, with a 50% ULSD + 50% Jet mixture giving a conductivity value of 163 pS/m and a 10% ULSD + 90% Jet mixture giving 128 pS/m. These results indicated that there is no obvious reaction between ULSD and jet fuel that might cause the increases of conductivity value in jet fuel. However, it does indicate that a proportionally small amount of the additised ULSD does increase the conductivity by more than would have been expected in a simple linear relationship.

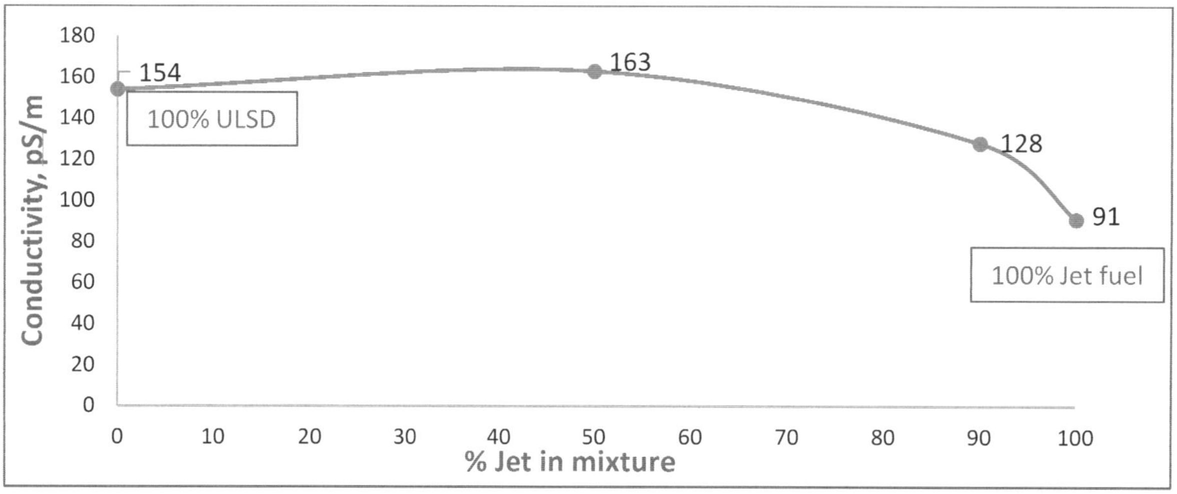

Figure 1: Interaction / reaction of Jet and ULSD

3- Additive trail back.

One of the potential mechanisms for higher than expected conductivity is trail-back into the Jet fuel of the additives from the preceding batch of ULSD which had passed through the pipeline immediately prior to the Jet fuel batch.

In this mechanism (Figure 2), when the ULSD is pushed through the pipeline, surface-active additives in the ULSD may adsorb onto the walls of the pipeline. When the following batch of jet fuel passes into the pipeline, those surface-active additives left by the ULSD are desorbed which may lead to an increase in the conductivity of the jet fuel.

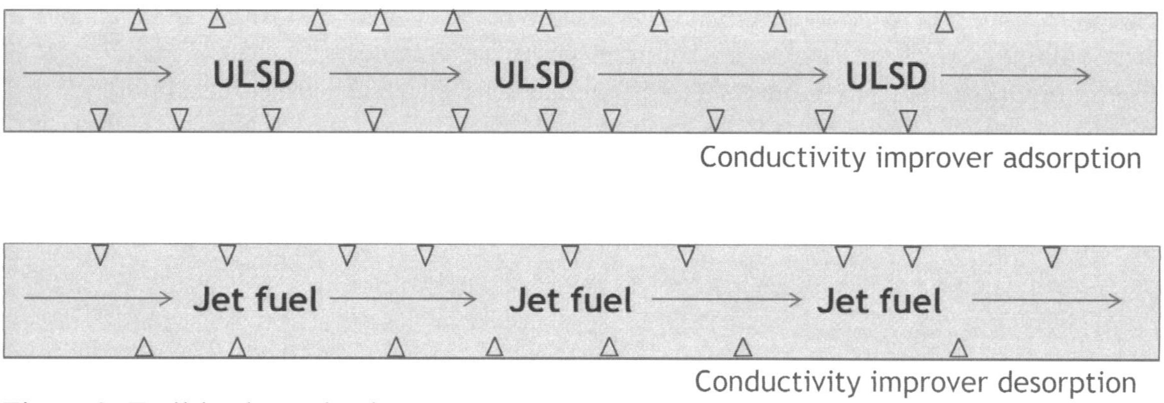

Figure 2: Trail-back mechanism

To establish if this may be the underlying mechanism, the conductivity was measured as the interface passed a fixed point in the pipeline. Normal practice in multi-product operation is for different batches of fuel to be transported through the pipeline without a physical barrier between them, recognizing that there will be small volume of mixed product where the two fuels meet. This volume is known as the interface and is segregated and handled / disposed of as required. Normally the interest is to ensure that the quality before and after the interface shows no cross contamination between the two fuels, but in this case, knowing what was happening within the interface was likely to be enlightening.

The purple line is the amount of ULSD in the fuel (left Y-axis) as it passes Bulk Plant A, the orange line being the conductivity of that fuel (right Y-axis), the X-axis being time.
It can be seen that prior to the interface arrival, the fuel is all ULSD, and after the interface has passed, the fuel is all Jet. Similarly, the conductivity before the interface is as per the ULSD, and after it is as per the main volume of Jet fuel.

However, looking inside the interface, it can be seen that the proportion of ULSD rapidly drops to 10% and beyond, but the conductivity has increased significantly, greatly exceeding that of either of the 'parent' fuels. It is interesting to compare this with the results shown in Figure 1, where the conductivity of a 10% mixture is somewhere between the two parent fuels, and very different to that measured in the interface.

Moving on to Bulk Plant B, this time the amount of ULSD is shown in green, and conductivity in grey. In this case, the amount of ULSD in the fuel before, during and after the interfaces passes is very similar to that seen at A. The conductivity within the interface is again very high, even more so than at A. However, at location B, it can also be seen that the elevated conductivity is persisting even after the interface, supporting the proposal that trail back of the ULSD additive may be affecting the following jet fuel product.

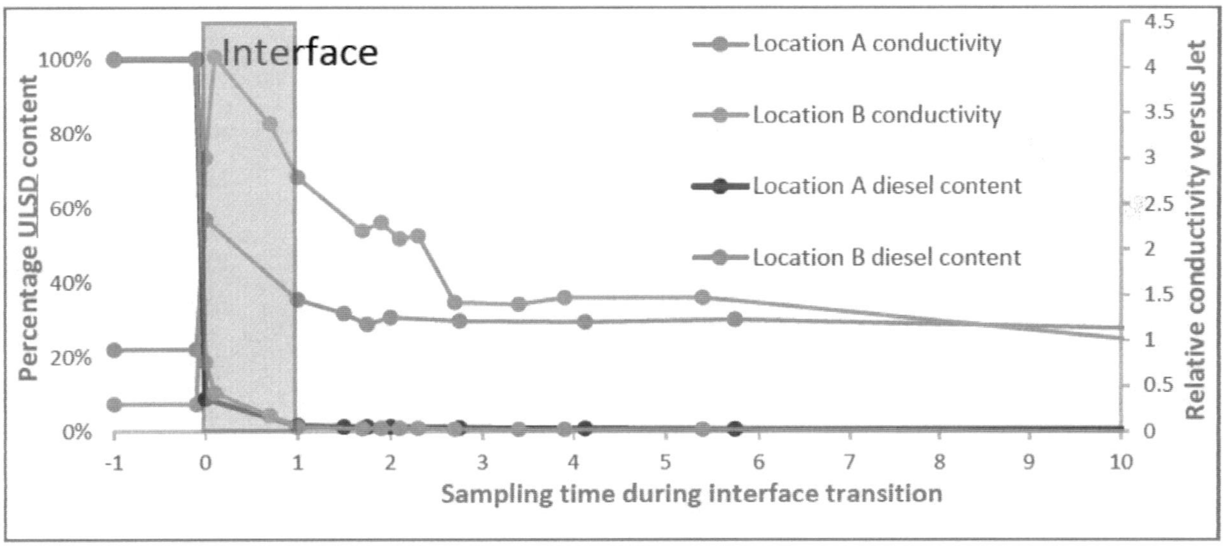

Figure 3: Interfaces - Bulk Plant A and Bulk Plant B

Trailback mechanism bench test design.

As there is no standard bench-test method to determine trail back tendency of additives in a pipeline, it was necessary to develop a procedure to simulate in a laboratory what happens inside the pipeline. To develop this experiment, 3 factors were considered:

ULSD residence time.
Due to the long pipeline and very large ULSD batches, there is significant opportunity and time for the ULSD conductivity improver to adsorb onto the pipe walls. Does a longer exposure time lead to a greater amount being adsorbed?

Jet fuel residence time.
How long does it take for the desorption of the ULSD additive into the Jet fuel to take place? Is the desorption likely to still be occurring after the interface has passed?

Conductivity improver type.
Does the type of conductivity improver in the ULSD have an effect of the trail back tendency?

To try address these questions, a simple bench-test procedure was developed to simulate what happens in the pipeline. The procedure uses stainless-steel beakers as a surrogate for the steel pipeline, with manual addition and removal of fuels.
The general approach was to fill the beaker with additized ULSD, seal it and allow to stand for an appropriate time, then empty it and refill with Jet fuel with no rinsing or washing in-between. The expectation was that the adsorbed surface-active additives might remain on the wall of the wet beaker and then desorb into the Jet fuel, increasing the conductivity, which would be measured to establish the possible effect of desorption duration. Realizing that there would be additised ULSD remaining as wetting of the beaker walls, the volume of that ULSD was established by adding a known amount fuel from a measuring cylinder to the beaker and then returning the fuel to the cylinder, the difference in volume being that retained by the beaker. In this case, approximately 1.5 ml was retained from a 350 ml fuel sample.

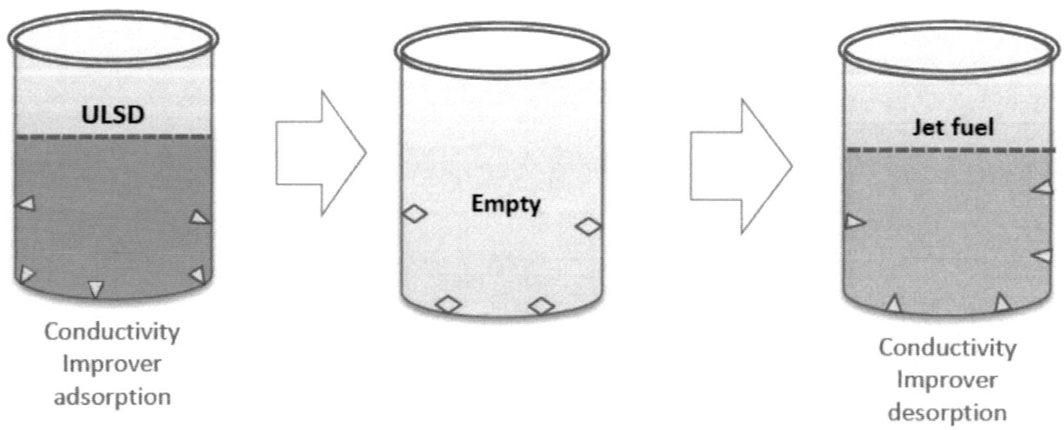

Figure 4: Trail-back Simulation - Basic Bench Method

Figure 5 shows the results of an initial proof of concept test, measuring the conductivity of a jet fuel sample after adding it to a beaker that had previously held ULSD for 96 hours. Prior to the addition to the beaker, the additised jet fuel had a conductivity of 62 pS/m. After adding to beaker, there is a small increase likely due in part to the additised ULSD remaining in the beaker. However, after waiting 2 hours, it was seen that the conductivity had then risen to 123 pS/m and continued to rise, peaking after 20 hours at 165 pS/m.

Figure 5: Jet fuel added to beaker that previously held ULSD for 96 hours

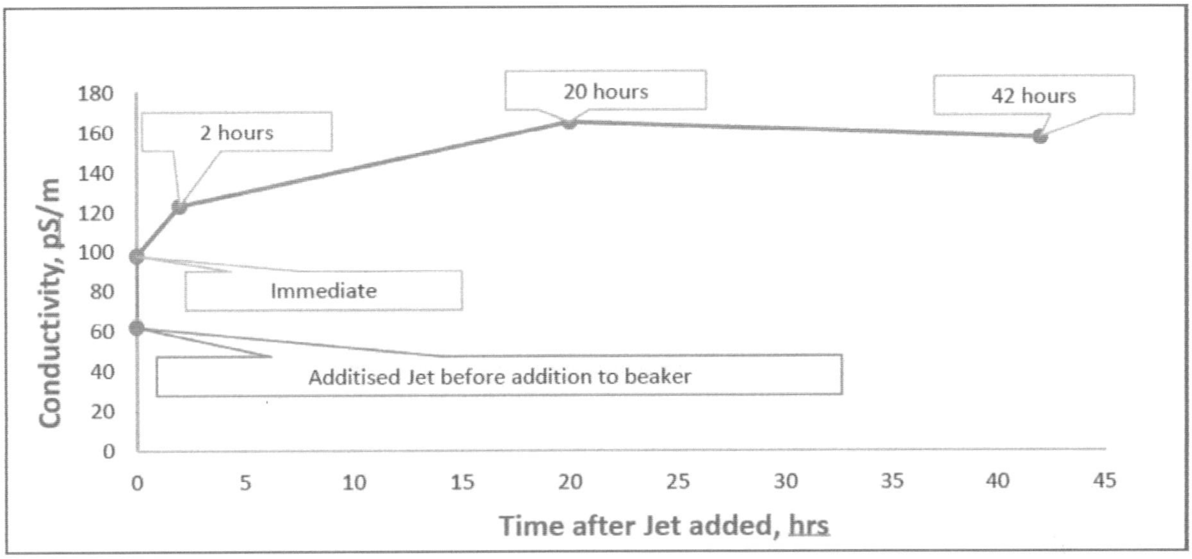

These results indicated that the procedure had merit and further tests were undertaken to expand on this.

The effect of the ULSD residence time was investigated by varying how long the ULSD was in the beaker before discarding, then adding the Jet fuel and measuring the conductivity after 24 hours, this being the time for maximum conductivity after refilling with Jet fuel, as per Figure 5.

In the first case, ULSD was added to beaker and then emptied almost immediately and replaced with the jet fuel. In the second case the ULSD was retained for 2 hours before replacement and in the third case, the ULSD has kept in the beaker for 72 hours

It was seen that there was no significant difference between the results of very short exposure, 155 pS/m and exposure for two hours, 148 pS/m. However, with the longer 72 hour exposure time, the conductivity had increased to 200 pS/m, suggesting that a short exposure period in a pipeline would not allow time for significant additive to be adsorbed.

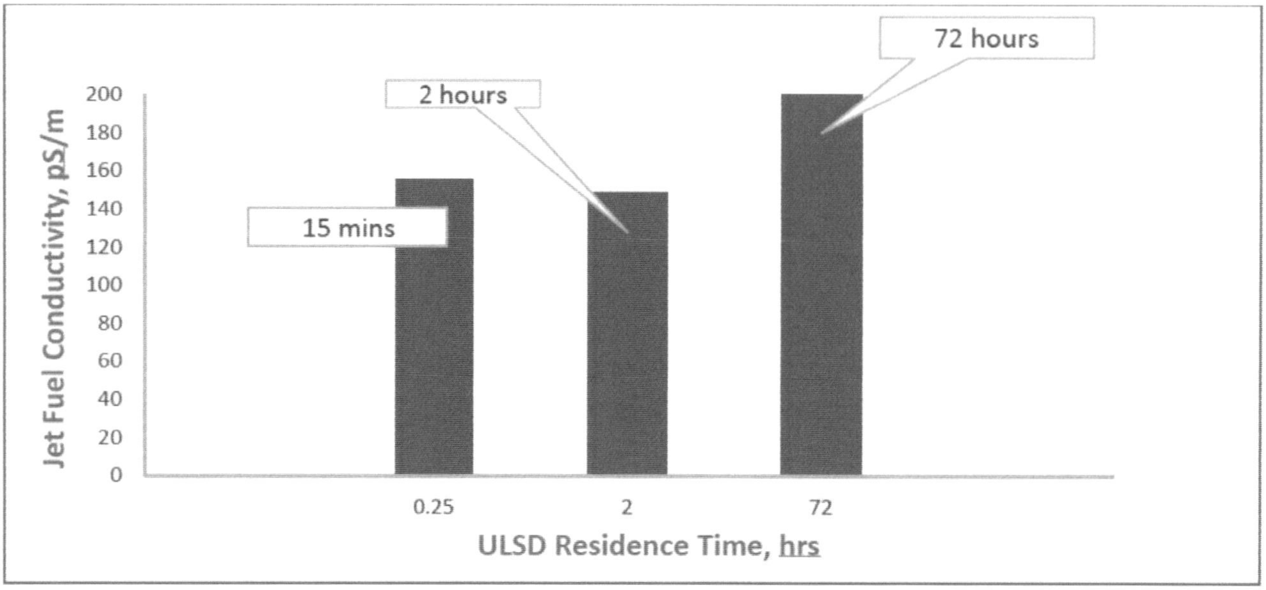

Figure 6: Jet fuel added to beakers that previously held ULSD for 0.25, 2 or 72 hours.

The third factor of the experiment was to consider if the type of conductivity improver in the ULSD had an effect. In this case, two different conductivity improvers were evaluated, one being the improver normally used in the ULSD, the other being that used in Jet fuel.

As per previous experiments, beakers were filled with additised ULSD, but in this case the ULSD samples were additised with one or other of the above conductivity improvers. The samples were held for 72 hours, then emptied and the contents replaced with additised jet fuel. As a control, a third beaker that had been previously empty was also filled with additised Jet. The results of this experiment are shown in Figure 7. For this experiment, the amounts of conductivity improver used for both ULSD samples was the same, even though the improvers are different.

The grey line shows conductivity measurement for the Jet fuel not exposed the ULSD beaker residue as a control. There is a gradual decline in conductivity, which is considered typical of such an additive

The green line shows conductivity measurement for the Jet fuel after adding to the beaker that had held ULSD with the Jet fuel conductivity improver. There is a minor conductivity increase at the beginning of the test, but the sample otherwise reflects the control.

The red line shows conductivity measurement for the Jet fuel after adding to the beaker that had held ULSD with the ULSD intent conductivity improver. As per previous experiments, after a short period, the conductivity increased for that additive and continued to do so, showing a significant increase with time. This is in contrast to the ULSD that used the improver more normally used for Jet fuel.

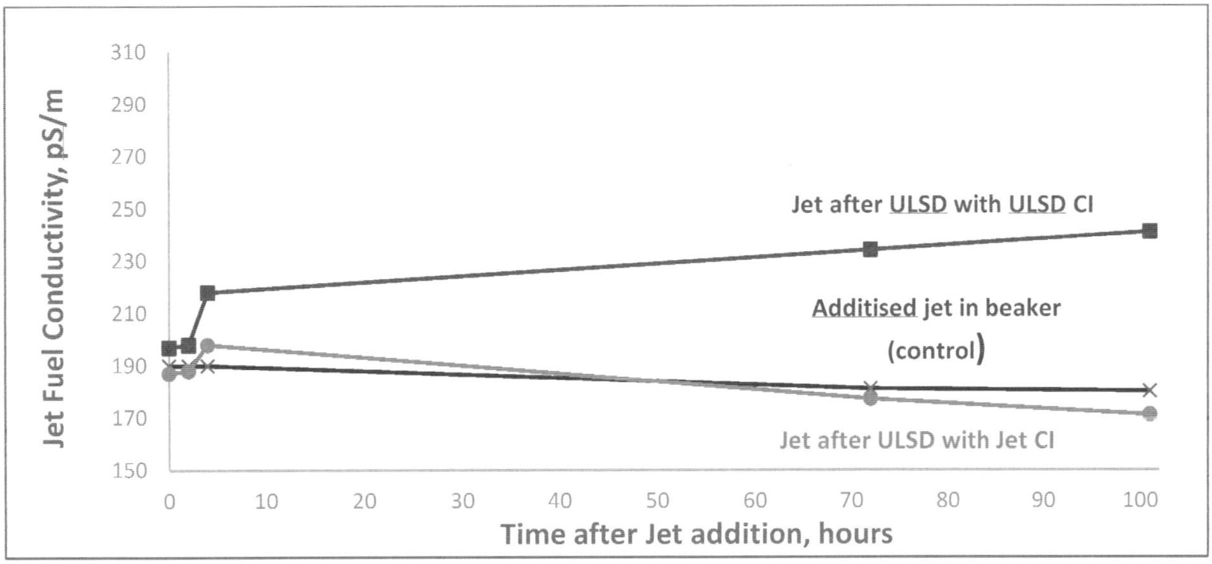

Figure 7: Jet fuel added to beakers that previously held ULSD for 72 hours. Different conductivity improvers in the ULSD's, same treat rate.

In the process of making the two ULSD's with the different additives, it was noted that the additive designed for ULSD use gave a stronger dose rate response. To ensure that the smaller response in Figure7 for the Jet intent improver as not due to this difference, the test was repeated, but this time with the amount of Jet improver being increased to give the same ULSD conductivity, Figure 8.

Despite the ULSD conductivity matching, it can be seen that the ULSD intent additive again gives a much stronger and sustained increase, whilst the Jet intent additive rapidly returns to and falls below the starting conductivity.

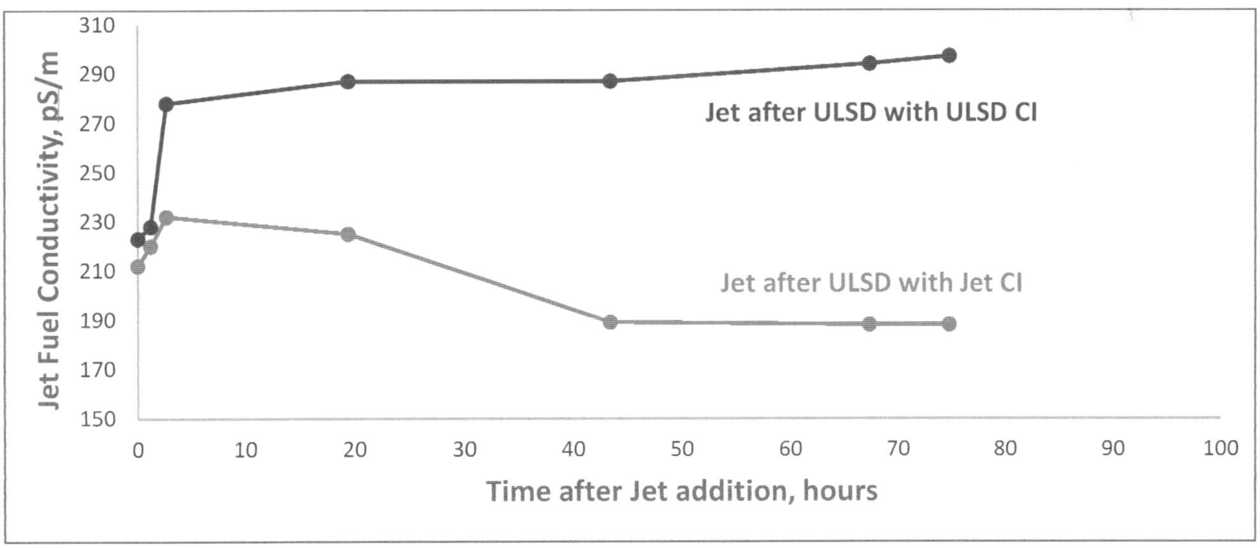

Figure 8: Conductivity of Jet fuel. Beaker previously held ULSD for 72 hours. Different conductivity improvers in ULSD's, same target conductivity.

Pipeline debris testing.

During the period of the study, part of the pipeline was scraped as part of routing maintenance and the opportunity was taken to try and identify if debris material within the pipeline might be having an effect on the increased conductivity of the jet fuel.

A sample of material scraped from within the pipeline was received, a liquid heavily contaminated with solid material. The solid material was filtered out using the procedure from ASTM D5452 but without the rinsing stage. The filter cake debris material was then mixed with unadditised Jet fuel and left overnight. This mixture was then filtered to remove the debris.
The filtrate was compared with Jet fuel that had not been exposed to debris and tested for water, salt (chlorides) content and conductivity, results shown in Table 2.

Table 2: Pipeline debris test

Sample Type	Water (mg/kg) ASTM D6304	Salt (mg/kg) ASTM D3230	Conductivity (pS/m) ASTM D2624
Jet fuel before exposure to scraping debris	79	0	3
Jet fuel after exposure to scraping debris	54	0	106

The results for water content in both two samples were broadly similar and at a low level. There was no salt content present in either of the samples, indicating that the scraped material had not contained any. However, there was a significant increase in the conductivity of the jet fuel sample that had been exposed the debris material. Whilst circumstantial, this further supports the proposition that surface-active material can be adsorbed onto the surfaces within the pipeline and increase the conductivity value of the Jet fuel, and it was also noted that immediately following the scraping, the increased in conductivity for Bulk Plant B had reduced.

CONCLUSIONS

An increase in the conductivity of Jet fuel arriving at a remote bulk plant fed by a multi-product pipeline was investigated, with findings as follows:

There was no external source of contamination causing increased conductivity

There was no reaction or interaction from mixing of the two products being transported through the (ULSD + Jet fuel) that might lead to an increased conductivity in the Jet fuel.

Trail-back (surface active additive adsorption-desorption) of the ULSD conductivity additive into the Jet fuel was identified as the likely root cause of the issue.

A test method was developed to investigate this issue and successfully demonstrated the root cause.

Trail-back was shown to be significantly increased by a long residence time for the preceding ULSD batch.

Long distance / duration distribution pipelines transporting very large discrete batches may be more prone to this issue.

Conductivity improver additive selection may affect the propensity to see this issue.

Pipeline debris may exaggerate the issue.

A tailback bench test method is being developed as a potential Saudi Aramco Lab Analytical Method (SALAM).

REFERENCES

Essa. H. M, Evaluation of The Conductivity of Ultra-Low Sulfur Diesel (ULSD), ChemIndix Conference in Bahrain, October 2019.

ASTM D6304 – 20 Standard Test Method for Determination of Water in Petroleum Products, Lubricating Oils, and Additives by Coulometric Karl Fischer Titration. ASTM International, West Conshohocken, PA. 2022

ASTM D3230 – 19 Standard Test Method for Salts in Crude Oil (Electrometric Method). ASTM International, West Conshohocken, PA. 2022

ASTM D7111 – 16 Standard Test Method for Determination of Trace Elements in Middle Distillate Fuels by Inductively Coupled Plasma Atomic Emission Spectrometry (ICP-AES). ASTM International, West Conshohocken, PA. 2022

ASTM D2624 – 15 Standard Test Methods for Electrical Conductivity of Aviation and Distillate Fuels. ASTM International, West Conshohocken, PA. 2022

ASTM D5452 – 20 Standard Test Method for Particulate Contamination in Aviation Fuels by Laboratory Filtration. ASTM International, West Conshohocken, PA. 2022

IASH 2022, the 17th INTERNATIONAL SYMPOSIUM ON STABILITY, HANDLING AND USE OF LIQUID FUELS
Dresden, Germany
September 11 - 15, 2022

RECENT DEVELOPMENTS AND VALIDATION OF A MORE ACCURATE TOOL FOR FUEL CLARITY BY D8148 SPECTROSCOPIC METHOD FOR HAZE IN FUELS

Ranzy Morgan*

Choice Analytical Inc., 1118 First Street, LaMarque, Texas 77568 United States

ABSTRACT

Providing fuel with the necessary cleanliness has been desirable since the earliest days of motorized transportation. Engine manufacturers expend significant effort to protect sensitive components of modern high pressure fuel systems from fuel contaminants. With an unprecedented number of new fuel types and feedstocks entering the worlds fuel supply, the industry is becoming increasingly concerned about fuel cleanliness and clarity.

Over the decades, engine manufacturers, fuel producers and fuel distributors, worked within ASTM to establish specification limits for fuel haziness and clarity. Because of the variability and subjectivity found with the current visible methods such as D4176 and because ASTM specifications may be applied anywhere within the fuel distribution systems, fuel pipelines and transport companies have traditionally not supported these efforts. During the last five years, major pipelines have worked closely to support the optics-based technology found in D8148.

ASTM D8148 Haze Clarity index is an instrumental objective technique which eliminates the subjectivity of the visual haze methods. D8148 measures and reports haze results in two formats 1) Instrumental Haze Rating (IHR) ranging from 1-6 which is calibrated to be equivalent to the D4176 Visual Haze rating, and 2) Haze Clarity Index (HCI) ranging from 50-100. HCI provides greater resolution and specificity of fuel cleanliness. With the large amount of data provided in the recent D8148 Interlaboratory Study, as well as CRC report 675, which compares D8148 to D4176 and actual amounts of water in fuel samples, it may be the time to seriously reconsider putting measurable limits around fuel cleanliness. This presentation will review the D8148 method, recent ILS results along with the data and conclusions presented in CRC report number 675.

*Corresponding Author: President Co-Founder Ranzy Morgan; rmorgan@choiceanalytical.com; 832-409-8150

IASH 2022, the 17th INTERNATIONAL SYMPOSIUM ON STABILITY, HANDLING AND USE OF LIQUID FUELS
Dresden, Germany
September 11 - 15, 2022

GRAPHENE OXIDE-BASED MICROBIAL FUEL FILTER: FROM LABORATORY BREAKTHROUGH TO FIELD DEMONSTRATION

Oscar N. Ruiz*

Fuels and Energy Branch, Aerospace Systems Directorate, U.S. Air Force Research Laboratory, 1790 Loop Rd. North, Wright-Patterson AFB, Ohio 45433-7103 USA

ABSTRACT

Biocontamination can damage fuel and fuel systems, thus, it is critical to develop methods to prevent biocontamination. Filtration is a very effective way to maintain fuel cleanliness. However, even the best aircraft fuel filters can only remove particles of about 1 μm in size. Most bacteria have a diameter of less than 0.5 μm requiring a 0.22 μm filter for effective filtration. Such filter would present a strong barrier to fuel flow and would lead to large pressure drops. Previously, we demonstrated the development of a graphene oxide (GO) nanomaterial fuel filtration technology proving 99.97% microbial filtration efficiency in the laboratory without the need of small filter pore sizes for microbial filtration. Now we show further development, maturation and validation of this technology into field-demonstrated microbial fuel filters for gas station fuel dispensers (diesel fuel pumps) and electric power generators. We describe the engineering, laboratory validation, and field testing that when into creating a fuel filter capable of providing >90% microbial filtration efficiency at fuel flow rates of up to10 GPM for about 60,000 gallons of diesel fuel. The filter has a delta-P of ≤4 psi, which is well below the maximum allowed for current filters. The filter is of modular design allowing easy scale up for applications such as jet fuel. The field demonstration showed the functionality and durability of the filter under real use conditions helping to mature and de-risk the technology one-step further into implementation.

*Corresponding Author: Principal Biological Scientist Oscar Ruiz; oscar.ruiz@us.af.mil; 1-937-255-6237

IASH 2022, the 17TH INTERNATIONAL CONFERENCE ON STABILITY, HANDLING AND USE OF LIQUID FUELS
Dresden, Germany
11 - 15 September 2022

THE POTENTIAL OF ORGANIC SOLVENT NANOFILTRATION A SUSTAINABLE SEPARATION METHOD TO ENHANCE FUEL QUALITY, STABILITY AND PERFORMANCE

Marie-Sophie Haverkamp[1] , Jan L.W. den Boestert [2] , Robin Wilmshöfer[3]

[1] *BORSIG Membrane Technology GmbH, Bottroper Straße 279, 45964 Gladbeck, Germany, m.haverkamp@borsig-mt.com*
[2] *ConSsepT®, Independent Consultancy - Special Separation Technology, The Netherlands*
[3] *BORSIG Membrane Technology GmbH, Bottroper Straße 279, 45964 Gladbeck, Germany, r.wilmshoefer@borsig-mt.com*

ABSTRACT

Within a rapidly changing world, liquid fuel is increasingly produced from a wide variety of feedstocks. Due to a depletion of conventional oil resources and the growing demand for renewable feedstock, properties vary significantly. Each processing route and feedstock causes specific contaminants within the process itself or in the fuel product that may negatively impact quality specifications, performance or stability aspects.

In particular, dissolved large-chain molecules such as gums, asphaltenes, metal trace components and reactive molecules cause quality, stability and fouling issues during production and/or application. Their removal using conventional technologies is difficult and mostly not economical or sustainable. New separation tools such as organophilic or Organic Solvent Nanofiltration[1] (OSN) may provide the remedy. The inherent OSN separation principle enables efficient separation of dissolved molecules in the molecular weight above 300 Da, as well as organo-metal compounds and water traces from hydrocarbon-based mixtures. In the last decade, OSN has proven its industrial applicability and robustness in large-scale (petro-) chemical and hydrocarbon processing. In addition, various solvent and fuel-related applications have been successfully demonstrated and implemented providing evidence that OSN may contribute to ensure and enhance fuel product quality.

The objective of this paper is to inform about OSN as a sustainable separation method for a range of fuel and feedstock-related applications. In addition, to show that the achieved fuel properties can have a sustainable impact, such as fuel saving or reduced maintenance.

Keywords: Membrane, Organic Solvent Nanofiltration, shale oil, SAF, metals, fuel, stability, condensate, asphaltenes, gum, military, automotive, jetfuel, diesel.

INTRODUCTION

The early developments in membrane technology between the 60-70ties were mainly related to water applications. Meanwhile, these are state of the art technologies e.g., Reverse Osmosis[1] utilized for desalination applications. However, the applied membranes were barely solvent-resistant. Applications in relation to solvent and fuel-related applications were at an infant stage. Especially, when facing challenges in relation to chemical robustness and the exposure to fouling-promoting components, such as particulates, asphaltenes and sticky gum-like molecules. Although in an early R&D stage, the potential of siloxane-based materials had already proven stability amongst others in jet fuel applications for e.g. metal removal[2], mitigating the undesired gum formation caused by the metal-catalysed jet fuel thermal oxidation. One of the first large commercial membrane applications for oil refining concerned the successful implementation for the recovery of Methyl-ethyl-ketone and Toluene within a solvent-based dewaxing process[3], developed by Mobil Oil and WR Grace in the nineties (MAX-DEWAX ™). This was soon followed up with application developments for the removal of sulphur molecules from gasoline[4].

Meanwhile, significant progress was made with the commercialization of suitable membrane material as well as membrane module developments to enhance performance such as to be less prone to fouling effects[5]. A significant co-development of BORSIG in collaboration with Shell was achieved in the clean-up of so-called Dark Condensates[6,7], which are heavily asphaltene-contaminated naphtha-like feedstock. Once decontaminated, these low-cost feedstocks can for example be fed to naphtha-cracker. In addition to the retention of asphaltenes, Mercury removal from Natural Gas Condensates [8,9] was demonstrated. With respect to improving feedstock stability, the purification and stabilization of Shale Oil condensate[10,11] was a significant achievement. This challenging application concerned a chemically unstable naphtha-like feed material, which even under nitrogen-blanketing conditions forms undesired gum-like components. Nanofiltration proved to be a successful and sustainable separation method for purification and stabilization due to the retention of the unstable, reactive molecules. The relevance to fuel quality, handling and stability are provided by several experimental examples in the section below. All examples relate to typical poly-siloxane based membrane materials, which have been commercialized by BORSIG Membrane Technology and its subsidiary company GMT Membrantechnik GmbH, Germany.

SEPARATION PRINCIPLE

Organic Solvent Nanofiltration (OSN) membranes are specifically designed to be stable in organic solvents. BORSIG manufactures and provides polymeric membranes made of different layers, so called Thin-Film Composite (TFC) membranes. The upper layer is a dense layer with a thickness of a few microns and forms the selective and semi-permeable separation barrier of the membrane. Since the selective layer is considered to be dense, no pores are available. Hence, no pore plugging occurs when sub-micron or small particulates occur in the feed. The mass transport through the membrane is by solution-diffusion[1,14]. Larger molecules in the range of 300 to 1.000 Da are retained while smaller molecules such as solvents permeate through the membrane. The membranes are operated in cross-flow mode. A pressure difference from feed-retentate to the permeate is the driving force. Typical pressures are in the range between 10 and 35 bar.

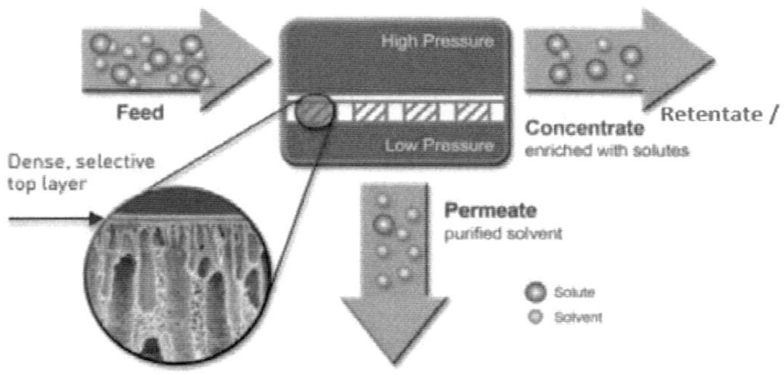

Figure 1. Schematic separation mechanism and membrane structure

CASE STUDIES AND APPLICATION EXAMPLES

The case studies and application examples have been chosen from open literature with a focus on fuel improvement objectives like product quality, stability and handling, such as for kerosene, Naphtha/condensate, diesel and Shale oil. The tested membranes are either BORSIG OSN or early-development versions which are chemically similar to the commercial BORSIG membranes.

Example 1, Quality and performance

In relation to thermal stability, the metal removal from a kerosine stream is of significance, since the catalytic effect of metals during combustion are the cause of undesired gum formation. Lower metal contents will result in an improved JFTOT performance (Jetfuel Thermal Oxidation Test, ASTM 3241). The following example from a jet fuel-related patent[2] forms a good basis to show metal separation effect. A kerosene feed fraction had a boiling range of from 120 to 250 °C and contained 920 parts per million by weight (ppmw) of water and 174 ppbw of metals, of which 160 ppbw was copper and 14 ppbw was iron. This kerosene fraction was treated with a (non-commercialized) non-porous membrane at a pressure of 25 bar and a temperature of 25 °C. The experimental results and its retention effect are shown in Table 1;

Kerosine Component	Conc. Feed (ppb)	Conc.Permeate (ppb)	Avg. Retention (%)
Fe	14	7	50
Cu	160	67	60
H_2O	920	138	85

Table 1: Retention results Kerosene Nanofiltration test

The reported degree of retention is defined as follows;

$$\text{Retention}^{1}\ (R) = 1 - \frac{\text{Conc. Permeate}}{\text{Conc. Feed}} * 100\%$$

With decreased metal levels, the JFTOT will improve in line with the retention rate. Within its development stage (early 90ties), both the membrane material and porous support were not optimal and not yet commercialized, which most-likely is the reason that the application never commercialized. Typical permeabilities at ambient temperature are between 1 to 3 kg/m2.hr.bar. The current commercial membranes are expected to show higher retention due to material improvements.

Example 2, Stabilization and storage improvement:

In relation to fuel stability and to demonstrate the fouling robustness of OSN membranes, reference to a case study with Shale Oil is made. This feedstock is derived from an in-situ conversion process (ICP)[10,11]. ICP Shale oil derived condensates are considered to be a worst-case feed material due to its instability through reactive precursor molecules which form gum/asphaltene-related solids. The basic feed material is black (Figure 2).

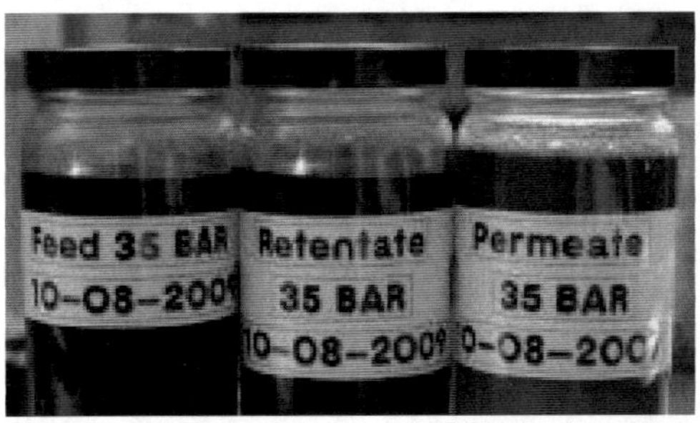

Figure 2. Image of an ICP Shale-oil feedstock, retentate and permeate sample[11]

A field trial was conducted, utilizing a 4" spiral wound membrane element from BORSIG. Permeation was executed at typical pressure regime of 10 to 25 bar and at ambient temperature. Permeate material was collected as a clear and bright product, which showed stability even without Nitrogen blanketing. To prove the effect of OSN treatment on stability and fouling tendency an Alcor® test was done, which is a thermally-based fouling promotion test. The heat-transfer loss is measured as an indicator of the fouling behaviour of the test material. A lower

Alcor reading reflects a reduction in heat transfer, which means a higher fouling tendency for the material. The results of this testing method showed that the permeate materials (A) had a far lower, in fact unmeasurable, fouling tendency in comparison to the original feed (B) and the retained fraction, as is shown in Figure 3.

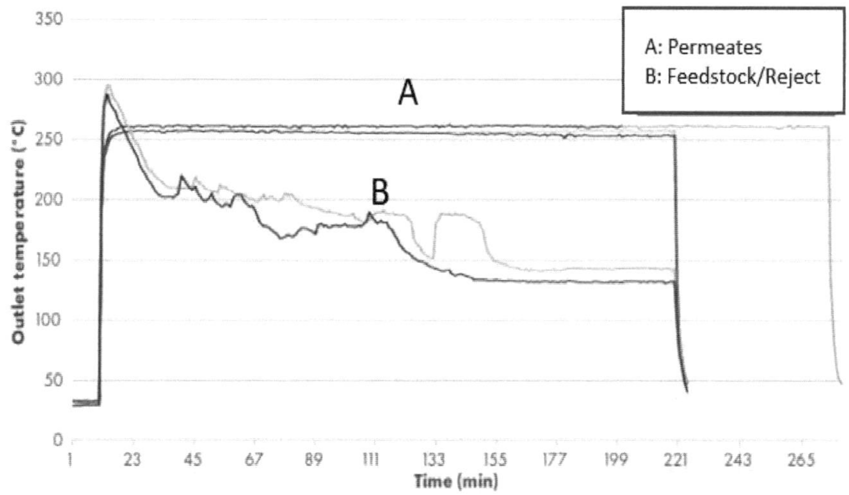

Figure 3. Alcor Test® results of the produced permeates versus Feed/Retained material

ICP shale-oil fouling behaviour is significant and fouling mitigation of such feed materials was not achieved by conventional filtration techniques. In contrast to conventional filtration processes, the dense nanofiltration membrane did not suffer from irreversible fouling because there was no pore plugging. This has resulted in a patent[10]. Through that capability, also high permeate recoveries larger than 90% were achieved. The non-permeating material and dirt is kept on the membrane surface and washed-off by the applied cross-flow. The resulting turbulence contributes to fouling prevention.

Example 3, Upgrading conventional and alternative feedstocks:

Natural Gas related condensates are by-products from natural gas production. When contaminated with mercury and /or the presence of asphaltenes (Dark Condensates) such condensates are marketed at substantial discounts. Shell investigated the use of OSN membranes for treatment of natural gas condensate[7,8,9]. These condensates are Naphtha-like, black feedstocks (typically boiling-range, 35 to 350°C, density 720-770 kg/dm^3). Trials with OSN membranes generated a clear and bright permeate sample. The Patent publication[8] on sweet condensate (in absence of H_2S, the Mercury is most-likely to be present in metallic and/or ionic form) underlines the efficiency for mercury removal by showing the mercury reduction from 1.000 ppb in the feedstock to less than 25 ppb in the permeate.

The mercury removal from natural gas condensate was also successfully demonstrated in a scale-up with the demonstration plant[9], shown in Figure 6.

Figure 6: In-situ module testing / Demonstration plant[9]

RESULTS AND DISCUSSION

The aforementioned examples emphasize the potential of utilizing OSN membranes.
The process can be optimized with regard to permeance and selectivity by choosing the right membrane. Commercial membranes of different "tightness" are available. In the membrane world the expression Molecular Weight Cut Off (MWCO) is used to characterize a membrane. The MWCO is the molecular weight that is retained by 90% by the specific membranes at defined conditions and in a defined feed system.

BORSIG OSN membranes are roughly covering the following MWCOs:

oNF1, MWCO* approx. 600 g/mol (Da)
oNF2, MWCO* approx. 350 g/mol (Da)
oNF3, MWCO* approx. 900 g/mol (Da)
*subject to test conditions and feed mixture/ only for orientation

The ability of water retention as shown in example 1 for Jet fuel is related to the hydrophobic behaviour of the selective membrane layer. Beside using optimized and commercial membranes, the overall membrane performance can be optimized by adjusting the operating conditions such as pressure and temperature to the process optimum. The pressure difference between the feed-retentate and permeate side is the key driving force, no further energy input is required. Hence, the nanofiltration systems are low energy (electricity) consuming and may contribute to sustainable processing.

To give an indication of the required footprint comparable to fuel applications, photos of two industrial OSN units are provided in Figure 5.

Figure 5: Examples of technical OSN applications
(Left) Solvent Polishing 25000 kg/hr - (Right) Used Oil Application 1000 kg/hr

The design is fully modular and therefore easy to stack in case of restricted space as well as extendable in case of capacity creep. Dependent on the application, the nanofiltration systems can be a stand-alone device or combined e.g. with specific filtration systems or even conventional unit operation.

The variety of the presented examples may give rise to explore other application options, such as in tank farm options to fine-tune desired fuel specification (water, metals etc.) or to deliver on specific high-purity demand, such as for military. For example, an improved thermal stability would result in a significant jet-engine maintenance (cost) reduction and the removal of hydrophilic molecules could reduce water pick-up and improve water shedding, which support to improve longer-term and strategic fuel storage.

Meanwhile, other innovative options have been published such as the in-situ removal of metals from diesel in a so-called 'kidney-loop configuration'[13] where Zinc retentions of 99,9% were achieved. This configuration has been recognized as a method for reducing injector deposits within the combustion engine, improving fuel efficiency, thermal stability, engine cleanliness and fuel economy. In addition, this would reduce the use of expensive detergents in fuel compositions. Also, for these concepts the typical OSN membrane materials are being applied. Again, advantages that contribute to sustainability aspects within the automotive industry.

CONCLUSIONS

Based on the technical results as well as from implementation experiences, OSN can be seen as a mature and robust separation method with high Technical Readiness Level (TRL). Its modularity gives flexibility for capacity creep, but also for mobility. Due to its pressure-driven separation mechanism, the technology can be seen as sustainable. In fuel applications, the resulting fuel quality aspects, such as thermal stability and fuel economy, like for diesel, also contribute to the sustainability aspects. The provided separation examples may give rise to scout similar application options.

ACKNOWLEDGEMENT

The author and BORSIG Membrane Technology GmbH would like to acknowledge the good collaborations with SHELL Projects and Technology through the various application development. Especially, Shell's Lead Technology Expert Prof. Arian Nijmeijer and Senior Technologist Special Separation Jan den Boestert, who recently retired from SHELL. Meanwhile, Mr. den Boestert is active as a Separation Technology consultant and contributed to this technical paper. Furthermore, we thank Evren Unsal who's the Executive Editor of Shell TechXplorer Digest magazine and facilitated the technical support documentation and several figures being used for this technical paper. The 2021 edition of Shell TechXplorer Digest can be read here: www.shell.com/energy-and-innovation/shell-techxplorer-digest-2021.html Last but not least, we have valued the professional steer of Dr Joanna Bauldreay, Principal and Director of Bauldreay Jet Fuel Consulting Ltd.

REFERENCES

1. Prof. Dr Ir. Marcel Mulder, Basic Principles of Membrane Technology – ISBN 0-7923-4248-8
2. J.G.A. Bitter, J.L.W. den Boestert and R.H. Clark; Process for reducing the Metal content of a hydrocarbon mixture; Patent Application No. 5031881 – 1991
3. R.M. Gould et al. Lubricating Oil Dewaxing with membrane separation. Patent Application No. US 5651877
4. L.S. White, R.F. Wormsbecher and M. Leseman; Membrane separation for Sulphur reduction. Patent Application No. US 6,896,796.
5. R.P. Cossee et al. Process for purifying a liquid hydrocarbon product; Patent Application No. US6488856 - 2001
6. Chandrasekharan et al. Producing light olefins from a contaminated liquid hydrocarbon stream by means of thermal cracking. Patent Application No. WO1999027036
7. J.L.W. den Boestert and J. van Westrenen; Process for separating colour bodies and/or asphaltenic contaminants from a hydrocarbon mixture Patent Publication Number 20090014359
8. J.L.W. den Boestert, A. Nijmeijer, K. Smit; Method for reducing the Mercury content of a Natural Gas Condensate. Patent Application No. WO 2008/116864 and US 20100147745
9. Shell TechXplorer Digest 2021; Developments in suitable separation methods for condensates. < https://www.shell.com/energy-and-innovation/shell-techxplorer-digest-2021.html>
10. W.M. Bond, J.L.W. den Boestert and V. Nair; Methods of filtering a liquid stream produced from an in-situ heat treatment process. Patents number WO2007050446 and US20070131428
11. Shell TechXplorer Digest 2021; Low-energy nanofiltration for abating alternative feedstock < https://www.shell.com/energy-and-innovation/shell-techxplorer-digest-2021.html >
12. E.S. Tarleton, J.P. Robinson and J.S. Low; 2009. Nanofiltration, a technology for selective solute removal from fuels and solvent. Chemical Engineering Research and Design, 87 (3), pp. 271-279.
13. R.W.M. Wardle, J.L. den Boestert, H. Jones and N.J. Rounthwaite; Process for reducing injector deposits. Patent Application WO 2021/058537
14. P. Vandezande, L.E.M. Gevers and I.F.J. Vankelekom; Solvent resistant nanofiltration; separating on a molecular level. Advanced article in Chemical Society Reviews. web version 8th October, 2007
15. Heng (2011): *Method and system for purifying used oil.* WO 2011/105966 A1. Int. Filling date 24.02.2011

IASH 2022, the 17th INTERNATIONAL SYMPOSIUM ON STABILITY, HANDLING AND USE OF LIQUID FUELS
Dresden, Germany
September 11 - 15, 2022

DEVELOPMENT OF A MONITORING SYSTEM FOR DIRECT ANALYSIS OF MICROBIAL CONTAMINANTS IN FUEL

Dr. Jiri Snaidr*, Dr. Claudia Beimfohr

Vermicon AG, Zeppelinstraße 3, 85399 Hallbergmoos, 85399 Germany

ABSTRACT

A significant and yet often underestimated problem in the storage of fuels is contamination with microorganisms. The growth of mainly bacteria in the tanks leads to fuel losses and considerable quality degradation. When used in engines and power plants, these can sometimes have catastrophic consequences. For example, contaminated kerosene can cause damage or, in the worst case, engine failure due to clogged filters.

Establishing an effective monitoring of microorganisms is very important. Standard investigations are essentially based on cultivation methods in which the bacteria are grown on artificial media and can thus be identified and quantified. However, this falls short as they can not analyze whole biocenoses. Therefore appropriate technologies should take a holistic approach.

The VIT® gene probe technology is a powerful tool here. With its help, bacteria and entire populations can be identified, quantified and visualized directly in the sample. So-called gene probes, small pieces of genetic material specific to a bacterial species or group, are coupled with a dye and introduced directly into the completely preserved bacteria. Only in bacteria that have a corresponding target site, the gene probe and thus also the dye is bound and cannot be washed off: the bacteria specifically start to glow in the sample and can thus be analyzed.

Using different fuel samples, we demonstrate how entire biocenoses in fuel can be analyzed using VIT technology. Based on this, we have built a innovative monitoring system that follows a holistic approach and effectively controls microbial contaminations in fuels.

*Corresponding Author: CEO Jiri Dr. Snaidr; snaidr@vermicon.com; +4989158820

IASH 2022, the 17TH INTERNATIONAL CONFERENCE ON STABILITY, HANDLING AND USE OF LIQUID FUELS
Dresden, Germany
11 - 15 September 2022

FAME – QUALITY, APPLICATION AND CHALLENGES

Katharina Friedrich[1] and Dr. Richard Wicht[1].

[1]*AGQM Association Quality Management Biodiesel, Am Weidendamm 1A, 10117 Berlin, GERMANY. info@agqm-biodiesel.de*

KEYWORDS
Biodiesel, FAME, quality, engine oil dilution, greenhouse gas emissions, metals, non-metals

ABSTRACT

The Association Quality Management Biodiesel (AGQM) was founded to ensure the highest quality of biodiesel, especially in Germany. Therefore, a separate QM system was developed which applies to all members of AGQM producing and trading biodiesel. The most important quality-determining task within the QM system is the unannounced sampling of production sites and warehouses of AGQM members. The results which account for approx. 75% of German production capacity and a large proportion of Austria's capacity, are published anonymously in an annual report and show that the quality of AGQM biodiesel is guaranteed, also for parameters that are critical for the exhaust gas aftertreatment system, such as alkaline and alkaline earth metals, sulphur, and phosphorus. In the area of research and development AGQM actively cooperates with the mineral oil, automotive and additive industries in various research projects and working groups. In particular, to prove the suitability of FAME in modern and future vehicles and ships under the current demanding technical boundary condition and to enable the use of higher FAME or as pure fuel. The current AGQM approval lists for commercial vehicles and inland navigation show that biodiesel in various blending quantities or as pure fuel can already be used to reduce greenhouse gas emissions significantly.

INTRODUCTION

The Association Quality Management Biodiesel (AGQM) was founded in 1999 as a quality assurance initiative by leading biodiesel producers and traders. The members represent the whole production and supply chain. After a few years an own quality management system (QM system) was introduced and implemented that guarantees the high-quality standard of biodiesel, especially in Germany.

The QM system specifies at least three unannounced samplings are carried out at the members' sites per year as one of the requirements. The results of these samplings will be published anonymously as an annual report. To ensure the highest quality, there are stricter AGQM limits for critical parameters compared to the FAME standard DIN EN 14214[1]. For the water content the limit is 0.027 % (m/m) instead of 0.050 % (m/m) according to DIN EN 14214 and for the parameter total contamination the AGQM limit is only 20 mg/kg.[1,2] Despite the unannounced samplings annual AGQM audits to check the standards of the members, round robin tests FAME and Vegetable & Used Cooking Oils as well as the quality check material (QC material) are part of the QM work.

According to EN 590[3], the maximum share of biodiesel in diesel fuel is currently limited to 7%. Additionally, the German Federal Immission Control Ordinance 10th BImSchV[4] currently only permits B7 and B100 in Germany. But there are also already the European standards EN 16734[5] and EN 16709[6] published which define the requirements for B10 and B20, B30. In addition, the Fuel Quality Directive (FQD II)[7] is currently being revised, possibly the new standard fuel grade could be B10 with the protection class of B7. So that the basis for higher blends is given in principle.

Furthermore, the Renewable Energy Directive (RED II)[8] provides for a continuous increase of the greenhouse gas (GHG) quota from 6% to 25% in 2030 in Germany. However, the transport sector is lagging behind the GHG reduction targets.[9,10] To achieve these goals, a higher content of biodiesel is the most obvious option, because biodiesel is a currently available, renewable fuel, the most cost-efficient CO_2 abatement option, can also achieve great greenhouse gas savings in the existing fleet (average GHG-savings of 81%[11]) and can use as drop-in component or neat fuel.

RESULTS AND DISSCUSSION

If we want to use higher biodiesel blends, some technical aspects have to be taken into account. For the filling stations the specifications for separator systems for light liquids are laid down in the European standards EN 858-1[12] and EN 858-2[13]. The responsible national German committee assumes that no major additional measures will have to be carried out on the majority of existing systems.[14] So that the dispending of FAME and FAME diesel blends (B10/ B20/ B30) at public filling stations can be classified as unproblematic.

Additionally, various projects were carried out on the subject of material compatibility. In the DGMK project 780, various metallic materials and polymers were examined for their resistance in B20 blends and as result could be classified as resistant.[15] In another project, the material and component compatibility during longer storage times was examined with B10 fuels. No degradation was found in the parts and components stored with B10 and tested for function compared to the B0 tests.[16] Consequently, for the use of B10 or higher biodiesel blends, it is not expected that adjustments will have to be made to the materials used.

In terms of emissions, the raw emissions of fuels containing FAME are more beneficial than those of a diesel fuel without FAME admixture with the exception of nitrogen oxides (lower hydrocarbons, particulates and CO emissions).[17,18] However, the raw emissions of the fuels are hardly relevant, since the total amount emitted, measured at the exhaust, must be reduced completely with the exhaust gas aftertreatment system. The concentration of sulphur, phosphorus and the alkali and alkaline earth metals sodium, potassium, calcium and magnesium could be critical and is regulated by EN 14214.[19] Some vehicle manufacturers point out that the legal limits of EN 14214 for metals and non-metals are probably not sufficient to safely guarantee the function of the exhaust aftertreatment system over the entire life cycle when using B100.
To overcome this aspect, the average values of the alkali and alkaline earth metals of the AGQM members are determined in the three annual unannounced sampling campaigns according to DIN EN 14538[20]. Figure 1 shows, that the average value of sodium and potassium content is below 1 mg/kg for the period 2014 to 2021, and thus clearly below the standard limit of 5 mg/kg[19]. This also applies to the calcium and magnesium content, which is around 0.1 mg/kg for the entire period (see Figure 1).

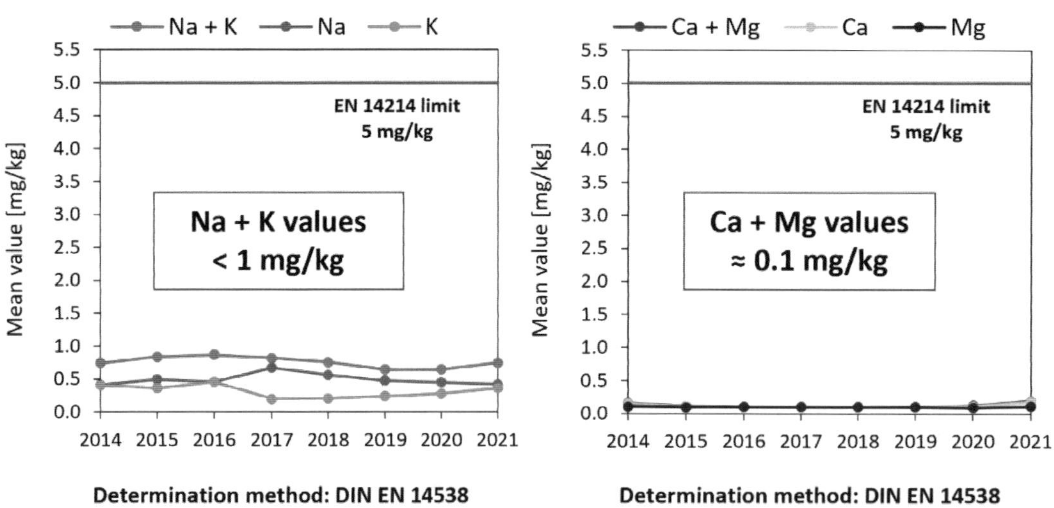

Figure 1: Average values of sodium, potassium, calcium, and magnesium contents of AGQM members

for the period 2014 to 2021. The results are determined within the annual unannounced samplings.

For the non-metals sulphur and phosphorus, the average values of the AGQM members were also determined according to DIN EN ISO 20846[21] and DIN EN 14107[22]. In Figure 2 it can be shown that the sulphur content is on average below 4 mg/kg, and thus well below the standard limit value of 10 mg/kg[19]. For phosphorus content, the mean value is even below 0.9 mg/kg and thus also well below the limit value of 4 mg/kg[19] (see Figure 2).

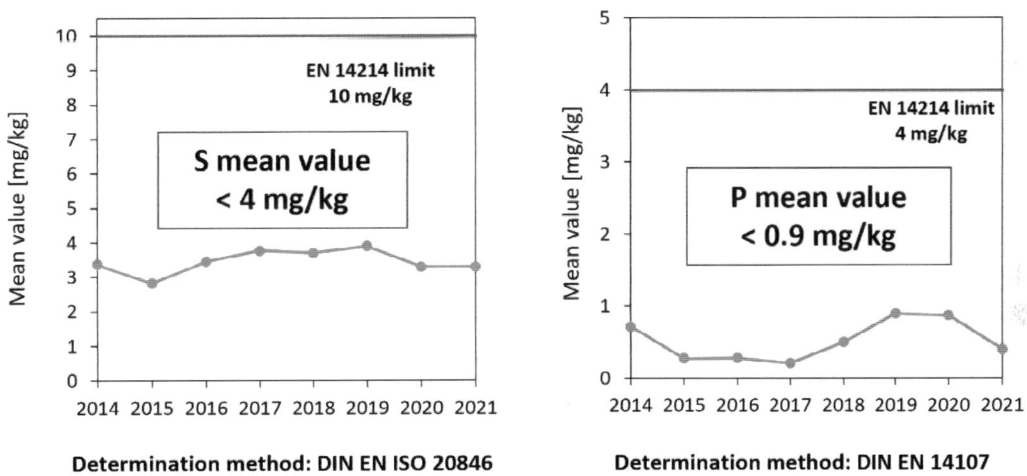

Figure 2: Average values of the non-metals sulphur and phosphorus of AGQM members for the period 2014 to 2021. The results are determined within the annual unannounced samplings.

With these results, the content of metals and non-metals can be classified as uncritical and that a higher proportion of biodiesel or the use as pure fuel would not be a problem in this aspect. Regarding the topic of engine, in the project "deposit formation and deposit prevention biodiesel" of the University of Rostock (Germany) it could be investigate that the use of higher FAME blends compared to B7 and B0 leads to a lower deposit formation in the injector-critical low temperature range. A cleaning effect through higher FAME blends cloud also be shown.[23]

With projects and approval lists, for example in the area of inland navigation and commercial vehicles, many vehicles and engine manufacturer demonstrate that a higher amount of biodiesel is technically possible already now.[24,25]

However, there is still a challenge to overcome when using B10/B20/B30, namely the issue of engine oil dilution. Engine oil dilution can occur when late fuel post-injection is used to regenerate the diesel particulate filter. Diesel fuel without FAME content can evaporate again form the engine oil, while diesel fuel with FAME content is not completely able to do so due to its higher boiling point. This leads to a reduction in viscosity and faster ageing of the engine oil.[26] In the past, this was particularly relevant for short-distance operation and usually resulted in shortened engine oil change intervals form B20 on.

The aspect of engine oil dilution is being examined again in two AGQM projects. One of the projects will be the B10 car fleet test, in which the practicality of a B10 will be investigated in non-approved passenger cars. For this purpose, realistic driving tests of EURO 5 and EURO 6 vehicles will be carried out under different conditions (city traffic, ultra-short distance, and mixture of city/ highway/ country road). The second project is about B100 engine oil testing of commercial vehicles. Therefore, the engine oils of buses that are and have been operated with B100 are to be tested. It is expected that it can be shown that the engine oil dilution and ageing is unproblematic when using B10 and B100 compared to B7 in modern vehicles. As a consequence, the oil change intervals can ideally be increased for commercial vehicles form approx. 35,000 km to 70,000 km or more.

CONCLUSIONS

In the near future, higher FAME blends can be used in passenger cars, commercial vehicles and maritime shipping because the legal framework is already in place and the general technical questions have already been answered. Furthermore, the use of biodiesel saves a large amount of greenhouse gases even in the existing fleet and is a currently available and renewable fuel.

REFERENCES

1 DIN EN 14214 - Flüssige Mineralölerzeugnisse - Fettsäure-Methylester (FAME) zur Verwendung in Dieselmotoren und als Heizöl - Anforderungen und Prüfverfahren; Deutsche Fassung EN 14214:2012+A2:2019.

2 Association Quality Managment Biodiesel (AGQM), Quality Management System - Requirements for Procution, Transport and Storage of Biodiesel, **18th February 2022**.

3 DIN EN 590, Automotive fuels - Diesel - Requirements and test methods.

4 10. BImSchV, Zehnte Verordnung zur Durchführung des Bundes-Immissionsschutzgesetzes (Verordnung über die Beschaffenheit und die Auszeichnung der Qualitäten von Kraft- und Brennstoffen).

5 EN 16734, Automotive fuels - Automotive B10 diesel fuel - Requirements and test methods.

6 EN 16709, Automotive fuels - High FAME diesel fuel (B20 and B30) - Requirements and test methods.

7 DIRECTIVE 2009/30/EC OF THE EUROPEAN PARLIAMENT AND OF THE COUNCIL of **23 April 2009** amending Directive 98/70/EC as regards the specification of petrol, diesel and gas-oil and introducing a mechanism to monitor and reduce greenhouse gas emissions and amend.

8 RED II, DIRECTIVE (EU) 2018/2001 OF THE EUROPEAN PARLIAMENT AND OF THE

COUNCIL of **11 December 2018** on the promotion of the use of energy from renewable sources.
9 Projektionsbericht 2021 für Deutschland, Gemäß Artikel 18 der Verordnung (EU) 2018/1999 des Europäischen Parlaments und des Rates vom 11. Dezember 2018.
10 Statistisches Bundesamt (Destatis). [Online] [Zitat vom: **30.09.2022**.] https://www.destatis.de/Europa/EN/Topic/Environment-energy/CarbonDioxideRoadTransport.html.
11 Bundesanstalt für Landwirtschaft und Ernährung - Referat 523 - Nachhaltige Biomasse, Evaluations- und Erfahrungsbericht für das Jahr 2020.
12 EN 858-1, Separator systems for light liquids (e. g. oil and petrol) - Part 1: Principles of product design, performance and testing, marking and quality control.
13 EN 858-2, Separator systems for light liquids (e.g. oil and petrol) - Part 2: Selection of nominal size, installation, operation and maintenance.
14 AGQM Biodiesel, [Online] [Zitat vom: 30.09.2022.] https://www.agqm-biodiesel.de/application/files/1816/5096/0197/RZ_AGQM_0221_Merkblaetter_Broschuere_DE_web-neu_neu.pdf.
15 DGMK-Research Report 780, Development of a test method for assessment of material resistance in applications with middle distillate.
16 FVV- Project no. 1253, Investigation of interactions between fuels and fuel leading components in Plug-in Hybrid Electric Vehicles (PHEV).
17 Landbauforschung Völkenrode - FAL Agricultural Research, Systematische Untersuchungen der Emissionen aus der motorischen Verbrennung von RME, MK1 und DK, **2003**.
18 J. Hunicz, P. Krzaczek, M. Gęca *et al*, Comparative study of combustions and emission of diesel engine fuelled with FAME and HVO, *Combustion Engines*, **2021** 184(1), 72-78.
19 EN 14214, Liquid petroleum products - Fatty acid methyl esters (FAME) for use in diesel engines and heating applications - Requirements and test methods.
20 DIN EN 14538, Fat and oil derivatives - Fatty acid methyl ester (FAME) - Determination of Ca, K, Mg and Na content by optical emission spectral analysis with inductively couled plasma (ICP OES).
21 DIN EN ISO 20846, Petroleum products - Determination of sulfur content of automotive fuels - Ultraviolet fluorescence method.
22 DIN EN 14107, Fan and oil derivates - Determination of phoshporus content by inductively coupled plasma (ICP) emission spectrometry.
23 Universität Rostock; Projektnummer 550_2018_1; Belagsbildung und Belagsvermeidung Biodiesel (DIN EN 14214).
24 AGQM, MVaK, ufop, VDB, Approvals in inland navigation.
25 AGQM, MVaK, ufop, VDB, Approval list of engine and commercial vehicle manufactures.
26 FUELS JOINT RESEARCH GROUP, Handlungsfelder und Forschungsbedarf bei Biokraftstoffen. Göttingen : CUVILLIER VERLAG, **2019**.

IASH 2022, the 17TH INTERNATIONAL CONFERENCE ON STABILITY, HANDLING AND USE OF LIQUID FUELS
Dresden, Germany
11-15 September 2022

MONITORING BIODIESEL BLENDS IN HEATING APPLICATIONS – EFFECTS OF EXPOSURE CONDITIONS

Ryan Kerr[1], Thomas Butcher[1], Neehad Islam[1], and Marty Haverly[2]

[1]*National Oilheat Research Alliance (NORA), Plainview, N.Y. 11803*
tbutcher@noraweb.org
[2]*Renewable Energy Group (REG), Inc., Ames, IA 50010*

Keywords: heating oil; biodiesel; copper; oxidation reserve; oxidative stability; Rancimat

ABSTRACT

Petroleum distillate fuel has been used for many years for building heating applications. This market is now increasingly using biodiesel blended with petroleum fuel. While copper is not recommended for conventional diesel engines and fuel systems, the use of copper lines in the heating oil industry in the U.S. is common due to their lower cost and ease of manipulation and installation. This study focused on the potential impacts of copper on the fuels used in today's home heating oil systems (i.e. ultra low sulfur No. 2 petroleum distillates and their blends with biodiesel) and showed the effects depend largely on the surface to volume ratio of the exposure to copper, access of the copper-exposed fuel to oxygen, residence time of the fuel in contact with copper, and the overall time before the fuel is burned. The practical implications of this research are that the presence of copper in heating systems should not present significant formation of acids or filterable insolubles that could adversely affect system operation of over time with either ultra-low sulfur No. 2 petroleum distillates or their biodiesel blends. This is likely due to the specific exposure conditions to copper in normal heating oil systems and the relatively small amounts of fuel exposed to both copper and oxygen for extended periods of time. To the extent conditions do exist where changes over time could occur, the changes observed with biodiesel blends were found to be similar or less than those with conventional No. 2 petroleum distillate fuel oil. However, exposure to copper can dramatically impact the measured oxidative reserve and this has important implications when sampling for monitoring fuel quality in this application.

INTRODUCTION

The heating oil market is an important consumer of distillate fuel in the U.S. with about 5 million homes using this fuel. Most of this consumption is in the Northeast. The heating oil market is currently in a period of transition to ultralow sulfur fuel (i.e. less than 15 parts per million sulfur) and increased use of biodiesel in blends with petroleum distillate (No. 2 heating oil)[1]. Some fuel marketers have taken the initiative to blend 20% biodiesel into No. 2 heating oil. Still others are exploring higher blend levels. Other biofuels are also being explored[2], but biodiesel is the most widely available and the lowest cost, so it is currently of the most commercial interest to the heating oil market in the U.S.

Figure 1 provides an illustration of a typical, residential fuel storage and burner supply system. This shows a storage tank indoors and at about the same elevation as the burner. Fuel storage tanks can also be outdoors, above ground or, less commonly, buried under ground. Tanks are most commonly made of carbon steel and have a typical storage volume of 1041 liters (275 gallons). This figure also shows the fuel line running from the tank to the burner under the foundation and coated with plastic. However, it is also common for the fuel line to be uncoated and to run inside the living or basement space. Copper fuel lines are most often 3/8-inch (9.5 mm) nominal diameter.

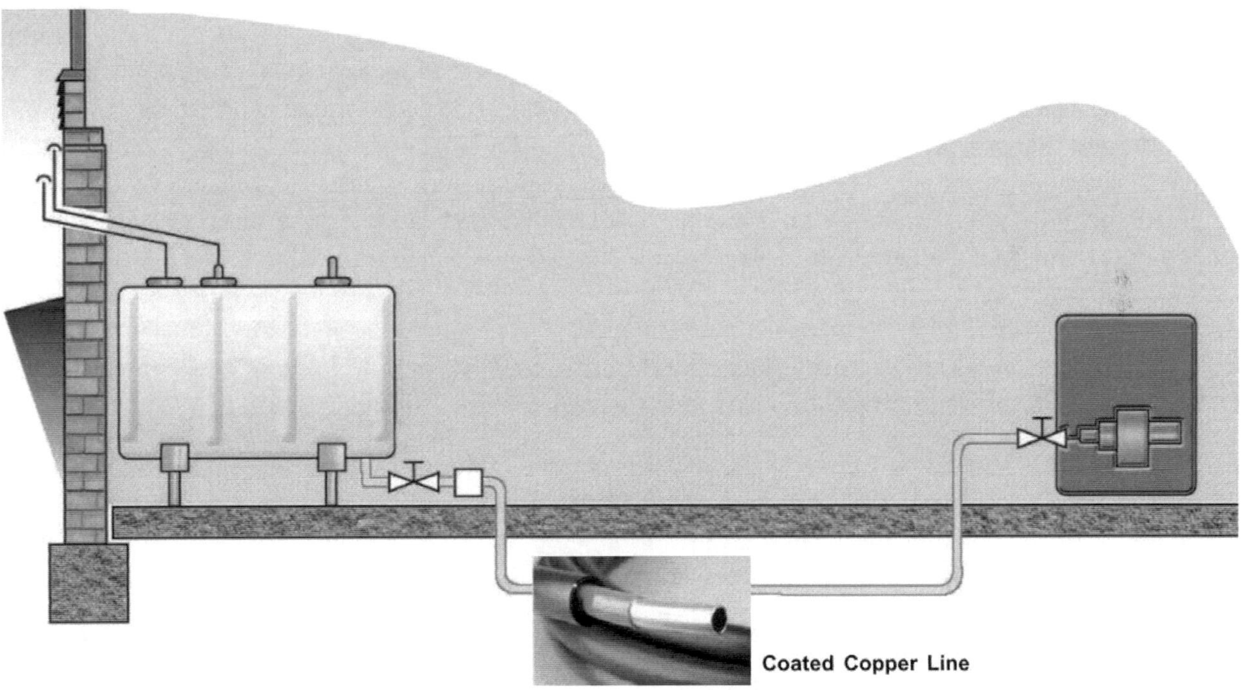

Figure 1. Illustration of a typical residential, oil-fired heating system fuel arrangement.

During normal winter operation, fuel flows from the tank to the burner periodically as the burner fires. When the burner is off, fuel is stagnant in the line, but during regular operation the residence time in the copper fuel line is only a few seconds.

Some boiler-based heating systems also operate during the summer for domestic hot water service. In this case, the fuel has the potential to be exposed to copper in the lines for longer time periods, but this rarely exceeds a few hours. In systems which are fully out of service during the summer, such as warm air furnaces, fuel can remain in the copper fuel line for up to 3-6 months.

The system in Figure 1 is termed a "one-pipe system". The burner fuel pump has an integral pressure regulator which bypasses excess fuel from the pump discharge internally back to the pump suction. In an alternative configuration, this bypass flow is directed back to the storage tank in a second copper line that runs parallel to the supply line. Fuel in these "two-pipe systems" can have greater exposure of the fuel to copper than the one-pipe systems. Two-pipe systems are typically only used when the fuel storage tank is at an elevation significantly below the burner and there is potential for air ingress into the fuel system due to the higher pump suction needed for these applications. In the U.S. market, it is estimated that 80 percent of the installations are of the one-pipe configuration, while the remaining are two-pipe.

Within common burner systems, fuel is directed at high pressure from the discharge of the pump through a short, small-diameter copper line to the atomizing fuel nozzle. The nozzle is the highest temperature component in the system. While stainless steel nozzles are available, most nozzles are brass. Nozzles are typically replaced annually but fuel lines are rarely replaced.

In the storage and handling of distillate fuel oil, it is often recognized that exposure to metals and the presence of dissolved metals in the fuel can accelerate oxidative degradation[3,4,5,6,7,8]. Some products of oxidation are relatively harmless in fuel while others that can form such as gels, particulates, or acids can be problematic. Copper is one such metal and it is generally recommended that copper should not be used in distillate fuel systems. The ASTM D975 standard for on/off road diesel states, "The formation of degradation products can be catalyzed by dissolved metals, especially copper and zinc. When dissolved copper and zinc are present it can be deactivated with metal deactivator additives." It goes on further to state, "Copper, copper-containing alloys, and zinc-coated or galvanized equipment should be avoided. Copper can promote fuel degradation and can produce mercaptide gels. Zinc coatings can react with water or organic acids in the fuel to form gels that rapidly plug filters." In general, copper is not recommended for either distillate fuel oils nor for biodiesel due to these factors.

In oil-fired home heating systems, however, copper fuel lines are commonly used due to their lower cost and ease of manipulation and installation. With the introduction of higher blends of biodiesel into home heating oil systems, questions have arisen about any potential differences in fuel performance of biodiesel blends vs. those of conventional petroleum-based distillates with the higher presence of copper in U.S. home heating oil systems vs. those commonly used for on- and off-road diesel fuel.

Currently, the ASTM D396 standard specification for fuels commonly used in home heating systems in the U.S. allows for up to 5% biodiesel by volume in the No. 1 and No. 2 grades, and also provides specifications for 6% to 20% biodiesel blends as heating fuels under the B6-B20 grade[9]. All three grades contain a specification for copper strip corrosion using the ASTM D130 test method of No. 3 maximum rating, as does the ASTM D6751 standard specification for B100. According to ASTM D396, "The corrosion test serves to indicate the presence or absence of materials that could corrode copper, brass, and bronze components of the fuel system." This parameter serves to help control potential corrosion of metal by the fuel, but it does not, however, control potential adverse impacts of metal contact on the properties of the long term storage of the fuel.

The Rancimat Induction Period (IP) provides a rapid measure of fuel oxidation reserve and is specified as the indicator of storage stability in the ASTM D396 fuel specification for biodiesel blends. The oxidative reserve values mentioned throughout the remainder of this paper are those provided by the fuel version of the Rancimat test, EN15751 as per ASTM D396.

In a recent study[10], the impact of copper exposure on diesel fuel with 7% biodiesel was investigated at both low (12°C) and high (30°C) storage temperatures. One-liter samples were stored for up to 4 weeks in borosilicate glass flasks with and without copper plates. At one-week intervals samples were taken for evaluation of oxidative reserve or fuel property changes by three methods: oxidation reserve/Induction Period (EN15751), rapid small-scale oxidation test (EN 16091), and a gravimetric method (EN ISO 12205, the European version of ASTM D2274). All three measures indicated more changes that may impact fuel performance over time for the fuel samples stored with copper compared to those without copper. This effect was most pronounced at the highest storage temperature. The copper content of the fuel exposed to copper after the 4-week test was measured using an ICP-OES technique and found to increase only slightly from <0.1 ppm to 0.2 ppm for the fuel at 12°C, and 0.4 ppm for the fuel at 30°C.

In another study[11], the impact of different metal-containing additives on the measured oxidation reserve/Induction Period was studied over a range of biodiesel blend levels. Oxidation reserve was found to be reduced for metal contents up to 2.5 ppm, with copper having the most significant impact. Above 2.5 ppm, oxidation reserve was found to be unaffected by metal concentration. Antioxidant (storage enhancing) additives were found to improve the oxidation reserve for metal containing fuel blends. Similar results were found by Sarin et. al. for different biodiesel types[12,13]. Knothe and Steidly[14] extended studies of this type to a much wider range of metals and confirmed that copper stands out as having the most impact on oxidation reserve and potential changes in the fuel over time.

Öl-Wärme-Institute (OWI) in Aachen, Germany studied the long-term stability of samples of No. 2 heating oil and 20% biodiesel/80% heating oil blends with and without copper wire coils and additives. Their results showed a consistent decrease in oxidation reserve after exposure to copper. These tests were done in two series at 40°C, each lasting for 24 months. Storage enhancing additives were found to be effective in reducing the apparent rate of degradation. Gum formation was found on the copper coils for the samples that had them[15]. Chendynski et al.[16] and Kugelmeier et. al.[17] also demonstrated a strong reduction in the measured oxidative reserve of biodiesel fuels with exposure to copper. Squissato et. al. developed a new method for the measurement of free Cu (II) ions in biodiesel and found a strong negative correlation with oxidation reserve.

Myers et al, evaluated the response of different petroleum distillate fuels to copper exposure. This team concluded that the property changes of low-sulfur fuels was impacted by copper to a greater degree than high-sulfur fuels[18]. Storage additives were found to be effective in reducing the impact of copper exposure.

Despite the potential for adverse effects, residential oil-fired heating systems in North America almost universally use copper pipe to connect the fuel tank to the burner and for some components in the burner itself. The authors of this study have a strong interest in understanding the quality of fuel actually being used in these small heating systems and the impact that the growing use of biodiesel has on this quality. The goal of this study was to evaluate the practical impact of the presence of copper on these modern fuels and the implications this impact has on fuel sampling for quality monitoring.

EXPERIMENTAL

A variety of fuel experiments were conducted to evaluate copper exposure impact. This included exposure of small fuel samples to copper under static conditions at ambient temperature, exposure and storage under warm conditions, and circulation of larger fuel samples in a tank-pump-tubing loop arrangement. The conditions were designed to simulate common conditions of exposure level of the fuel to copper in common fuel oil systems in the U.S.

Petroleum No. 2 heating oil samples meeting ASTM D396[9] were obtained from local heating oil distributors. Each sample was checked for biodiesel content using an InfraCal 2 ATR-B Biodiesel-in-Diesel Analyzer. It is common, presently, for heating oil to have some biodiesel in it and all of the samples referred to as petroleum No. 2 heating oil samples used in this work contained 2-4% biodiesel by volume. The 100% biodiesel used in this study for producing blends was a commercial fuel produced by Renewable Energy Group, Inc. (REG) and met the requirements of ASTM D6751[19]. Table 1 provides a summary of selected properties the biodiesel fuel used. The production facility for this fuel confirmed that additives were used with this fuel, as is common practice, to ensure acceptable stability.

Blends of the petroleum No. 2 heating oil and biodiesel mentioned above were prepared to determine the impact of copper exposure on fuel blends. The fuels were blended on a volumetric basis such that 20% and 50% biodiesel was blended into petroleum No. 2 heating oil.

The Induction Period (IP or "oxidative reserve") of fuel blends was measured to evaluate oxidative stability in storage, following EN 15751 using a Metrohm 893 Professional Biodiesel Rancimat instrument. In addition, long term storage impacts of fuel blends were measured following ASTM D4625[20]. In this test, 400 ml of sample were placed in an incubator set to 43 ± 1°C. The samples were left in the incubator, open to the air, for a predetermined amount of time. Heating the samples to 43°C increases degradation approximately fourfold compared to degradation at ambient temperature such that 1 week in the incubator is equal to approximately 1 month at ambient temperature. After the planned amount of time, samples were removed from the incubator and both filterable and adherent insolubles were measured. Filterable insolubles were measured by forcing each sample through a nylon membrane filter with a nominal pore size of 0.8 μm and determining the mass of filtered insolubles. The adherent insolubles, which are the materials remaining on the sample container walls, were then measured by washing the container twice with a solvent comprised of equal parts acetone, methanol, and toluene. The solution was then heated to 160°C to drive off the solvent. The residual mass is taken as that of the gum. Fuel acid numbers were measured via manual titration following ASTM D664[21].

Trace metals analysis of the fuel samples before and after exposure to copper was conducted at REG using a Spectro Arcos inductively coupled plasma optical emission spectrophotometer (ICP-OES). The ICP-OES was calibrated to measure copper below 1 ppm levels. A 2-ppm check standard was included intermittently between each fuel sample to check for instrument drift.

Table 1. Properties of the Biodiesel Fuel (B100) Used in this Study

Property	Value	Units	Test Method
Cloud point	-1.1	°C	D7397
Free Glycerin	0.01	% mass	D6584
Total Glycerin	0.11	% mass	D6584
Acid Number	0.34	mg KOH/g	D664
Oxidation Reserve (IP)	9.0	hours	EN 15751
Cold Soak Filtration	102	seconds	D7501
Sulfur	0.6	mg/kg	D5453
Sodium & Potassium Combined	0.2	mg/kg	EN 14538
Calcium & Magnesium Combined	0.0	mg/kg	EN 14538
Kinematic Viscosity at 40°C	4.13	mm^2/sec	D445
Distillation at 90% Recovered	352	°C	D1160

Exposure of fuel blends to copper was done using several methods as follows:

Exposure Method 1 – A section of copper tubing commonly used in fuel systems with a length of 291 mm, was capped on one end and positioned in a vertical orientation. The tubing has an inside diameter of 6.76 mm and an outside diameter of 9.36 mm. A volume of 14 ml of fuel was then added to the tubing and maintained at ambient temperature for 24 hours. This arrangement provided a surface area to volume ratio of 5.92 cm^2/cm^3.

Exposure Method 2 – A section of the standard copper tubing described in Exposure Method 1, was cut into short (8 mm) pieces. For each test, 50 pieces were combined with 400 ml of each fuel blend. In this case the fuel was exposed to the inside, outside, and end of each piece. Exposure time to copper was again 24 hours at ambient temperature. The ratio of surface area of copper tubing to volume of fuel was 0.59 cm^2/cm^3. This exposure level is more than found in the bulk fuel tank in Exposure Method 4 but provides a means to accelerate the potential impacts of copper.

Exposure Method 3 – This is the same as in Exposure Method 2 except the total number of pieces was reduced to just 8 but the pieces were longer, giving the same surface to volume ratio as in Exposure Method 2. This provides a slightly different avenue for accelerating the impacts of copper.

Exposure Method 4 – This method involved a much larger volume of fuel and an exposure arrangement which more closely emulated field conditions. A commercially available plastic fuel storage tank with a total volume of 379 liters (100 gallons) was used in combination with a standard oil burner pump configured to simulate a two-pipe arrangement. The pump was connected to the storage tank with two coils of copper tubing, each with a total length of 7.86 m. A pump operated continuously for a period of two months. Based on the total volume of oil in the tank and the total internal surface area of the copper tubing, the surface area to volume ratio in this arrangement was 0.018 cm^2/cm^3.

Exposure Method 5 – This method was designed to simulate a summer shutdown field situation, where fuel could be stationary in a copper line for many months. A copper tube with a total length of 9.14 m was filled with a volume of 420 ml of fuel. This copper tube was then capped at both ends to prevent air exposure. The copper tube with the fuel inside was then left at ambient temperature for 6 months. The ratio of surface area of copper tubing to volume of fuel was 5.92 cm^2/cm^3.

RESULTS

Table 2 shows the results of a series of tests with different fuel blends exposed to copper using Method 1. All of these tests were conducted at the NORA laboratory in Plainview, N.Y. A sample of each fuel was held in a glass beaker without exposure to copper to serve as a control. Table 2 compares the IP of the control samples and those exposed to copper using Method 1. Each fuel showed a significant drop in oxidation reserve after copper exposure compared to the control sample with no copper. Specifications for the properties of heating oil blends with 6 to 20% biodiesel content in the ASTM D396[9] standard require a minimum oxidation reserve of 6 hours at the time of manufacture which has been shown to provide a consistent lack of fuel degradation products for a minimum of at least 6 months. All of the No. 2 heating oil, B20, and B50 fuels tested using this exposure method had measured IP below 6 hours after exposure to copper. Similarly, the B100 sample also has measured IP below the 3-hour value for as-manufactured B100 in ASTM D6751[19].

To confirm these results, a copper tube was prepared per Exposure Method 1 and sent to the REG lab in Iowa for repeat testing at a B50 blend level using different samples of No. 2 heating oil and biodiesel. REG also left portions of the B50 blend in both the copper tube and a glass beaker for 24 h at room temperature. In addition to IP, each sample was analyzed by ICP-OES to measure trace metals. The results from these tests are shown in Table 3. The IP values from REG are close to the B50 induction times obtained in the tests at the NORA lab even though the fuels were different. Results of the ICP test also show a small but measurable increase in the amount of copper in the exposed sample compared to the unexposed sample, with the copper value increasing more than that of lead or zinc.

Table 2. Oxidation Reserve (Rancimat Induction Period) for Fuel Samples Held in a Vertical Copper Tube and the Control Samples held in a Glass Beaker without Copper Exposure. Each was held for 24 hours at room temperature (~22°C) at the NORA Lab.

Copper Exposure	No.2 Oil	B20	B50	B100
Not Exposed to Copper	18.29	8.16	7.82	6.13
Exposed to Copper	3.07	0.98	0.11	0.11

Table 3. Oxidation Reserve (Rancimat Induction Period - hours) and Selected Trace Metals for a B50 Blend Exposed and not Exposed to Copper in a Vertical Tube for 24 hours at Room Temperature (22°C) at the REG lab.

	B50 (control)	B50 (copper)
Induction Period (hr)	7.6	0.18
Copper Content (ppm)	0.1	4.1
Lead Content (ppm)	0.1	1.3
Zinc Content (ppm)	0.1	1.6

The first long term-storage stability test following ASTM D4625 involved eight samples, two control samples of No. 2 heating oil not exposed to copper, two samples of No. 2 heating oil exposed to copper, two control samples of a B20 blend not exposed to copper, and two samples of a B20 blend exposed to copper. The exposure in this test followed Exposure Method 2. Fuels were separated from the copper prior to starting the long-term storage test. All of the samples were filtered through a 0.8 micron filter before testing to remove any initial particles. The samples were removed after 9 weeks and IP, acid number, filterable insolubles and adherent insolubles were measured. Results are shown in Table 4. When comparing the control samples to those that were exposed to copper, the oxidation reserve values were reduced to similar values for both the No. 2

heating oil samples and the B20 samples. After aging the control samples all remained above 3 hour Induction Period which has been used in previous research as an acceptable performance indicator for aged fuel. Similar results were observed for acid number; the samples exposed to copper showed elevated acid numbers compared to the control samples although all control results remained low and all aged results remained below 1 mg KOH/g. Filterable insolubles and adherent insolubles also increased noticeably for the samples exposed to copper, although the increase was about 5 times less for the B20 compared to No. 2 oil. This could be the result of the presence of oxidation reserve specifications for B100 and for B20, leading to introduction of stability-additives during commercial production, while none exist for No. 2 oil.

Table 4. Results of Long-Term Storage Test #1. Four Samples of No. 2 oil and Four Samples of B20 Blends. Half of the Samples were Exposed to Copper for 24 hours Before Starting the Long-Term Storage Test.

Sample ID	IP (hr)	Acid Number (mg KOH/g)	Filterable Insolubles (mg/100 mL)	Adherent Insolubles (mg/100 mL)
No. 2 heating oil #1, (initial)	30.58	-	-	-
No. 2 heating oil #1, (control)	10.66	0.01	0.5	5
No. 2 heating oil #1, (copper)	0.02	0.81	12	93.25
No. 2 heating oil #2, (control)	10.92	0.05	0.25	6
No. 2 heating oil #2, (copper)	0.03	0.75	5.5	95
B20 #1, (control)	3.58	0.13	0.25	6.5
B20 #1, (copper)	0.02	0.87	5.25	17.75
B20 #2, (control)	4.49	0.13	0.25	6.5
B20 #2, (copper)	0.01	0.87	6	12.75

In the second long-term test, eight samples of No. 2 heating oil were exposed to copper for 24 hours using exposure Method 3 in duplicate before starting the long-term storage test. Again, all samples were filtered through a 0.8 micron filter before testing to remove any initial particulate contamination. Every three weeks from the beginning of the test (at weeks 3, 6, 9 and 12) one set of duplicates was removed and analyzed for oxidation reserve, acid number, filterable insolubles and adherent insolubles. A control sample of No. 2 heating oil, stored in a glass beaker and not exposed to copper, was also held in the incubator for the full 12 weeks. Results are listed in Table 5. All of the samples exposed to copper showed a large decrease in oxidation reserve with respect to the control sample. The initial oxidation reserve of the No. 2 heating oil was 23.6 hours. By the end of 12 weeks, the oxidation reserve fell below 0.06 hours for the samples exposed to copper. The acid number of the samples exposed to copper also increased with time, with the samples from week 12 going slightly over 1 mg KOH/g. The filterable insolubles amount increased for each sample as the test progressed up to week 9 but fell slightly in the samples removed at 12 weeks. The adherent insolubles increased substantially through the 12-week period. In contrast, the control sample showed only minor changes for the four quality metrics analyzed.

Table 5. Results of Long-Term Storage Test #2. Eight Samples of No. 2 Heating Oil were Exposed to Copper for 24 Hours Using Exposure Method 3 in Duplicate. Each Set of Duplicates were Removed at Weeks 3, 6, 9 and 12. One Control Sample of No. 2 Heating Oil was Stored in a Glass Beaker and not Exposed to Copper. This Sample was Held for the Full 12 Weeks.

Sample ID	Duration (weeks)	IP (hr)	Acid Number (mg KOH/g)	Filterable Insolubles (mg/100 mL)	Adherent Insolubles (mg/100 mL)
No. 2 heating oil (initial)	0	23.59			
No. 2 heating oil (copper)	3	1.86	0.10	0.8	9.5
No. 2 heating oil (copper)	3	0.03	0.01	1.3	9
No. 2 heating oil (copper)	6	0.06	0.13	5.5	14.5
No. 2 heating oil (copper)	6	0.06	0.29	12.5	16.25
No. 2 heating oil (copper)	9	0.02	0.40	13.8	31
No. 2 heating oil (copper)	9	0.02	0.37	13.3	27.8
No. 2 heating oil (copper)	12	0.02	1.14	5.5	317.5
No. 2 heating oil (copper)	12	0.02	1.10	7.5	764.3
No. 2 heating oil (control)	12	10.61	0.06	0.5	14.3

The oxidation reserve measurements taken for the two-pipe system of Exposure Method 4, are shown in Figure 2. In this exposure technique, the fuel is constantly recirculating, so it is not stagnant. Samples were pulled from the pump bleed valve about every 7 days. The oxidation reserve of the No. 2 heating oil in the tank was 24.9 hours before the start of the test. For a sample taken from the pump bleeder, which in this case is a good representation of the fuel in the bulk tank since it is constantly recirculating, one day after the start of the test, the oxidation reserve fell to 19.8 hr. The next collection at day 8 fell again to 15.0 hours, and all collections after this were relatively similar to this measurement, varying from 14.2 to 16.3 hours. The final sample was collected at 57 days. Interestingly, ICP-OES indicated that no significant amount of copper was present in this sample.

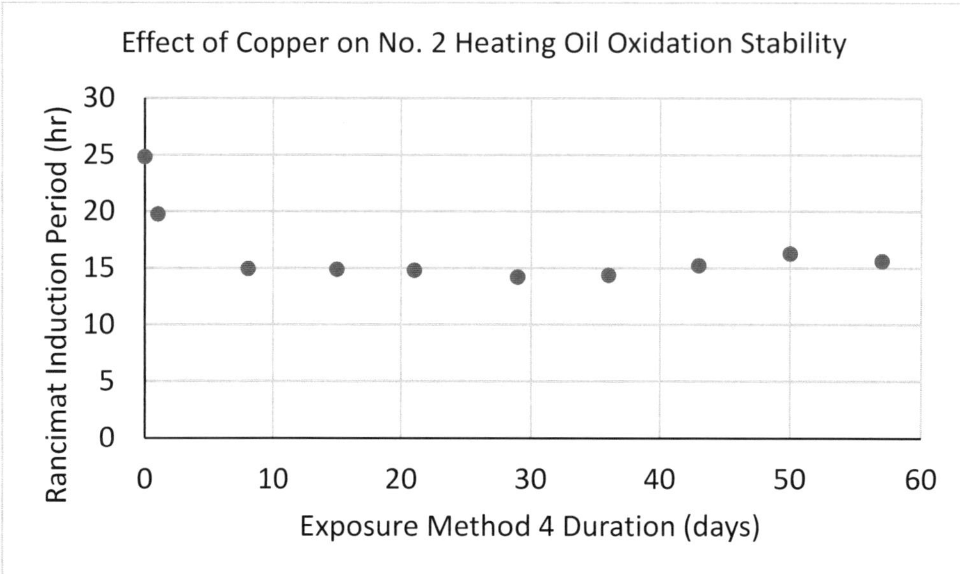

Figure 2. Results of periodic Rancimat IP measurements of No. 2 heating oil exposed to copper for 57 days using Exposure Method 4.

As shown in Figure 2, this test exhibited some reduction in oxidation reserve over the two-month test period but showed no decrease after the first week. After this test was completed, a sample of the fuel was removed and exposed to copper at much higher surface to volume ratio using Exposure Method 1. After 24 hours at room temperature in the vertical tube, the Rancimat Induction Period fell to 1.4 hours, confirming results of the earlier tests.

After the completion of the two-pipe system test, the pump was shut down and remained off. Fuel remained stagnant in the copper pipes for 2 months to simulate leaving fuel in the copper pipes of a 2-pipe system at the end of a heating season. In this case, the fuel inside the copper pipes is not being constantly exposed to fresh oxygen. A fuel sample was taken at the end of the 2-month period and analyzed for IP, acid number, and copper content. IP was 13.5 hours, similar to the IP measured for samples in the two-pipe system test. The acid number was 0.03 mg KOH/g, and the copper concentration was 0.2 ppm.

After the test was concluded, the fuel removed from the 2-pipe set-up was left at room temperature for 1 month in an open-top glass beaker where it was exposed to fresh oxygen. After this month it was observed that a large amount of adherent insoluble had formed on the bottom of the glass beaker. The filterable and adherent insoluble were measured. The filterable insolubles were measured as 8.5 mg/100 ml, while the adherent insolubles were measured at 16.5 mg/100 ml.

Based on these results, a test protocol was developed to study the effect of summer shutdown on a fuel. A sample of No. 2 oil was prepared in copper according to Exposure Method 5 and was left at room temperature for 6 months. In this exposure technique fuel inside a long copper line is held without exposure to oxygen, as would be the case in the field. An identical sample of No. 2 heating oil was also stored in a sealed epoxy-lined metal can for 6 months to serve as a control. The IP of this fuel was measured before the start of the test. Both samples were filtered through a 0.8 micron filter before the start of the test. At the end of the 6 months, both fuels were removed from storage and analyzed for IP, acid number, and filterable particulates. The results for this are shown in Table 6. These results show that the only major difference between the two samples is the IP, which fell heavily for the sample exposed to copper.

After testing was complete on both samples, each was stored in a glass beaker with an open top. This was done to see how the fuels would degrade over time when exposed to air. Both samples were left at ambient temperature for 15 weeks. After this time, both samples were again analyzed for IP, acid number, filterable particulates and adherent insolubles. These results are shown in Table 7. While both samples saw a decrease in IP and an increase in acid number and filterable particulates, the change for the fuel stored in the copper tubing was much greater. The amount of adherent insolubles formed between the two samples was also very different, with the sample exposed to copper forming much more particulates on the glass beaker.

Table 6. Results of Simulated Summer Shutdown Test. One Sample of No. 2 Oil Exposed to Copper Using Exposure Method 5. One Sample of No. 2 Oil Stored in an Epoxy-Lined Metal Can and Not Exposed to Copper.

Sample ID	IP (hr)	Acid Number (mg KOH/g)	Filterable Insolubles (mg/100 mL)
No. 2 heating oil, (initial)	17.8		
No. 2 heating oil, (control)	15.2	0.025	0.2
No. 2 heating oil, (copper)	1.2	0.02	0.2

Table 7. Results of Simulated Summer Shutdown Test #2. One Sample of No. 2 Heating Oil Exposed to Copper Using Exposure Method 5, Then Left for 15 Weeks. One Sample of No. 2 Oil Stored in an Epoxy-Lined Metal Can and Not Exposed to Copper, Then Left for 15 Weeks.

Sample ID	IP (hr)	Acid Number (mg KOH/g)	Filterable Insolubles (mg/100 mL)	Adherent Insolubles (mg/100 mL)
No. 2 heating oil #1, (control)	8.9	0.04	0.6	11.7
No. 2 heating oil #1, (copper)	0.01	0.55	2	230

DISCUSSION

In traditional diesel engine fuel systems, the presence of copper can be problematic. The presence of copper can catalyze the formation of degradation products and produce mercaptide gels in petroleum-based diesel fuels. For these reasons, copper is not normally used or recommended in diesel engine fuel systems with petroleum distillates or biodiesel blends. Copper tubing and brass nozzles are commonly used in heating oil, however, due to lower cost and ease of manipulation and installation. The adverse impacts of copper particularly pertinent to home heating oil fuel systems are potential increases in acid number and increases in filterable or adherent solids from degradation products or their complexes with sulfur.

This study showed that the potential impacts of copper on the fuels used in today's home heating oil systems (i.e. ultra low sulfur No. 2 petroleum distillates and their blends with biodiesel) depend largely on the surface to volume ratio of the exposure to copper, access of the copper-exposed fuel to oxygen, residence time of the fuel in contact with copper, and the overall time before the fuel is burned. Potential impacts of copper can also be mitigated by the presence of storage enhancing additive packages, metal deactivators, and natural or synthetic antioxidants although this study did not specifically control or evaluate them.

The conventional fuel specifications for petroleum distillates do not have controls for acid number or filterable insolubles other than those measured as sediment in the D2709 water and sediment test. Biodiesel (B100) on the other hand, has the same water and sediment specifications but it also has specifications for acid number, oxidative reserve, and cold soak filtration which is a measure of sediments after cold-soaking the fuel using a 0.7 micron filter.

Previous research used to set the ASTM standards for biodiesel and biodiesel blends indicated fuels with high Rancimat Induction Periods would not form problematic acids or filterable insolubles over time under normal use conditions with diesel engine fuel systems. These systems, however, do not normally contain copper. Low Rancimat values did not necessarily mean problematic acids or sediments would form over time, and Rancimat values twice as high may not mean storage life is doubled, but samples that met a relatively high value consistently had low acids and sediments. This made the fuel version of the Rancimat Induction Period test a conservative method—and one that was readily available and that provided relatively quick, precise, repeatable results—so it was adopted into the biodiesel specifications and is commonly used today. This previous research was used to inform and design the testing conducted in this study to understand the long-term implications of exposure to copper in modern heating oil systems.

This testing confirmed that copper can have adverse long-term effects if the fuel is exposed to copper under high surface-to-volume ratios and then has access to oxygen without being intermingled with a larger quantity of bulk fuel. In these cases, the high contact with copper can dramatically reduce the measured oxidation reserve and accelerate the formation of acids and filterable and adherent insolubles which could result in clogged strainers, coked nozzles, and problematic system deposits. The impacts observed under these conditions were similar or less for the biodiesel blends than the No. 2 petroleum distillates, likely due to oxidation reserve requirements for biodiesel and biodiesel blends that are not present in No. 2 fuel oil. Fuels in home heating oil systems, however, are not normally subject to these conditions.

Over a normal summer shutdown, the fuel is exposed to high surface-to-volume ratios inside the copper tubing but without access to oxygen for extended periods of time. The amount of copper tubing is approximately double for the two-pipe systems, but these systems are much less prevalent in the U.S. market. The fuel is also subject to the same high surface-to-volume ratio conditions inside copper lines during normal operation, although for much shorter periods of time. In these cases of high surface-to-volume copper exposure but without access to oxygen formation of potentially problematic acids or sediments inside the copper tubing were not substantially increased, even though the exposure to copper dramatically reduced the oxidation reserve value as measured by the Rancimat Induction Period. A location that experiences a summer shutdown and uses a 2-pipe system could see fuel stability issues, as the fuel that will rapidly degrade in the presence of oxygen will be recirculated to the tank.

Fuel in the bulk 275 gallon home heating oil bulk tank is exposed to oxygen but is not normally exposed to copper for a one pipe system. A notable exception to this would be the two-pipe system, where the fuel is subjected to the high surface-area-to-volume ratio in the copper tubing going to the burner as well as in the return line back to the tank. In tests simulating a one-pipe system, a two-pipe system under recirculation, and a two-pipe system under summer shutdown, fuel did not show problematic increases in acid number or filterable insolubles, or a substantial decrease in measured oxidative reserve. This is likely due to the dilution effect of the relatively small amount of high surface-area-to-volume fuel in the copper tubes co-mingling with the bulk fuel after summer shutdown, potential passivation of the copper tubing by the fuel in general, and the relatively low contact time with copper in the fuel lines during on-going operations.

CONCLUSIONS

The practical implications of this research are that the presence of copper lines in most home heating oil systems in the U.S. should not present significant formation of acids or filterable insolubles that could adversely affect system operation over time with either ultra low sulfur No. 2 petroleum distillates or their biodiesel blends. This is likely due to the specific exposure conditions to copper in normal heating oil systems and the relatively small amounts of fuel exposed to both copper and oxygen before being burned. To the extent conditions do exist where changes over time could occur, the changes observed with biodiesel blends were similar or less than those with conventional No. 2 petroleum distillate fuel oil. This is likely due to the oxidative reserve and acid number controls present in the biodiesel and biodiesel blend specifications that are not present in traditional heating oil specifications.

An additional practical implication of this research is the significant negative impact that high surface-to-volume contact with copper has on oxidation reserve testing methodology. Oxidation reserve, measured via the Rancimat Induction Period, is used with biodiesel and its blends as a convenient and conservative surrogate for deleterious fuel changes that could happen over time. Contact with copper, however, can cause a significant reduction in the oxidative reserve testing methodology with either No. 2 petroleum distillate or its blends with biodiesel even though problematic acids or filterable insolubles may not have formed. This can particularly be the case when sampling for troubleshooting purposes or periodic quality checks from a copper line or pump bleeder valve in the field. A low Rancimat Induction Period value from a sample from a copper line or pump bleeder may not represent problematic fuel in the line or in the bulk tank. In this case use of the acid number, filterable insolubles, or taking a sample of the bulk tank would be a better means to assess whether the fuel has changed to the point of being problematic.

FUNDING SOURCES AND ACKNOWLEDGEMENTS

This work was sponsored by the National Oilheat Research Alliance and the National Biodiesel Board. The authors would like to acknowledge contributions to the experimental work by Jonathan McCutcheon, John Levey, and Robert O'Brien. We would also like to thank Steve Howell for his careful review of the manuscript and many helpful suggestions.

References

1. *Five states reach deadline for ULS heating oil.* Indoor Comfort Marketing, 2018: p. 22.
2. Mante, O.D., et al., *Evaluation of biomass-derived distillate fuel as renewable heating oil.* Energy & Fuels, 2015. **29**(10): p. 6536-6543.
3. *ASTM International, ASTM D975-18 Standard Specification for Diesel Fuel Oils*. 2018, ASTM International, West Conschohocken, PA.
4. Petroleum, B. *Long Term Storage of Diesel.* Feb 10, 2005; Available from: https://www.bp.com/content/dam/bp-country/en_au/media/fuel-news/long-term-storage-diesel.pdf.
5. Chevron. *Diesel Fuels Technical Review.* 2007 6/12/18]; Available from: https://www.chevron.com/-/media/chevron/operations/documents/diesel-fuel-tech-review.pdf.
6. Litzke, W.L., *A Guide to Fuel Performance*. 2004, National Oilheat Research Alliance.
7. Batts, B.D. and A.Z. Fathoni, *A literature review on fuel stability studies with particular emphasis on diesel oil.* Energy & Fuels, 1991. **5**: p. 2-21.
8. White, E.W., *Storage Stability of Distillate Fuels for Ships*, in *Manual on Requirements, Handling, and Quality Control of Gas Turbine Fuels ASTM STP 531*. 1973, American Society for Testing and Materials. p. 143-166.
9. *ASTM International, ASTM D396-15c, Standard Specification for Fuel Oils*. 2015, ASTM International West Conshohocken, PA.
10. Bukrejewski, P. and D. Wardzinska, *The influence of copper on the oxidation stability of commercial diesel oil.* The Archives of Automotive Engineering - Archiwum Motoryzacji, 2016. **72**(2): p. 5-14.
11. Jain, S. and M.P. Sharma, *Effect of metal contents on oxidation stability of biodiesel/diesel blends.* Fuel, 2014. **116**: p. 14-18.
12. Sarin, A., et al., *Effect of metal contaminants and antioxidants on the oxidation stability of the methyl ester of Polgamia.* Journal of the American Oil Chemists Society, 2010. **87**: p. 567-572.

13. Sarin, A., et al., *Oxidation stability of palm methyl ester: effect of metal contaminants and antioxidants.* Energy and Fuels, 2010. **24**: p. 2652-2656.

14. Knothe, R. and K.R. Steidley, *The effect of metals and metal oxides on biodiesel oxidative stability from promotion to inhibition.* Fuel Processing Technology, 2019. **177**: p. 75-80.

15. van Rheinberg, O. and H. Dirks, *Untersuchungen zur Produktqualität von Mischungen aus Heizöl EL schwefelarm und 20% Fettsäuremethylester bei der Langzeitlagerung*. 2010, Öl-Wärme-Institut: Aachen, Germany.

16. Chendynski, L.T., et al., *Influence of Copper and Metallic Alloys on the Oxidation Reaction of Commercial Biodiesel Mixture with Natural Antioxidant.* J. Braz. Chem. Soc., 2019. **30**(1): p. 90-96.

17. Kugelmeier, C.L., et al., *Corrosion behavior of carbon steel, stainless steel, aluminum and copper upon exposure to biodiesel blended with petrodiesel.* Energy, 2021. **226**.

18. Myers, K.M., R.E. Morris, and T.N. Loegel. *Assessment of the role of copper contamination on diesel fuel storage stability (Poster Presentation).* in *IASH Annual Conference*. 2017. International Association for Stability, Handling, and Use of Liquid Fuels.

19. *ASTM International, ASTM D6751-15ce1, Standard Specification for Biodiesel Fuel Blend Stock (B100) for Middle Distiallate Fuels*. 2015, ASTM International West Conshohocken, PA.

20. *ASTM International, ASTM D4625-16, Standard Test Method for Middle Distillate Fuel Storage Stability at 43 °C(110 °F)*. 2016, ASTM International West Conshohocken, PA.

21. *ASTM International, ASTM D664-18, Standard Test Method for Acid Number of Petroleum Products by Potentiometric Titration*. 2018, ASTM International West Conshohocken, PA.

IASH 2022, the 17th INTERNATIONAL SYMPOSIUM ON STABILITY, HANDLING AND USE OF LIQUID FUELS
Dresden, Germany
September 11 - 15, 2022

AN OPTIONEERING EXERCISE INTO THE POTENTIAL FOR USING ALTERNATIVE, MORE ENVIRONMENTALLY FAVOURABLE FUELS THAN CLASS A2 OR CLASS D GAS OIL.

Carl Atkinson*

EDF Energy, Barnett Way Barnwood, Gloucester GL4 3RS United Kingdom

ABSTRACT

The 2020 UK HM Government financial budget statement included an intent to review the eligibility of applications using ‘red diesel’, on the premise that reducing the number of applications would lead to environmental benefits either by consumers using less of a more expensive fuel or using an alternative, more environmentally acceptable fuel. Following the announcement, under which EDF Energy would not qualify for tax-exempt status, an optioneering exercise was conducted to assess the potential for using alternative fuels/fuel-additive combinations as an alternative to Class A2 or Class D gas oils. The review was comprehensive and captured all aspects that a change in fuel would incur, including the practicality of change, the environmental benefits both short and long term, fuel availability and resilience both short and long term, and potential future trends in fuel development.

*Corresponding Author: Chemist, Engineering Division Carl Atkinson; carl.atkinson@edf-energy.com; +44 07790005353

IASH 2022, the 17th INTERNATIONAL SYMPOSIUM ON STABILITY, HANDLING AND USE OF LIQUID FUELS
Dresden, Germany
September 11 - 15, 2022

EVALUATION OF COLD FLOW RESPONSE OF ULTRA LOW SULFUR DIESEL (ULSD) IN COMBINATION WITH RENEWABLE DIESEL AND BIODIESEL

Jason A. Lennon*, Jessica N. Grabowski, Casey Bouissey

Innospec Fuel Specialties LLC, 200 Executive Dr, Newark, DE 19702.

ABSTRACT

The use of ultra low sulfur diesel (USLD) in cold regions is greatly affected by the composition of the fuel. The most common test method used to ensure low temperature vehicle operability is ASTM D6371 - Standard Test Method for Cold Filter Plugging Point (CFPP) of Diesel and Heating Fuels. ASTM D975 (Standard Specification for Diesel Fuel Oils) provides guidance on air temperatures in different regions of the United States during winter months, which may be used to inform CFPP targets. Interest in the use of green fuels (common names for these materials are biodiesel, renewable diesel, and hydrotreated vegetable oil (HVO)) has gained interest in order to limit the increased carbon intensity associated with the combustion of petroleum derived fuels. To date, there has not been an extensive study on the CFPP responsiveness of ULSD in combination with varying levels of both biodiesel and HVO; and their response to cold flow improver (CFI) additives. A study was performed to measure the impact of combinations of ULSD, HVO, and FAME at varying levels on CFPP depression with and without the use of additives. After performing this study, we found that fuels inclusive of biodiesel had diminished CFPP responsiveness to cold flow improver than fuels blended solely with HVO. However, in either case, blends made into ULSD with either of these fuel streams were responsive to CFPP depression with the use of cold flow additives.

*Corresponding Author: Performance Products Supervisor Jason Lennon; jason.lennon@innospecinc.com; +1-302-451-1356

IASH 2022, the 17th INTERNATIONAL SYMPOSIUM ON STABILITY, HANDLING AND USE OF LIQUID FUELS
Dresden, Germany
September 11 - 15, 2022

THE IMPACT OF CHANGES TO UK LEGISLATION AFFECTING THE ELIGIBILITY OF COMMERCIAL CONSUMER'S USE OF DUTY EXEMPT DIESEL FOR ESSENTIAL OR EMERGENCY PLANT SYSTEMS.

Carl Atkinson*

EDF Energy, Barnett Way Barnwood, Gloucester GL4 3RS United Kingdom

ABSTRACT

In the UK, to indicate that a diesel fuel has not attracted the standard rate of fuel duty (tax), diesel includes a red dye/markers to denote that it is tax exempt and this fuel is available for use in a range of specific applications such as agriculture, non-public road going vehicles, construction etc. The tax-exempt list also includes all applications on power stations used for emergency power generation, emergency pumping systems and steam generating boilers. The 2020 UK HM Government financial budget statement included an intent to review the eligibility of applications using 'red diesel', on the premise that reducing the number of applications would lead to environmental benefits either by consumers using less of a more expensive fuel or using an alternative, more environmentally acceptable fuel. Later, details emerged indicating that the changes, which remove almost all eligibility for diesel fuelled power station plant systems, would take effect on 1st April 2022 and that in advance of this date any procurement of fuel not expected to be consumed before this date, must meet the non-exempt requirement. This presentation describes the impact of this legislative change and assesses the practicality of transition from red duty exempt to white duty paid diesel, the resilience of fuel supply these changes invoke and the overall impact from an environmental perspective.

*Corresponding Author: Chemist, Engineering Division Carl Atkinson; carl.atkinson@edf-energy.com; +44 07790005353

IASH 2022, the 17th INTERNATIONAL SYMPOSIUM ON
STABILITY, HANDLING AND USE OF LIQUID FUELS
Dresden, Germany
September 11 - 15, 2022

THE VERSATILITY OF THE GC-VUV ANALYZER

Dan Wispinski*, Sean Jameson

VUV Analytics, Inc. 1500 Arrow Point Drive, Cedar Park Texas 78613 USA

ABSTRACT

The application of gas chromatography vacuum ultraviolet (GC-VUV) spectroscopy has gained acceptance as an analytical technique in the modern petroleum laboratory. The adoption of GC-VUV has been due to the flexibility and versatility of standard methods, the recognition by regulatory bodies and the adaptability to complex R&D functionality. In this rapidly changing world of liquid fuels the GCVUV technique is strategically positioned to encompass a broad range of conventional, non-conventional and renewable fuels.

The VUV Analyzer Platform provides an automated, fast and accurate approach to analyze products ranging from propane (C3) to triacontane (C30). The platform was developed to seamlessly switch between four standard analysis methods using the same hardware and software.

Standard method ASTM D8071 provides a regulatory approved automated, fast, and accurate approach to alternative multiple gasoline methods for benzene, ethanol, olefins and aromatics. Standard method ASTM D8267 provides an ASTM D1655 and DefStan 91091 specification approved method for aromatics in jet fuel. Standard method ASTM D8368 analyzes biodiesel and aromatic content in diesel fuels. The latest method, prEN17734 analyzes EN228 petrol for saturates, aromatics, benzene, and oxygenates including ethers. This paper will describe the latest ASTM and CEN developments for these four methods and how they integrate into one automated system - the VUV Analyzer platform.
In addition to the standardized methods, the determination of advanced chemical and physical properties of petroleum products by GCVUV will be introduced.

*Corresponding Author: Standard Methods Development Manager Dan Wispinski; dan.wispinski@vuvanalytics.con; +1 512 922-6400

IASH 2022, the 17th INTERNATIONAL SYMPOSIUM ON STABILITY, HANDLING AND USE OF LIQUID FUELS
Dresden, Germany
September 11 - 15, 2022

APPLICATION OF GCXGC-VUV FOR THE ANALYSIS OF SYNTHETIC FUELS

Dr. Max Jennerwein*, Dr. Klaus Tröster

ASG Analytik-Service AG, Trentiner Ring 30, Neusaess 86356 Germany

ABSTRACT

Synthetic fuels represent a genuine challenge for instrumental analysis methods, independent from their source (e.g. FT or MtG conversion, etc.) or their purpose (e.g. Gasoline, Diesel Fuel, Jet Fuel, etc.). The scope of application of established methods is often not suitable to deal with strongly diverging compositions of Synfuels or regarding the limits of detection of certain target compounds, respectively. Numerous analytical questions cannot even be answered since adequate methods do not yet exist. Thus, in order to analyze such novel fuels and feedstocks and to guarantee their quality, the development of adequate analytical techniques is inevitable. In the past years and decades techniques like comprehensive two-dimensional GC, mostly in combination with mass spectrometry or FID, and vacuum-UV detection for one-dimensional GC found their why in lab routine for the separation of complex samples. Unfortunately, those techniques are often not sufficient to answer certain questions completely. Some of these questions can be answered by the application of a hyphenation of both techniques. Previous publications and presentations have shown the successful application and advantages of GCxGC-VUV. In the present study the opportunities of GCxGC-VUV analysis methods and data processing using different types of software were investigated and evaluated.

*Corresponding Author: Dr. Max Jennerwein; max.jennerwein@asg-analytik.de; 082145042320

IASH 2022, the 17th INTERNATIONAL SYMPOSIUM ON STABILITY, HANDLING AND USE OF LIQUID FUELS
Dresden, Germany
September 11 - 15, 2022

SYNTHETIC AVIATION JET FUELS, RELATIVE PERMITTIVITY & AIRCRAFT GAUGING SYSTEMS

Ian Mylrea[1]*, Paul Pilkington[2]

[1]Stanhope-Seta Limited London Street, Chertsey, Surrey KT16 8 AP United Kingdom
[2]Airbus UK ltd. Pegasus House, Aerospace Avenue, Filton, Bristol BS34 7PA

ABSTRACT

Fuel gauging systems use permittivity to estimate the mass of fuel on board jet aircraft during flight. Gauge accuracy relies on the relationship between temperature, permittivity and density of the jet fuel being the same as the historic norm. For fuels derived from alternative sources this relationship can be different than for fuels derived from crude oil. Measuring permittivity is difficult, and as noted in the CRC report, there is need for a dedicated method to measure jet fuel. This paper describes some new fuels in terms of permittivity and describes a new test method published by the UK Energy Institute for measuring the property.

*Corresponding Author: Mr Ian Mylrea; im@stanhope-seta.co.uk; +447854044200

IASH 2022, the 17th INTERNATIONAL CONFERENCE ON STABILITY, HANDLING AND USE OF LIQUID FUELS
Dresden, Germany
September 11-15, 2022

USING A FIELD PORTABLE LC FOR DETECTING CORROSION INHIBITOR/LUBRICITY IMPROVERS AND METAL DEACTIVATOR ADDTIVIES

Thomas N. Loegel[1]*, Iwona Leska [2]

[1]U.S. Naval Research Laboratory, 4555 Overlook Avenue S.W., Washington, D.C. 20375 USA.
thomas.loegel@nrl.navy.mil
[2]NOVA Research, Inc. 1900 Elkin Street, Suite 230, Alexandria, Virginia 22308 USA

ABSTRACT

This presentation will show ongoing efforts focused on the development of a new high-performance liquid chromatograph (HPLC) method utilizing an ultra violet-visible light absorbance photodiode array (UV-vis PDA) detector for the determination of concentration of monomeric and dimeric fatty acids used as corrosion inhibitor/lubricity improvers (CI/LI) in fuels and presence and concentration of metal deactivator additive (MDA). Dimers of fatty acids, especially of linoleic and oleic acid, are a class of compounds that are widely used in many practical applications including paints and coating and can be used to prevent corrosion and/or decrease wear between metal-to-metal contact surfaces. The use of HPLC with a UV-Vis detector for the analysis of fatty acids is very challenging due the lack available chromophores in their molecular structures which typically would require the use of very low UV wavelengths (<200 nm) which is not practical for portable systems. The goal of this effort is to see if any CI/LI additives can be detect in their native form and to explore different derivatization strategies to add chromophores for these compounds along with various chromatographic conditions. We have shown improvements to the MDA separation method through the use of a mobile phase modifier trifluoroacetic acid. This chromatographic additives also shows promise in the separation of CI/LI additives for both monomeric and dimoraic formulations. Additionally, results of esterification or amidification derivatization strategies are shown.

*Corresponding Author: Thomas Loegel; thomas.loegel@nrl.navy.mil; 202-404-1157

IASH 2022, the 17TH INTERNATIONAL CONFERENCE ON STABILITY, HANDLING AND USE OF LIQUID FUELS
Dresden, Germany
11 – 15 September 2022

DETERMINATION OF HYDROPEROXIDES IN LIQUID FUELS: A SYSTEMATIC COMPARISON OF TITRATIONS AND ABSORPTION METHODS AND THEIR COUPLING TO HPLC

Ryma Benrabah[1], Zaki El Sayah[1], Pierre-Alexandre Glaude[1], Philippe Arnoux[1], Baptiste Sirjean[1]
[1]*Université de Lorraine, CNRS, LRGP, F-54000 Nancy, FR.* ryma.benrabah@univ-lorraine.fr

ABSTRACT

Hydroperoxides are major products in the autoxidation of fuels and play a central role in oxidation kinetics. Their presence affects the oxidation rate, even at low concentrations, and, in turn, the thermal stability of fuels. Standard techniques for the quantification of hydroperoxides (ROOH) in fuel involve the reduction of ROOH by I^-, yielding iodine (I_2), which can be quantified by titration. These ASTM methods are known to have poor reproducibility, inadequate lower detection limits and limited dynamic ranges. In this work, several analytical methods for the quantification of hydroperoxides in surrogate fuels have been systematically compared for a known concentration of ROOH: titration of iodine with thiosulfate (ASTM D3703) or triphenylphosphine (TPP) solutions with colorimetric and potentiometric methods and quantification of iodine by UV-visible absorption. *n*-Decane was used as the surrogate fuel and tert-butyl hydroperoxide as the controlled ROOH content. Our systematic study shows that the ASTM method based on colorimetric titration using thiosulfate in aqueous solution is the least accurate because of the non-miscibility of water and *n*-decane, which even makes the UV-visible spectroscopy method ineffective. The other titration method in the literature, using TPP dissolved in *n*-decane as the titrant does not have any miscibility problem. Potentiometric titration is found to be more reproducible and accurate than colorimetric titration. The UV-visible absorption method allows reaching lower detection limits. We show that the combination of the potentiometric titration of iodine with TPP and UV-visible absorption allows for determining ROOH concentrations with low detection limits and a large dynamic range. The advantages of both methods have been used to develop a new analytical method based on their coupling in HPLC.

KEYWORDS: hydroperoxides, HPLC, autoxidation

INTRODUCTION

The thermal oxidation stability of fuels is a key factor in the transportation and the energy fields. This stability can be jeopardized by the slow oxidation of fuels occurring during their storage, manipulation, and use in engines. One of the main chemical markers of fuel oxidation is the formation of hydroperoxides[1,2]. The hydroperoxide content is therefore a key factor to characterize the degree of oxidation of a fuel and its ability to reach the expected specifications for a given engine application. An ATSM standard (D3703) has been used for several years but has been shown to be poorly reproducible.[3,4,5] Indeed, this norm quantifies the total hydroperoxides content of an organic phase (hydrocarbons and organic components) with a method using an aqueous phase, which could lead to miscibility problems. Moreover, many works have been recently performed to determine new methods to better quantify hydroperoxides. These approaches are still based on iodometry.[1,6,7]

Recently, *Roohi and Rajabi* [6] proposed an iodometric titration, based on a visual change of the color to find the endpoint (colorimetric approach). Triphenylphosphine (TPP) is the titrant for the I_2 produced from the reaction of ROOH with iodide ions I^- in an acidic organic phase. In the first step, ROOH is reduced with iodide ions from potassium iodide in absolute ethanol in presence of acetic acid to produce a yellow solution of I_2 as presented in the equation below:

$$ROOH_{(decane)} + 2H^+_{(organic)} + 2I^-_{(organic)} \rightarrow I_{2(organic)} + ROH_{(organic)} + H_2O \quad (1)$$

Then, triphenylphosphine is added to react with I_2 producing an uncolored complex $[TPPI]^+[I]^-$ as shown below:

$$I_{2(organic)} + TPP_{(decane)} \rightarrow [TPP]^+[I]^-_{(aq)} \quad (2)$$

The iodine concentration can be determined from the quantity of TPP added and the endpoint is found when the mixture turns transparent.

Roohi and Rajabi[6] proposed a colorimetric titration, but the equivalence point can also be found using potentiometry.[7] In this work, we used this method (TPP titrant/potentiometry) for the first time in the literature.

The ASTM D3703 standard uses thiosulfate in aqueous phase as the titrant that reacts with I_2, according to:

$$I_{2\,(organic)} + 2\,S_2O_3^{2-}{}_{(aq)} = S_4O_6^{2-}{}_{(aq)} + 2\,I^-_{(aq)} \quad (3)$$

The equivalence point can be determined by colorimetry and potentiometry titrations. The titration methods offer many advantages, including simplicity, not being time and labor-consuming, accuracy at relatively low concentrations, and, in most cases, easy endpoint determination[3,8].

Alternatively, the absorbance of the solution formed in equation (1) can be measured by spectroscopy. Indeed, triiodide ion I_3^- is formed in the presence of I_2 and I^- (equation 4). In absolute ethanol, triiodide has one of a maximal absorbance at a wavelength of 362 nm.

$$I_2 + I^- \rightleftharpoons I_3^- \quad (4)$$

The absorbance depends on the dilution factor of the solution.[8] Since reaction (1) is total, $[I_2]$=[ROOH]. The Beer-Lambert law is used to link the absorbance of I_3^- and its concentration:

$$A = \varepsilon \, l \, C \qquad (5)$$

A: absorbance of the solution at 362 nm
ε: molar extinction coefficient of I_3^- (cm^{-1} L mol^{-1})
l: optical length of quartz spectroscopy cell (cm)
C: concentration of I_3^- (mol L^{-1})

All the above-mentioned methods are usually performed in laboratory flasks (for titration) and cuvettes (UV-Vis absorption). In this study, we will show that reaction (1) and UV-Vis absorption can be combined in a post-column reactor within a High-Performance Liquid Chromatography (HPLC) device. A photodiode array (PDA) detector can also be used to follow the absorbance and the transmission of a colored solution at different wavelengths.[9,10] The high added value of HPLC in this method is its ability to separate the ROOH from other compounds in the oxidized fuel.[10]

This study aims to determine the detection limits of each method and compare their reproducibility and performance for the determination of the total hydroperoxide content of a surrogate fuel containing a known quantity of hydroperoxides.

EXPERIMENTAL SECTION

Reagents

Potassium iodide (KI), anhydrous sodium thiosulfate ($Na_2S_2O_3$; 0.1 N in aqueous solution), and absolute ethanol were purchased from VWR Chemicals. Butylated hydroxytoluene (BHT) (≥ 99 %), triphenylphosphine (TPP; ≥99 %), acetic acid (≥ 99.8 %), tert-butyl hydroperoxide in *n*-decane (TBHP / C_{10}; 5.5-6.5 M) 2,2,4-trimethylpentane (isooctane; ≥ 99.5 %) were purchased from Sigma Aldrich. *n*-Decane was provided by ACROS Chemicals. The surrogate fuel is *n*-decane (C_{10}), which is a usual surrogate of diesel and jet fuels[11,12], and the total ROOH content is represented by the addition of TBHP. To evaluate the performance of the different analytical methods, TBHP/*n*-decane solutions were prepared with decreasing concentrations of TBHP obtained by dilutions.

Preparation of the solution

The first reaction used for all types of titrations and the UV-vis spectroscopy is based on the protocol from *Roohi and Rajabi*[6]: 0.5 g of KI and 0.5 mL of acetic acid in 25 mL of absolute ethanol were heated for 5 min in slow agitation (400 rpm) to 68°C. Then, a 1 mL sample of TBHP/*n*-decane blend was added and the mixture was heated at 68°C, at the same agitation, for 15 min to ensurethat reaction (1) was complete. The solution was kept warm during the subsequent manipulations.

Colorimetric titration

A usual setup for titration was used. When TPP is the titrant, a 25 mL burette filed with a 0.08M solution of 2.1g of TPP and 0.01g of BHT in *n*-decane is used to titrate iodine and the endpoint of titration was determined when the solution became transparent. When $Na_2S_2O_3$ (0.025M) in water is the titrant, the equivalence point is also reached when the solution becomes transparent. Each experiment was done twice.

Potentiometric titration

An automated potentiometric titrator (Mettler Toledo G10S Compact Potentiometric Titrator) was used, equipped with a Pt indicator electrode vs. an AgCl / Ag reference electrode. For analysis, 1 ml of the TBHP / C_{10} solution was put in the cell stirred at 20%, Two titrants were used: TPP (0.08M-8mM-0.8mM) in *n*-decane and $Na_2S_2O_3$ (0.025M) in water. Each experiment was repeated 3 times.

UV-spectrophotometric titration

A sensitive UV-visible double beam spectrophotometer UV-3600 Shimadzu was used to determine the precise amount of peroxides in the TBHP/C_{10} solutions. The absorbance of triiodide formed by I_2 with excess iodide ions was measured at 362 nm.[10] Measurements were done in a 4 mL quartz cell of dimensions 10×10×40 mm.

HPLC titration

An HPLC device combining a separation column in the normal phase (silica) and a post-column reactor was adapted from the work of *Cullis et al.*[10] A HPLC Shimadzu Nexera XR was used at a flow of 1 mL per minute of solvent (isooctane: absolute ethanol 95:5). 1 μL of the sample TBHP/C_{10} was injected using an autosampler through the separative column. At the exit of this column, the fractions were mixed with a 10% acetic acid in absolute ethanol (0.2 mL/min) and a 0.125 g/L solution of NaI in absolute ethanol (0.2 mL/min). This mixture passed through a reactive column packed with 30 μm glass beads, heated at 68°C with a residence time of 80 seconds.[10] A chromatogram was obtained at every wavelength between 190 nm and 800 nm, which allows hydroperoxide quantification in the sample after calibration. Figure 1 presents a schematic representation of the HPLC device.

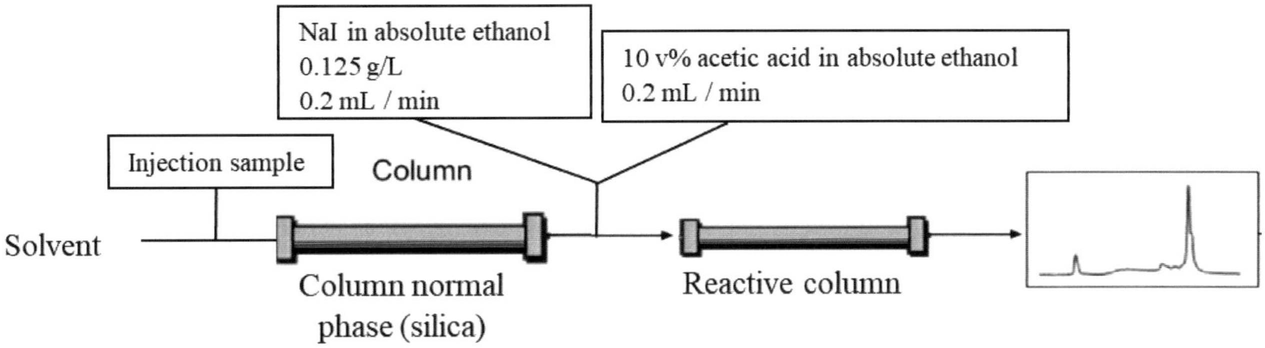

Figure 1: Representation of the HPLC device used in this work

RESULTS AND DISCUSSION

The colorimetric and the potentiometric methods using two different titrants are compared, as well as UV-spectrophotometric measurement, to determine the concentration of total peroxide content in the TBHP / C_{10} solutions. HPLC using a PDA detector (spectroscopy UV-vis) is also used.

Comparison and limits of colorimetric titration using TPP and aqueous $Na_2S_2O_3$

Figure 2 presents a comparison between the experimental and the theoretical (vendor certifications for the stock solution) concentrations of TBHP in *n*-decane. TBHP concentrations were measured by colorimetry with two titrants: a solution of TPP in *n*-decane and a solution of sodium thiosulfate in the aqueous phase. The results of Figure 2 are also given in Tables 1 and 2.

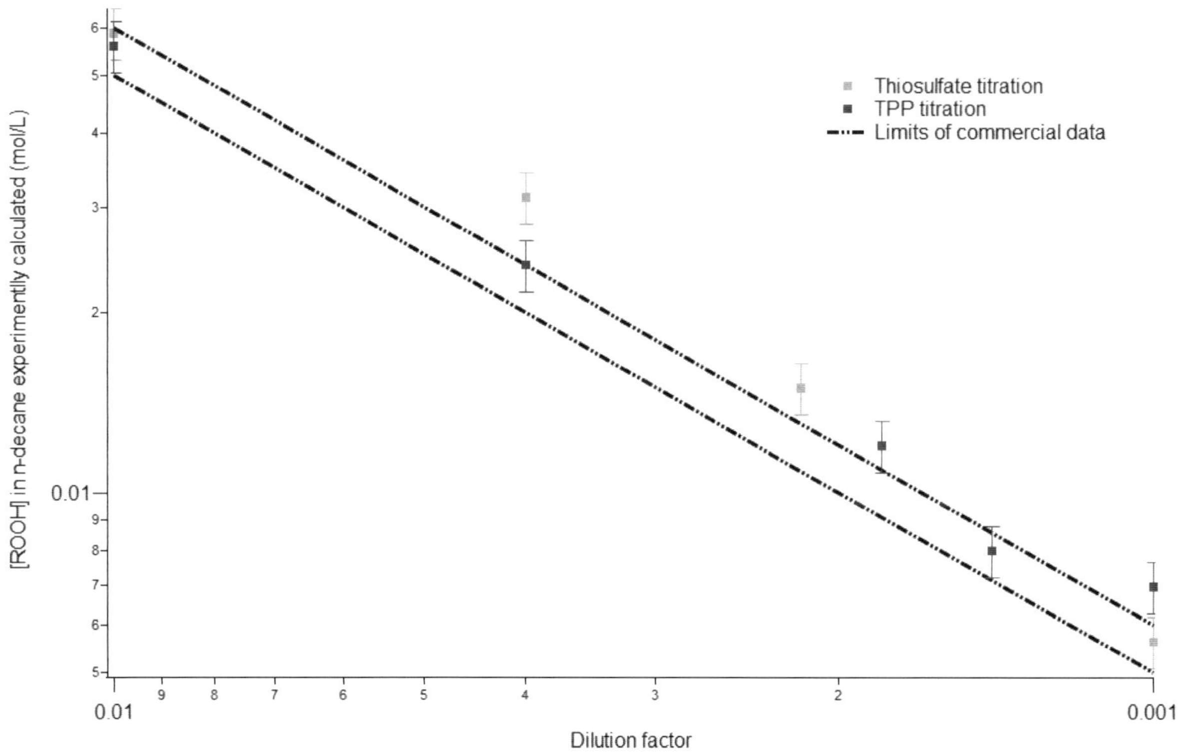

Figure 2: Comparison of experimental and theoretical (commercial certifications) TBHP (ROOH) concentrations as a function of dilution. Experimental method: colorimetric titration with thiosulfate in water and TPP in *n*-decane titrants.

Sigma-Aldrich indicates that the concentration of TBHP in *n*-decane (our stock solution) ranges between 5.5 and 6.5M. It is worth noting that their measure of the TBHP concentration is performed by iodometry using the thiosulfate and colorimetric method. Therefore, we decided to represent these limits on the graph (black dashed straight line) to ensure that the concentration found with each solution remains in these data.

The comparison between the results obtained by colorimetric titration with TPP in *n*-decane (mean relative deviation (MRD) with the theoretical value of 14.8%) and sodium thiosulfate in water (MRD of 20.7%) shows better results between the experimental and commercial concentrations in the case of the titration using TPP. All the colorimetric titration results seem to show that the TBHP concentration is within the upper limit of the vendor certifications. It is important to note that the main difference in the titration with an aqueous titrant and an organic titrant lies in the formation of two phases which brings accuracy issues in the colorimetric (and potentiometric) methods. Figure 3 illustrates the non-miscibility problems observed when thiosulfate is used as the titrant.

Table 1: Molar concentration and titration volume of ROOH in TBHP / C_{10} after iodometric titration with TPP in *n*-decane titrant.

TBHP / C_{10} Dilution factor	[TBHP / C_{10}] buffer (mol. L^{-1})	[TPP/C_{10}] (mol. L^{-1})	$V_{titration}$ (mL)	[ROOH] (mol. L^{-1})
100	5.50E-02	0.08	0.80	6.40E-02
250	2.20E-02	0.08	0.30	2.40E-02
550	1.00E-02	0.08	0.15	1.20E-02
700	7.86E-03	0.08	0.10	8.00E-03
1000	5.50E-03	0.08	8.7E-02	6,97E-03

Table 2: Molar concentration and equivalence point volume of ROOH in TBHP / C_{10} after iodometric titration with $Na_2S_2O_3$ / water solution.

TBHP / C_{10} Dilution factor	[TBHP / C_{10}] buffer (mol. L^{-1})	[$Na_2S_2O_3$ / water] (mol. L^{-1})	$V_{titration}$ (mL)	[ROOH] (mol. L^{-1})
100	5.50E-02	0.025	4.70	5.88E-02
250	2.20E-02	0.025	2.50	3.13E-02
460	1.00E-02	0.025	1.20	1.50E-02
700	7.86E-03	0.025	0.64	8.00E-03
1000	5.50E-03	0.025	0.45	5.63E-03

$Na_2S_2O_3$

TPP

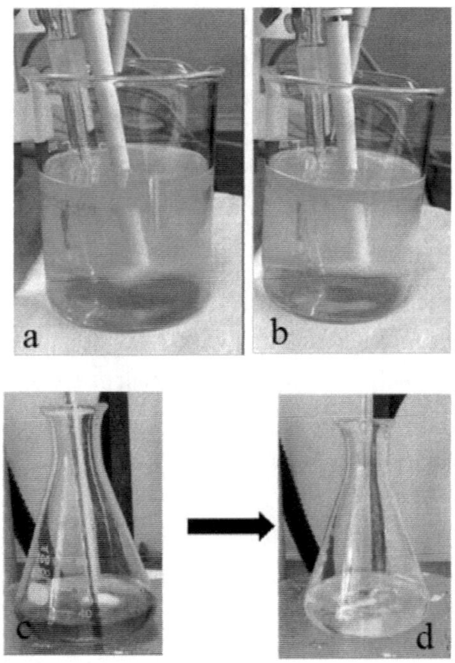

Figure 3: Pictures representing the different states of colorimetric titration (before the equivalence volume (a, c) and after (b,d) of iodine using an aqueous solution of $Na_2S_2O_3$ and using TPP.

Figures 3b and 3c clearly show that two phases are involved when water is involved in the titration. The colorimetric determination of the endpoint can even be challenging as illustrated in Figure 3c.

Comparison and limits of potentiometry using TPP and aqueous $Na_2S_2O_3$

Potentiometry has also been used for the determination of the equivalence point in thiosulfate and TPP titrations. The results using two different titrants are shown in Figure 4 and detailed in Table 3. TPP in *n*-decane titrant (0.08 M-8mM and 0.08mM) was used since the volume of a droplet in the potentiometer is small compared to the one in the burette.

Table 3: Molar concentration and equivalence point volume of ROOH in TBHP / C_{10} after potentiometric titration with organic TPP / C_{10} solution.

TBHP / C_{10} Dilution factor	**[TBHP / C_{10}] buffer (mol. L^{-1})**	**[TPP/C_{10}] (mol. L^{-1})**	**$V_{titration}$ (mL)**	**[ROOH] (mol. L^{-1})**
10	5.50E-01	0.08	7.79	6.23E-01
100	5.50E-02	0.08	8.0E-01	6.40E-02
250	2.20E-02	0.08	2.9E-01	2.32E-02
550	1.00E-02	0.08	1.6E-01	1.28E-02
700	7.86E-03	0.08	1.00E-02	8.18E-03
1000	5.50E-03	0.08	8.72E-02	6.97E-03
2500	2.40E-03	8.00 E-03	6.16E-01	4.93 E-03
5000	1.20E-03	8.00 E-03	3.60E-01	2.88 E-03
10000	6.00E-04	8.00 E-03	1.07E-01	7.17 E-04
500000	1.20E-04	8.00 E-04	2.54	2 E-04
700000	8.57E-05	8 E-04	2.00	7.14 E-05
1000000	6E-05	8 E-04	1.00	5.0 E-05

The experimental results of the quantification of peroxides in the different samples of TBHP in *n*-decane and the commercial concentrations agree well (MRD of 32.7%), as shown in Table 3. The limit of detection of this method is considerably lower than the colorimetric approach with the same reagents and titrants. Figure 2 presents the variation of the different concentrations obtained by potentiometric titration with TPP titrant, within the range of concentration given by the vendor.

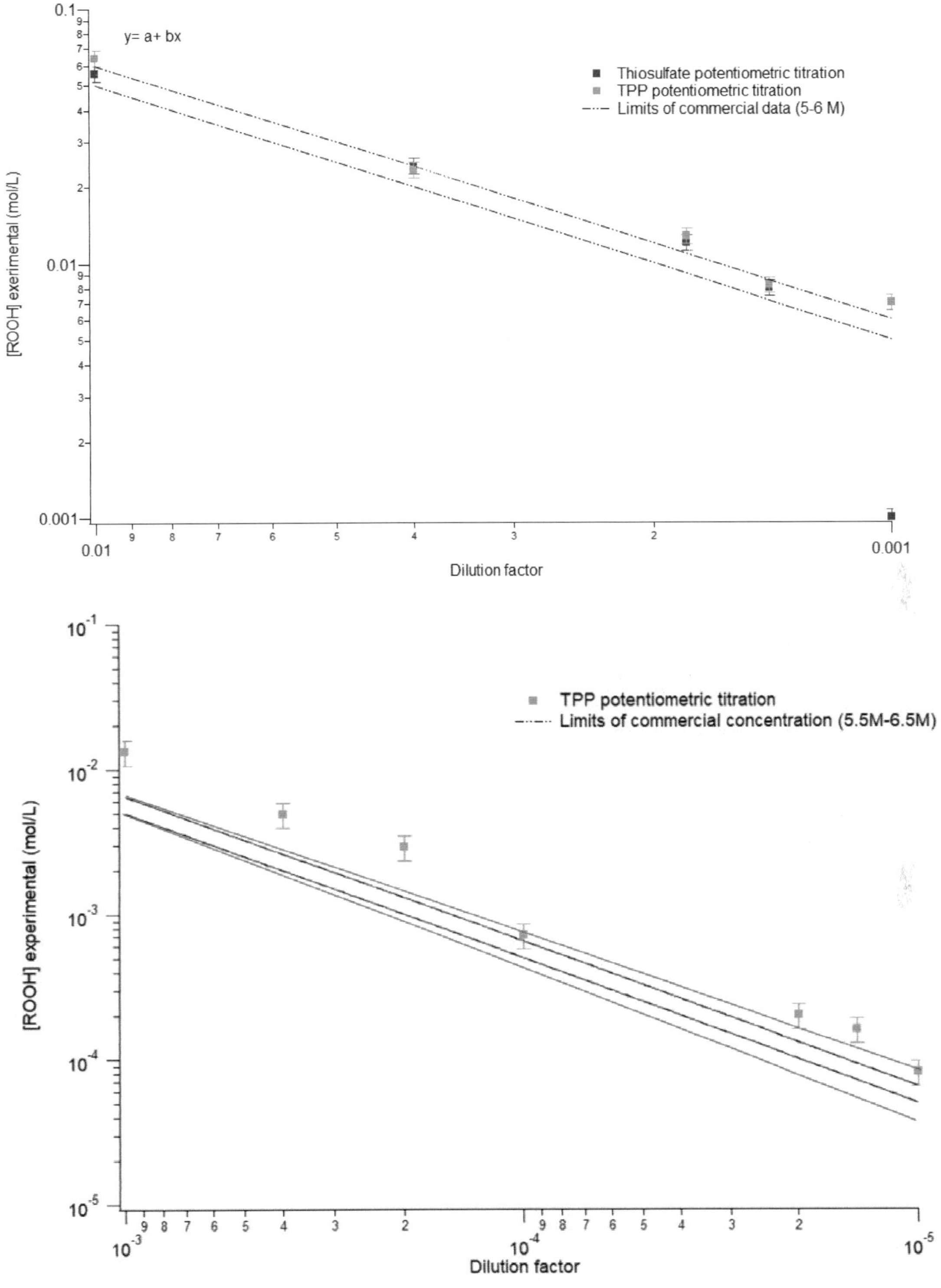

Figure 4: Comparison between experimental and theoretical (vendor certification) TBHP concentrations. Potentiometric titration with TPP and thiosulfate titrants. Theoretical concentrations (black lines 5,5-6,5M) and the corrected concentration due to repeated dilutions (red lines) are represented.

Figure 3 shows the agreement between the concentrations obtained experimentally and theoretically of the different TBHP in *n*-decane, highlighting the efficiency of the potentiometric titration method with TPP in *n*-decane for the quantification of hydroperoxides in an organic phase. ROOH concentrations as low as 5 10^{-5} mol L^{-1} can be reached.

The same potentiometric method can be used with $Na_2S_2O_3$ (0.025 mol. L^{-1}) in water. It was done in this work to see the effect of an aqueous phase on the potentiometry. The results are shown in Table 4.

Table 4: Molar concentration and titration volume of ROOH in TBHP / C_{10} after potentiometric titration with $Na_2S_2O_3$ in water.

TBHP / C_{10} Dilution factor	**[TBHP / C_{10}] buffer (mol. L^{-1})**	**[$Na_2S_2O_3$ / water] (mol. L^{-1})**	**$V_{titration}$ (mL)**	**[ROOH] (mol. L^{-1})**
100	5.50E-02	0.025	4.44918	5.56E-02
250	2.20E-02	0.025	2.34941	2.94E-02
460	1.19E-02	0.025	1.9332	2.42E-02
550	1.00E-02	0.025	1.097	1.37E-02
1000	5.50E-03	0.025	0	0

The results of the potentiometric titrations of 4 samples of TBHP represented in Table 4 show convergent experimental and theoretical concentrations. However, only a high concentration of peroxides can be measured as the measure fails at ROOH concentration below 1 10^{-2} mol L^{-1}. This is due to the non-miscibility observed in Figure 3.

Comparison and limits of potentiometry and UV-vis spectroscopy

Absorbance is known to poorly follow linearly Beer-Lambert law when OD=1. The results presented in Figure 5 show that the absorbance of the mixture containing I_3^- is linear with its dilution so its concentration for the sample diluted more than 10^3. The linearity is shown with a correlation coefficient of 0.99751. Despite this absorbance, the concentration calculated thanks to the potentiometric titration is less linear. It can be explained by the different uncertainty of repeated dilutions of TBHP stock solution (5.5-6.5M).

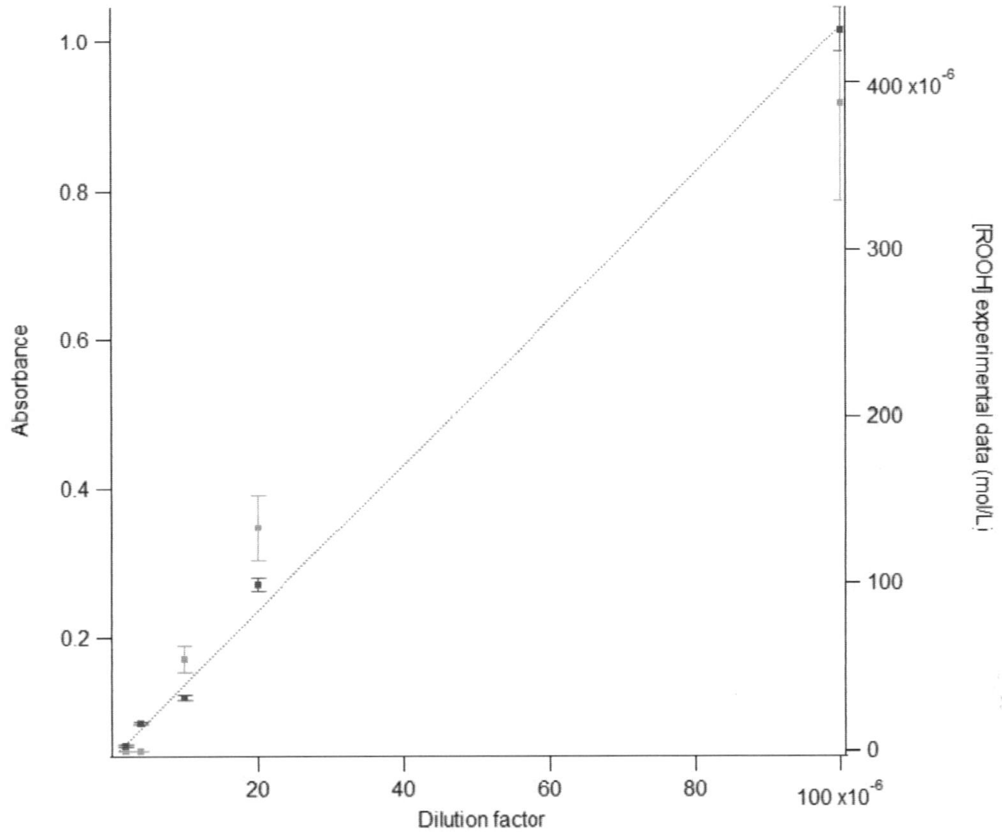

Figure 5: Comparison of absorbance of the reactive solution and the concentration obtained using potentiometric titration with TPP (green) and the Beer-Lambert law (purple).

Table 5: Experimental concentration of ROOH in TBHP / C_{10} after potentiometric titration with TPP solution and the absorbance of the solution before the titration

TBHP / C_{10} Dilution factor	**[ROOH] experimental TPP/C_{10} potentiometry (mol/L)**	**Absorbance**
1.00E-03	1.97E-03	2.95E+00
4.00E-04	1.64E-03	2.93E+00
2.00E-04	9.61E-04	2.26E+00
1.43E-04	2.85E-04	9.80E-01
1.00E-04	3.87E-04	1.02E+00
2.00E-05	1.33E-04	2.71E-01
1.00E-05	5.45E-05	1.19E-01
4.00E-06		8.60E-02
2.00E-06		5.50E-02
1.43E-06		4.10E-02
1.00E-06		4.40E-02

The absorbance technique seems to be more reproducible than potentiometric since the standard deviation of each point is less than 5%. The error in the successive dilution of the stock solution of TBHP seems to be linear enough to not affect the linearity of absorbance due to Beer-Lambert law.

Results of Table 5 show that lower concentration (up to 1.0 10^{-6} mol L^{-1}) of TBHP can be measured experimentally compared to the potentiometric titration with TPP. We also see that a higher concentration of ROOH cannot be measured by this absorption method because the absorbance exceeds 1 for concentrations above about 3.0 10^{-4} mol L^{-1}.

Quantification of hydroperoxides using HPLC device

Experiments consisted of the injection of 10 μL or 40 μL of a solution of TBHP in eluent (isooctane: absolute ethanol 95:5) inside the separative column. PDA detector records a chromatogram shown in Figure 6. Absolute ethanol was used because of its resistance to autoxidation. Isopropanol, which is commonly used in this type of experiment, has the property of easily autoxidizing which produces hydroperoxides[6]. A new solution of absolute ethanol was used every day and the baseline check was done each time. For these results, the baseline check showed an error ± 0.497 mAU. We can determine a peak if its height is at least 10 times higher than this error.

Table 6: Experimental absorption of ROOH in TBHP / iso-C_8 after injection in HPLC after the reactive column.

TBHP / iso-C_8 Dilution factor	Volume of injection (μL)	Area (mAU)	Intensity (mAU)
1.00E-04	10	8639.526	643.323
1.00E-06	40	8855.279	201.097
1.00E-07	40	3767.283	114.99
1.00E-08	40	165.309	8.2725

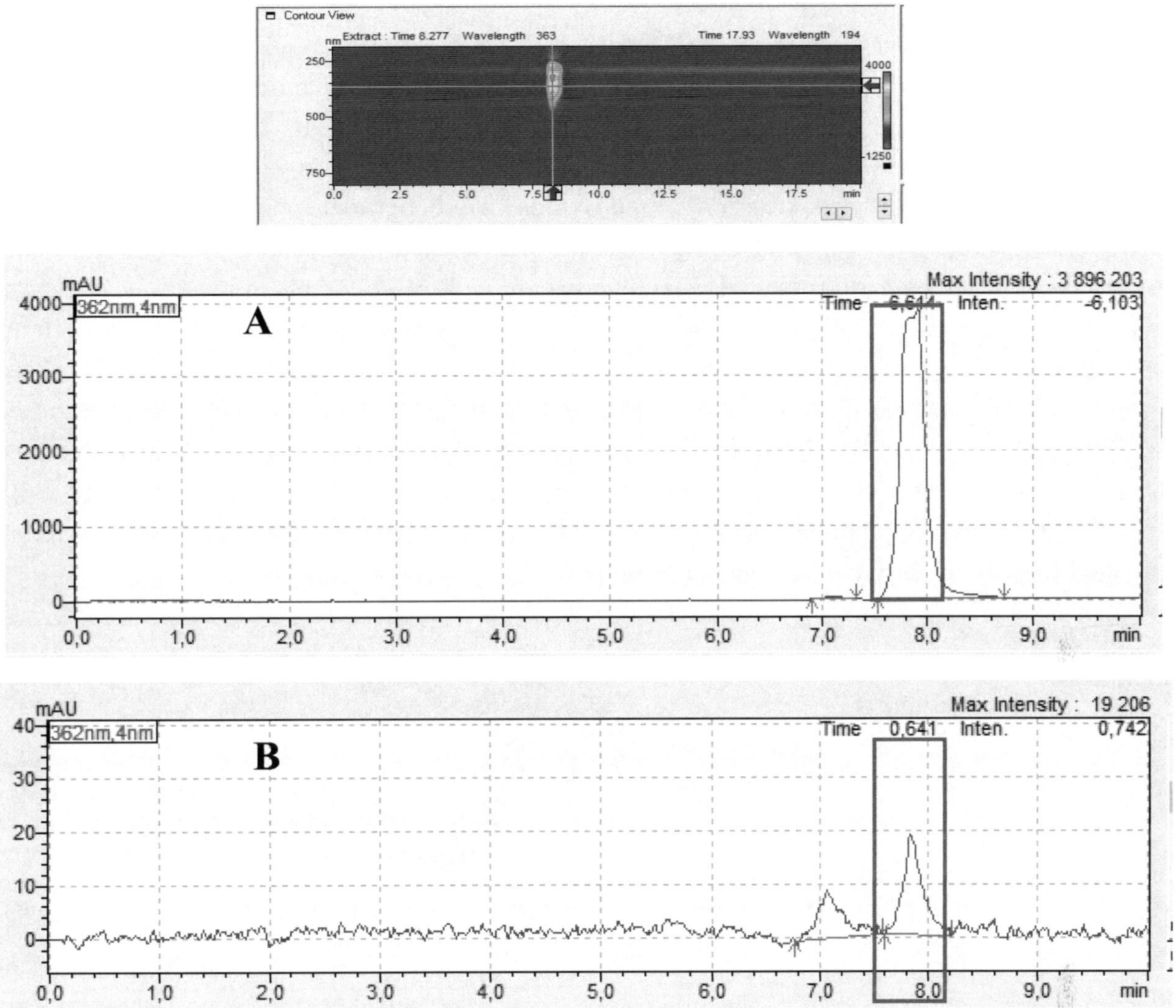

Figure 6: Chromatogram of TBHP (framed in blue) at 362nm A: diluted 10^4 times (10 µL injected at 1 mL/min for 9.2 min) B: diluted 10^8 times (40 µL injected at 1 mL/min for 15min)

The results presented in Figure 6 show that the HPLC set-up used can detect I_3^- which result from a reaction between TBHP and iodide. We can detect a TBHP solution diluted 10^8 fold (Table 6).

The presence of peaks at different retention times results from the difference of the quantity of TBHP injected in the HPLC. Moreover, this detection is reproducible, each experiment was repeated at least 2 times and the deviation is generally 1.45 mAU which is adequate. In Table 6, we can observe that the area and the height of the peaks decrease with the dilution factor. However, this decrease is not linear, which can be explained by the uncertainty of multiple dilutions of TBHP the same as the uncertainties using potentiometric quantification of hydroperoxide.

This HPLC setup quantifies ROOH in solution by determining the concentration of I_3^- using an easy and reproducible method. This method will be improved to have a stable retention time for ROOH to be able at term to isolate and quantify ROOH in a mixture of oxidized components.

CONCLUSIONS

This work aimed to determine the detection limits of different techniques quantification of hydroperoxides in fuels. The standard ASTM method is known to be little reproducible. Colorimetry, potentiometry and UV adsorption spectroscopy were compared to quantify iodine formed during the iodometric titration of peroxides. To better determine the limits of each technique, a standard solution of TBHP in *n*-decane was diluted. Potentiometric and colorimetric titrations were performed at 70°C. These techniques are well reproducible at high concentrations. The potentiometric method gives reproducible results until a concentration of 10^{-4} mol/L. Below that value, experimental uncertainties take a larger place and biased the results. On the other hand, UV-vis absorption spectroscopy at 362 nm allows to quantify iodine and indirectly hydroperoxides in a reproducible way until 10^{-6} mol/L. The results obtained with the HPLC, coupled with a post-colomn iodometric reactor, are very promising because we can detect concentration until 10^{-8} mol/L. HPLC is also interesting because of the possibility of this device to separate hydroperoxides in an aged fuel and quantify their concentration in a more complex mixture.

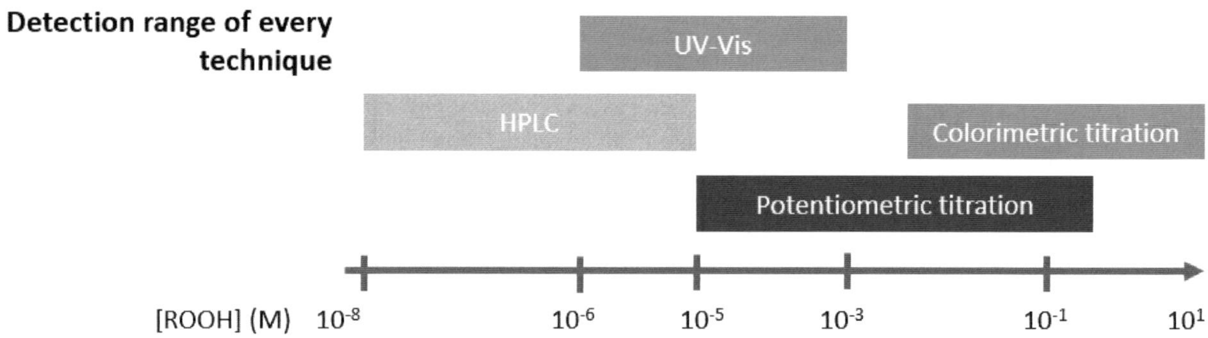

Figure 7: Detection range of each method in organic media mentioned in this study

ACKNOWLEDGMENTS

The project leading to this application has received funding from the European Research Council (ERC) under the European Union's Horizon 2020 research and innovation program (grant agreement No 101003318). This work was also financially supported by the BioACe project, grant ANR-18-CE05-002 of the French National Research Agency, and the Grand-Est region.

REFERENCES

1. West, Z. J. Studies of Jet Fuel Autoxidation Chemistry: Catalytic Hydroperoxide Decomposition & High Heat Flux Effects. , PhD dissertation, University of Dayton, Dayton, US, 2011

2. Jia T, Zhang X, Liu Y, et al. A comprehensive review of the thermal oxidation stability of jet fuels. *Chem Eng Sci*. 2021;229.

3. West, Z. J.; Zabarnick, S.; Striebich, R. C. Determination of Hydroperoxides in Jet Fuel via Reaction with Triphenylphosphine. Ind. Eng. Chem. Res. 2005, 44 (10), 3377–3383.

4. ASTM Standard D3703, " Standard - Standard Test Method for Hydroperoxide Number of Aviation Turbine Fuels, Gasoline, and Diesel Fuels" ASTM International, 2017, DOI: 10.1520/D3703-18, www.astm.org.

5. ASTM Standard E298-17a, " Standard Methods for Assay of Organic Peroxides.," ASTM International, 2017, DOI: 10.1520/E0298-17A, www.astm.org.

6. Roohi H, Rajabi M. Iodometric Determination of Hydroperoxides in Hydrocarbon Autoxidation Reactions Using Triphenylphosphine Solution as a Titrant: A New Protocol. *Ind Eng Chem Res*. 2018;57(20):6805-6814.

7. Gimeno MP, Mayoral MC, Andrés JM. A potentiometric titration for H2O2 determination in the presence of organic compounds. *Anal Methods*. 2013;5(6):1510-1514.

8. Stein RA, Slawson V. Spectrophotometric Hydroperoxide Determination by the Use of Triphenylphosphine. *Anal Chem*. 2002;35(8):1008-1010.

9. Pinkernell U, Effkemann S, Karst U. Simultaneous HPLC Determination of Peroxyacetic Acid and Hydrogen Peroxide. *Anal Chem*. 1997;69(17):3623-3627.

10. The oxidation of decane in the liquid and gaseous phases, Cullis and *al.*, *Proc R Soc London A Math Phys Sci*. 1981;375(1763):543-563.

11. Mohammed Hichem Hakka, Pierre-Alexandre Glaude, Olivier Herbinet, Frédérique Battin Leclerc. Experimental study of the oxidation of large surrogates for diesel and biodiesel fuels. Combustion and Flame, Elsevier, 2009, 156 (11), pp.2129-2144. 10.1016/j.combustflame.2009.06.003, hal-00724812.

12. El-Sayah, Z.; Glaude, P.-A.; Fournet, R.; Sirjean, B.; Baptiste, S. Comparison of the Effects of Different Biofuels on the Oxidation Stability of a Hydrocarbon Fuel. 2020. https://doi.org/10.4271/2020-01-2101ï.

IASH 2022, the 17th INTERNATIONAL SYMPOSIUM ON STABILITY, HANDLING AND USE OF LIQUID FUELS
Dresden, Germany
September 11 - 15, 2022

USING HIGH MAGNIFICATION OPTICS AND SHAPE RECOGNITION CAPABILITIES OF DYNAMIC IMAGING TO ACHIEVE A MORE ACCURATE ANALYSIS OF FUELS

Colin Dalton*, Tod Canty

Canty, 6100 Donner Road, Lockport, New York 14094 USA

ABSTRACT

High magnification optical imaging techniques for characterizing hard particles, water and soft organic particles in fuels is crucial for managing the operation of equipment and extending operational life cycles.

Traditional particle counting methods are unable to characterize between particles, hard and soft and misrepresent water and bubbles as particles. They are unable to detect particles <4µm, which can have a significant impact on operational equipment.

They are often limited to off line lab environments where a process sample is taken to a lab and timely control is difficult. These methods do not provide continual analysis of the fuels.
High magnification imaging systems lend themselves well to the analysis of fuels as they clearly differentiate between hard and soft particles, water and droplets down to 1µm in size based on the high resolution images and shape information obtained. This reduces the likelihood of characterization errors and enhances reproducibility, repeatability and accuracy. The visual capability allows the user to input his or her knowledge of the process to work in making meaningful analyses decisions. Imaging filters can be applied to the analysis that can distinguish solids from water droplets thereby doubling the capability of the instrument over non-visual equipment which cannot make this distinction.

This paper describes a high magnification dynamic imaging system used for measuring particle (Hard & Soft) concentration and size in addition to water concentration based on providing a true 2-dimensional size and shape analysis. The paper will also outline the critical steps in measuring, sampling and the analysis of fuel for both lab and in-line systems.

*Corresponding Author: Colin Dalton; colind@jmcanty.com; +353860456982

IASH 2022, the 17th INTERNATIONAL SYMPOSIUM ON STABILITY, HANDLING AND USE OF LIQUID FUELS
Dresden, Germany
September 11 - 15, 2022

THE FAST PEROXIDE ANALYZER

Bas de Jonge*, Larissa Ram, Lou Cheng

Da Vinci Laboratory Solutions, Sydneystraat 5, Rotterdam 3047BP Netherlands

ABSTRACT

Peroxides are reactive compounds that may impact product quality, promote unwanted side effects such as autopolymerisation, and present safety hazards such as the risk of explosion. Traditional methods often make use of iodometric titration methods like ASTM D2340 and D5799. These methods are susceptible to oxygen interference, causing false positives, expose the analyst to hazardous substances, and are time consuming. Furthermore, most iodometric titrations have lower quantitation limits of 1 mg/kg active oxygen with high uncertainties at this level. Da Vinci Laboratory Solutions has developed an automated alternative to the current titration method based on a flow injection system: the Fast Peroxide Analyzer (FPA). The FPA is based on the chemistry used in iodometric titration. HPLC hardware is used to pump iodide and acetic acid reagents through a dedicated reaction module after a sample has been injected by an automated liquid sampler or sampling valve. Any peroxides present in the sample will react with iodide to form iodine, which is detected by a UV-Vis detector. The system addresses the common challenges associated with the traditional titration methods by eliminating oxygen interference and chemical exposure due to the closed system, while simultaneously lowering detection limits, improving reproducibility, and lowering analysis time. The system can analyze liquid streams like; styrene monomer, alfa-olefins, and knocking engine reference fuels; liquified hydrocarbon streams like 1,3-butadiene; and solid hydrocarbon streams like dicyclopentadiene.

*Corresponding Author: Application Specialist Development Bas de Jonge;
larissa.ram@davinci-ls.com; +31 6 50967360

IASH 2022, the 17th INTERNATIONAL SYMPOSIUM ON STABILITY, HANDLING AND USE OF LIQUID FUELS
Dresden, Germany
September 11 - 15, 2022

GCXGC FOR HYDROCARBON TYPE ANALYSIS OF AVIATION FUELS AND COMPARISON OF DIFFERENT MODES OF OPERATION

Richard C. Striebich*, Linda M. Shafer, Zachary West, Steven Zabarnick

University of Dayton Research Institute, 300 College Park, Dayton, OH USA 45469-0043

ABSTRACT

Hydrocarbon type analysis using GCxGC has become a popular way to characterize jet fuel (and diesel fuel) to determine quantitative levels of n-paraffins, iso-paraffins, mono-, di- and tri-cylcloparaffins, alkylbenzenes, cycloaromatics, di- and tri-aromatics. Like few other techniques, this method can quickly and accurately determine the mass percentage of each of these hydrocarbon categories. As its development progresses, comparisons of results within the same laboratory (repeatability) and comparisons of results between other laboratories running exactly the same method with the same equipment (reproducibility) were conducted for a limited number of samples. Because there are so many different modes of operating GCxGC (reverse phase vs normal phase and flow modulation vs thermal modulation, primarily), comparisons of results were conducted between a laboratory operating with normal phase flow modulation and another laboratory operating with reverse phase thermal modulation. By examining these comparisons, we hope to get an indication of the accuracy and precision of this technique, when collaborators use exactly the same instrumentation and conditions, and when both instrument and conditions may differ in process or approach. While larger comparative studies with other collaborators and more fuels are in-process, the comparison described in this contribution focuses on GCxGC accuracy and precision between two labs running normal phase flow modulation GCxGC and then two labs running different modes of GCxGC.

*Corresponding Author: Mr. Richard Striebich; Richard.striebich.ctr@us.af.mil; 1-937-255-7295

IASH 2022, the 17th INTERNATIONAL CONFERENCE ON STABILITITY, HANDLING AND USE OF LIQUID FUELS
Dresden, Germany
11 – 15 September 2022

CHEMICAL CHARACTERIZATION OF NITROGEN- AND OXYGEN-CONTAINING COMPOUNDS IN LIQUID FUELS BY USING ELECTROSPRAY IONIZATION COUPLED TO A LINEAR QUADRUPOLE ION TRAP/ORBITRAP MASS SPECTROMETER

Mark Romanczyk[1*] and Thomas Loegel[1]

[1]US Naval Research Laboratory, Washington, District of Columbia 20375, United States. *mark.romanczyk@nrl.navy.mil, orcid.org/0000-0002-6907-9613

ABSTRACT: The ability to detect and chemically characterize heteroatom-containing compounds (HCCs) in liquid fuels is critical as these compounds adversely affect fuel storage and thermal stability. To address these needs, Electrospray Ionization (ESI) coupled to a linear quadrupole ion trap/orbitrap mass spectrometer was used to chemically characterize oxygen- and nitrogen-containing compounds (OCCs and NCCs, respectively) in aviation and diesel fuels and a marine gas oil. After methanol extraction of the liquid fuel, the methanol layer was directly injected into the ESI probe. The OCCs and NCCs were ionized via deprotonation and protonation, respectively, with little to no fragmentation. Kendrick Mass Defect (KMD) values were determined for the protonated and deprotonated ions, and the values were plotted against their respective nominal Kendrick masses to facilitate comprehensive characterization. For example, up to ten individual homologue ion series of ionized NCCs (*i.e.,* $C_nH_{(2n+4)}N$, $C_nH_{(2n+2)}N$, $C_nH_{(2n)}N$, etc.) were detected in the aviation fuels with an overall mass range of m/z 100 – 314. Several proposed chemical classifications of the NCCs include: alkylamines, pyrroles, pyridines, indolines, tetrahydroquinolines, indoles, quinolones, carbazoles and triphenylamines. Further, up to twelve distinct homologue ion series of OCCs (*i.e.,* $C_nH_{(2n-7)}O$, $C_nH_{(2n-9)}O$, $C_nH_{(2n-9)}O_2$) were detected in the aviation fuels with an overall mass range of m/z 89 – 283. Proposed classifications of the OCCs include: saturated carboxylic acids, phenols, benzoic acids, indenols, hydroxytetralins and naphthols. The Kendrick Mass Defect plots showed clear differences when the disparate fuel types were compared and among the same type of fuel (*i.e.,* aviation fuels). Unknown ions derived from an aviation fuel and ionized model compounds were subjected to collision-activated dissociation (CAD) to facilitate further structural analysis. Several of the CAD spectra of unknown compounds were nearly identical to the model compounds. The information collected from this method may aid in providing more through chemical characterization of fuels, which may be used to provide stronger links between fuel composition and the propensity of fuels to fail storage and thermal stability.

Keywords: heteroatom-containing compounds, nitrogen-containing compounds, oxygen-containing compounds, chemical characterization, mass spectrometry, analytical methods, storage and thermal stability, orbitrap mass spectrometer

INTRODUCTION

The chemical compositions of liquid fuels (*i.e.,* aviation and diesel fuels) are exceptionally complex, with each fuel containing thousands of individual compounds.[1] In liquid fuels, hydrocarbons are the most abundant and are classified into one of three major groups: alkanes, cycloalkanes and aromatic compounds. The three hydrocarbon classes influence many important physical and chemical properties of fuels (*i.e.,* seal swell, emissions, viscosity, density, net heat combustion, etc.).[2-7]

Although significantly less abundant than hydrocarbons, heteroatom-containing compounds (HCCs) are also present in fuels. The HCCs include nitrogen-, oxygen- and sulfur-containing compounds (NCCs, OCCs, and SCCs, respectively). The chemical classes of the HCCs found in fuels are vast and include phenols, indoles, anilines, tetrahydroquinolones, quinolones, carbazoles, thiols, thiophenes, and benzothiophenes.[8-15] Further, the abundances of HCCs in fuels fluctuate; one study demonstrated a range from 10 – 2400 ppm in total HCC content in several different aviation fuels.[8] More specifically, reported concentrations of OCCs and NCCs in aviation fuels ranged from 1 – 1900 ppm and ~ 1 – 750 ppm, respectively.[8,9.] Additional studies supported that total sulfur concentrations in aviation fuels varied from 400 – 1400 ppm.[10,11] Hence, fuels have considerable variation of HCC content.

As observed with hydrocarbons, HCCs influence physical and chemical properties of fuels, including storage and thermal stability.[3,8,16-19] In the context of this paper, stability refers to the resistance of a fuel to degrade in overall quality, germane to fuel stored at ambient conditions over extended periods (*i.e.,* storage) or exposed to high temperatures for short durations (*i.e.,* thermal stability). The mechanisms that contribute to instability are highly intricate, but likely involve free radical autoxidative chain mechanisms (**Figure 1**).[18-20] Briefly, the first step involves the formation of an alkyl radical, R•, derived from an initiator. The R• species reacts with dissolved oxygen to produce an alkylperoxy radical, RO_2•. Then, the alkylperoxy radical reacts with a fuel molecule to form hydroperoxide, RO_2H, and an additional radical R•. The presence of the additional radical promotes chain propagation. In addition, autoacceleration reactions may also occur upon the homolytic cleavage of RO_2H forming two radicals, •OH and RO•.[18]

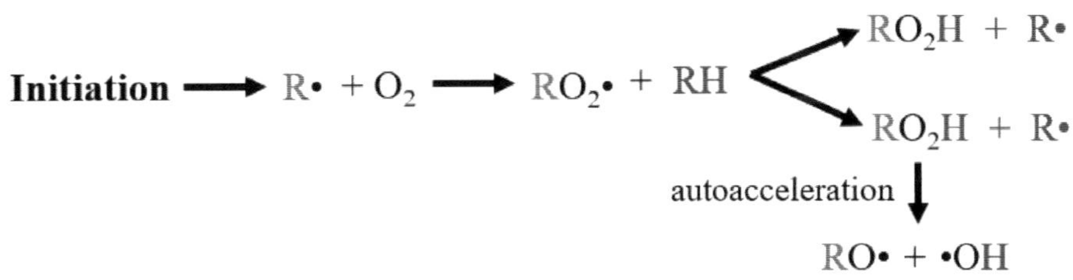

Figure 1. Initial and propagation steps for the autoxidative pathway that contribute to instability in fuel. Initiation may occur through several means including homolysis of compounds with heteroatom-heteroatom σ bonds upon exposure to light and/or heat. Autoacceleration is shown as an ancillary step after the formation of hydroperoxide.[18]

Throughout the autoxidation chain mechanism, HCC radical species may react to form dimers, referred to as soluble macromolecular oxidative reactive species (SMORS).[19,21] As SMORS continue to react with additional HCCs, large insoluble molecular weight sediments and gums form. Upon exposure to high temperatures, the harmful by-products may produce coking and fouling materials. Collectively, these insoluble materials are of great concern as they may obstruct fuel filters and/or ignition injectors, resulting in fuel starvation, decreased fuel spray efficiency and/or attenuated engine performance of the vehicle. The insoluble species formed in fuels by autoxidative reactions pose major risks to routine operations, incur high costs for maintenance-related issues and adversely affect mission readiness.[22] Hence, it is critical to detect HCCs in fuels.

Past studies provided in-depth classification and quantitation of HCCs in fuels and used a variety of different instrumentation.[8-16,23-26] Several of these studies used quadrupole mass spectrometers (QMS) to facilitate detection and qualitative analysis for the unknown HCCs in fuels.[8,11-13,16,23,24,26] QMS have limited resolution and low mass accuracy when compared to high resolution mass spectrometers such as orbitraps and FT-ICR detectors.[27] Time-of-Flight (TOF) mass spectrometers were also used to enable identification of unknown HCCs.[9,10,14] TOF instruments have improved resolution and mass accuracy when compared to QMS, but do not surpass orbitraps or FT-ICR detectors. Further, electron ionization (EI) was used to ionize the HCCs in the fuels measured.[8-14,16,23,24,26] EI is a hard ionization technique that induces extensive fragmentation, which attenuates the signal of molecular ions resulting in misidentification and/or low match score from a mass spectrometer library. Further, the absence of HCCs in libraries prevents the ions identification. Therefore, the above methods have limitations for HCC characterizations in fuels.

The limitations of the methods described above affords an opportunity to evaluate new instrumentation to determine the effectiveness for compositional analysis of HCCs in liquid fuels. Previously, the orbitrap mass spectrometer demonstrated excellent resolution and mass accuracy for unknown ions derived from complex petroleum mixtures.[28] Further, Electrospray Ionization (ESI) is selective for ionizing analytes with polar functionalities and is a soft ionization technique that limits fragmentation of molecular / pseudo-molecular ions.[27] Therefore, the aim of this study was to evaluate the applicability of (+) and (-) ESI coupled to a linear quadruple ion trap/orbitrap (LQIT/orbitrap) to chemically characterize NCCs and OCCs derived from different fuels (for a general schematic of the instrument see **Figures S1** and **S2**). Further, the information collected in the spectra was converted into Kendrick Mass Defect (KMD) plots to provide comparison of the fuels measured. Finally, several unknown ions derived from fuels and known model compounds were subjected to collision-activated dissociation to facilitate structural elucidation and class determination.

RESULTS AND DISCUSSION

In this study, NCCs and OCCs from several fuels were ionized by using ESI and detected with a LQIT/orbitrap mass spectrometer. The high resolution MS^1 measurements provided useful information for the classification of NCCs and OCCs in the fuel samples. Detailed qualitative analysis of NCCs and OCCs is critical as these compounds adversely affect fuel stability.[3,8,16-19] Hence, establishing stronger links between the NCCs and OCCs composition in fuels to thermal and storage stability may advance the current understanding of the mechanisms that contribute to failures and provide useful insight into the overall propensity of fuels to fail.

Chemical Characterization of Nitrogen-Containing Compounds in Liquid Fuels

The MS^1 spectra of the aviation fuels, and a diesel fuel and marine gas oil (MGO) are shown in **Figures 2** and **3**, respectively. The detection of ions with even m/z values indicates that the NCCs were ionized via protonation $(MH)^+$. This result was anticipated as the proton affinity / gas-phase basicity values of NCCs indicates that protonation is highly favorable (**Equation S1**). When the MS^1 spectra of the three aviation fuels are compared to each other, clear differences in the distribution of NCCs are observed. Jet fuel (A) has the largest distribution of NCCs with a mass range of m/z 100 – 314. Jet fuels (B) and (C) have mass ranges of m/z 122 – 272 and m/z 136 – 260, respectively. From the m/z values provided, the number of carbon atoms in the detected NCCs derived from jet fuels range from 6 – 22.

When the MS^1 spectra of the diesel fuel and MGO are compared to each other, the diesel fuel has a wider mass range corresponding to m/z of 122 – 372. The MGO has a mass range of m/z 186 – 316. The number of carbons in NCCs derived from diesel and MGO range from 8 – 27 and 13 – 22 carbons, respectively.

For a completed a list of the ionized NCCs detected in the individual fuel samples, see **Table 1**. The table includes the m/z and ring and double bond equivalents (RDB) values for the detected ions. For each series of ions corresponding to the same chemical formula (*i.e.,* a homologue ($C_nH_{2n-x}N$)), the average carbon number was provided and proposed classification(s) for the ion series were made. The average carbon number is a weighted average of the carbons in the ions within the homologue series. The mass accuracy of the NCCs are all less than 10 ppm (**Table S1**).

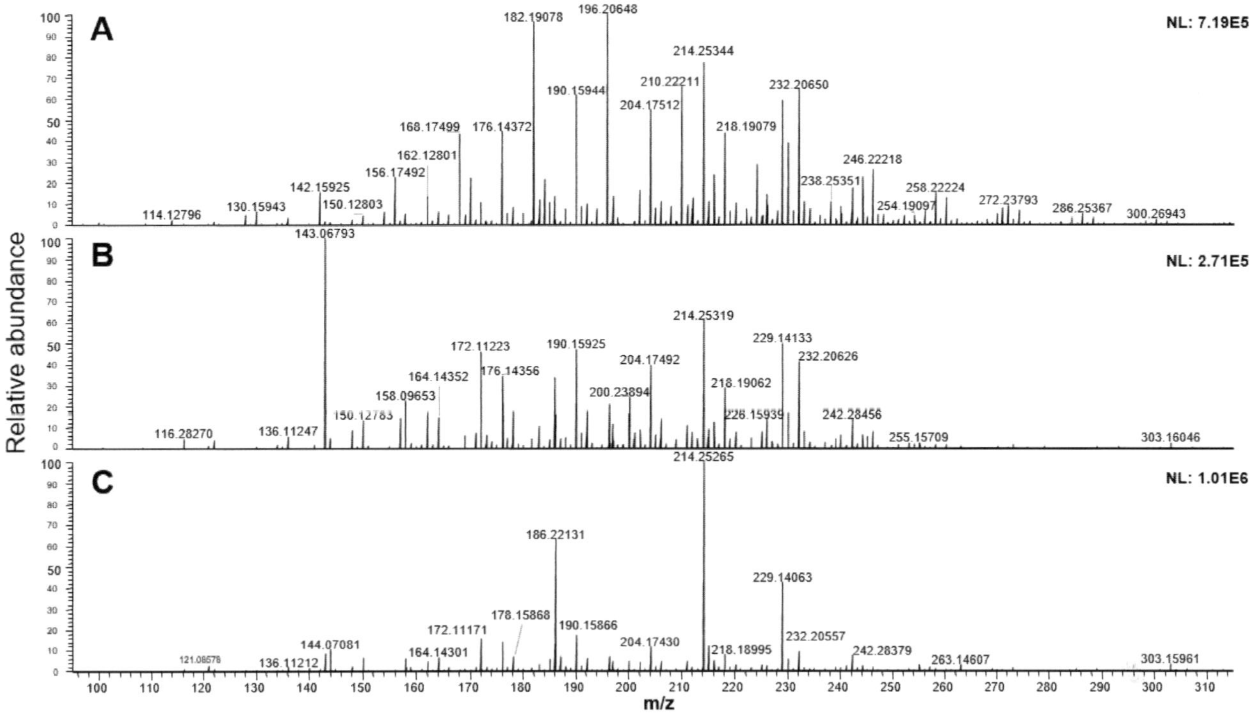

Figure 2. (+) ESI MS[1] spectra for (**A**) jet fuel (A), (**B**) jet fuel (B) and (**C**) jet fuel (C). Ions of m/z 143 derived from jet fuel (B) have the elemental composition $[C_5H_{12}O_3]Na^+$, likely corresponding to sodium adducts of the fuel system icing inhibitor, 2-(2-methoxyethoxy)ethan-1-ol (DiEGME).

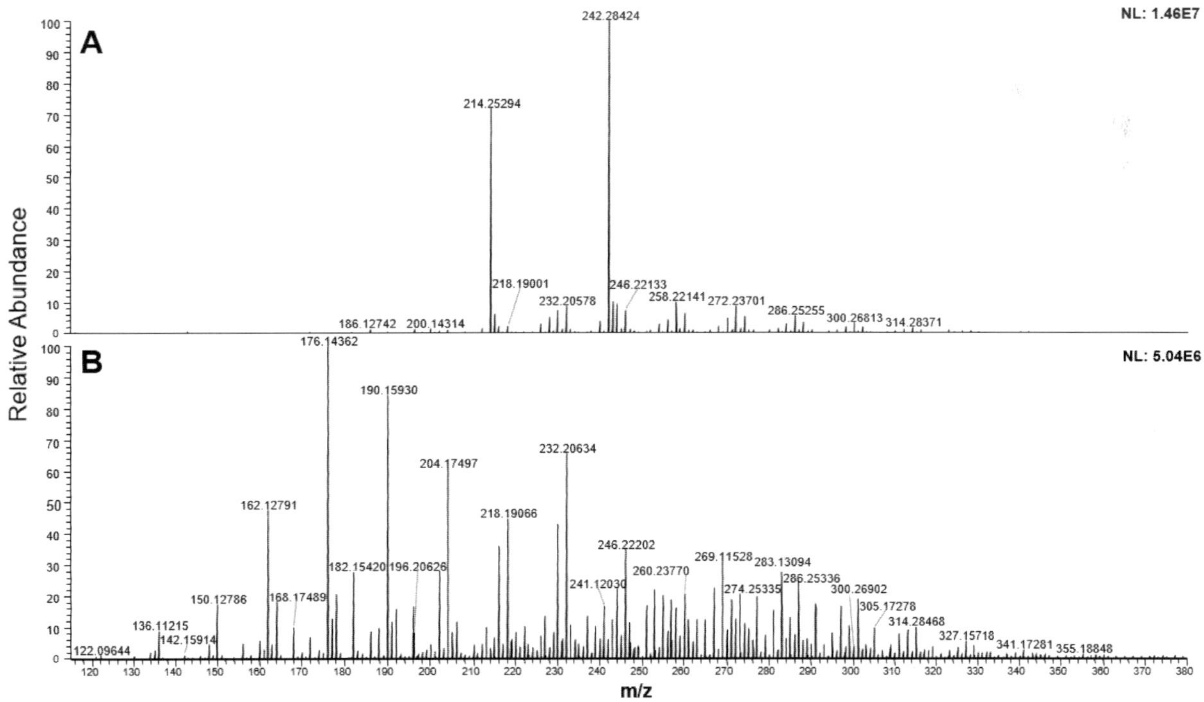

Figure 3. (+) ESI MS[1] spectra for the (**A**) marine gas oil and **(B)** diesel fuel.

Table 1: The m/z, average carbon number, RDB values and empirical formulas for ionized NCCs derived from fuels corresponding to each homologue ion series. Proposed classifications for each homologue series were made and assigned a group number, in **blue bold text** within the parenthesis. This number is referred to when comparing the compositions of fuels to each other. For the m/z column, increments of 14 Da correspond to the numerical range provided in the table (*i.e.,* ions of m/z 130-214 is equal to m/z 130, 144, 158, 172, 186, 200 and 214).

Fuel	Ions detected (m/z)	Average carbon number of ions detected	RDB value	Empirical formula of ions (homologue series)	Proposed classification(s) and group number
Jet 1	130, 186, 214, 242	(14)	-0.5	$C_nH_{(2n+4)}N$	Alkylamines (**1**)
Jet 2	130, 186, 200, 214, 242, 270	14			
Jet 3	158, 186, 214, 242	13			
Diesel	ND	-			
Jet 1	100-268	12	0.5	$C_nH_{(2n+2)}N$	Piperidines / pyrrolidines (**2**)
Jet 2	ND	-			
Jet 3	ND	-			
Diesel	142-184	11			
Jet 1	154-266	13	1.5	$C_nH_{(2n)}N$	Decahydroquinoline / decahydroisoquinoline (**3**)
Jet 2	ND	-			
Jet 3	ND	-			
Diesel	168-238	13			
Jet 1	166-264	14	2.5	$C_nH_{(2n-2)}N$	Pyrroles (**4**)
Jet 2	166-180	11			
Jet 3	166-180	13			
Diesel	208-236	15			
Jet (A)	122-290	14	3.5	$C_nH_{(2n-4)}N$	Pyridines / anilines (**5**)
Jet (B)	122-234	12			
Jet (C)	136-234	13			
Diesel	122-276	14			
MGO	290	18			
Jet (A)	134-302	14	4.5	$C_nH_{(2n-6)}N$	Indolines / tetrahydroquinolines / tetrahydroisoquinolines (**6**)
Jet (B)	134-260	14			
Jet (C)	162-260	14			
Diesel	134-316	14			
MGO	204-316	18			

Jet (A)	160-314	16	5.5	$C_nH_{(2n-8)}N$	Indoles / pyrrolines (**7**)
Jet (B)	146-272	15			
Jet (C)	188-258	16			
Diesel	146-328	16			
MGO	216-314	18			
Jet (A)	144-298	15	6.5	$C_nH_{(2n-10)}N$	Quinolines / isoquinolines / naphthylamines (**8**)
Jet (B)	144-256	13			
Jet (C)	158-256	13			
Diesel	158-326	17			
MGO	186-312	17			
Jet (A)	184-282	16	7.5	$C_nH_{(2n-12)}N$	Diphenylamines (**9**)
Jet (B)	184-254	16			
Jet (C)	212-240	16			
Diesel	198-324	18			
MGO	212-240	17			
Jet (A)	196-280	18	8.5	$C_nH_{(2n-14)}N$	Carbazoles (**10**)
Jet (B)	ND	-			
Jet (C)	ND	-			
Diesel	252-294	19			
MGO	266	19			
Jet (A)	ND	-	10.5	$C_nH_{(2n-18)}N$	Naphthylbenzylamine (**11**)
Jet (B)	ND	-			
Jet (C)	ND	-			
Diesel	234-360	21			
MGO	ND	-			
Jet (A)	ND	-	11.5	$C_nH_{(2n-20)}N$	Triphenylamine (**12**)
Jet (B)	ND	-			
Jet (C)	ND	-			
Diesel	246, 288-358	23			
MGO	ND	-			

To facilitate qualitative comparison of the ionized NCCs derived from the fuels, the complex spectra were converted to Kendrick Mass Defect (KMD) plots (**Figures 4** and **5**). KMD plots convert the high resolution mass spectra into easy to view compositional maps with chemical groupings of data points corresponding to ions with the same chemical formula (*i.e.,* $(C_xH_{2n-x})N$). Please note that the group numbers assigned from **Table 1** are the same in **Figures 4** and **5**.

When the three aviation fuels are compared to each other, ten homologue ions series were detected in jet fuel (A) and seven were detected in jet fuels (B) and (C). Hence, at least ten different classes representative of NCCs were found in the aviation fuels. The ionized NCCs are representative of aromatic and non-aromatic compounds. Compounds with aromaticity have up to three fused rings (*i.e.,* carbazole). The most abundant chemical groups in jet fuel (A) are **1**, **3**, and **6**. For jet fuel (B), groups **1**, **6** and **8** are the most abundant and for jet fuel (C), group **1** is the most abundant. The presence of NCCs in groups **2** (piperidines / pyrrolidines) and **3** (decahydroquinoline / decahydroisoquinoline) in jet fuel (A) is a remarkable difference when compared to the other jet fuels. These observed chemical differences may play crucial roles in determining the propensity of a fuel to fail thermal and/or storage stability.

When the ions of the same homologue series were detected in all three jet fuels, the average carbon number for the detected NCCs are similar. Hence, the molecular weight distribution for these groups are similar.

From **Table 1** and **Figure 4**, the mass ranges for the NCCs detected in the three fuels were considerably wider than compared to the NCCs identified in other jet fuels from a previous work.[15] This implies that the number of carbons in alkyl groups attached to the aromatic core (*i.e.,* indole, pyridine, quinolone cores) of NCCs in these fuels are greater than compared to the NCCs detected in the previous study.[15] The presence of more carbons in chains attached to the core structures may influence chemical reactions germane to stability-related reactions. For example, NCCs exhibiting greater steric hindrance may attenuate reaction efficacy in autoxidative reactions and/or the production of SMORS species.

The chemical classes of NCCs indentified in fuels in this study were also identified in other aviation fuels measured with different instrumentation. For example, one study utilized (+) ESI coupled to a Time-of-Flight (TOF) mass spectrometer and detected NCCs in groups **3**, **4**, **5**, **6**, and **7**.[15] These five chemical groups were present in the three fuels measured in this study. The additional five groups of NCCs in the fuels measured for this work include groups **1**, **2**, **7**, **9**, and **10**. Additional studies using solid phase extraction high performance liquid chromatography gas chromatography mass spectrometry[8] and multidimensional gas chromatography-(TOF) mass spectrometry[9] reported NCC classes of **5**, **7**, **8** and **10** in several aviation fuels. Both studies utilized solid phase extractions prior to analysis. However, a simple methanol extraction was carried out prior to analysis in this study. Hence, future efforts should compare the data collected in this study to a method incorporating an SPE procedure completed on jet fuels (A), (B) and (C) prior to analysis. If the results are similar, the simple methanol extraction process can improve overall robustness and throughput when measuring NCCs and OCCs than compared to past SPE methods.

With exception of group **1** (alkylamines), all other NCC groups were detected in the diesel fuel (**Table 1** and **Figure 5**). Furthermore, NCCs with the largest RDB values (groups **11** and **12**) were detected in diesel fuel, which was anticipated when compared to aviation fuels as diesel fuels have greater distillation temperatures. Ion in chemical groups **6** and **7** were the most abundant in the diesel fuel. Despite the wide mass ranges for the NCCs in groups **6** and **7**, the average carbon number for ions in this homologue series corresponded to 14 and 16 carbons, respectively, which were very similar to that of aviation fuels. Therefore, the most abundant NCCs detected in groups **6** and **7** have a comparable number of carbons in alkyl chains attached to indoline / tetrahydroquinoline / tetrahydroisoquinoline, and indole / pyrroline cores, respectively, as the NCCs detected in jet fuels.

The MGO contained six NCC chemical groups, the fewest of all fuels measured (**Table 1** and **Figure 5**). Ions in groups **1**, **5**, **6**, **7**, **8** and **9** were detected in MGO, with group 1 (alkylamines) being the most abundant. Specifically, ions of m/z 214 and 242 were the most abundant (**Figure 3**). It is noteworthy that the most dominant group in the MGO and jet fuel (C) are group 1, but the other detectable groups of NCCs are far less abundant.

As previously demonstrated, NCCs from multiple chemical classes (*i.e.,* pyrroles, isoquinolines, alkylamines, indoles) adversely influence fuels by chemical reactions that result in insoluble sediments, gums and thermal deposits.[8,29-33] The simple methanol extraction of fuel followed by (+) ESI LQIT/orbitrap measurements provided detailed characterization of NCCs in the fuels and showed several major compositional differences when the fuels were compared to each other. With the qualitative NCC information obtained from this method, future performance and ASTM testing (*i.e.,* ASTM D3241, D4625, D6468, D6748) should be completed to determine if links are present between the NCC composition and thermal and storage stability properties (*i.e.,* breakpoint) of fuels.

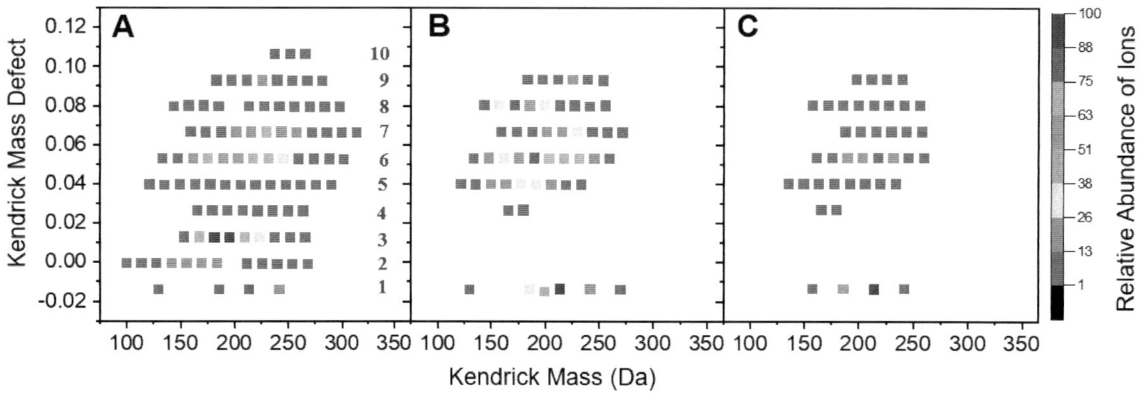

Figure 4. Kendrick Mass Defect plots for ionized NCCs derived from (**A**) jet fuel (A), (**B**) jet fuel (B) and (**C**) jet fuel (C). Numbers 1 – 10 indicate the proposed individual classes of NCCs: alkylamines **1**, piperidines / pyrrolidines **2**, decahydroquinoline / decahydroisoquinoline **3**, pyrroles **4**, pyridines / anilines **5**, indolines / tetrahydroquinolines / tetrahydroisoquinolines **6**, indoles **7**, quinolines / isoquinolines / naphthylamines **8**, diphenylamines **9**, and carbazoles **10**.

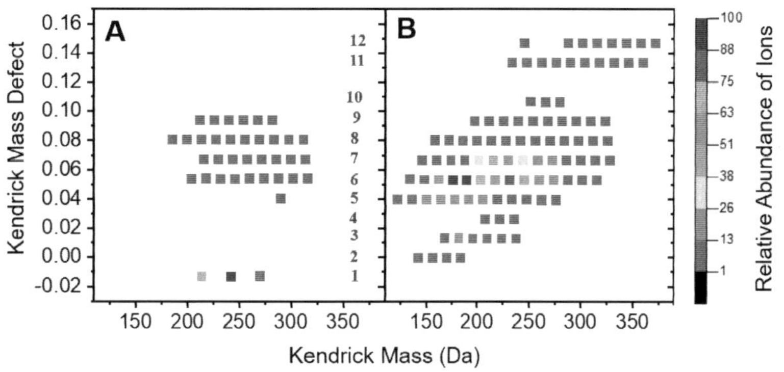

Figure 5. Kendrick Mass Defect plots for ionized NCCs derived from (**A**) MGO and (**B**) diesel fuel. Numbers 1-12 indicate the proposed individual classes of NCCs. Numbers 1-10 were indicated previously in Figure 4. Proposed classes corresponding to **11** and **12** are naphthylbenzylamines and triphenylamines, respectively.

Collision-activated Dissociation of Nitrogen-Containing Compounds

Collision-activated dissociation (CAD) provides useful structural information and possible identification for unknown ions.[27,34] Within the LQIT, ions of interest are first isolated in the ion trap, and are activated through collisions with helium (**Figure S2**). During the activation step, the ions gain internal energy and eventually fragment at the weakest bonds within the ion. The fragment ions detected in the tandem mass spectra (MS^2) reveal important information used to elucidate the structures of unknown ions. As can be observed from **Table 1**, several different classes of NCCs may be representative of a chemical group with the same formula (*i.e.,* Group **5** can be representative of anilines and/or pyridines). Therefore, CAD was attempted on known model compounds and compared to unknown ions of the same m/z value derived from jet fuel (B) to determine if the class of NCCs could be identified.

When protonated quinolone (m/z 130) and unknown ions of m/z 130 were subjected to CAD, the MS^2 spectra look very similar to each other (**Figure 6**). As the homologue series increases by units of 14 Da, unknown ions of m/z 144 and 158 were also subjected to CAD and compared to known protonated quinolines with the same m/z value. The spectra for the unknown ions and known protonated quinoline ions also look similar. Hence, the unknown ions of m/z 130, 144 and 158 are likely representative of the quinolone/isoquinoline class, but not the naphthylamines (to confirm, model compounds of naphthylamine are currently being purchased).

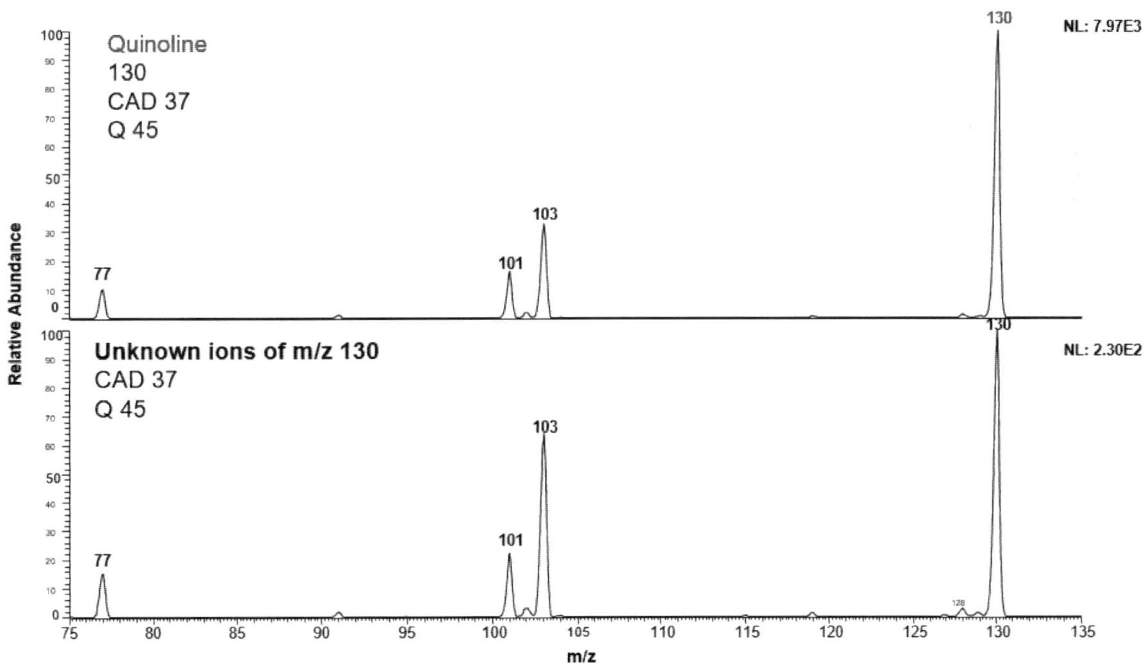

Figure 6. MS^2 spectra of protonated quinolone and unknown ions of m/z 130 derived from jet fuel (B). The similarity of the spectra support that the unknown ions are likely protonated quinoline.

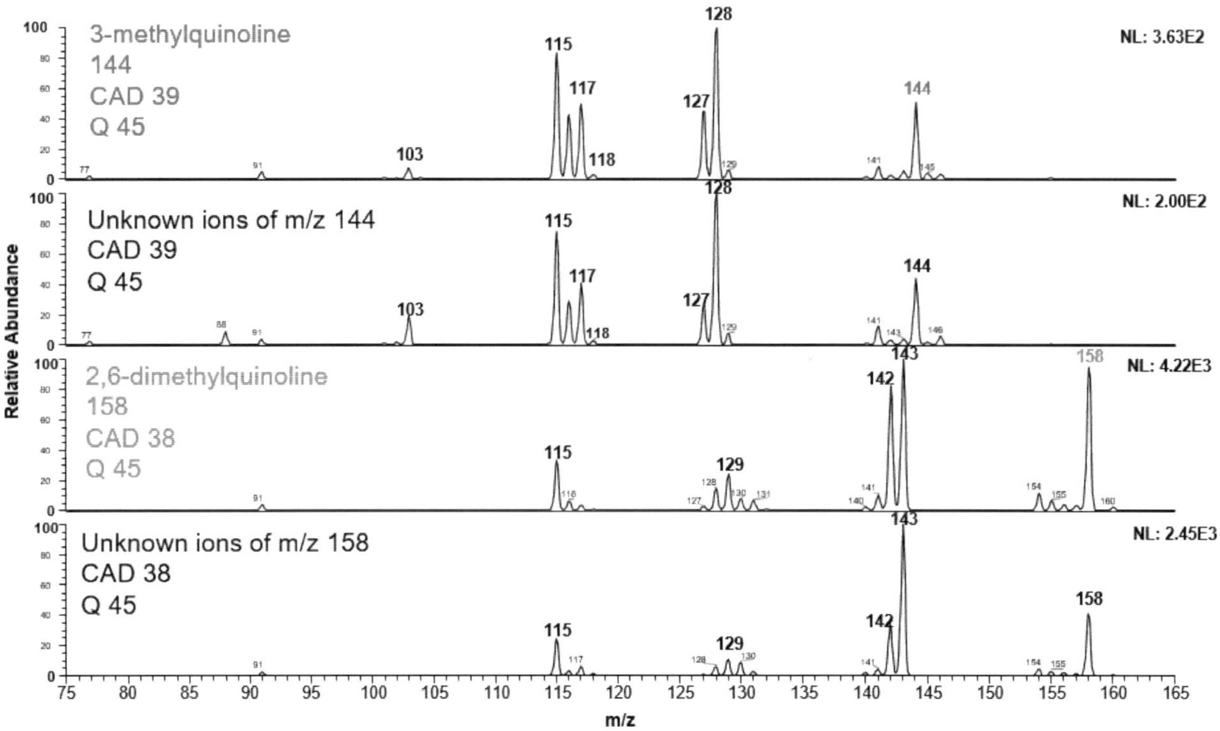

Figure 7. By descending order, the MS^2 spectra of protonated 3-methylquinoline, unknown ions of m/z 144, protonated 2,6-dimethylquinoline and unknown ions of m/z 158. The spectra of known and unknown ions are similar, but not identical to each other. Hence, unknown ions of m/z 144 and 158 may be representative of several structural isomers of methyl- and dimethylquinolines.

Additional CAD measurements were completed on unknown ions of m/z 122; this class may be representative of anilines or pyridines or a combination of both classes. The MS^2 spectra of unknown ions of m/z 122 and protonated 3,5-dimethylaniline and 2,4,6-trimethylpyridine are shown in **Figure 8**. The MS^2 spectra of protonated 2,4,6-trimethylpyridine and the unknown ions are similar to each other. Specifically, fragment ions of m/z 77, 79, 81 and 94 are observed in both spectra. The results support that the unknown ions may be classified as pyridines. Additional model compounds (*i.e.,* dimethylpyridines and ethylanilines) are currently being purchased. If the spectra do not include the fragment ions of m/z 77, 79, 81 and 94 the unknown ions are likely representative of trimethylpyridines.

The ability to differentiate classes of NCCs from each other with the same chemical formulas may have critical applications for fuel stability properties. As demonstrated, imbalances of basic (derivatives of pyridine) and non-basic (derivatives of pyrroles) NCCs contributed to stability-related failures.[16] With evidence of distinguishing different ionized classes with the same chemical formula from each other, CAD will likely play an important role in future chemical characterization methods.

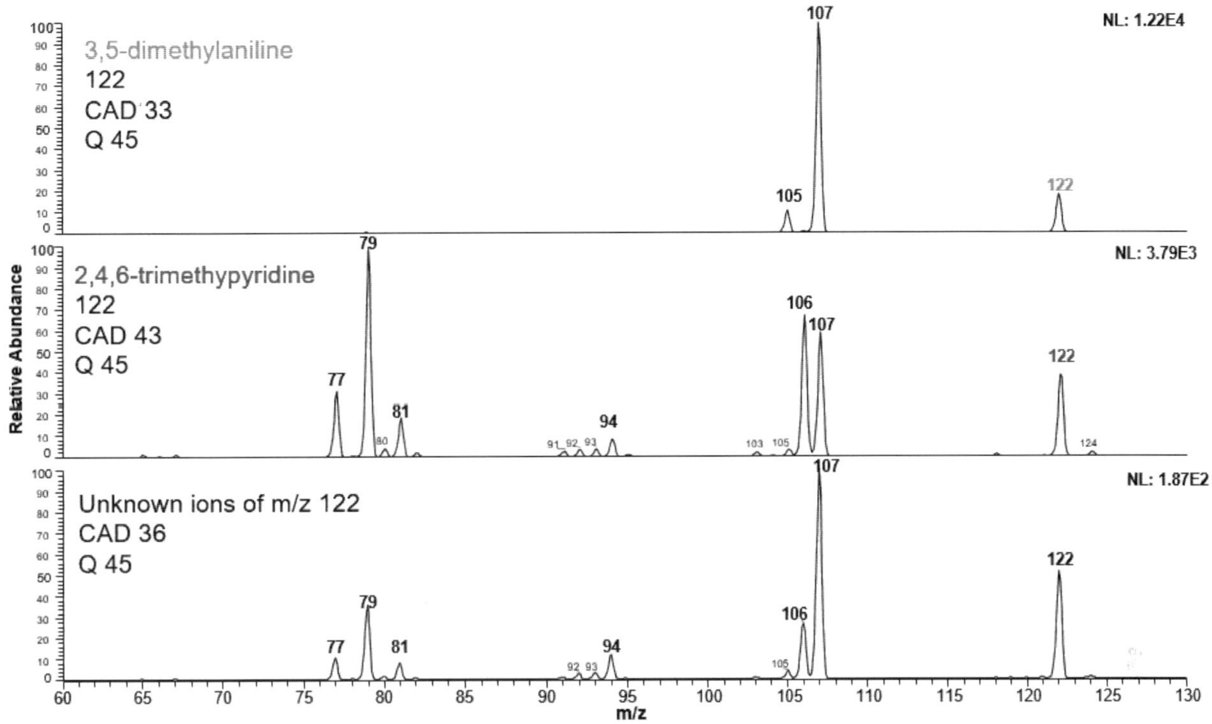

Figure 8. By descending order, the MS^2 spectra of protonated 3,5-dimethylaniline, protonated 2,4,6-trimethylpyridine and unknown ions of m/z 122 derived from jet fuel (B).

Chemical Characterization of Oxygen-Containing Compounds in Liquid Fuels.

The MS^1 spectra for aviation fuels, and a diesel fuel and marine gas oil (MGO) are shown in **Figures 9** and **10,** respectively. The detection of ions with odd m/z values specifies that the OCCs were ionized via deprotonation $(M-H)^-$, indicating energetically favorable removal of protons from the oxygen-containing functionalities of the OCCs (**Equation S2**). The mass accuracy of the OCCs in the fuel samples are all less than 10 ppm (**Table S2**). Noteworthy, many of the detected OCCs contain more than one oxygen atom with ranges of 1 to 6. For OCCs with at least four oxygen atoms, an adverse peroxide functionality may be present, which are known to contribute to free radical reactions.[35] The detection of such OCCs demonstrates the excellent sensitivity of the LQIT/orbitrap mass spectrometer. Consequently, the method has practical importance for the industry as a possible screening tool for harmful compounds in fuels, which may aid in identifying fuels as marginally stable and/or a scale of the fuels' propensity to fail stability during its life cycle.

When the MS^1 spectra of the three aviation fuels are compared to each other, differences in the distribution of the OCCs are observed. Detected OCCs derived from jet fuel (A) have a mass range of m/z 89 – 269. Jet fuel (B) has the smallest mass range from m/z 121 – 226. Jet fuel (C) has a mass range of m/z 89 – 283, which was the largest among the three aviation fuels. The overall range of carbon atoms in the detected OCCs in the aviation fuels is 3 – 18.

When the MS[1] spectra of the diesel fuel and MGO are compared to each other, the diesel fuel has a wider mass range of m/z 133 – 393. The MGO has a mass range of m/z 143 – 311. The range of carbons atoms in the detected OCCs in diesel and MGO are 9 – 22 and 8 – 20, respectively.

For a complete a list of the ionized OCCs derived from the fuel samples, see **Table 2**. The table includes the m/z and RDB values for the detected OCCs. For each series of ions with the same chemical formula (*i.e.*, $C_nH_{2n-x}O_y$), the average carbon number was provided and proposed classification(s) for the ion series were made. As stated previously, the average carbon number is a weighted average of the carbons in the ions within the homologue series. **Table 3** includes the same information for OCCs detected only in the diesel fuel; however, proposed classifications were not assigned to the ion series as the number of possibilities was considered too broad. Future CAD measurements will be used to elucidate the structures and presence of oxygen-containing functionalities to aid in narrowing down the number of possible classifications.

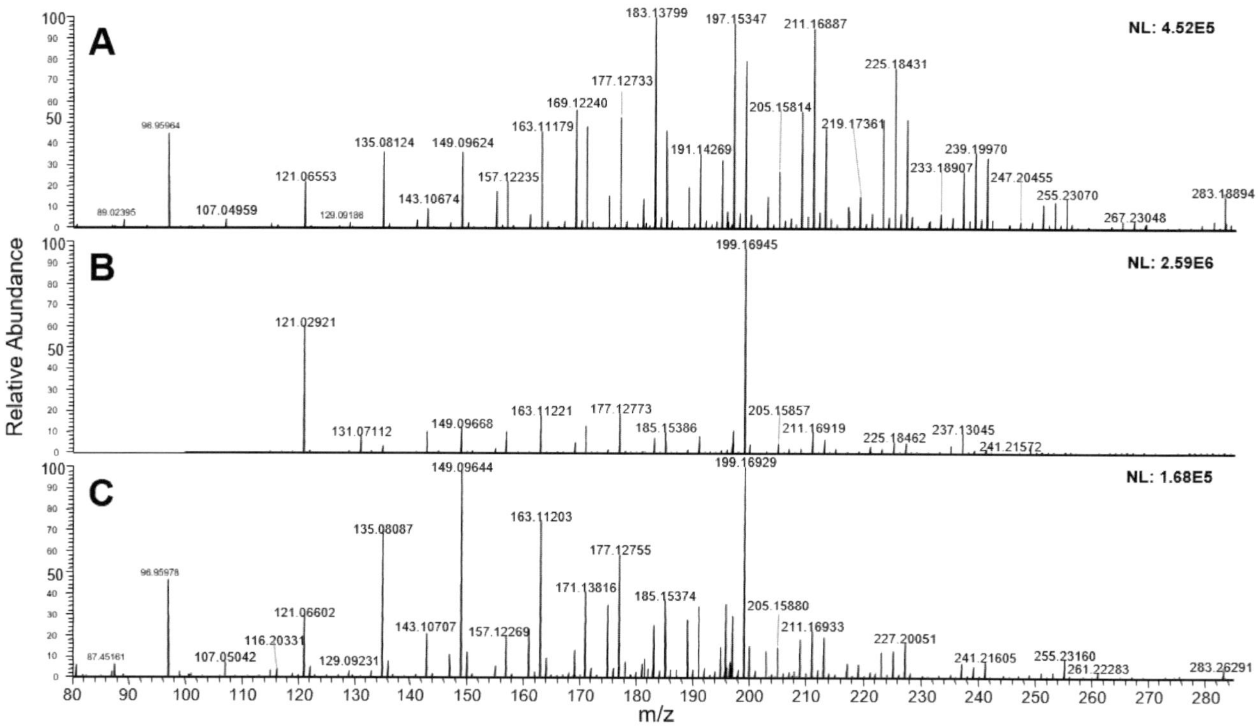

Figure 9. (-) ESI MS[1] spectra for (**A**) jet fuel (A), (**B**) jet fuel (B) and (**C**) jet fuel (C).

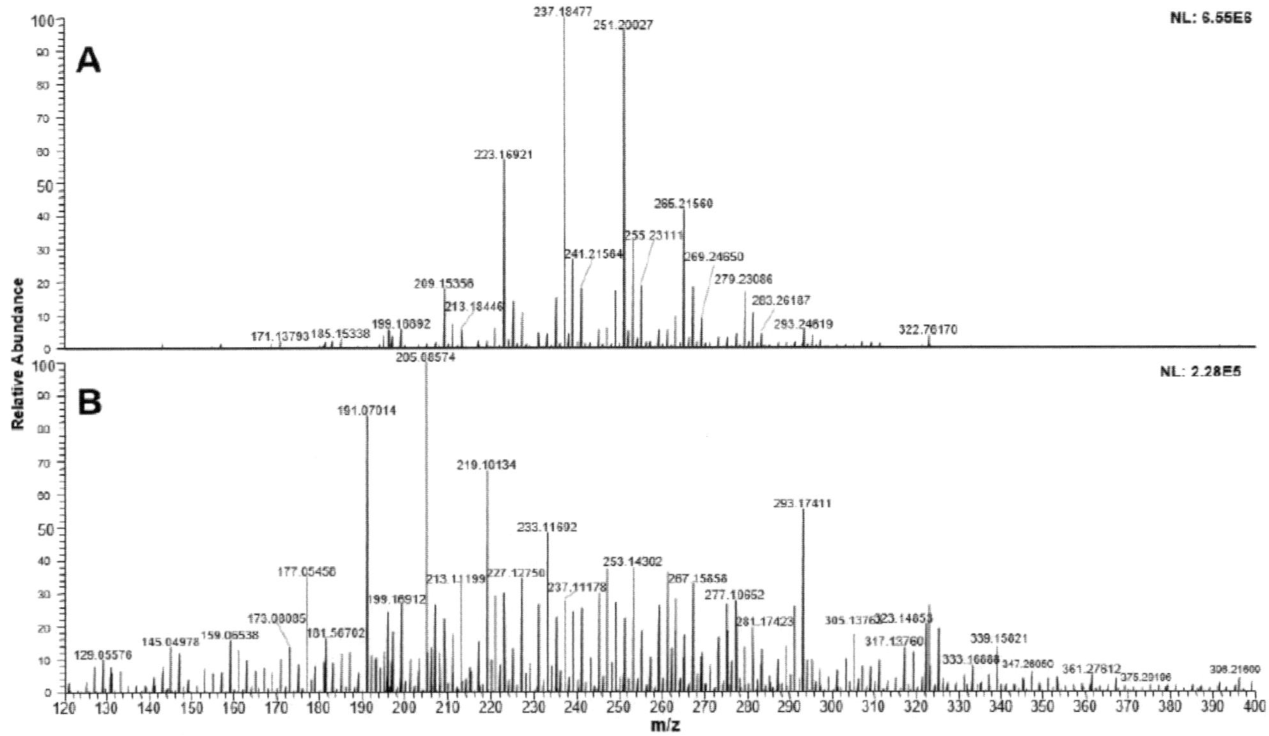

Figure 10. (-) ESI MS[1] spectra for (**A**) marine gas oil and (**B**) diesel fuel.

Table 2. The m/z, average carbon number, RDB values and empirical formulas for ionized OCCs derived from fuels corresponding to individual homologue ion series. Proposed classifications for each homologue ion series were made and assigned a group number in **blue bold text** within the parenthesis. This number is referred to when comparing the compositions of fuels to each other. For the m/z column, increments of 14 Da correspond to the numerical range listed in the second column (*i.e.,* ions of m/z 143 – 227 is equal to m/z 143, 157, 171, 185, 199, 213, 227).

Fuel measured	Range of ions detected	Average carbon number of ions detected	RDB value	Empirical formula of ions (homologue series)	Proposed classification(s)
Jet (A)	115-269	12	1.5	$C_nH_{(2n-1)}O_2$	Saturated carboxylic acids (**1**)
Jet (B)	143-227	11			
Jet (C)	101-283	11			
MGO	143-311	15			
Diesel	143,157,199,255, 283	13			
Jet (A)	141-267	12	2.5	$C_nH_{(2n-3)}O_2$	Cyclic carboxylic acids (**2**)
Jet (B)	155-225	12			
Jet (C)	141-253	12			
MGO	183-309	16			
Diesel	ND	-			

Jet (A)	93-233	11	4.5	$C_nH_{(2n-7)}O$	Phenols (**3**)
Jet (B)	121-219	12			
Jet (C)	107-233	11			
MGO	205-247	16			
Diesel	ND	-			
Jet (A)	89-103	3	1.5	$C_nH_{(2n-1)}O_3$	Hydroxy acids (**4**)
Jet (B)	131-145, 215	6			
Jet (C)	89	3			
MGO	ND	-			
Diesel	ND	-			
Jet (A)	167-265	13	3.5	$C_nH_{(2n-5)}O_2$	Bicyclic carboxylic acids / napthoquinones (**5**)
Jet (B)	181-209	12			
Jet (C)	167-265	13			
MGO	195-307	16			
Diesel	167-195	11			
Jet (A)	147-217	13	5.5	$C_nH_{(2n-9)}O$	Hydroxtetralins (**6**)
Jet (B)	175-203	13			
Jet (C)	133-217	12			
MGO	203-231	15			
Diesel	133-217	12			
Jet (A)	193-235	13	4.5	$C_nH_{(2n-7)}O_2$	Benzenediols / oxyphenols (**7**)
Jet (B)	ND	-			
Jet (C)	137-249	12			
MGO	207-291	16			
Diesel	ND	-			
Jet (A)	187-201	14	6.5	$C_nH_{(2n-11)}O$	Indenols (**8**)
Jet (B)	ND	-			
Jet (C)	173-215	14			
MGO	ND	-			
Diesel	145-285	13			
Jet (A)	121, 149-219	11	5.5	$C_nH_{(2n-9)}O_2$	Benzoic acids (**9**)
Jet (B)	121, 149-191	7			
Jet (C)	191-219	12			
MGO	219-275	16			
Diesel	135-289	13			
Jet (A)	157-171	12	7.5	$C_nH_{(2n-13)}O$	Naphthols (**10**)
Jet (B)	ND	-			
Jet (C)	ND	-			
MGO	ND	16			
Diesel	185-283	16			

Jet (A)	189-203	13	6.5	$C_nH_{(2n-11)}O_2$	Cinnamic acids (**11**)
Jet (B)	ND	-			
Jet (C)	ND	-			
MGO	217-287	16			
Diesel	161-301	15			
Jet (A)	189-217	14	8.5	$C_nH_{(2n-15)}O$	Diphenylmethanols (**12**)
Jet (B)	ND	-			
Jet (C)	169-225	14			
MGO	ND	-			
Diesel	197-239	15			
Jet (A)	ND	-	7.5	$C_nH_{(2n-13)}O_2$	Naphthalenediols (**13**)
Jet (B)	ND	-			
Jet (C)	ND	-			
MGO	243-271	17			
Diesel	173-299	16			
Jet (A)	ND	-	6.5	$C_nH_{(2n-11)}O_3$	Hydroxycinnamic acids / acetyl benzoic acids (**14**)
Jet (B)	ND	-			
Jet (C)	ND	-			
MGO	ND	-			
Diesel	163-331	13			

When examining the OCCs in aviation fuels, up to twelve distinct homologue ions series were detected. Therefore, at least twelve different classes of OCC are found in these fuels. Individually, jet fuels (A), (B) and (C) have 12, 7 and 11 detected OCC homologue ion series, respectively. The detected OCCs are representative of aromatic and non-aromatic compounds and may contain one, two or three oxygen atoms. Compounds exhibiting aromaticity contained up to two fused rings (*i.e.*, naphthols). The KMD plots for the aviation fuels are shown in **Figure 11**.

When the aviation fuels are compared to each other, the OCC composition of jet fuel (B) is considerably different when compared to jet fuels (A) and (C) (**Figure 11**). The most abundant chemical groups in jet fuel (B) are **1** and **3**. The most abundant chemical groups in jet fuel (A) are **1**, **2**, **3**, and **5**. The most abundant groups in jet fuel (C) are **1** and **3**. The presence of phenols (group **3**) as an abundant OCC class in aviation fuels is strongly supported from previous works.[8,9,12,15,24]

In general, jet fuels (A) and (C) have similar m/z ranges for ions in homologue series identified in both fuels. However, the relative abundances of the individual ions in the series for the two fuels differ. For OCCs in jet fuels (A) and (C), ions greater than 25 % abundance have mass ranges of m/z 149 – 241 and m/z 134 – 198, respectively. Hence, the most abundant ions (*i.e.*, > 25 %) derived from jet fuel (A) have more carbon atoms in alkyl chains attached to the cores than compared to the most abundant ions derived from jet fuel (C). As stated previously for the NCCs, the differences in the relative sizes and number of alkyl groups attached to the OCCs may affect chemical reactions necessary that contribute to stability-related mechanisms. When examining jet fuel (B), the saturated carboxylic acids (m/z 199) and benzoic acids (m/z 121) are the most abundant. All other ions detected are less than 25 % relative abundance.

When comparing the OCC compositions of MGO to diesel fuel, major differences are observed **Tables 2-3** and **Figure 12**). Most notably is the number of homologue ion series present in the diesel fuel, eighteen in total, compared to MGO with only nine. Further, the number of oxygen atoms present in the OCCs in diesel fuel range from 1-6. Conversely, only 1 or 2 oxygen atoms are present in the OCCs in MGO. These differences may influence the reactivity of the OCCs in the fuels. For example, acyl peroxides (OCC with four oxygen atoms) undergo homolytic bond cleavages upon exposure to light and heat.[35] Radicals derived from the acyl peroxides may serve as initiators in the autoxidation chain pathway and react with oxygen (**Figure 1**).

The most abundant ions derived from MGO were in groups **2** and **5**; assigned as cyclic carboxylic acids, and bicyclic carboxylic acids / napthoquinones, respectively. The most abundant ions derived from the diesel fuel were in groups **11** and **14**, assigned as cinnamic acids, and hydroxycinnamic acids / acetyl benzoic acids, respectively. It is noteworthy that the diesel fuel did not have any detected OCCs in groups **2**, **3**, **4**, and **7**; with exception of group **4** not present in MGO, all other fuel samples had detected OCCs in these chemical groups. As classified as an ultra low sulfur diesel, the groups may have been removed due to refinery processes.

Table 3. Additional m/z values for OCCs detected only in the diesel fuel, the average carbon number, RDB values and chemical formulas of the ions for each ion series.

Range of ions detected	Average carbon number of ions detected	RDB value	Chemical formula of ions
171-325	13	3.5	$C_nH_{(2n-5)}O_4$
259-357	15		$C_nH_{(2n-5)}O_6$
183-337	14	4.5	$C_nH_{(2n-7)}O_4$
195-349	26	5.5	$C_nH_{(2n-9)}O_4$
193-347	15	6.5	$C_nH_{(2n-11)}O_4$
281-393	17		$C_nH_{(2n-11)}O_6$
189-315	15	7.5	$C_nH_{(2n-13)}O_3$
205-331	15		$C_nH_{(2n-13)}O_4$
249-347	16		$C_nH_{(2n-13)}O_5$
185-283	16	8.5	$C_nH_{(2n-15)}O_2$
181-265	15	9.5	$C_nH_{(2n-17)}O$

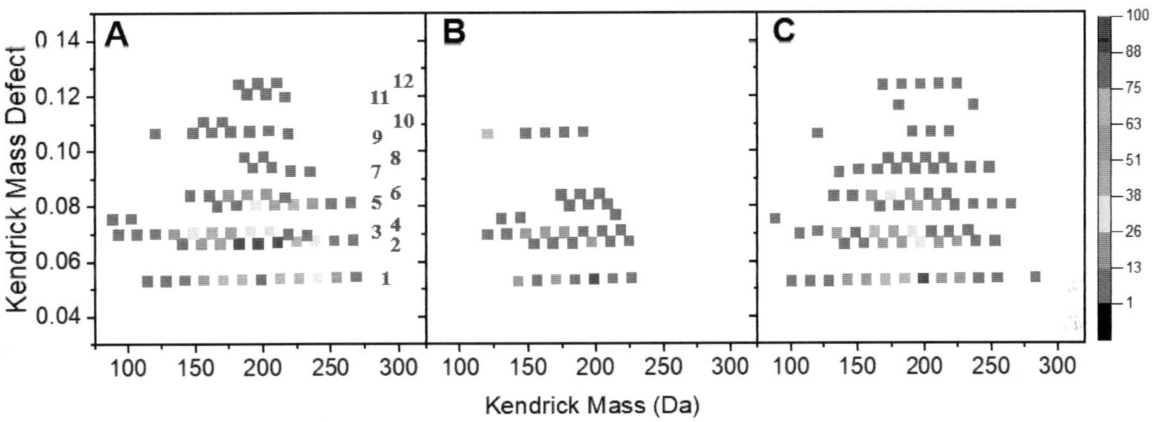

Figure 11. Kendrick Mass Defect plots for ionized OCCs derived from (**A**) Jet (A), (**B**) Jet (B) and (**C**) Jet (C). Numbers 1-12 indicate the proposed individual classes of OCCs: saturated carboxylic acids **1**, cyclic carboxylic acid **2**, phenols **3**, hydroxy acids **4**, bicyclic carboxylic acids **5**, hydroxytetralins **6**, benzenediols / oxyphenols **7**, indenols **8**, benzoic acids **9**, naphthols **10**, cinnamic acids **11**, and diphenylmethanols **12**.

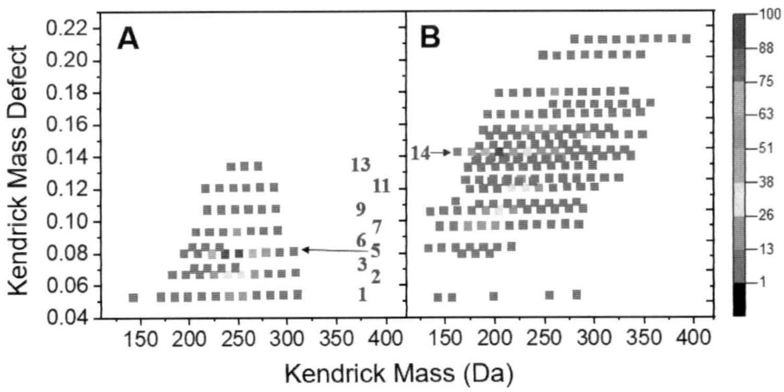

Figure 12. Kendrick Mass Defect plots for ionized OCC derived from (**A**) MGO and (**B**) diesel fuel. The numbers indicate the proposed classes of OCCs. saturated carboxylic acids **1**, cyclic carboxylic acids **2**, phenols **3**, bicyclic carboxylic acids / napthoquinones **5**, hydroxytetralins **6**, oxyphenols / benzenediols **7**, benzoic acids **9**, cinnamic acids **11**, naphthalenediols **13**, and hydroxycinnamic acids / acetyl benzoic acids **14**. If a blue group number is omitted, the class was not detected.

Other differences in the OCC chemical composition of MGO and diesel fuel include the m/z ranges for homologue ion series identified in both samples. For ion groups **9**, **11** and **13**, the m/z range is much greater for OCCs in diesel fuel than the MGO. Further, as the Kendrick Mass Defect increases for the ion series in the diesel fuel, the ions' starting and ending m/z values generally increase (the series shift from left to right with increasing KMD values, **Figure 12 B**). This observation implies the molecular weight for the ions within that series also increase. Hence, ions with the greatest Kendrick Mass Defect are also the largest molecules detected in the diesel fuel.

Collision-activated Dissociation of Oxygen-Containing Compounds

To aid in structural analysis and possible class differentiation of OCCs, CAD measurements were completed on unknown ions derived from jet fuel (C) and compared to model compounds. Deprotonated 2-ethylphenol (m/z 121) and 4-isopropylphenol (m/z 135) were subjected to CAD and compared to the MS^2 spectra of unknown ions of m/z 121 and 135, respectively **(Figure 13)**. Deprotonated 2-ethylphenol and unknown ions of m/z 121 predominantly fragment by loss of 15 Da (methyl radical) and the MS^2 spectra are similar to each other. The MS^2 spectra of deprotonated 4-isopropylphenol and unknown ions of m/z 135 are not similar. If unknown ions of m/z 121 and 135 are representative of phenols, multiple structural isomers are possible. Hence, the lack of similarity in CAD spectra between 4-isopropylphenol and unknown ions of 135 does not preclude the presence of phenols. Additional compounds will be purchased (*i.e.*, tri-methylated phenols, propylphenols) and subjected to CAD to elucidate possible structures of unknown ions of m/z 135 and further probe if the ion series are phenols.

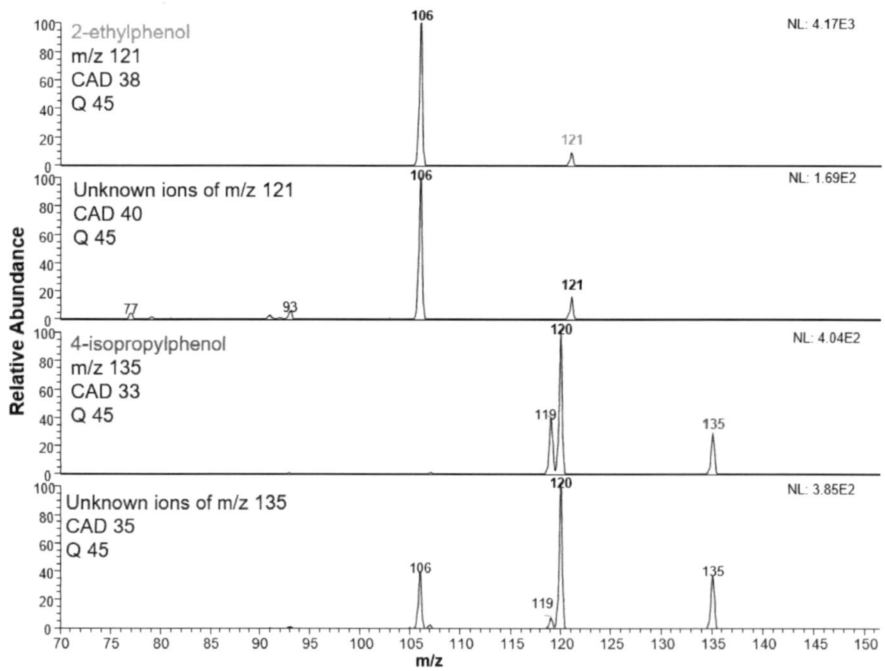

Figure 13. By descending order, the MS^2 spectra of deprotonated 2-ethylphenol, unknown ions of m/z 121, deprotonated 4-isopropylphenol and unknown ions of m/z 135.

Deprotonated 1-octanoic acid (m/z 143) and 1-dodecanoic acid (m/z 199) and unknown ions of m/z 143 and 199 were also subjected to CAD (**Figure 14**). For both model and unknown compounds, losses of 18 Da and 44 Da were observed and correspond to water and CO_2 losses, respectively. Hence, the unknown ions of m/z 143 and 199 likely contain two oxygen atoms and an carboxylic acid functionality. Although the fragment ions (*i.e.,* m/z 125 and 181) are smaller in abundance in the MS^2 spectrum for unknown ions than the model compounds, the spectra are fairly similar. Hence, the class of compounds representative of the chemical formula $C_nH_{(2n-1)}O_2$ are likely saturated carboxylic acids.

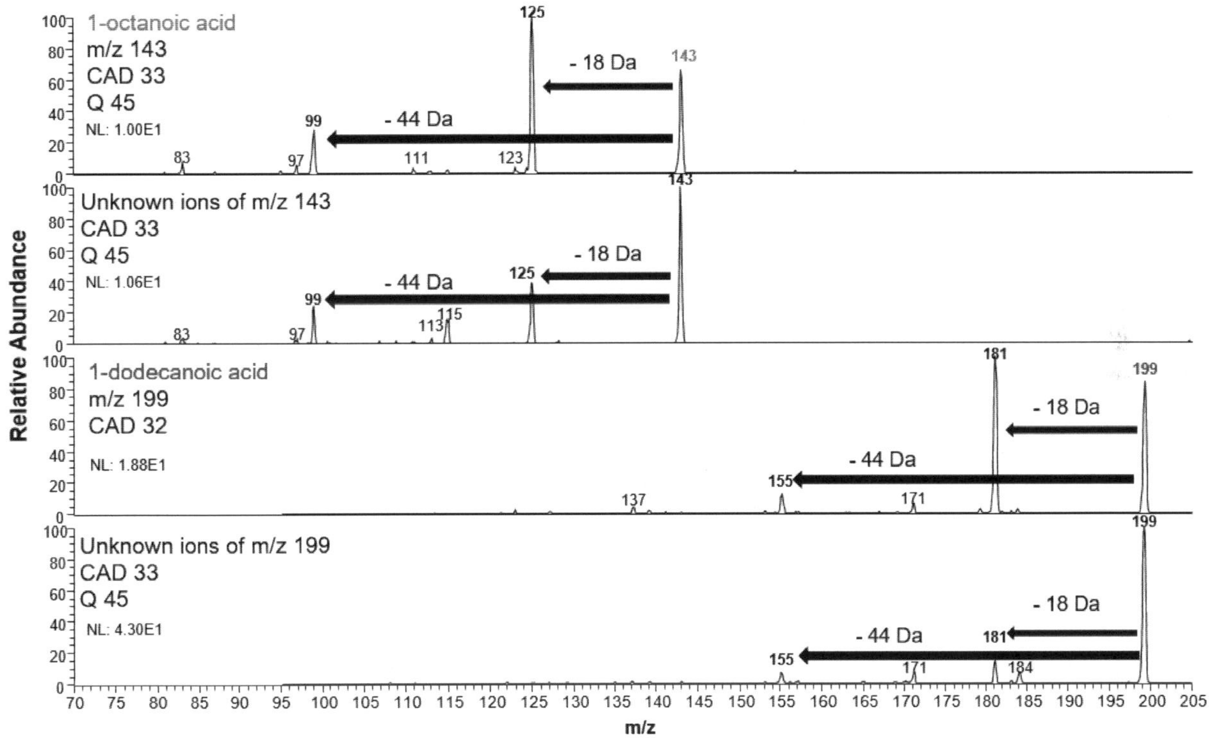

Figure 14. By descending order, the MS^2 spectra of deprotonated 1-octanoic acid, unknown ions of m/z 143, deprotonated 1-dodecanoic acid and unknown ions of m/z 199.

The importance of differentiation classes of OCCs from each other with the same chemical formulas will likely have critical applications for stability properties. Hence, future efforts will include purchasing additional model compounds to aid in elucidating unknown structures and functionalities for OCCs found in fuels.

Quantitative Considerations from (+)/(-) ESI LQIT/Orbitrap

Accurate qualitative data is essential to understanding how HCCs contribute to thermal and storage stability failures. Equally important is obtaining quantitative data for the HCCs in fuels. ESI presents cogent concerns over ionization bias. Consequently, the results obtained by using ESI should not be considered as relative concentrations as certain compounds have greater efficacy at ionization due to their gas-phase basicity / acidity values.[27] The values differ for classes of analytes with specific functionalities and also within the same class of compounds.[15] Therefore, when analyzing MS^1 spectra, it is difficult to extrapolate quantitative data when resorting to direct inject ESI. Methods that incorporate chromatography, flame ionization and/or chemiluminescence detectors with acceptable equimolar responses and linear dynamic ranges show promise at providing accurate quantitative data.[9,36-38] Hence, hybrid analytical methods that provide both accurate quantitative and qualitative data should be considered complementary to each other. Collectively, comprehensive chemical characterization of fuels provided by hybrid methods may provide essential scientific knowledge for understanding the stability of liquid fuels in their production, transportation, storage and use.

CONCLUSIONS

A method using (+) and (-) ESI coupled to LQIT/orbitrap mass spectrometer was developed for the qualitative characterization of OCCs and NCCs in several liquid fuels. The method demonstrated excellent mass accuracy with ppm values less than 10 and exhibited excellent sensitivity. These characteristics are needed to evaluate fuel as a function of time to determine if changes in fuels occur in relatively short durations. Further, individual Kendrick Mass Defect plots were made for both NCCs and OCCs derived from the fuel samples. The plots clearly show compositional differences when the fuels are compared to each other. These qualitative differences in composition are likely critical to understanding stability-related issues in fuels. Additionally, CAD was performed on several unknown ions and compared to known model compounds with the same m/z values. The results support CAD as a reliable tool that provided further structural elucidation of unknown ions and chemical class differentiation. The method also demonstrated high throughput as a simple methanol extraction was only needed prior to analysis. Although not qualitative, the method shows critical importance and superior ability to classify HCCs likely not possible when using mass spectrometers with less resolution and mass accuracy. Hence, the ESI-LQIT/orbitrap mass spectrometer method should be considered complementary to other established methods that provide qualitative data (*i.e.,* two-dimensional gas chromatography coupled to flame ionization detector, two-dimensional gas chromatography / (electron ionization) high resolution TOF spectrometry and two-dimensional gas chromatography coupled to chemiluminescence detectors). Together, the chemical information obtained from such methods will likely aid in developing stronger links between fuel composition and storage and thermal stability.

ACKNOWLEDGEMENTS

The authors would like to acknowledge the Defense Logistics Agency - Energy and project manager Yan Guo for their support and guidance of this work.

REFERENCES

1. Maurice, L.Q.; Lander, H.; Edwards, T.; Harrison, W.E. Advanced Aviation Fuels: A Look Ahead Via a Historical Perspective. *Fuel*, **2001**, *80* (5) 747-756.
2. Romanczyk, M.; Ramirez Velasco, J.H.; Xu, L.; Vozka, P.; Dissanayake, P.; Wehde, K.E.; Roe, N.; Keating, E.; Kilaz, G.; Trice, R.; Luning Prak, D.J.; Kenttämaa, H. I. The Capability of Organic Compounds to Swell Acrylonitrile Butadiene O-rings and Their Effects on Mechanical Properties. *Fuel* **2019**, *238*, 483 – 492.
3. Corporan, E.; Edwards, T.; Shafer, L.; DeWitt, M.J.; Klingshirn, C.; Zabarnick, S.; West, Z.; Striebich, R.; Graham, J.; Klein, J. Chemical, Thermal Stability, Seal Swell, and Emissions Studies of Alternative Fuels. *Energy Fuels* **2011**, *25*, 955 – 966.
4. Smith, B.L.; Bruno, T.J. Composition-Explicit Distillation Curves of Aviation Fuel JP-8 and a Coal-Based Jet Fuel. *Energy Fuels* **2007** *21*, 2853 – 2862.
5. Berrier, K.L.; Freye, C.E.; Billingsley, M.C.; Synovec, R.E. Predictive Modeling of Aerospace Fuel Properties Using Comprehensive Two-Dimensional Gas Chromatography with Time-of-flight Mass Spectrometry and Partial Least Squares Analysis. *Energy Fuels* **2020**, *34*, 4084 – 4094.
6. Shi, X.; Li, H.; Song, Z.; Zhang, X.; Liu, G. Quantitative Composition-property Relationship of Aviation Hydrocarbon Fuel Based on Comprehensive Two-Dimensional Gas Chromatography with Mass Spectrometry and Flame Ionization Detector. *Fuel* **2017**, 100, 395 – 406.
7. Lovestead, T.M.; Burger, J.L.; Schneider, N.; Bruno, T.J. Comprehensive Assessment of Composition and Thermochemical Variability by High Resolution GC/QToF-MS and the Advanced Distillation-Curve Method as a Basis of Comparison for Reference Fuel Development. *Energy Fuels* **2016**, 30, 10029 – 10044.
8. Balster, L.M.; Zabarnick, S.; Striebich, R.C.; Shafer, L.M.; West, Z.J. Analysis of Polar Species in Jet Fuel and Determination of Their Role in Autoxidative Deposit Formation. *Energy Fuels* **2006**, *20*, 2564 – 2571.
9. Striebich, R.C.; Contreras, J.; Balster, L.M.; West, Z.; Shafer, L.M.; Zabarnick, S. Identification of Polar Species in Aviation Fuels Using Multidimensional Gas Chromatography-Time of Flight Mass Spectrometry. *Energy Fuels* **2009**, *23*, 5474 – 5482.
10. Timko, M.T.; Schmois, E.; Patwardhan, P.; Kida, Y.; Class, C.A.; Green, W.H.; Nelson, R.K.; Reddy, C.M.. Response of Different Types of Sulfur Compounds to Oxidative Desulfurization of Jet Fuel. *Energy Fuels* **2014**, *28*, 2977 – 2983.
11. Link, D.D.; Baltrus, J.P.; Rothenberger, K.S. Class- and Structure-Species Separation, Analysis, and Identification Techniques for the Characterization of the Sulfur Components of JP-8 Aviation Fuel. *Energy Fuels* **2003**, *17*, 1292 – 1302.
12. Link, D.D.; Baltrus, J.P.; Zandhuis, P.; Hreha, D.C. Extraction, Separation, and Identification of Polar Oxygen Species in Jet Fuel. *Energy Fuels* **2005**, *19*, 1693 – 1698.
13. Lee, I.C.; Ubanyionwu, H.C. Determination of Sulfur Contaminants in Military Jet Fuels. *Fuel* **2003**, *87*, 312 – 318.

14. van der Westhuizen, R.; Ajam, M.; De Coning, P.; Beens, J.; de Villiers, A. Comprehensive Two-dimensional Gas Chromatography for the Analysis of Synthetic and Crude-derived Jet Fuels. *J. Chromatogr. A* **2011**, *1218*, 4478 – 4486.

15. Adams, R.K.; Zabarnick, S.; West, Z.J.; Striebich, R.C.; Johnson, D.W. Chemical Analysis of Jet Fuel Polar, Heteroatomic Species via High-Performance Liquid Chromatography with Electrospray Ionization Mass Spectrometric Detection. *Energy Fuels* **2013**, *27*, 2390 – 2398.

16. Bauserman, J.W.; Mushrush, G.W. Hardy, D.R. Organic Nitrogen Compounds and Fuel Instability in Middle Distillate Fuels. *Ind. Eng. Chem. Res.* **2008**, *47*, 2867 – 2875.

17. R. Epping, S. Kerkering and J.T. Anderson. Influence of Different Compound Classes on the Formation of Sediments in Fossil Fuels During Aging. *Energy Fuels* **2014** *28*, 5649 – 5656.

18. Heneghan, S.P.; Zabarnick, S. Oxidation of Jet Fuels and the Formation of Deposit. *Fuel* **1994**, *739*, 35 – 43.

19. Beaver, B.; Gao, L.; Burgress-Clifford, C. Sobkowiak, M. On the Mechanisms of Formation of Thermal Oxidative Deposits in Jet Fuels. Are Unified Mechanisms Possible for Both Storage and Thermal Oxidative Deposit. *Energy Fuels* **2005**, *19*, 1574 – 1579.

20. Kuprowicz, N. J.; Zabarnick, S.; West, Z. J.; Ervin, J. S. Use of measured species class concentrations with chemical kinetic modeling for the prediction of autoxidation and deposition of jet fuels. *Energy Fuels* **2007**, *21* (2), 530−544.

21. Hardy, D. R.; Wechter, M. A. Characterization of soluble macromolecular oxidatively reactive species (SMORS) from middle distillate diesel fuels: Their origin and role in instability. *Energy Fuels* **1994**, *8*, 782 – 787.

22. MIL-STD-3004D. Department of Defense Standard Practice Quality Assurance/Surveillance for Fuels, Lubricants and Related Products. https://assist.dla.mil, 2016.

23. Link, D.D.; Baltrus, J.P. Isolation and Identification of Nitrogen Species in Jet Fuel and Diesel Fuel. *Energy Fuels* **2007**, *21*, 1575 – 1581.

24. Yuan, S.Y.; Li, H.; Liu, Z.; Wang. Y.; Wang, L.; Zhang. X. Measurement of Non-hindered and Hindered Phenolic Species in Aviation Fuels via Tandem-SPE. With Comprehensive GCxGC-MS/FID. *Fuel* **2021**, *287*, 119561, 1 – 8.

25. Sharma, P. Spectroscopic Analysis of Jet A-1 Heteroatomic Components. *Chem. Eng. Sci.* **2019**, *207*, 588 – 599.

26. Liu, G.; Cao, Y.; Jiang, R.; Wang, L.; Zhang, X.; Mi, Z. Oxidative Desulfurization of Jet Fuels and Its Impact on Thermal-Oxidative Stability. *Energy Fuels* **2009**, *23*, 5978 – 5985.

27. Hoffman, E.; Stroobant, V. *Mass Spectrometry Principles and Applications*. John Wiley & Sons, Hoboken, 2007.

28. Yerabolu, R.; Kotha, R.R.; Niyonsaba, E.; Dong, X.; Manheim, J.M.; Kong, J.; Riedeman, J.S.; Romanczyk, M. Johnston, C.T.; Kilaz, G.; Kenttämaa, H.I. Molecular Profiling of Crude Oil by Using Distillation Precipitation Fractionation Mass Spectrometry (DPF-MS). *Fuel* **2018**, *234*, 492-501.

29. Frankenfeld, J.W.; Taylor, W.F. *Fundemental Synthetic Fuel Stability Study. U.S. Department of Energy Report No. DOE/BC/10045-12*; U.S. Department of Energy, 1981.

30. Frankenfeld, J.W.; Taylor, W.F.; Brinkman, D.W. Storage Stability of Synfuels from Oil Shale. 1. General Features of Sediment Formation in Model Fuel Systems. *Ind. Eng. Chem. Prod. Res. Dev.* **1983**, *22*, 608 – 614.

31. Frankenfeld, J.W.; Taylor, W.F.; Brinkman, D.W. Storage Stability of Synfuels from Oil Shale. 2. Effects of Nitrogen Compound Type and the Influence of other nonhydrocarbons on Sediment Formation in Model Fuel Systems. *Ind. Eng. Chem. Prod. Res. Dev.* **1983**, *22*, 615 – 621.

32. Batts, B.D.; Fathoni, A.Z. A Literature Review on Fuel Stability Studies with Particular Emphasis on Diesel Oil. *Energy Fuels* **1991**, *5*, 2 – 21.

33. Astolfi, P.; Greci, L.; Rizzoli, C.; Sgarabotto, P.; Marrosu, G. Radical Intermediates in the Peroxidation of Indoles. *J. Chem. Soc., Perkin Trans.* **2001**, *2*, 1634 – 1640.

34. Busch, K.L.; Glish, G.L.; McLuckey, S.A. (1988) *Mass Spectrometry/Mass Spectrometry Techniques and Applications of Tandem Mass Spectrometry*; VCH Publishers Inc: New York.

35. Grossman, R. B. (2003) *The Art of Writing Reasonable Organic Reaction Mechanisms* (2nd Edition). Springer: New York.

36. Romanczyk, M.; Deese, R.D.; T. Loegel, T.N.; Metz, A.E.; Morris, R.E.; Kamin, R.; McDaniel, A.; Peretich, M. Examination of two-dimensional gas chromatography with a nitrogen chemiluminescence detector to facilitate quantitation and characterization of nitrogen-containing compounds in petroleum-derived fuels. *Energy Fuels* **2021**, *35 (7)*, 5867 – 5878.

37. Deese, R.D.; Morris, R.E.; Metz, A.E.; Myers, K.M.; Johnson, K.; Loegel, T.N. Characterization of Organic Nitrogen Compounds and Their Impact on the Stability of Marginally Stable Diesel Fuels. *Energy Fuels* **2019**, *33 (7)* 6659 – 6669.

38. Striebich, R.C.; Shafer, L.M.; Adams, R.K.; West, Z.J.; DeWitt, M.J. Zabarnick, S. Hydrocarbon Group-Type Analysis of Petroleum-Derived and Synthetic Fuels Using Two-Dimensional Gas Chromatography. *Energy Fuels* **2014**, *28*, 5696 – 5706.

SUPPORTING INFORMATION

CHEMICAL CHARTERIZATION OF NITROGEN- AND OXYGEN-CONTAINING COMPOUNDS IN LIQUID FUELS BY USING ELECTROSPRAY IONIZATION COUPLED TO A LINEAR QUADRUPOLE ION TRAP/ORBITRAP MASS SPECTROMETER

Mark Romanczyk[1*] and Thomas Loegel[1]

[1]US Naval Research Laboratory, Washington, District of Columbia 20375, United States. *mark.romanczyk@nrl.navy.mil, orcid.org/0000-0002-6907-9613

Table S1. Representative raw data collected for NCCs derived from jet fuel (A) with the chemical composition formula of $C_nH_{(2n-2)}N$.

m/z	Relative abundance	Mass accuracy (ppm)	RDB value	Elemental Composition of ions
166.15937	4.91	2.04	2.5	$C_{11}H_{20}N$
180.17509	5.81	2.28	2.5	$C_{12}H_{22}N$
194.19083	7.67	2.59	2.5	$C_{13}H_{24}N$
208.20649	8.86	2.47	2.5	$C_{14}H_{26}N$
222.22219	7.43	2.53	2.5	$C_{15}H_{28}N$
236.23785	4.57	2.42	2.5	$C_{16}H_{30}N$
250.25350	2.14	2.3	2.5	$C_{17}H_{32}N$
264.26930	1.06	2.72	2.5	$C_{18}H_{34}N$

Table S2. Representative raw data collected for OCCs derived from jet fuel (C) with the chemical composition formula of $C_nH_{(2n-7)}O$.

m/z	Relative abundance	Mass accuracy (ppm)	RDB value	Elemental Composition of ions
107.04976	4.23	5.82	4.5	C_7H_7O
121.06465	1.16	-1.15	4.5	C_8H_9O
135.08082	56.56	2.76	4.5	$C_9H_{11}O$
149.09638	80.13	1.92	4.5	$C_{10}H_{13}O$
163.11195	66.28	1.28	4.5	$C_{11}H_{15}O$
177.12746	50.63	0.40	4.5	$C_{12}H_{17}O$
191.14308	29.26	0.21	4.5	$C_{13}H_{19}O$
205.15871	12.2	0.10	4.5	$C_{14}H_{21}O$
219.17430	5.75	-0.18	4.5	$C_{15}H_{23}O$
233.18980	1.24	-0.81	4.5	$C_{16}H_{25}O$

Figure S1. General schematic of the Thermo Fisher Scientific LQIT/orbitrap mass spectrometer with germane parts and concepts labeled.

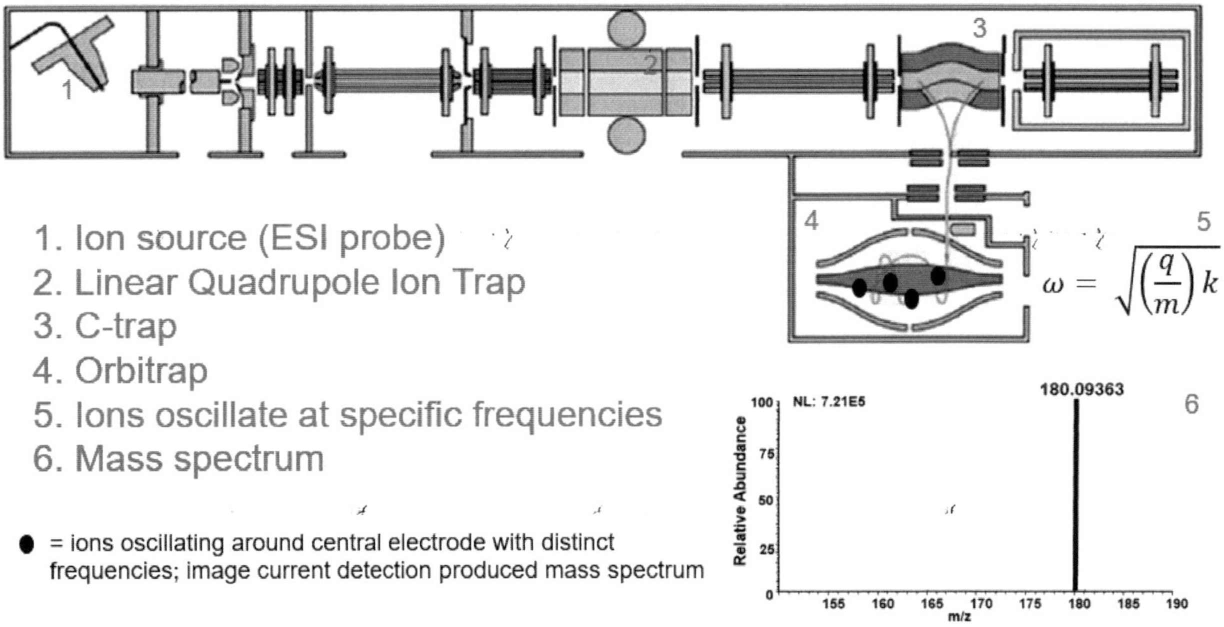

Figure S2. Location of the ion trap in the Thermo Fisher Scientific LQIT/orbitrap mass spectrometer where the CAD process occurs.

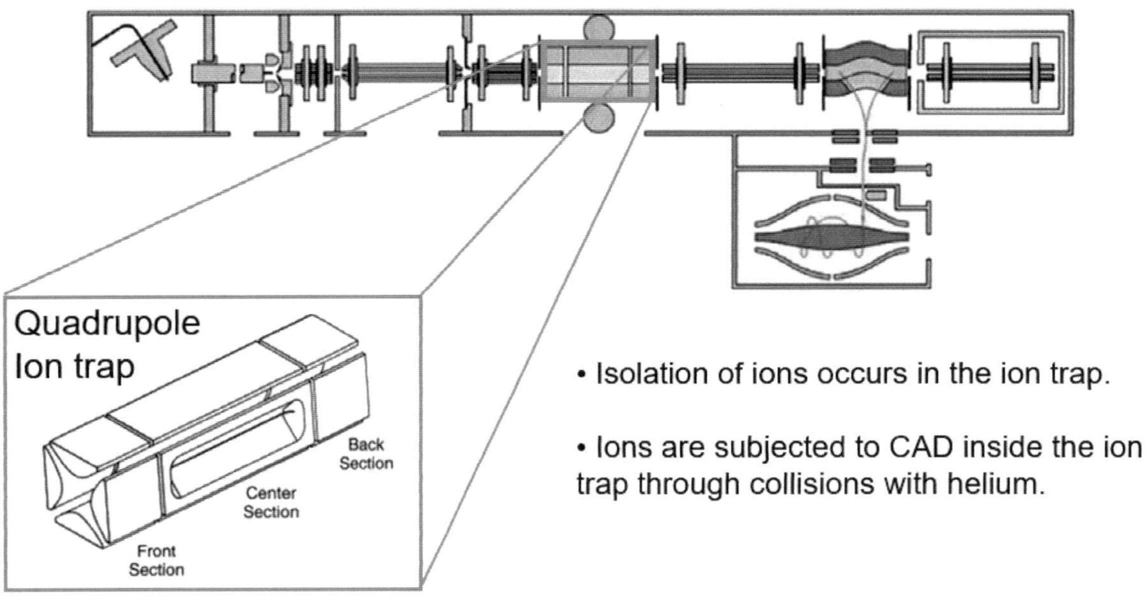

Equation S1. Proton affinity and gas phase basicity are defined as the negative enthalpy change or negative Gibbs free energy change, respectively, for the reaction shown.

$$M + H^+ \rightleftharpoons MH^+$$

Equation S2. The gas phase acidity of a reaction is defined as the negative Gibbs free energy change for the removal a proton from a molecule.

$$MH \rightleftharpoons M^- + H^+$$

IASH 2022, the 17th INTERNATIONAL CONFERENCE ON STABILITY, HANDLING AND USE OF LIQUID FUELS
Dresden, Germany
September 11-15, 2022

THE IMPACT OF NON-FERROUS CORROSION INHIBITORS ON REACTIVE SULFUR SPECIES AND THE DOCTOR TEST

Matthew T. Barnes[1], Sankaran Murugesan[2], Brandon Vittur[3]

[1]*Subject Matter Expert: Finished Fuel Additives, Baker Hughes, 12645 West Airport Blvd., Sugar Land, TX 77478, matthew.barnes@bakerhughes.com*

[2]*Project Manager, Baker Hughes, 12645 West Airport Blvd., Sugar Land, TX 77478, sankaran.murugesan@bakerhughes.com*

[3]*Project Manager, Baker Hughes, 12645 West Airport Blvd., Sugar Land, TX 77478, brandon.vittur@bakerhughes.com*

ABSTRACT

It is well known that the presence of hydrogen sulfide in fuels can result in failure of copper and silver corrosion tests, and can be detected by several methods, including the doctor test. Additionally, the presence of elemental sulfur is known to cause non-ferrous corrosion issues but can be more difficult to detect and quantify. The presence of mercaptans alone will not typically cause tarnish to silver, but in combination with elemental sulfur, can form hydrogen sulfide and polysulfide compounds that corrode and tarnish silver. Additives containing dialkyl dithiothiadiazoles are commonly used to correct copper and silver corrosion problems. In certain circumstances, these same additives were being tested in field locations and reportedly correcting failures of the Doctor test. It was not immediately apparent what mechanism might be allowing a fuel to become Doctor negative following non-ferrous corrosion inhibitor treatment. This paper presents a detailed experimental study to understand the nature of the interactions between non-ferrous corrosion inhibitor additives and active sulfur species. Reaction mechanisms are proposed by which the corrosion inhibitor may be affecting the Doctor test results.

Keywords: mercaptans, sulfur, Doctor test, non-ferrous corrosion, corrosion inhibitor, disulfide, thiol-disulfide exchange

INTRODUCTION

It is well known that the presence of hydrogen sulfide in fuels can result in failure of copper and silver corrosion tests. Hydrogen sulfide can be detected by several methods, including lead acetate paper, the Doctor test, gas detection tubes, gas chromatography and potentiometric titration. Hydrogen sulfide is highly toxic, corrosive to ferrous and non-ferrous metals, and presents an odor concern.

The presence of elemental sulfur (as well as polysulfides) is also known to cause non-ferrous corrosion issues but typically is more difficult to detect and quantify. Elemental sulfur can form via oxidation of hydrogen sulfide in the fuel. The Doctor test may detect elemental sulfur.

Mercaptans also contribute to non-ferrous corrosion issues. Mercaptans can be detected by various methods, such as the Doctor test, gas detection tubes, gas chromatography and potentiometric titration. The corrosiveness to copper coupons from mercaptans is well documented[1]. In addition to their contribution to non-ferrous corrosion, mercaptans tend to have objectionable odors.

Short-chain mercaptans (generally C1-C4 range) tend to be the most reactive, at least with respect to reactivity towards a scavenging chemical. Prior silver corrosion studies by Baker Hughes demonstrate that mercaptans alone do not tend to produce visible tarnish on silver test coupons. When mercaptans are combined with sulfur in a fuel sample (as sulfur powder), the solution is corrosive to the silver coupon. In this respect, the mercaptans may interact with sulfur to form disulfides, polysulfides or hydrogen sulfide[1,2]. Elemental sulfur alone in solution, at sufficient concentration, will tarnish both copper and silver coupons.

Additives containing dialkyl dithiothiadiazoles are commonly used to correct copper and silver corrosion problems. It is generally accepted, and supported in literature[3], that these types of non-ferrous corrosion inhibitors (NFCI) work by forming a protective film on the metal surface. These inhibitors are sometimes described as "fooling" the corrosion test in this manner. It is also thought that the NFCI may react with and possibly scavenge elemental sulfur from fuels. This study reinforces the notion that other mechanisms may be at work, and that the concept of the corrosion inhibitor filming the metal surface may not be entirely descriptive of the mechanism of action.

The Doctor Test is used to detect the presence of mercaptans, hydrogen sulfide and elemental sulfur in gasoline and other fuels. For the reasons previously stated, the presence of these reactive sulfur compounds in finished fuels is undesirable. Finished fuels may have a requirement to be Doctor negative.

Baker Hughes became aware that NFCI additives were being tested by field operations and reportedly found to correct failures of the Doctor Test in gasoline. It was not immediately apparent what mechanism might be allowing a fuel to become Doctor negative following non-ferrous corrosion inhibitor treatment. In response, a detailed experimental study was undertaken to understand the nature of the interactions between NFCI additives and reactive sulfur species. Several mechanisms are proposed by which the corrosion inhibitor may be affecting the Doctor test results.

EXPERIMENTAL

All work detailed in this study (where a fuel is indicated) was conducted in a surrogate gasoline-like test fuel composed of a mixture of 75% iso-octane (2,2,4-trimethylpentane) and 25% toluene, reagent-grade materials obtained from Sigma-Aldrich. Toluene was included to ensure dissolution of sulfur[4]. The fuel was analyzed for mercaptan sulfur and subjected to the Doctor test. It was shown to contain no mercaptan and was Doctor negative.

Silver corrosion testing was carried out according to ASTM D7671: Standard Test Method for Corrosiveness to Silver by Automotive Spark-Ignition Engine Fuel – Silver Strip Method. Silver coupons were polished according to the test method guidelines, and exposed to 30 mL samples of the described fuel solutions for a period of three hours at a temperature of 50°C. Upon test conclusion, the coupons were rated on a scale of 0 to 4, with 0 indicating no tarnish, and 4 indicating blackening of the coupon surface.

Doctor tests were carried out according to ASTM D4952: Standard Test Method for Qualitative Analysis for Active Sulfur Species in Fuels and Solvents (Doctor Test). In a test tube, 10 mL of the fuel sample to be tested was mixed with 5 mL of the sodium plumbite solution (obtained from Ricca Chemical), followed by addition of ~20 mg of sulfur (sulfur powder, precipitated, 99.5%, obtained from Alfa Aesar). Observations of color change to the solution or the sulfur powder were made to determine a passing or failing result.

Potentiometric titrations were carried out according to UOP163: Hydrogen Sulfide and Mercaptan Sulfur in Liquid Hydrocarbons. Fuel samples were titrated against 0.0141N silver nitrate solution (obtained from Ricca Chemical) to quantify hydrogen sulfide and mercaptans.

Raman analysis was conducted using a handheld Raman Spectrometer, from Bruker Optics. Nuclear Magnetic Resonance analysis was conducted using a 400 MHz NMR instrument, also from Bruker Optics.

RESULTS AND DISCUSSION

To understand the effect of mercaptans on corrosion of silver coupons, solutions containing ethanethiol, n-propanethiol and n-butanethiol in the test fuel were evaluated. Silver was chosen for evaluation rather than copper, as silver has been noted to be more sensitive to common reactive sulfur species[5]. Note that where "90 ppm mercaptans" is indicated, a mixture containing 30 ppm each of ethanethiol, n-propanethiol and n-butanethiol was used. All coupons shown in Figure 1 were rated "0" for silver corrosion.

We observed that addition of mercaptan to a sample containing sulfur does not necessarily change the corrosion rating relative to that of a sample containing the same amount of sulfur. Instead, the color of the tarnish can appear different relative to a sample containing sulfur only. It has been reported that mercaptans can mask the corrosiveness of other sulfur species; however, mercaptans do not seem to mask the corrosiveness from sulfur. In this case, the corrosion rating of the coupon tested in the fuel containing 2.5 mg/L sulfur rated "0", while addition of mercaptans to the fuel containing 2.5 mg/L sulfur produced corrosion ratings of "2". Increasing sulfur to 5 mg/L resulted in a "3" rating. Addition of mercaptans along with 5 mg/L did not change the rating, but differences in color were observed.

Acceptable corrosion test results (rating "1" or higher) could be obtained for each of these conditions with addition of 5 ppm of NFCI.

Figure 1. Silver Corrosion Test Coupons.

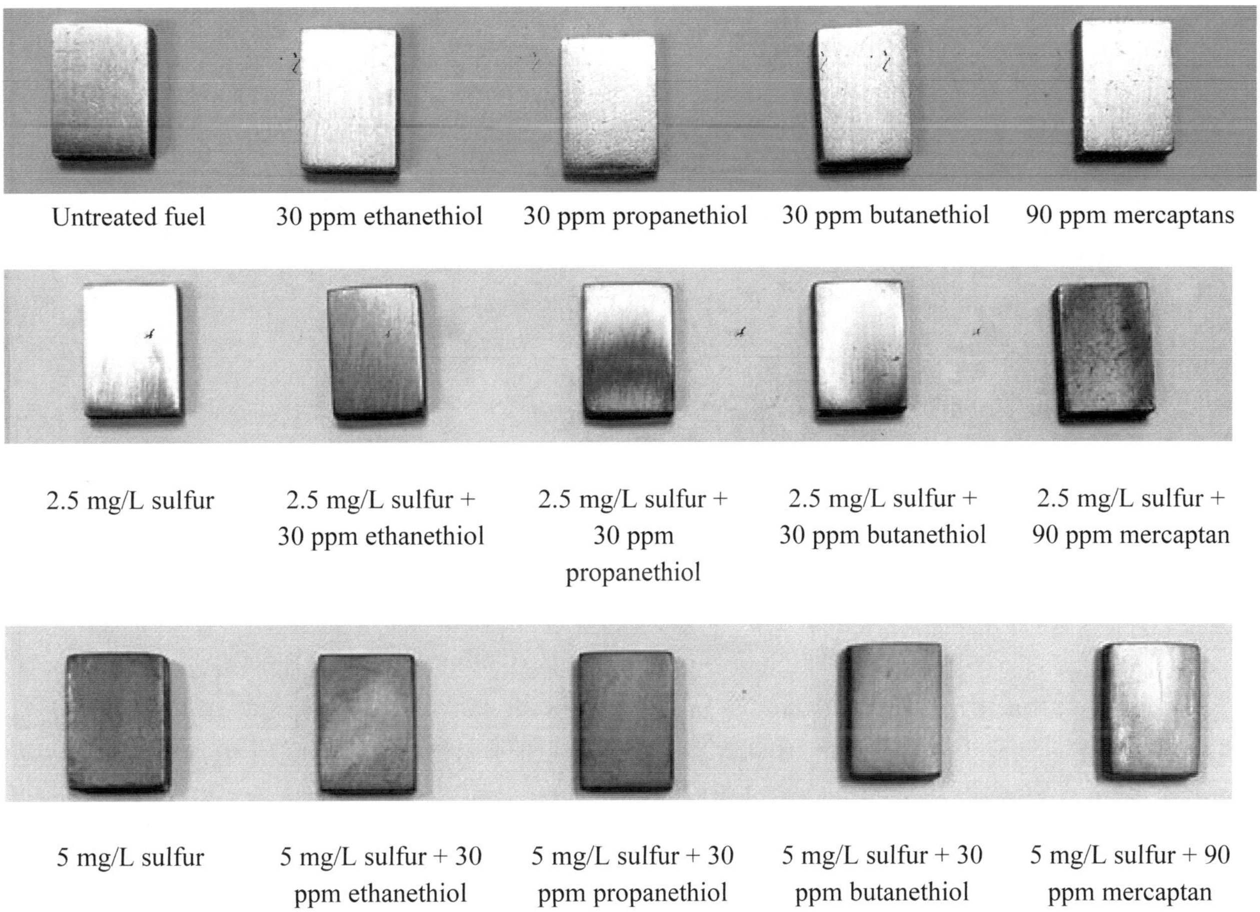

A series of experiments was conducted to determine possible reaction mechanisms between the NFCI and sulfur compounds in the fuel. We considered a possible reaction between the NFCI and mercaptans, between the NFCI and elemental sulfur, and between the NFCI and hydrogen sulfide.

Mercaptans were considered first for their contribution to failure of the Doctor test. We determined the approximate quantity of mercaptan sulfur required to produce a positive result in the Doctor Test. Small quantities of 1-propanethiol were added to the test fuel and the Doctor Test was performed until the point of failure was determined to occur at ~13-14 ppm S. Mercaptans were quantified by potentiometric titration.

Using the sample of fuel with propanethiol that failed the doctor test (which was determined to contain 13.4 ppm mercaptan sulfur), the NFCI in successively higher concentrations to determine the concentration of the NFCI required to produce a passing result. The solutions were mixed on the stir plate overnight, for a period of 18 hours, after which the mercaptan sulfur concentration was determined and Doctor Test results were recorded. A concentration of 25 ppm was found to elicit a negative Doctor Test result, as shown in Table 1.

Table 1: Determination of the Quantity of NFCI Required to Correct Failing Doctor Test Result.

Sample Description	Mercaptan, as ppm S	Doctor Test Result
Test fuel + 1-propanethiol	10.8	Positive
Test fuel + 1-propanethiol + 5 ppm NFCI	10.8	Positive
Test fuel + 1-propanethiol + 25 ppm NFCI	10.4	Negative

A similar test was conducted using 1-butanethiol and found that a sample containing 13.7 ppm mercaptan sulfur resulted in a positive Doctor Test result. The NFCI was added at a dosage of 50 ppm, and after 18 hours, results of Doctor Test and mercaptan titration were recorded. Treatment with NFCI at 50 ppm produced a negative Doctor test result. See Table 2.

Table 2: Determination of the Quantity of NFCI Required to Correct Failing Doctor Test Result.

Sample Description	Mercaptan, as ppm S	Doctor Test Result
Test fuel + 1-butanethiol	14.1	Positive
Test fuel + 1-butanethiol + 50 ppm NFCI	10.4	Negative

Another study was conducted with ethanethiol in the test fuel, targeting 30 ppm S. The NFCI was added to the solution at 1:1 and 4:1 ratio (NFCI to ethanethiol). Samples were taken for mercaptan sulfur analysis and Doctor Test at 1 hour, 2 hours and 24 hours. Mercaptan titration results are given in Table 3, in ppm S, with Doctor Test results in parenthesis next to the sulfur values. In this case, the mercaptan concentration was not sufficiently reduced within 24 hours to produce a negative Doctor Test result.

Table 3. Meracptan Sulfur and Doctor Test Results for Test Fuel with Ethanethiol and NFCI.

Sample Description	Initial	1 hour	2 hours	24 hours
Test fuel + ethanethiol	29.82 (+)	25.89 (+)	26.20 (+)	26.10 (+)
Test fuel + ethanethiol + 1:1 molar ratio NFCI	-	21.55 (+)	19.60 (+)	15.12 (+)
Test fuel + ethanethiol + 4:1 molar ratio NFCI	-	15.87 (+)	14.45 (+)	13.37 (+)

The next part of the investigation focused on analysis to determine the mechanism responsible for reduction in mercaptan sulfur levels.

A solution was prepared by adding butanethiol to a sample of neat NFCI. The sample was mixed and allowed to react prior to analysis via ^{13}C Nuclear Magnetic Resonance Spectroscopy (NMR).

While the reaction was proceeding, we observed the formation of a yellow precipitate that started to form within the first hour.

Samples of the mixture were taken over a period of several hours and analyzed via NMR. Those scans were compared to a scan of the NFCI product. Figure 2 shows the NMR spectrum of the NFCI product. The circled peak corresponds to the two carbons in the central ring structure of the corrosion inhibitor's active component.

Figure 2. ^{13}C NMR Spectrum of NFCI.

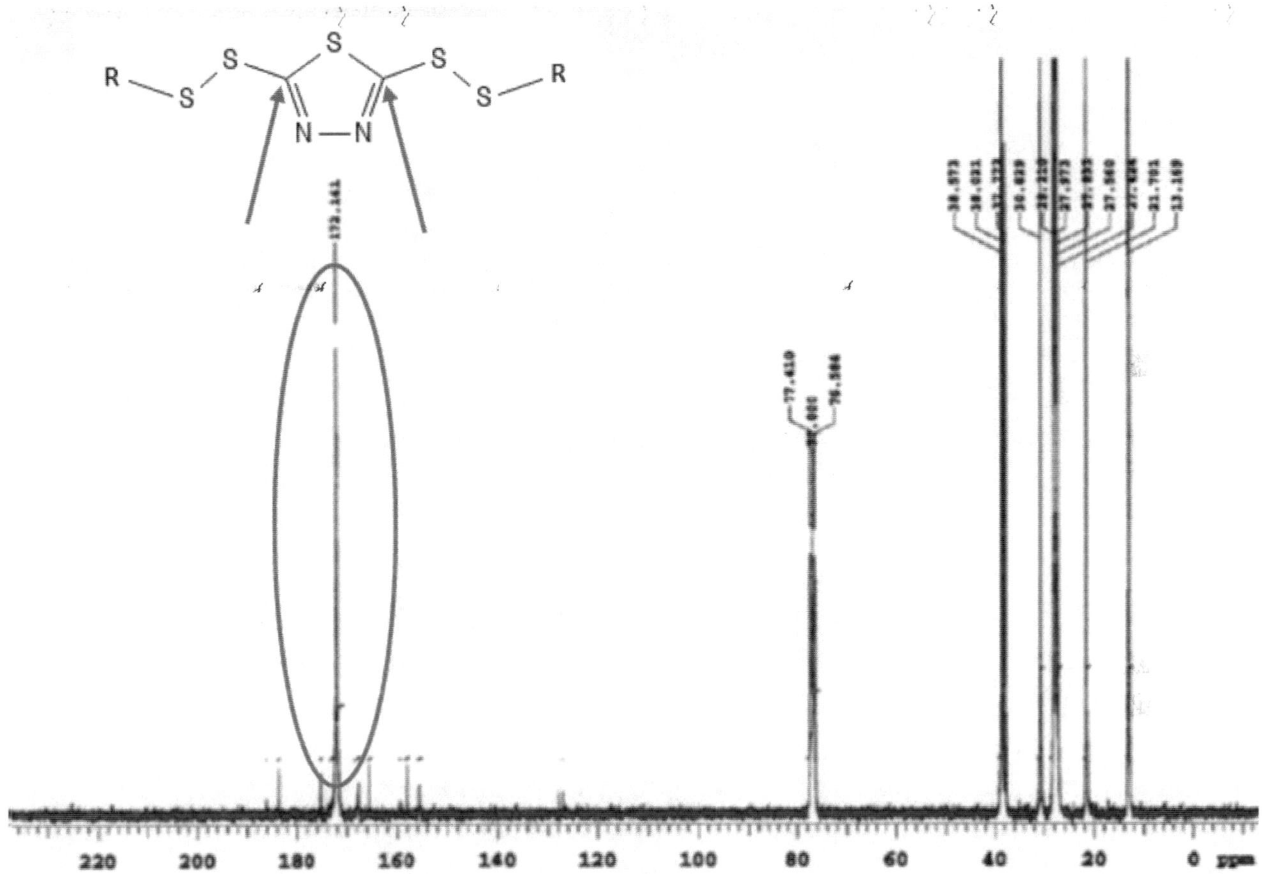

In the proposed reaction scheme, a mercaptan (in this case, butanethiol) reacts with the corrosion inhibitor, combining with the R group from the NFCI to form a disulfide, in a thiol-disulfide exchange or thiol-disulfide interchange mechanism[6]. Figure 3 shows the proposed reaction pathway by which the concentration of available mercaptan in solution may be reduced, facilitating passing Doctor Test results. The reaction mechanism is similar to a thiol-pyridyl disulfide exchange described by Altinbasak et al[7] (see figure 4). The rate of reaction is governed by the pK_a of the thiol compound, thus in general a shorter chain mercaptan is expected to be more reactive. Their research indicates that this type of reaction can occur under mild conditions, in a variety of solvents, and over a range of temperatures and pH levels. However, higher pH facilitates the reaction. It should be noted that no heat was required to facilitate this reaction. The reactions were carried out at laboratory ambient temperatures (~70-72°F).

The resulting disulfide compound may act as an inhibitor of silver tarnishing[1,8], representing an additional potential mode of action for the inhibitive effect of the NFCI.

Using NMR to study the reaction, we observed a change from the dialkyl dithiothiadiazole to an asymmetric molecule (see Figure 3, labeled A) with a single alkyldithio group, resulting in the appearance of two new peaks. See Appendix 1. These two peaks correspond to the same two carbon atoms in the ring structure of the corrosion inhibitor.

Figure 3. Proposed Thiol-Disulfide Exchange Reaction Mechanism.

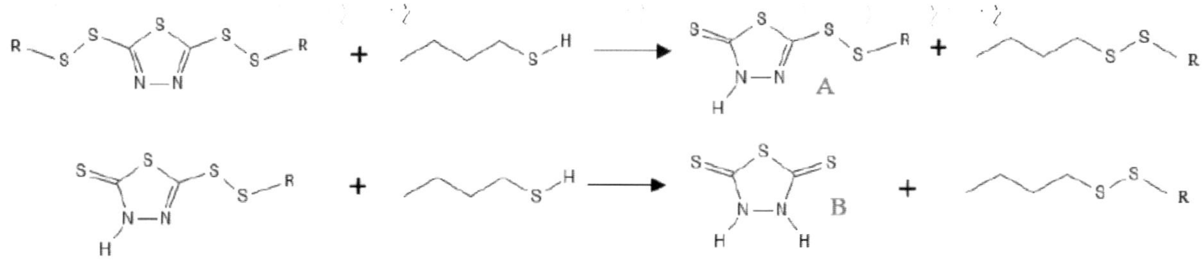

Figure 4. Thiol-Pyridyl Disulfide Exchange Mechanism.

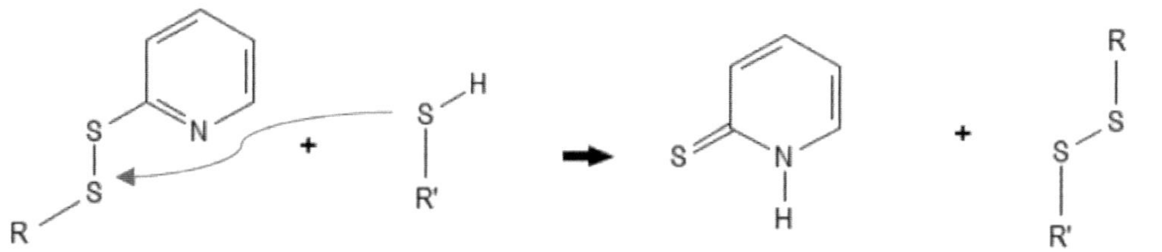

With increasing concentration of butanethiol, we observed a decrease in the concentration of the product A, and an increase in the resulting disulfide compound. These spectra can be found in Appendix 1. The NFCI appears to be steadily consumed, eventually leaving a mixture of disulfides, thiols, and solid precipitate. Integration data for peaks corresponding to the mono-alkyldithio thiadiazole intermediate (component A) and the primary disulfide are also given in Appendix 1. A comparison of the NMR spectra of the NFCI before and after these reactions (1:1, 2:1 and 4:1 ratios of butanethiol to NFCI) is given in Appendix 2.

We did not observe the molecule labeled B in Figure 3 on these NMR scans. As previously mentioned, a solid yellow precipitate was formed, which is insoluble in the hydrocarbon solvents used. This precipitate was isolated, analyzed, and found to be a very close spectral match to 2,5-dimercapto-1,3,4-thiadiazole (DMTD). The comparative NMR spectra between DMTD and the precipitate are shown in Appendix 3. Elemental analysis of the precipitate, as compared to theoretical composition and analysis of a sample of 98% DMTD, obtained from Sigma-Aldrich, support the conclusion that DMTD is the precipitate from this reaction. See Table 4. NMR peak integration data for the spectra in Appendix 3 is given in Table 5.

Table 4. Elemental Analysis Comparison of Purified Precipitate with DMTD.

Sample Description	Wt.% C	Wt.% H	Wt.% N	Wt.% S
Theoretical DMTD	16	1.3	18.7	64
98% Pure DMTD	15.9	1.2	18.4	64.4
Precipitate Material	16.7	1.3	18.1	62.5

Table 5. NMR Comparison of Purified Precipitate with DMTD.

98% DMTD Peaks	Integration	Precipitate Peaks	Integration
190.68	20.34	190.37	40.95
171.73	100	171.74	100
155.19	22.57	154.97	35.04
86.80	23.85	86.69	27.65

Further evidence of reaction between the NFCI and mercaptan was generated by examining several solutions via Raman Spectroscopy, as shown in Figure 5. A solution containing 10 mL of the test fuel containing 10% by volume of butanethiol was reacted with 10 mL solutions containing increasing concentrations of NFCI (from 10% to 50%) in the test fuel solution, such that the final concentration of butanethiol was 5% and the concentration range of the NFCI between 5-25%. The consumption of mercaptan with higher concentration of NFCI is clear in the observed spectra, via the peak at ~2580 cm^{-1}.

Figure 5. Raman Spectra of Solutions Prepared by Reacting NFCI and Butanethiol.

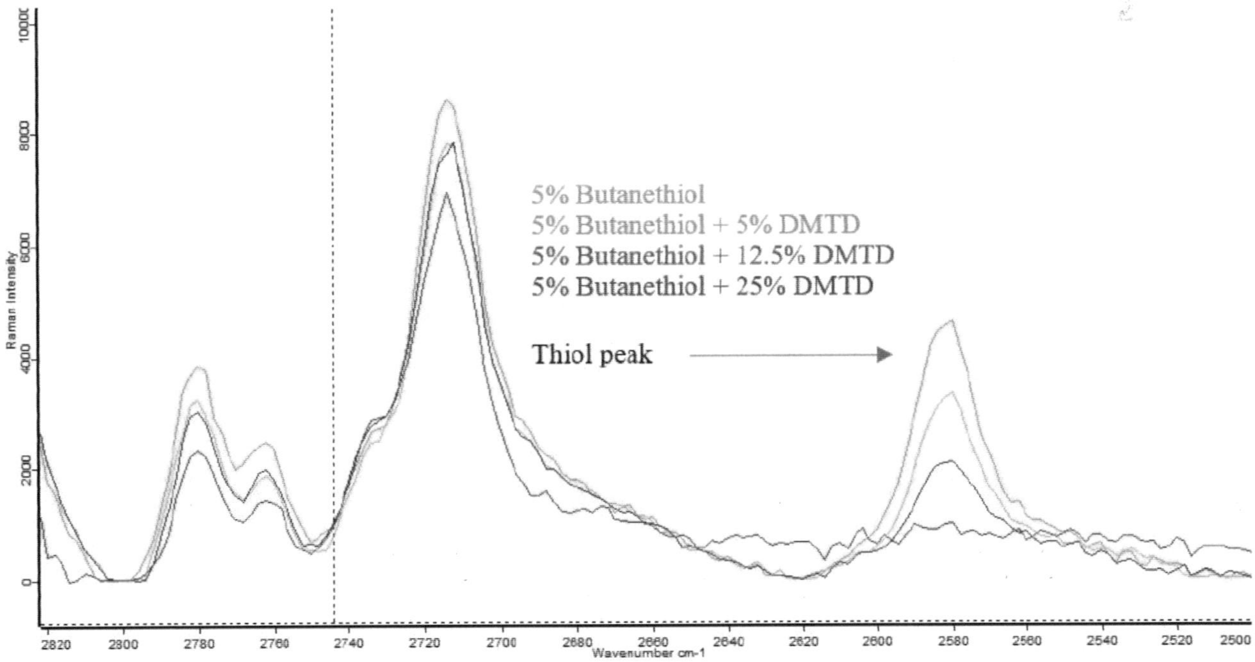

After confirming the reaction between NFCI and mercaptans, we turned our attention to elemental sulfur. A series of solutions was prepared by dissolving increasing amounts of sulfur (50 mg, 100 mg and 200 mg) in 10 mL of toluene and combining those solutions with 10-mL samples of the neat NFCI. Scans of the resulting solutions were taken using Raman Spectroscopy, and the results are shown in Figure 6.

The proposed reaction between the NFCI and sulfur results in incorporation of sulfur into a polysulfide compound, like that shown in Figure 7. Raman spectra indicate a shift away from disulfide during the reaction, as the amount of sulfur added increases. This conclusion is supported by Luo et al[9], where evidence of the formation of a trisulfide is presented. The peak (~476 cm^{-1}) adjacent to the disulfide peak (~520 cm^{-1}), though the signal is weak, seems to correspond to trisulfide[10], and increases somewhat upon addition of higher amounts of sulfur.

Figure 6. Raman Spectra of Solutions Prepared by Reacting NFCI and Sulfur.

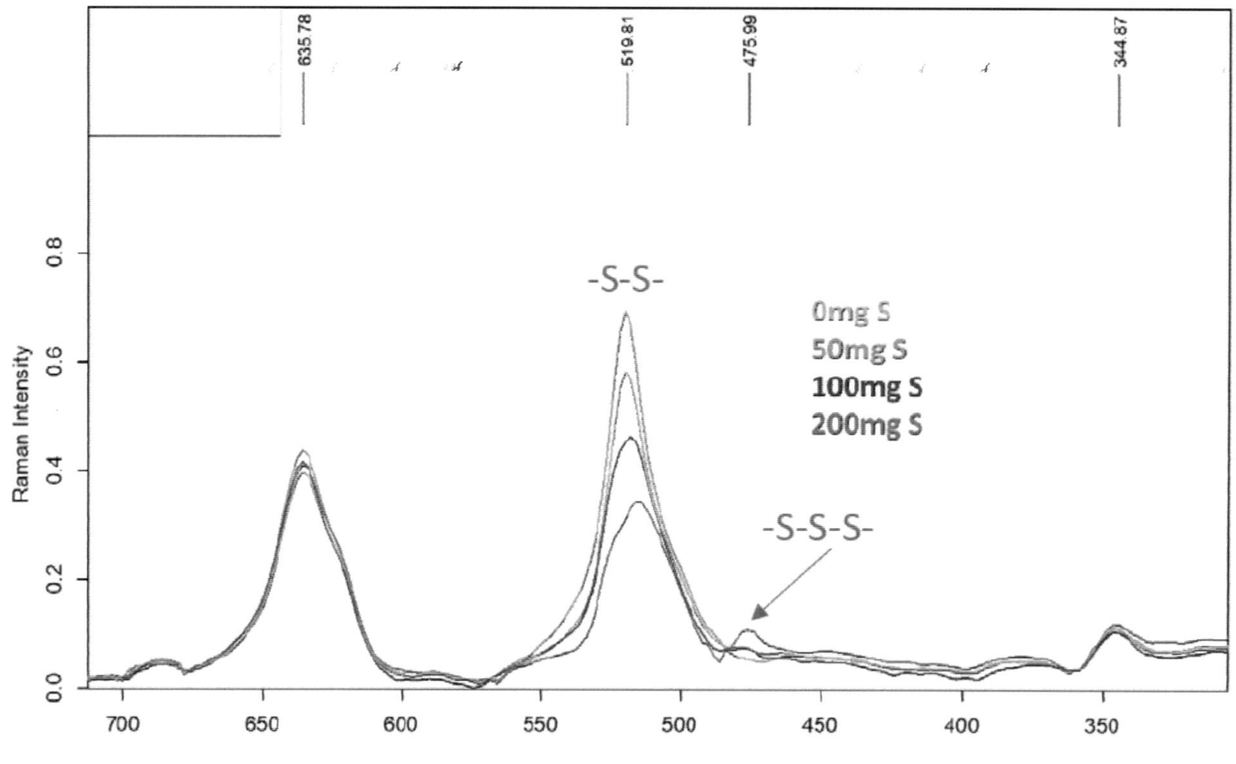

Figure 7. Possible Reaction Mechanism: NFCI with Sulfur.

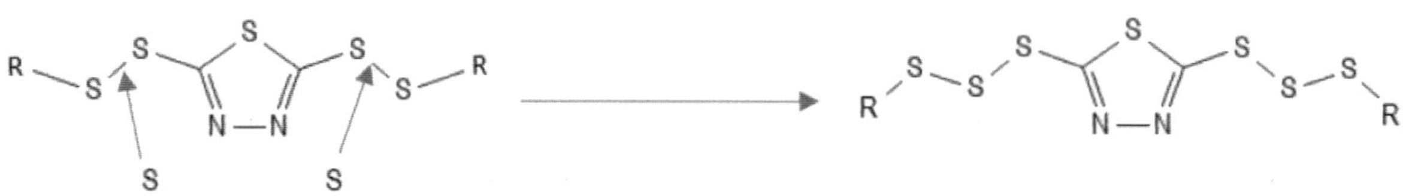

Another possible reaction was considered between the sulfur and the carbon-nitrogen double bond of the NFCI, which may form an isothiocyanate[11]. The literature suggests that hydrogen sulfide would be evolved during this reaction. No hydrogen sulfide was detected in any of the prepared samples. Furthermore, NMR scans did not indicate this change, nor did Raman spectra. The molecular structure of that possible reaction product could not be determined from the available data, nor can that reaction be confirmed by this data. No other significant changes to the spectra were observed, apart from the disulfide and apparent trisulfide/polysulfide peaks, suggesting that incorporation of sulfur to the disulfide linkage in the NFCI is the primary pathway.

A reaction between hydrogen sulfide and the NFCI was also considered. A solution of the NFCI at 50% concentration in isoparaffinic solvent was prepared. A second solution, also 50% concentration of the NFCI, was prepared in isoparaffinic solvent that contained H_2S. To prepare this sour solvent, H_2S gas (1% in nitrogen) was bubbled through the solvent for 4 hours. Raman scans of the two solvents were compared, but no significant differences were noted. No visible evidence of reaction was observed, and no precipitate formed. Further to this test, the 1% H2S in nitrogen gas mixture was bubbled through a sample of neat NFCI for 4 hours. Again, no visible changes were observed, nor were any significant changes noted in NMR spectra.

Most of these experiments were conducted with concentrated solutions, where reactions proceeded rapidly at laboratory ambient temperatures. It is unclear how readily these reactions occur when the reactants are present at low concentration (i.e. parts-per-million), as would typically be found in fuels. It is difficult to detect the minute quantities that would be present in a typical fuel sample by spectroscopic analysis. The authors postulate that an excess of the NFCI relative to other reactive sulfur species may facilitate the reactions, which may explain why higher dosage of additive is required to produce a negative Doctor Test result relative to typical dosage rates required for corrosion inhibition. Titration and Doctor Test results from dosing a fuel containing light mercaptans with the NFCI in the iso-octane/toluene test fuel suggests that these reactions can occur, but they appear to proceed slowly at low concentrations.

CONCLUSIONS

This work demonstrates that a reaction mechanism may exist whereby a non-ferrous corrosion inhibitor (of the dialkyl dithiothiadiazole type) reacts with mercaptans in a fuel sample. One of the products of this reaction is a disulfide compound comprised of a combination of the R group from the NFCI and that of a representative mercaptan molecule. The effective reduction in mercaptan concentration in the fuel may allow that fuel sample to pass the Doctor Test where that same sample had previously failed. The work also demonstrates that a threshold concentration of mercaptan sulfur is required to produce a positive Doctor Test result, such that complete elimination of mercaptans from the fuel is not required to achieve a negative result. In this manner, the NFCI may exert both metal surface filming and scavenging capabilities to achieve passing results in copper or silver corrosion tests as well as the Doctor Test.

Similarly, a possible reaction mechanism between an NFCI and sulfur has been demonstrated. These reactions could explain how a sample may pass the Doctor test after treatment with the NFCI and likely demonstrate an additional mechanism by which the NFCI inhibits corrosion of non-ferrous metals by incorporating free sulfur from the fuel.

Despite this evidence, the application of a non-ferrous corrosion inhibitor for the purposes of achieving a negative Doctor test result produces inconsistent results. With respect to mercaptans, results will depend upon speciation and concentration of mercaptans in the fuel. It should be noted that the dosage required to affect Doctor test results (based on the observed test results from this study and from reported field results) may exceed recommended dosage rates for an NFCI. Specialized mercaptan scavengers are more appropriate for this type of application. There are also methods for extraction of mercaptans combining chemical and mechanical solutions. A suitable scavenger may be combined with typical recommended dosages of NFCI, as needed, if copper or silver corrosion is still an issue after appropriate scavengers have been applied.

The use of additives for purposes other than those for which they have been designed, or at dosage rates above those recommended by the manufacturer or supplier, may result in negative impacts to other fuel properties. For instance, any additive with surface active properties, such as a corrosion inhibitor, if used at dosages higher than typical recommendations, may result in fuel water tolerance issues. High corrosion inhibitor dosages may loosen solids from pipe walls, along with the possibility in this specific case of forming solid precipitate. Additionally, non-ferrous corrosion inhibitors of the type discussed in this paper contain high sulfur concentrations. Therefore, caution must be exercised in the use of these additives to avoid exceeding fuel sulfur specifications.

REFERENCES

1. García-Antón, J., Monzó, J. and Guiñón, J.L. Effect of Elemental Sulfur and Mercaptans on Copper Strip Corrosion and Use of the ASTM D130 Test Method. Corrosion Engineering, NACE International, July 1995, 558-566.

2. Vineyard, B.D. Mercaptan-Sulfur Reaction. Alkyl Trisulfides. J. Org. Chem. 1966, 31, 2, 601–602.

3. Liu, L., Pan, X., Xing, J. Qian, J. Anti-corrosion behavior of thiadiazole derivatives for silver strip in hydrogen sulfide solutions. Anti-Corrosion Methods and Materials, Vol. 62, Iss. 6, November 2015, Pages 353-362.

4. Ren, Y. Shui, H., Peng, C., Liu, H. and Hu, Y. Solubility of elemental sulfur in pure organic solvents and organic solvent–ionic liquid mixtures from 293.15 to 353.15 K. Fluid Phase Equilibria. Fuel and Energy Abstracts. Volume 312, 25 December 2011, Pages 31-36.

5. Tripathi, R.P., Gulati, I. B., Pandey, S.S. and Rohatgi, H.S. Copper and Silver Corrosion by Aviation Turbine Fuels. Ind. Eng. Chem. Prod. Res. Develop., Vol. 12, No. 3, 1973.

6. Singh, R. and Whitesides, G.M., 1993. Thiol-disulfide interchange. Sulphur-containing functional groups (Chapter 13), pp.633-658.

7. Altinbasak, I., Arslan, M., Sanyal, R. and Sanyal, A. Polymer Chemistry. Pyridyl disulfide-based thiol–disulfide exchange reaction: shaping the design of redox-responsive polymeric materials. 2020,11, 7603-7624

8. Smith, V. and Carpenter, J.H. United States Patent US3398003A. Silver Polish-Tarnish Retarder Containing a Dialkyl Disulfide Having from 8 to 20 Carbon Atoms in Each Alkyl Radical. Filed Apr. 26, 1965, Ser. No. 451,070.

9. Luo, Y., Zhong, B. and Zhang, J. The Mechanism of Copper-Corrosion Inhibition by Thiadiazole Derivatives. Journal of the Society of Tribologists and Lubrication Engineers, Lubrication Engineering, Volume 51, 4, 293-296.

10. Rosario-Alomar, M.F., Quiñones-Ruiz, T, Kurouski, D, Sereda, V., Ferreira, E.B., De Jesus-Kim, L, Hernandez-Rivera, S., Zagorevski, D.V., Lopez-Garriga, J. and Lednev, I.K. Hydrogen Sulfide Inhibits Amyloid Formation. J. Phys. Chem. B., 2015, 119, 1265-1274.

11. Mansfield, R.C. Action of Sulfur on Primary Amine-Formaldehyde Condensates. J. Org. Chem. 1959, 24, 9, 1375–1376.

Raman spectral correlations:

12. University of California-Irvine. https://www.chem.uci.edu/~dmitryf/manuals/Raman%20correlations.pdf

13. Adar, F. Introduction to Interpretation of Raman Spectra Using Database Searching and Functional Group Detection and IdentificationSpectroscopy-07-01-2016, Volume 31, Issue 7, Pages: 16–23.

Appendix 1. NMR Scans Showing Increase in Molar Ratios Between the Thiol and the NFCI.

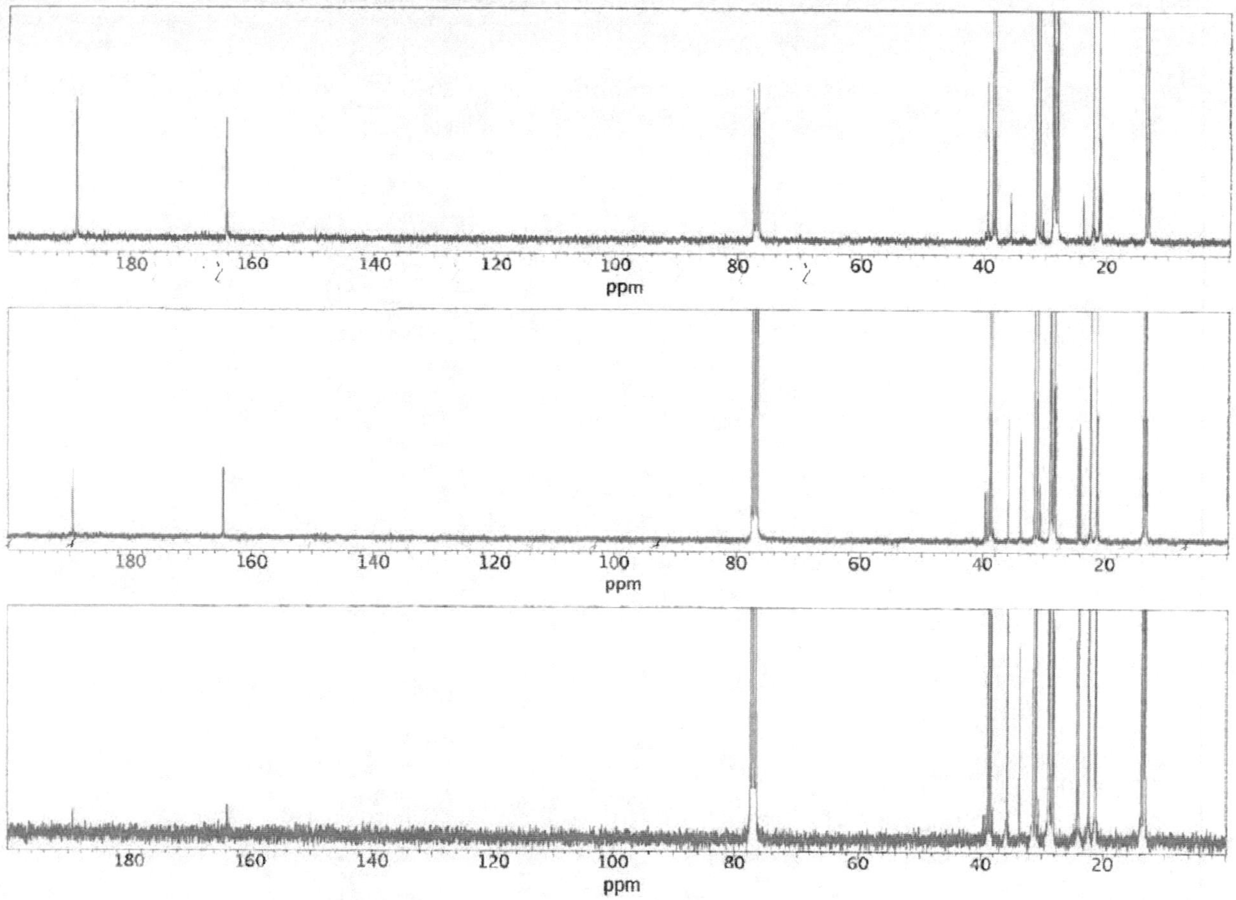

The blue spectrum is the result of reaction with a 1:1 molar ratio butanethiol to NFCI.

The red spectrum is the result of reaction with a 2:1 molar ratio butanethiol to NFCI.

The green spectrum is the result of reaction with a 4:1 molar ratio butanethiol to NFCI.

Peak (assignment)	**Integration @ 1:1**	**Integration @ 2:1**	**Integration @ 4:1**
189 (mono-alkyldithio thiadiazole intermediate)	35.12	21.3	3.01
164 (mono-alkyldithio thiadiazole intermediate)	33.68	21.66	3.71
38.42 (primary disulfide)	58.23	67.75	80.49
30.93 (primary disulfide)	37.14	57.22	73.88

Appendix 2. Comparison of NMR Scans of NFCI Before and After Reaction with Butanethiol.

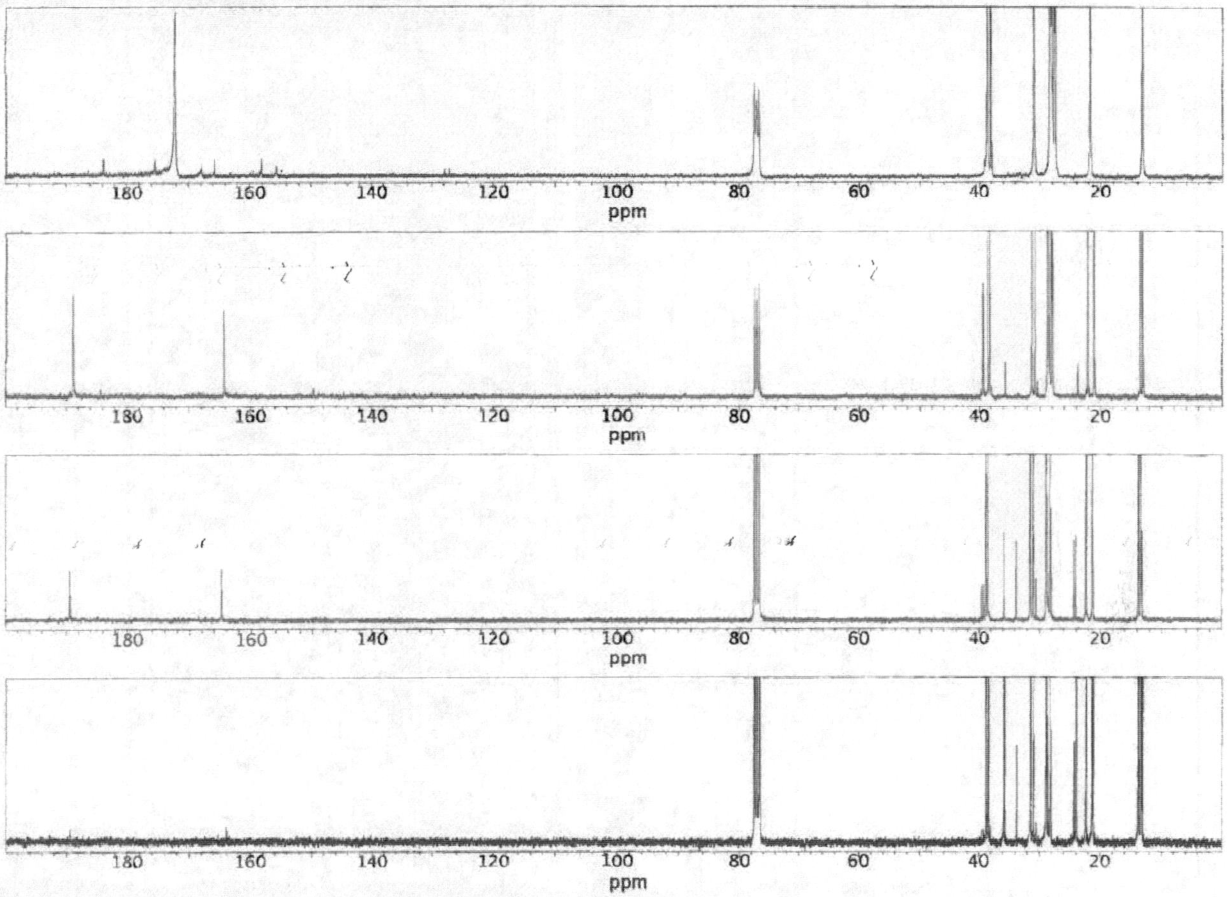

The blue spectrum represents the NFCI.

The red spectrum is the result of reaction with a 1:1 molar ratio butanethiol to NFCI.

The green spectrum is the result of reaction with a 2:1 molar ratio butanethiol to NFCI.

The purple spectrum is the result of reaction with a 4:1 molar ratio butanethiol to NFCI.

Appendix 3. ^{13}C NMR Spectral Comparison of DMTD (Top) with Purified Precipitate (Bottom).

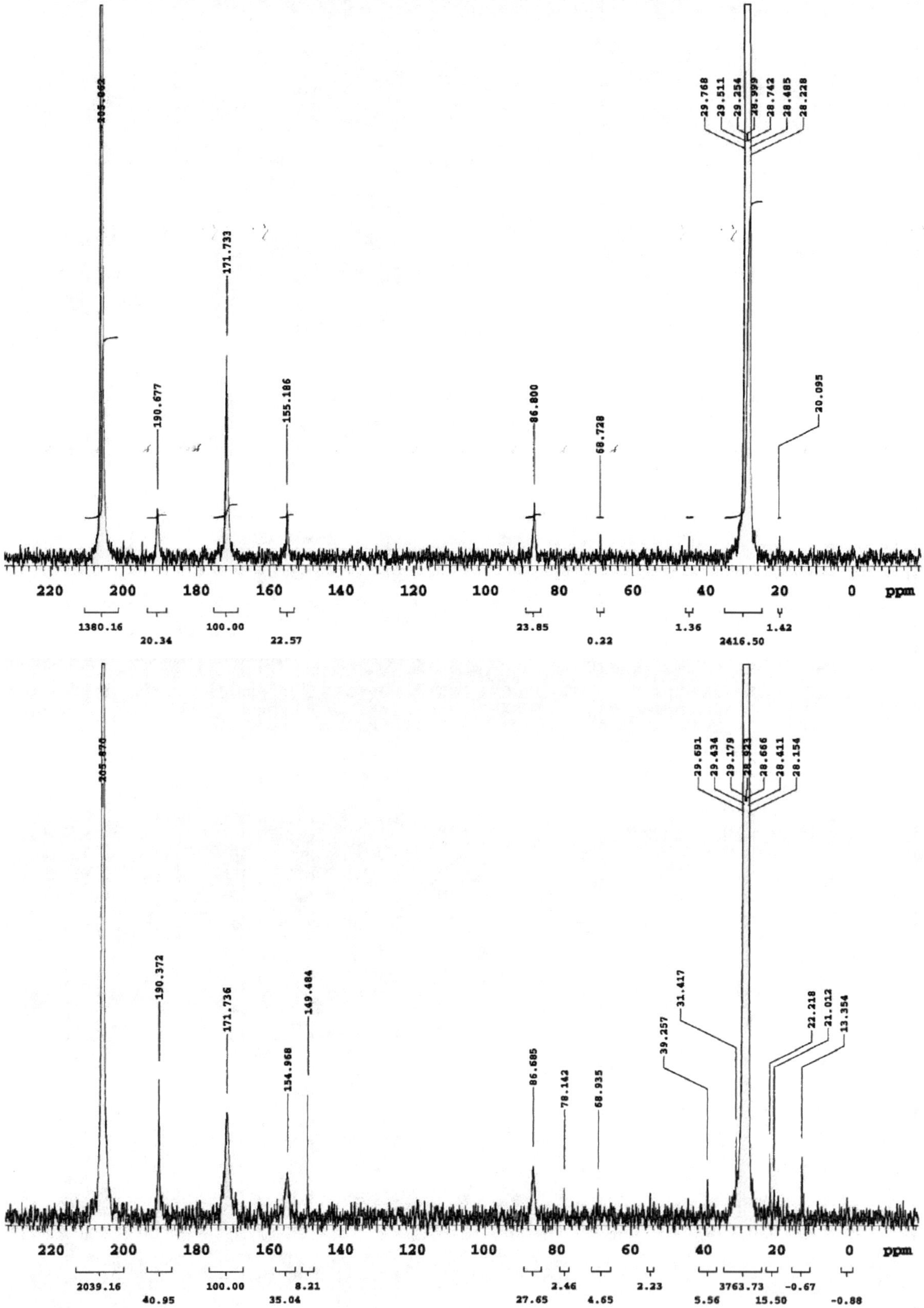

IASH 2022, the 17TH INTERNATIONAL CONFERENCE ON STABILITY, HANDLING AND USE OF LIQUID FUELS
Dresden, Germany
11 - 15 September 2022

GENOMIC PROFILES OF MICROBIAL POPULATIONS IN DIESEL FUEL MICROCOSMS

Frederick J. Passman[1], Katarzyna H. Kucharzyk[2], Danika Nicoletti[3], Jordan Schmidt[4]

[1]Biodeterioration Control Associates, Inc., PO Box 3659, Princeton, NJ 08543, USA, fredp@biodeterioraion-control.com

[2]Battelle Memorial Institute, 505 King Avenue, Columbus, OH 43201, USA, kucharzyk@battelle.org

[3] LuminUltra Technologies Limited, 520 King Street, Fredericton, New Brunswick, E3B 6G3, Canada, danika.nicoletti@luminultra.com

[4]LuminUltra Technologies Limited, 520 King Street, Fredericton, New Brunswick, E3B 6G3, Canada, jordan.schmidt@luminultra.com

KEYWORDS

Diesel, fuel, genomics, microbiology, microbiomes.

ABSTRACT

The relationship between microbial bioburdens detected in fuel and fuel-associated water, and evidence of biodeterioration is unclear. Fuel systems with negligible, detected bioburdens show evidence of substantial biodeterioration. Conversely, systems assessed to harbor heavy bioburdens sometimes show no evidence of biodeterioration. One hypothesis is that the types of microbes present in a system is a critical factor in determining whether a contaminant population is likely to be biodeteriogenic. Genomic testing was performed on aqueous-phase samples of microcosms used in a Coordinating Research Council sponsored study that was commissioned to evaluate principal factors contributing to mild steel corrosion in diesel fuel underground storage tanks. The results presented in this paper indicated that initially, microbiome taxonomic profiles are dynamic but that after three-months they remain stable for at least 18 months.

INTRODUCTION

Since 2007, when ultralow sulfur diesel (ULSD – sulfur concentration <15 $\mu g\ g^{-1}$) became the mandated fuel for U.S. on highway use, the number of reported diesel fuel dispensing system corrosion incidents has increased. It has not been established whether increased reporting reflected increased occurrence, greater awareness, increased sensitivity of modern dispensing equipment, or a combination of these factors. Common hypotheses posit that the reduced sulfur concentration, use of fatty acid methyl ester (FAME – biodiesel) blends, or increased risk of fuel contamination have been responsible for increased diesel fuel corrosivity. The results of two previous investigations [1, 2] suggested that fuel system corrosion was more pervasive than commonly thought. In 2017, The Coordinating Research Council (CRC) designed and commissioned a laboratory study to evaluate the impact of 11 controlled variables on diesel fuel corrosivity. The results of that study were published in 2021[3]. Among the factors tested, only water correlated consistently with corrosion. Although microbial contamination was intended to be one of the controlled variables, by the end of the study's fourth week, bioburdens in uninoculated microcosms were comparable to those in inoculated systems [4, 5]. One question not addressed by the CRC study was the impact of fuel chemistry on microcosm microbiomes. This paper reports the genomic profiles of the microcosms that contained an intentional aqueous phase.

MATERIALS AND METHODS

Experimental design and microcosm details have been described perviously[3, 4, 5]. Briefly, a fractional factorial experiment was designed to evaluate the relationships between 11 controlled variables and fuel corrosivity. Although the study ended after three months (T_{3mo}), microcosms were retained in the dark at room temperature (20 ± 2 °C) for 18 months (T_{18mo}).

Duplicate inoculum samples were tested as detailed in the CRC report[3]. At T_0 and T_{3mo} aqueous phase samples were collected, and DNA was extracted by the Meta-G-Nome™ DNA Isolation Kit (Epicentre, Madison, WI). The manufacturer's protocol for direct extraction from biomass captured on nitrocellulose filters was followed. Post-extraction cleanup for all specimens was performed using OneStep™ polymerase chain reaction (PCR) Inhibitor Removal Kit (Zymo Research Corp., Irvine, CA). Purified DNA specimens were analyzed with an ultraviolet (UV) absorbance (NanoDrop™ 200 spectrophotometer, Thermo Scientific, Waltham, MA), Qubit® dsDNA HS Assay Kit, and SYBR® Gold Nucleic Acid Gel Stain according to the manufacturer's protocols (Invitrogen/LifeTechnologies, Grand Island, NY).

At T_0 and T_{3mo}, genomic DNA was fragmented using a Covaris™ S220 Sonicator (Covaris, Inc., Woburn, MA) to approximately 300 base pairs (bps). Fragmented DNA was used to synthesize indexed sequencing libraries using the TruSeq DNA Sample Prep Kit V2 (Illumina, Inc., San Diego, CA), according to the manufacturer's recommended protocol. Cluster generation was performed on the cBOT using the TruSeq PE Cluster Kit v3 – cBot – HS (Illumina). Libraries were sequenced with an Illumina HiSeq 2000 at Nationwide Children's Hospital (NCH) Biomedical Genomics Core (Columbus, OH) using the TruSeq SBS Kit v3 reagents (Illumina) for paired end sequencing with read lengths of 100 bps (200 cycles). Primary analysis (image analysis and base calling) was performed using HiSeq Control Software version 1.5.15.1 and Real Time Analysis version 1.13.48. Secondary analysis (demultiplexing) was performed using Illumina CASAVA Software v1.6 on the NCH computer cluster. Sequence data (.fastq files) and QC reports for library construction were delivered to Battelle via an external hard drive.

The specimens were then analyzed through BLAST using the NCBI 16S database v. June 11, 2017. Post BLAST, additional filtering was used to remove false identifications and low-quality identifications. BLAST results were filtered to remove all results with less than 97% identity over less than 80% of the read length. Following this filter, a second filter, designed to remove false positive identifications was applied. False positive identifications were defined as hits that comprise less than 0.01 percent of the identifications in the results. False positive hits had too few reads identified as belonging to a given organism to be considered strong identifications.

At T_{18mo}, DNA extracted from the samples using a phenol chloroform extraction protocol. The DNA was then subjected to PCR amplification using sample-specific barcode-labeled versions of the primers *515F* [6] + *806R* [7], following the protocol of Caporaso *et al.* [8]. After quantification of the purified products using a Qubit High Sensitivity Fluorometer kit (Invitrogen, Carlsbad, CA, USA) amplicons from each sample were combined at equimolar concentrations into an amplicon library. Finally, the DNA concentration of the library was measured by Qubit once again, diluted to a concentration of 2 nM, and then sequenced using an Illumina MiSeq sequencer with v2 500 cycle reagent kit (Illumina, San Diego, CA, USA). Sequences were received for each sample as de-multiplexed FASTQ files representing forward and reverse paired end reads. Paired-end reads were merged and combined reads with a total expected error threshold above 0.5 were discarded using USEARCH (v10.0.240) [9] and the "classify.seqs" and "remove.lineage" commands in Mothur (v. 1.36.1) were used to identify and remove potential mitochondrial, chloroplast, archaeal and eukaryote contaminants [10]. Remaining sequences were grouped into operational taxonomic units (OTUs) based on a 3% sequence dissimilarity cut-off using the UPARSE greedy algorithm in USEARCH, during which chimeric sequences were also removed [9]. Subsequently, the taxonomic identity of each OTU was determined with the SINTAX algorithm implemented in USEARCH [11] in conjunction with the SILVA SSU ARB database version 1.32 [12].

The Shannon-Wiener Index (H)[13], for determining microbial diversity was calculated using Equation 1:

$$H = -\sum_{i=n}^{S} p_i \ln p_i \qquad \textbf{(1)}$$

Where: i_n is an individual species (S – operational taxonomic unit – OTU – genetic sequence that is determined [probability ≥ 95 %] to be unique), p_i = proportion of total sample represented by a given OTU, and S = number of OTU in the sample (i.e., *species richness*). Evenness – the observed diversity index relative to the theoretically possible diversity – was computed as Shannon's Equitability (E_H) per Equation 2. E_H values range between 0 and 1, with 1 being complete evenness/diversity. As the number of detected OTU decreases, H approaches 0.

$$E_H = H/ln\,S \qquad (2)$$

Analysis of variance (ANOVA), and diversity calculations were computed using Excel® (Excel is a registered trademark of Microsoft, Redmond, WA). Other microcosm populations comparisons of were computed using R, and Korona plots were developed using *KronaTools4* (National Institutes of Health, Bethesda, MD).

RESULTS

Repeatability – During the CRC study, duplicate samples were collected from the inoculum microcosm at T_0 and test microcosms 45 (uninoculated B5 ULSD, augmented with monoacid lubricity additive at 200 mg kg^{-1}, cold flow improver at 200 mg kg^{-1}, and 9.0 ± 1.0 mg kg^{-1} corrosion inhibitor) and 81 (inoculated B0 LSD, augmented with monoacid lubricity additive at 200 mg kg^{-1} and corrosion inhibitor at 2.5 ± 0.5 mg kg^{-1}). Both test microcosms included a fiber reinforced polymer (FRP) coupon. As shown in Figure 1, the genomic profiles of duplicate specimens were indistinguishable. At T_{18mo}, duplicate samples of the microcosm 121's interface (invert emulsion – I.E.) phase differed in both the number of taxa detected and their relative abundances (Figure 2 a and b). This reflected the interface's greater heterogeneity.

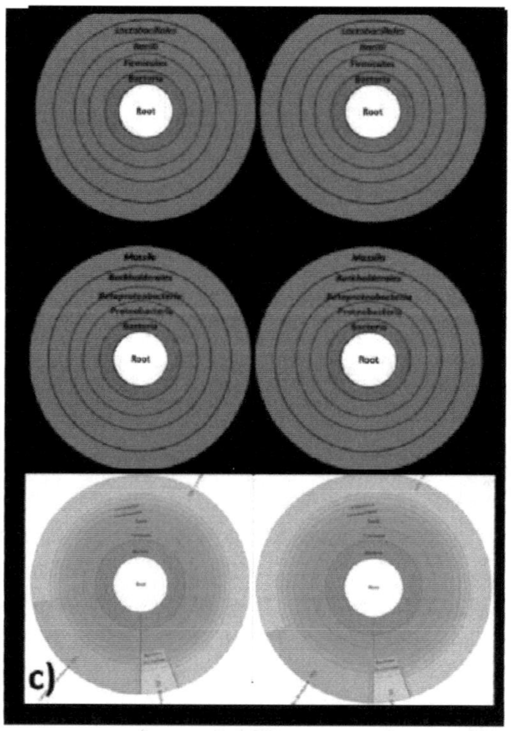

Figure 1. Korona plots of aqueous phase, NGS genomic profile repeatability– a) Inoculum microcosm @ T_0; b) Microcosm 45 @ T_{3mo} ; c) Microcosm 81@ T_{3mo} .

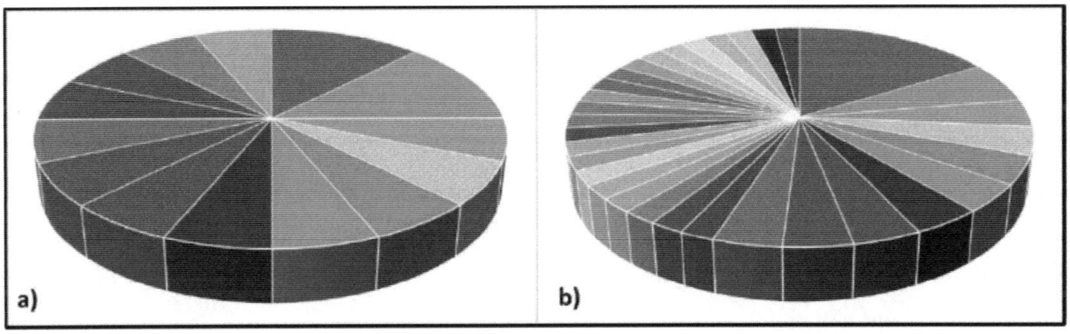

Figure 2. NGS genomic profile repeatability in Microcosm 121 interface phase – a) replicate 1; b) replicate 2.

Temporality – Table 1 lists the taxa detected in the inoculum. Table 2 is a heat map of the predominant taxa recovered from inoculated microcosms at T_{3mo}. In the inoculum, *Pseudomonas* spp. (37 %) and *Acidimicrobium* spp. (28 %) were the predominant taxa among five detected. Except for *Pseudomonas* spp. representing 1 % of the microcosm 28 and 32 microbiomes and 3 % of the microcosm 48 microbiome, the taxa recovered from the inoculum specimens were absent from T_{3mo} aqueous phase specimens. The possibility of these differences reflecting experimental error cannot be dismissed.

Table 1. Microbial taxa detected in inoculum suspension. Numbers indicate percentage of relative abundance. Colors indicate the OTU's relative abundance - high (≥ 5 % shades of blue, darkening with increased %) to low (red).

Organism	**%**
Pseudomonas	37
Acidimicrobium	28
Acidiphilium	10
Rhodomicrobium	9
Rhodobacter	10

Table 2. Microbial taxa detected in specimens from challenged microcosms. Numbers indicate percentage of relative abundance. Colors indicate the OTU's relative abundance - high (≥ 5 % shades of blue, darkening with increased %) to low (red)[3].

Organism	Microcosm 4	21	28	32	46	48	81
Acinetobacter	0.01					0.07	0.03
Aspergillus	0.09	0.01	0.06			0.02	0.02
Bacillus	31.1	2.0	0.18		0.08	0.11	0.05
Haemophilus		0.01	0.06				
Lactobacillus	0.68	12.1	0.40	0.15	0.42	83.9	94.8
Oerskovia						6.13	
Porphyromonas		0.01	0.08				0.01
Pseudomonas			0.01	0.01		0.03	
Stenotrophomonas				0.01			
Streptococcus							0.01

Although T_{3mo} and T_{18} sampling and NGS were performed by different teams, using different protocols, the aqueous phase microbiome profiles were quite similar in specimens from microcosms 46, 47, 56, 127 (Table 3). These observations indicated that data from NGS testing performed by different teams can be compared. They also indicated that under some conditions, after an initial period of selectin and adaptation, the microbiome remains stable for prolonged periods. However, substantial microbiome changes were observed in microcosms 40, 45, 81, and 102 (Table 4).

Table 3. Microbial taxa detected at T_{3mo} and T_{18mo} in specimens from microcosms in which the population remained stable. Numbers indicate percentage of relative abundance. Colors indicate the OTU's relative abundance - high (≥ 5 % shades of blue, darkening with increased %) to low (red).

Taxa	Microcosm 46		47		56		127	
	T_{3mo}	T_{18mo}	T_{3mo}	T_{18mo}	T_{3mo}	T_{18mo}	T_{3mo}	T_{18mo}
Lactobacillus (G)								
Lactobacillales (O)								
Acinetobacter (G)								
Enterobacteriaceae (F)								
Pseudomonas (G)								
Methylobacterium (G)								
Stenotrophomonas (G)								
Sphingomonas (G)								
Enhydrobacter (G)								
Bosea (G)								
Paenibacillaceae (F)								
Acidiphilum (G)								
Acidimicrobium (G)								
Rhodopseudomonas (G)								
Rhodobacter (G)								
Rhodomicrobium (G)								
Pseudogulbenkiania (G)								
Other								

Table 4. Microbial taxa detected at T_{3mo} and T_{18mo} in specimens from microcosms in which the population changes substantially. Numbers indicate percentage of relative abundance. Colors indicate the OTU's relative abundance - high (≥ 5 % shades of red, darkening with increased %) to low (pale green).

Taxa	Microcosm 40		45		81		102	
	T_{3mo}	T_{18mo}	T_{3mo}	T_{18mo}	T_{3mo}	T_{18mo}	T_{3mo}	T_{18mo}
Lactobacillus (G)								
Bacillales (O)								
Lactobacillales (O)								
Lactobacillaceae (F)								
Acinetobacter (G)								
Enterobacteriaceae (F)								
Pseudomonas (G)								
Burkholderia-Caballeronia-Paraburkholderia (G)								
Methylobacterium (G)								
Stenotrophomonas (G)								
Ralstonia (G)								
Burkholderiaceae (F)								
Betaproteobacteriales (O)								
Sphingomonas (G)								
Gammaproteobacteria (C)								
Massilia (G)								
Enhydrobacter (G)								
Pseudomonadaceae (F)								
Psychrobacter (G)								
Chryseobacterium (G)								
Escherichia-Shigella (G)								
Acidiphilum (G)								
Acidimicrobium (G)								
Rhodopseudomonas (G)								
Rhodobacter (G)								
Rhodomicrobium (G)								
Pseudogulbenkiania (G)								
Other								

Two (47 and 56) of the four microcosms with stable microbiomes had not been inoculated. Similarly, two (40 and 45) of the four microcosms with evolving microbiomes had not been inoculated. These observations suggested that indigenous and inoculum population dynamics were similar in terms of the likelihood of population succession.

The T_{3mo} and T_{18mo} microbiome profiles were evaluated using analysis of similarities (ANOSYM) per equation 3:

$$R = \frac{rB - rW}{\left(\frac{M}{2}\right)} \tag{3}$$

Where R is the similarity coefficient, r_B is the average of rank similarities of sample pairs from different groups (microcosms), r_W is the average of rank similarities of sample pairs from within groups (T_{3mo} v. T_{18mo}), and $M = [n(n-1)] \div 2$, when n is the number of microcosms. The value of R can range from -1 to +1. When R is positive the similarities between pairs (e.g., microcosm X at T_{3mo} and T_{18mo}) is greater than that among microcosms. When R is negative, the similarity among microcosms is greater than that between pairs. If R = 0 then the similarities between pairs and among microcosms are equal. For T_{18mo} versus T_{3mo} microbiome profiles, R = -0.094 (P = 0.51; if P >0.05 then R is not statistically significant). Consequently, although microbiome profiles continued to change in some microcosms but not others during the 12-month period between T_{3mo} and T_{18mo}, neither time nor microcosm conditions were significant factors.

Inoculation effect – To further test whether inoculation affected microcosm microbiomes, one-way analysis of variance was used to compare T_{3mo} and T_{18mo} aqueous phase population diversities (S), Shannon- Wiener indices (H) diversity indices, and evenness (E). As shown in Table 5, the none of the three statistics were significantly different between inoculated and uninoculated microcosms at either time point.

Table 5. ANOVA summary – impact of inoculation on diversity parameters S, H, and E at T_{3m} and T_{18mo}.

	Parameter					
Statistic	**S**		**H**		**E**	
	T_{3mo}	**T_{18mo}**	**T_{3mo}**	**T_{18mo}**	**T_{3mo}**	**T_{18mo}**
AVG_{Inoc}	11	26	0.69	1.19	0.29	0.38
AVG_{Uninoc}	23	31	0.85	1.24	0.34	0.34
F_{Obs}	1.40	0.30	0.69	0.008	0.54	0.07
F_{crit}[a]	4.49	4.75	4.49	4.75	4.49	4.75
P	0.25	0.59	0.42	0.93	0.47	0.79

Note a – At T_{3mo} $F_{crit\,[1,16;\,\alpha = 0.05]}$ = 4.49; at T_{18mo} $F_{crit\,[1,12;\,\alpha = 0.05]}$ = 4.75.

Aqueous versus invert emulsion microbiomes – At T_{18mo}, aqueous phase and interface specimens were collected from seven microcosms. A total of 43 taxa were detected (Table 6) with 35 taxa recovered from aqueous phase specimens and 38 from I.E. specimens. As indicated by the Table 6 heat map, *Lactobacillus* spp. were detected in all aqueous and five I.E. specimens. *Acinetobacter* spp. were detected in all aqueous and three I.E. specimens. *Pseudomonas* spp. were detected in six aqueous and four I.E. specimens. One-way analysis of variance was used to compare T_{18mo} aqueous phase versus I.E. S, H, and E. As summarized in Table 7, overall diversity among aqueous phase specimens was greater than that in I.E. specimens. Conversely, evenness was greater in I.E. than in aqueous phases specimens. The Shannon-Wiener diversity indices were statistically indistinguishable among specimens from the two phases. ANOSYM analysis yielded $R = -0.11$ ($P = 0.88$), indicating that microbiome diversity differences between phases was statistically indistinguishable from that among microcosms.

Table 6. Microbial taxa detected at T_{18mo} in aqueous and invert emulsion (I.E.) phase specimens from microcosms. Numbers indicate percentage of relative abundance. Colors indicate the OTU's relative abundance - high (≥ 5 % shades of red, darkening with increased %) to low (pale green).

Taxa	Microcosm													
	4		44		45		46		56		57		127	
	Aq.	I.E.	Aq.	I.E.	Aq.	I.E.	Aq.	I.E.	Aq.	I.E.	Aq.	I.E.	Aq.	I.E.
Methylobacterium (G)														
Lactobacillus (G)														
Pseudomonas (G)														
Chryseobacterium (G)														
Psychroglaciecola (G)														
Cloacibacterium (G)														
Enterococcus (G)														
Bacteria (K)														
Thermomicrobiales (O)														
Micrococcus (G)														
Mycobacterium (G)														
Massilia (G)														
Marmoricola (G)														
Dolosigranulum (G)														
Peptoniphilus (G)														
Actinobacillus (G)														
Alcaligenes (G)														
Enhydrobacter (G)														

Table 6. continued.

Taxa	Microcosm

	4		44		45		46		56		57		127	
	Aq.	**I.E.**	**Aq.**	**I.E.**	**Aq.**	**I.E.**	**Aq.**	**I.E.**	**Aq.**	**I.E.**	**Aq.**	**I.E.**	**Aq.**	**I.E.**
Escherichia-Shigella (G)														
Bosea (G)														
Sphingomonas (G)														
Acinetobacter (G)														
Chryseomicrobium (G)														
Flavobacterium (G)														
Burkholderia-Caballeronia-Paraburkholderia (G)														
Burkholderiaceae (F)														
Methylophilaceae (F)														
Bacillus (G)														
Corynebacterium (G)														
Leptotrichia (G)														
Leuconostoc (G)														
Saccharimonadales (O)														
Bacillales (O)														
Brachybacterium (G)														
Anaerobacillus (G)														
Flavobacteriales (O)														
Sandaracinobacter (G)														
Oligoflexales (O)														
Polynucleobacter (G)														
Staphylococcus (G)														
Psychrobacter (G)														
Alicyclobacillus (G)														
Thermicanus (G)														

DISCUSSION

Examining aviation fuel tank sump samples from 11 U.S. Air Force bases around the United States, Rauch and her coworkers [14] were the first to report using genomic methods to characterize microbial contaminants in fuel systems. However, the U.S. Air Force survey used genomics to characterize isolates obtained by culture methods.

Table 7. ANOVA summary – impact of matrix on diversity parameters S, H, and E from aqueous phase and invert emulsion specimens, where $F_{crit\,[1.12;\,\alpha = 0.05]} = 4.75$.

Parameter	AVG Aq.	I.E.	F_{obs}	P
S	37	19	9.17	0.011
H	1.13	1.28	0.097	0.76
E	0.31	0.43	4.75	0.415

In 2011, White *et al.* [15] used denaturing gradient gel electrophoresis (DGGE) and massively parallel signature sequencing (MPSS) to characterize fuel system microbiomes from 15 samples collected from retail and bulk storage tanks containing various fuel grades – including "automotive diesel" (equivalent to ULSD) and "automotive biodiesel" (B5 ULSD). White et al. reported that the dominant classes included taxa from *Betaproteobacteria* (42.8%), *Gammaproteobacteria* (30.6%) and *Firmicutes* (19.2 %). In this study, the genera *Bacillus* and *Lactobacillus* – which were dominant in several microcosms – are members of the Class *Firmicutes*. Suflita et al. [16] also reported high abundances of Firmicutes in ULSD samples. Also dominant in one or more microcosms, *Burkholderia* and *Paraburkholderia* genera are members of the Class *Gammaproteobacteria*. Restrepo-Flórez *et al.* [17] monitored genomic trends in ULSD and B100 microcosms. They found that the microbiome changed substantially during100-days of incubation. *Proteobacteria* (they did not differentiate among α, β, and γ-Proteobacteria) were dominant (96. 4 %) at T_0 in ULSD and at T_{100} in B100 (92.8 %), but at T_{100} in ULSD, they represented 23.9 % of the microbiome. At T_{100} in ULSD, *Rhodococcus* (Genus within the Class *Actinomycetia*) was the predominant taxon. No representatives of *Actinomycetia* were detected in the CRC inoculum or T_{3mo} or T_{18mo} microcosms.

Soriano *et al.* [18] investigated the impact of FAME source and concentration, and inoculation in low sulfur diesel (LSD). They used B5 and B100 soy and beef tallow FAME and collected specimens for genomic testing after 7, 28, and 42 days. Genomic profiles were affected by FAME chemistry. Soriano's team used a known inoculum mixture comprised of 10^5 MPN mL^{-1} each of "*Bacillus amyloliquefaciens*, *Bacillus* sp., *Micrococcus* sp., *Candida* sp., *Candida dubliniensis*, *Candida viswamathii*, and *Pichia anomala*. At T_{28d}, the investigators recovered *B. amyloliquefaciens* in uninoculated and inoculated B100 beef tallow FAME, but not in uninoculated or inoculated soy FAME. *Conidia* sp. was detected only in LSD (T_{28d} and T_{42d}). In soy B5 and B100, *C. viswamathii* was detected at T_{28d} but not T_{42d}. The absence of most inoculum taxa from inoculated microcosms was similar to what occurred in the CRC study microcosms.

Another study [19] examined the impact of sulfur concentration ([S]) on bacterial biomass (mg mL^{-1}) and taxonomic diversity in microcosms containing one of four fuel grades (ULSD, LSD, high sulfur diesel, HSD – [S] = 400 mg L^{-1}, and ultra-high sulfur diesel, UHSD – [S] = 1,400 mg L^{-1})

over 10 mL Bushnell-Hass mineral medium. The investigators reported that interface biomass varied with [S]: UHSD, 45 ± 5 mg mL^{-1}; HSD, 23 ± 2 mg mL^{-1}, LSD, 19 ± 4 mg mL^{-1}, and ULSD, 7 ± 2 mg mL^{-1}. Conversely, taxonomic diversity decreased as [S] increased. In all microcosms the dominant taxa were Firmicutes, Bacteroidetes, Actinobacteria and Ascomycota. At T_{40d} in ULSD and LSD, respectively, Firmicutes represented 65 % and 54 % of the total microbiome. Bacteroidetes represented 25 % and 20 % of the ULSD and LSD microbiomes, respectively. In the current study, Firmicutes were dominant in four of seven inoculated and all four inoculated microcosms from which T_{3mo} specimens for NGS testing were obtained. No representatives of Bacteroides were detected.

Focusing on fuel tank sludge and bottoms-water samples at different altitudes González-Benítez and her co-workers [20] reported that microbiomes were affected by tank altitude, total organic carbon (TOC), and phosphorous concentrations. The predominant phyla were similar to those identified by de Azambuja *et al.* [19]. In order of average relative abundance, González-Benítez's team recovered OTU representatives from Thermotogae, Spirochaetes, Firmicutes, Bacteroidetes, and Proteobacteria.

The similarities and differences among fuel system microbiome profiles reflect a combination of environmental and protocol factors. Consensus NGS protocols do not exist. Consequently, most labs use variants of previously published DNA extraction, amplification, and identification methods. Until genomic testing is standardized, conclusions regarding the impact of environmental factors on microbiome profiles will be suspect. Our observations that none of the controlled variables in our study – including intentional inoculation versus proliferation of contaminant microbes introduced with the test fuels – contributed significantly to variations of either taxonomic diversity or microbiome profiles, were consistent with the conclusion drawn by Azambuja *et al.* [19]: "there was an apparent lack of deterministic process in structuring microbial communities."

CONCLUSIONS

Although NGS test results are repeatable, it is not clear that they are reproducible. Results from samples collected 12-months apart and tested by two different labs, using different extraction, amplification, and identification protocols indicated that similarities are sufficient to support high-order comparisons among data sets from different research teams. Within fuel systems there are microbiome differences among phases. In diesel fuel over water microcosms taxonomic diversity tends to be greater in the aqueous than in the invert emulsion phase.

ACKNOWLEDGEMENTS

All microcosm work through the completion of the T_{3mo} testing was funded by the Coordinating Research Council, under project CRC DP-07-16-01and performed at Battelle, Columbus, OH. The T_{18mo} sampling and sample preparation work was performed at Marathon Petroleum Company LLC's Refining and Development Center, Catlettsburg, KY. LuminUltra Technologies Ltd, Fredericton, New Brunswick, provided all supplies and reagents for the T_{18mo} tests. The authors gratefully acknowledge the efforts of technical staff at Battelle, LuminUltra, and Marathon for their respective efforts in support of this study.

REFERENCES

1. Battelle, 2012. Corrosion in Systems Storing and Dispensing Ultra Low Sulfur Diesel (ULSD), Hypotheses Investigation, Study No 10001550 Final Report, Battelle Memorial Institute, Columbus. 146 pp. ULSDStoringSystemCorrosion.pdf (clean-diesel.org).

2. U.S. EPA, 2016. Investigation Of Corrosion-Influencing Factors In Underground Storage Tanks With Diesel Service, EPA 510-R-16-001, 68 pp, United States Environmental Protection Agency, Washington, DC. https://www.epa.gov/sites/default/files/2016-07/documents/diesel-corrosion-report_0.pdf

3. CRC, 2021. Identification of Potential Parameters Causing Corrosion of Metallic Components in Diesel Fuel Underground Storage Tanks, CRC Report No. DP-07-16-1, Coordinating Research Council, Alpharetta, GA. 401 pp. DP-07-16-01-FINAL-REPORT-REV-30JUL21-COMPLETE.pdf (crcao.org).

4. Passman, F.J., Schmidt, J, Lewis, R.P. and Christian, P., 2019. The Relationship between Planktonic and Sessile Microbial Population Adenosine Triphosphate in Diesel Fuel Microcosms. In: R. E. Morris. Ed. Proceedings of the 16th International Symposium on the Stability and Handling of Liquid Fuels, 08 to 12 September 2019, Long Beach, USA. http://iash.conferencespot.org/69070-iash-1.4569809/t001-1.4569870/f0010-1.4570090/a036-1.4570109/ap103-1.4570112

5. Passman, F.J., Schmidt, J, Lewis, R.P. and Christian, P., 2019. The Relationship between Microbial Community Vitality and ATP Bioburdens in Bottoms-waters under Fuel Microcosms. In: R. E. Morris. Ed. Proceedings of the 16th International Symposium on the Stability and Handling of Liquid Fuels, 08 to 12 September 2019, Long Beach, USA. http://iash.conferencespot.org/69070-iash-1.4569809/t001-1.4569870/f0010-1.4570090/a037-1.4570103/ap106-1.4570106

6. Parada, A. E., Needham, D. M., & Fuhrman, J. A., 2016. Every base matters: assessing small subunit rRNA primers for marine microbiomes with mock communities, time series and global field samples. Environmental Microbiology, 18(5), 1403–1414. http://doi.org/10.1111/1462-2920.13023

7. Apprill, A., McNally, S., Parsons, R., & Weber, L., 2015. Minor revision to V4 region SSU rRNA 806R gene primer greatly increases detection of SAR11 bacterioplankton. Aquatic Microbial Ecology, 75(2), 129–137. http://doi.org/10.3354/ame01753

8. Caporaso, J. G., Lauber, C. L., Walters, W. A., Berg-Lyons, D., Huntley, J., Fierer, N., Owens, S. M., Betley, J., Fraser, L., Bauer, M., Gormley, N., Gilbert, J. A., Smith, G., & Knight, R., 2012. Ultra-high-throughput microbial community analysis on the Illumina HiSeq and MiSeq platforms. ISME J 6, 1621–1624. http://doi.org/10.1038/ismej.2012.8

9. Edgar, R.C., 2013. UPARSE: highly accurate OTU sequences from microbial amplicon reads. Nature Methods, 10, 996. http://doi.org/10.1038/nmeth.2604.

10. Schloss P.D. ,Westcott S.L., Ryabin, T., Hall, J. R., Hartmann, M., Hollister, E. B., Lesniewski, R. A., Oakley, B. B., Parks, D. H., Robinson, C. J., Sahl, J. W., Stres, B., Thallinger, G. G., Van Horn, D. J., & Weber, C. F., 2009. Introducing mothur: open-source, platform-independent, community-supported software for describing and comparing microbial communities. Applied and Environmental Microbiology, 75, 7537-7541. https://doi.org/10.1128/AEM.01541-09.

11. Edgar, R.C., 2016. SINTAX, a simple non-Bayesian taxonomy classifier for 16S and ITS 513 sequences. http://dx.doi.org/10.1101/074161.

12. Quast, C., Pruesse, E., Yilmaz, P., Gerken, J., Schweer, T., Yarza, P., Peplies, J., Glöckner, F.O., 2013. The SILVA ribosomal RNA gene database project: improved data processing and web-based tools. Nucl. Acids Res. 41 (D1): D590-D596. https://doi.org/10.1093/nar/gks1219.

13. Spellerberg, I.F. and Fedor, P. J., 2003. A tribute to Claude Shannon (1916–2001) and a plea for more rigorous use of species richness, species diversity and the 'Shannon–Wiener' Index. Global Ecology and Biogeography, 12(3):177-179. https://doi.org/10.1046/j.1466-822X.2003.00015.x

14. Rauch, M. E., Graef, H.W., Rozenzhak, S.M., Jones, S.E., Bleckmann, C.A., Kruger, R.L., Naik, R.R., Stone, M.O., 2006. Characterization of microbial contamination in United States Air Force aviation fuel tanks. J. Ind. Microbiol. Biotech. 33(1):29–36. https://doi.org/10.1007/s10295-005-0023-x.

15. White, J., Gilbert, J., Hill, G., Hill, E., Huse, S.M., Weightman A.J., Mahenthiralingam, E., 2011. Culture-Independent Analysis of Bacterial Fuel Contamination Provides Insight into the Level of Concordance with the Standard Industry Practice of Aerobic Cultivation. Applied and Environmental Microbiology 77(13): 4527–4538. https://doi.org/10.1128/AEM.02317-10.

16. Suflita, J.M., Aktas, D.F., Oldham, A.L., Perez-Ibarra, B.M., Duncan, K. 2012. Molecular tools to track bacteria responsible for fuel deterioration and microbiologically influenced corrosion. Biofouling 28(9): 1003–1010. https://doi.org/10.1080/08927014.2012.723695.

17. Restrepo-Flórez, J-M, Bassi, A., Rehmann, L., Thompson, M.R., 2013. Effect of biodiesel addition on microbial community structure in a simulated fuel storage system. Bioresource Technology 147:456–463. https://doi.org/10.1016/j.biortech.2013.08.068.

18. Soriano, A.U., Martins, L.F., Ventura, E.S.A., Teixeir, F.H., de Landa, G., de Araújo Valonia, É., Faria, F.R.D., Ferreira, R.F., Faller, M.C.K., Valério, R.R., de Assis Leite, D.C., do Carmo, F.L., Peixoto, R.S., 2015. Microbiological aspects of biodiesel and biodiesel/diesel blends biodeterioration, Internat. Biodeter. Biodeg. 99:102-114. https://doi.org/10.1016/j.ibiod.2014.11.014.

19. de Azambuja, A.O., Bücker, F., de Quadros, P.D., Zhalnina, K., Dias, R., Vacaro, B.B., Correa. C., Ferrão, M.F., Camargo, F.A.O., Triplett, E., Bento, F.M., 2017. Microbial community composition in Brazilian stored diesel fuel of varying sulfur content, using high-throughput sequencing. Fuel, 189:340-349. https://doi.org/10.1016/j.fuel.2016.10.108.

20. González-Benítez, N., Bautista L.F., Simarro, R., Vargas, C., Salmerón, A., Murillo, Y., Molina, M.C., 2020. Bacterial diversity in aqueous/sludge phases within diesel fuel storage tanks. World J. Microbiol. Biotechnol., 36(12), 180. https://doi.org/10.1007/s11274-020-02956-6.

IASH 2022, the 17th INTERNATIONAL SYMPOSIUM ON STABILITY, HANDLING AND USE OF LIQUID FUELS
Dresden, Germany
September 11 - 15, 2022

INVESTIGATING THE EFFECT OF THE CONCENTRATION OF SEAWATER ON THE ACTIVITY OF SULPHATE REDUCING BACTERIA

Gareth J. Williams*, Graham C Hill

ECHA Microbiology Ltd. Unit 22&23, Willowbrook Technology Park, Llandogo Road, Cardiff, CF3 0EF, WALES, UK

ABSTRACT

Microbial sulphide generation in seawater compensated fuel tanks of ships, other vessels and oil platforms and has been reported for over 40 years. Sulphate, is utilised as an electron acceptor by a range of sulphide generating microbes, and is in plentiful supply in seawater. Hence, tanks and systems prone to seawater ingress are vulnerable to consequent microbial sulphide generation and Microbiologically Influenced Corrosion (MIC). In some circumstances, it is feasible to utilise freshwater as compensate water instead of seawater. This could reduce the levels of nutrients available for microbial growth and also the amount of sulphate available as a terminal electron acceptor, so mitigating this detrimental microbial process. This paper describes a laboratory investigation of the influence of varying proportions of sea water and fresh water on microbial sulphide generation and corrosion in simulations of seawater compensated fuel tanks. A series of microcosms containing naval distillate fuel and water with various seawater:freshwater ratios were inoculated with a consortium of microorganisms, including Sulphate Reducing Bacteria (SRB), from field samples. The extent of microbial growth and physico-chemical parameters of the water phase were monitored. The study sought to determine the overall effectiveness of using freshwater in compensate tanks to mitigate MIC. It also sought to assess the amount of seawater ingress that might be tolerated without a corresponding significant increase in SRB activity. Under the study conditions, levels of microbial sulphide generation were found to remain low providing sea water made up less than 25% of the compensate water.

*Corresponding Author: Dr. Gareth Williams; gwilliams@echamicrobiology.com; +44 2920 365930

IASH 2022, the 17th INTERNATIONAL SYMPOSIUM ON STABILITY, HANDLING AND USE OF LIQUID FUELS
Dresden, Germany
September 11 - 15, 2022

ESTABLISHING AN ELECTROCHEMICAL BIOSENSOR FOR RAPID DETECTION OF FILAMENTOUS FUNGI AND YEASTS IN FUEL-CONTAINING ENVIRONMENTS

Osman Radwan[1]*, Michael C. Brothers[2], Victoria Coyle[2], Steve S. Kim[3]

[1]Environmental Microbiology Group, University of Dayton Research Institute, Dayton, OH

[2]UES Inc, Dayton, OH

[3]Airman Systems Directorate, 711 Human Performance Wing, Air Force Research Laboratory, Wright-Patterson AFB

ABSTRACT

Fuels are not commonly thought of as a microbial growth medium. However, certain microorganisms have evolved the capacity to use fuel hydrocarbons as carbon sources, enabling proliferation in fuel systems including fuel storage tanks. Early detection of biocontamination would enable more effective implementation of mitigation strategies to prevent fuel system biodeterioration. Therefore, we developed a comprehensive method to develop a novel, reagentless, real-time, microbial sensor platform that is also fuel resistant. Firstly, we identified an extracellular protein epitope conserved in fuel-degrading fungi. Secondly, we used the epitope to identify a suitable biorecognition element (BRE) through biopanning of a 7-mer phage displayed peptide library and demonstrated the BRE's affinity to fungi with a fluorescence assay. Thirdly, we incorporated the BRE into a novel, reagentless real-time electrochemical sensing platform leveraging a mixed self-assembled monolayer containing a peptide BRE, redox reporter, and fuel-resistant fluorocarbon blocking group. Finally, we incorporated the real-time electrochemical sensing platform into a microfluidic device to demonstrate sensing of Yarrowia lipolytica extracted fuel. We demonstrated the detection of Y. lipolytica using the biosensor down to 1E4 CFU in a bath cell. In the microfluidic cell mode, the biosensor was shown to function after been exposed to jet fuel. Therefore, this work describes a process to generate fuel-resistant biosensors for detection of microbial growth in fuel system that can be leveraged in other applications including for food and biomedical pathogen detection. The biosensor platform described effectively detects multiple groups of fuel-degrading fungi including both filamentous fungi and yeasts.

*Corresponding Author: Acting Group Leader Osman Radwan; oradwan1@udayton.edu; +19372294668

IASH 2022, the 17th INTERNATIONAL SYMPOSIUM ON STABILITY, HANDLING AND USE OF LIQUID FUELS
Dresden, Germany
September 11 - 15, 2022

INVESTIGATION OF HYDROGEN SULPHIDE GENERATION IN AN UNDERGROUND SALT CAVERN USED FOR BUTANE STORAGE

Gareth J. Williams[1]*, Edward W. English[2], Howard L. Chesneau[2], Graham C. Hill[1]

[1]ECHA Microbiology Ltd. Willowbrook Technology Park, Llandogo Road, Cardiff CF3 0EF United Kingdom
[2]Fuel Quality Services, Inc. P.O. Box 1380, Flowery Branch, Georgia 30542 USA

ABSTRACT

This paper describes the application of molecular microbiological techniques and chemical analysis to investigate problems of sulphide generation in salt caverns used to store butane, and the use of these methods for monitoring on-going risk. Following several months storage in a newly created underground salt cavern, butane gas and brine was found to contain elevated levels of hydrogen sulphide. This led to concerns over personnel safety and potential detrimental impacts on the infrastructure and product. An investigation was initiated to establish whether there was a microbial involvement in the problems being experienced, the potential long-term implications, and potential remediation plans. Samples of brine were obtained from the affected cavern and from an unconnected cavern that had been in service since the 1950s. A sample of water from the same well that was used to create the affected cavern was also collected. Samples underwent qPCR analysis to determine the abundance of bacteria, archaea and specific groups of microorganisms capable of generating sulphide (Sulphate Reducing Bacteria, Sulphate Reducing Archaea and bacteria belonging to the genus Halanaerobium). This was followed by Next Generation Sequencing (NGS) to identify the bacteria and archaea species detected in the samples. Chemical analysis of the samples was performed to ascertain the chemical and environmental conditions of the various systems, the availability of nutrients and chemical compounds needed to facilitate microbial metabolic processes, and the presence of specific metabolites associated with hydrogen sulphide production by microorganisms.

*Corresponding Author: Dr. Gareth Williams; gwilliams@echamicrobiology.com; +44 2920 365930

IASH 2022, the 17th INTERNATIONAL SYMPOSIUM ON STABILITY, HANDLING AND USE OF LIQUID FUELS
Dresden, Germany
September 11 - 15, 2022

A HOLISTIC APPROACH TO MANAGING BUNKER FUEL QUALITY

Joseph L Stark & Paul Biggerstaff

Innospec Fuel Specialties LLC, 200 Executive Dr, Newark, DE 19702

ABSTRACT

On January 1st, 2020 the International Marine Organization (IMO), in accordance with MARPOL Annex VI, adopted a new regulation for the maximum concentration of sulfur in fuel oil used on board commercial ships trading outside Emission Control Areas (ECAs). The new rule, commonly referred to as "IMO 2020", states the sulfur content in bunker fuel must not exceed 0.50% m/m unless the vessel is equipped with an exhaust gas cleaning system (EGCS).

Compliance with IMO 2020 effectively created a new fuel for the marine market, which is called VLSFO (very low sulfur fuel oil) and defined as a bunker fuel containing ≤ 0.50% sulfur. Although the implementation of IMO 2020 has been viewed largely as a success by the industry, fuel quality issues associated with the new VLSFO fuels have been prevalent throughout the marine supply chain.

Producers and blenders are incorporating a wide variety of component streams to make VLSFO blends in order to economically meet the regulation. The increased complexity of these blends has resulted in a wider variety, and increased frequency, of fuel quality issues. These quality issues not only include traditional asphaltene destabilization, but now include paraffin wax crystallization and precipitation, sludge formation from various degradation mechanisms and solids fouling from inorganic particulates, such as catalyst fines, that can contaminate blend streams from refinery cracking units.

While current test methods, primarily those found in the ISO 8217 fuel standard, help to identify fuel quality issues, one may not be able to determine the true root cause of the concern due to the complexity and variable nature of the fuels. Additionally, existing stability test methods were created before the onset of IMO 2020 and the creation of VLSFO. Current stability test methods only provide a snapshot in time of the stability of these fuels, they do not provide insight into longer term stability or operability. This is an important issue as the industry has learned from field experience that the quality of today's marine fuels can change over time.

However, by conducting a detailed evaluation of a bunker fuel's characteristics, its composition, and the operational dynamics of the application in which it will be used, a more complete, holistic understanding of the fuel's nature and behavior can be of achieved. This greater insight allows for improved economic decision-making during fuel design and blending, more efficient bunker operability in the field and greater success in predicting and mitigating any fuel quality concerns.

VLSFO Instability Causes & Concerns

IMO 2020 regulations require the maximum sulfur content of fuel oil used on board commercial ships traveling outside Emission Control Areas (ECAs) to be no more than 0.50% unless the vessel is equipped with an emission scrubber. These regulations resulted in a new type of marine fuel, officially designated as very low sulfur fuel oil (VLSFO), being produced by refineries and blenders. [1]

Despite persistent stability issues that have occurred across the fuel supply chain with VLSFO since the new sulfur regulations went into effect on January 1st, 2020, there has been very little additional technical guidance on how to design compliant fuel blends or how to better ensure good fuel quality.

ISO 8217 is an industry marine fuel standard that provides a framework for the quality of marine fuels. It was originally designed for marine distillates and high sulfur, heavy fuel oil (HSFO). The standard was created in 1987 and it has been modified several times over the years, with the latest occurring in 2017. However, it has not yet been modified to address the unique quality challenges of VLSFO.

Although the standard provides helpful guidance, the existing test methods and limits within the standard do not necessarily provide a complete picture of the stability, operability and handling characteristics of a VLSFO bunker. For example, the standard includes an evaluation of existent and accelerated stability, but it does not -consider the stability of fuel over an extended period of time. The industry is learning this distinction can be important, as the long-term storage stability of VLSFO has been determined to be significantly less than that of legacy HSFO. Traditionally, HSFO has been stable in the field for six months or longer. However, the industry shelf-life guidance for VLSFO has been reduced to three months, on average, with some VLSFOs remaining stable for only a few weeks. Additionally, if exposed to oxygen or heat, some streams could completely degrade in a matter of hours.[2]

Field instability issues with VLSFO have occurred even when the fuel met all the ISO 8217 quality standards when initially tested. Evaluating the baseline stability of a fuel based on the ISO 10307-2 test method for total sediment provides a good baseline; however a VLSFO could still degrade and become unusable in extended storage, or if a ship goes into layup, unless the fuel is properly protected beforehand.

Taking more of a holistic approach to understanding the composition and quality characteristics of VLSFO can provide greater insight into the handling, storage and operability behavior of the fuel. This approach would also help to better identify what issues may be associated with VLSFO and help to determine the best proactive strategies to mitigate any existing or potential issues in the field.

Asphaltene Instability

As implementation of IMO 2020 approached, there were concerns within the marine industry about fuel compatibility and the impact on stability, specifically asphaltene stability, when lower sulfur, distillate streams were blended with, or bunkered on top of, traditional HSFO streams. Although specific blend strategies were, and remain, tightly guarded proprietary recipes, many VLSFO blends appear to contain a high percentage of distillate components or other non-traditional bunker blend streams.

While distillate and catalytically cracked blend streams can help reduce viscosity, density and sulfur content, they generally contain lower levels of aromatics and higher concentrations of saturates and straight-chained (normal) paraffins compared to legacy HSFO. These differences in composition, as illustrated in Figure 1, can lead to asphaltene destabilization and precipitation if a fuel blend is not fully compatible.

Figure 1. Differences in Composition by Fuel Oil Type.

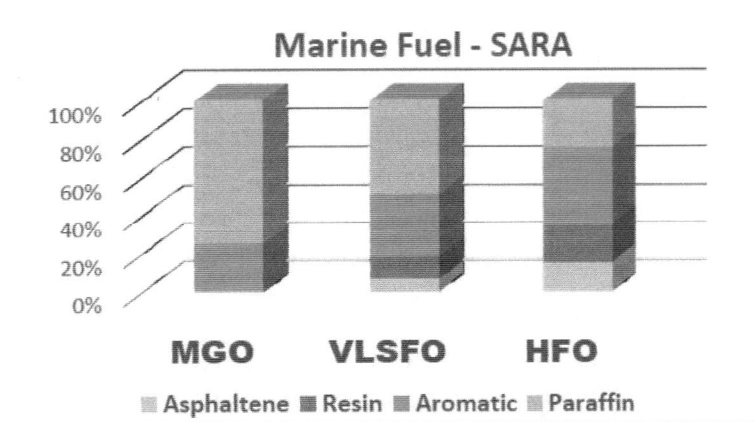

Asphaltene instability can impact fuel quality, resulting in reduced marketability and fuel value for traders and producers. It can also lead to sludge formation throughout the fuel supply chain, which can then contribute to storage, handling and operability issues. For example, sludge formation in transfer lines and storage tanks can cause a significant reduction in throughput and increased costs due to more frequent cleaning, maintenance and downtime. On board ships, sludge formation can contribute to system fouling, filter plugging, increased maintenance costs, poor combustion and higher emissions.

Figure 2. Theoretical Asphaltene Structure & Agglomerated Asphaltenes Under a Microscope.

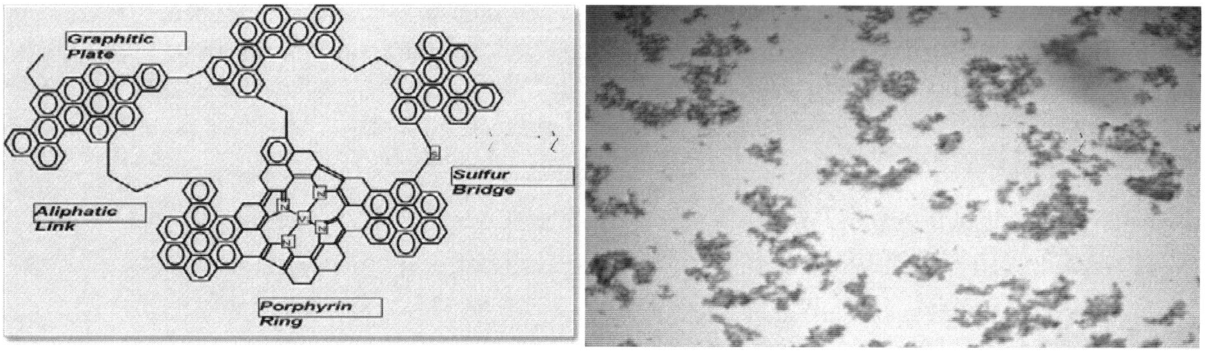

Thermal & Oxidative Instability

While the marine industry is accustomed to dealing with asphaltene-related issues, and it has a good understanding of how to mitigate their impact, asphaltene destabilization has not always been the primary fouling issue experienced with VLSFO. A new and unanticipated issue encountered within the industry has been instability with the distillate and intermediate (e.g., gas oil and slurry oil) streams that comprise some VLSFO blends.

Straight-run distillate (i.e., a stream straight off the distillation column that has not gone through hydroprocessing) and intermediate process streams can contain olefins, di-olefins, aldehydes, and reactive nitrogen compounds such as amines. Heavier process streams can contain acidic compounds and trace levels of metals. These constituents can all act as instability agents to drive various degradation reactions resulting in the formation of insoluble polymers and high molecular weight gums. Exposure to oxygen (oxidative) and heat (thermal) can accelerate these reactions leading, in some cases, to rapid degradation of the fuel.

Degradation can occur within a potential blend stream itself or it can occur when streams that exhibit good standalone stability are blended together to make VLSFO. As an example, one stream may be high in amines while another stream may be high in acidic components. Individually, the streams are stable. However, when blended together, the amines and acids can react to form salts. If heat is added, these salts can further react to form high molecular weight polymers and gums.

Figure 3. Sticky Oxidation Foulant.

Figure 4. Hard Thermal/ Oxidized Foulant.

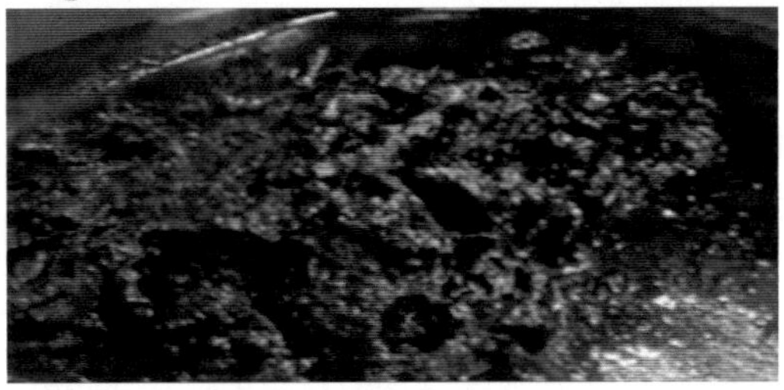

Paraffin Instability

Increasing levels of distillate and cracked streams entering the bunker market means there is a higher concentration and a broader carbon distribution of straight-chained paraffins being seen in VLSFO. The longer chain, higher carbon number paraffins can have melt points well above ambient temperature conditions. Thus, as the temperature of a fuel goes down, these waxes can crystallize and precipitate out leading to sedimentation and fouling. It is important to determine the wax appearance temperature (WAT) of a fuel and analyze its paraffin profile to better understand its temperature and flow behavior.

Increasing the temperature of a bunker fuel that contains a high level of paraffin will help to solubilize the wax crystals. However, higher temperatures may also have the unintended consequence of driving thermal degradation reactions in VLSFO resulting in increased sediment and sludge formation. Conversely, reducing operating temperatures to try and minimize degradation reactions can lead to an increase in fouling due to wax sedimentation. This is why understanding the true nature of a fuel is critical to maintaining fuel quality.

Paraffin-related issues can often be mistaken for other organic fouling issues. For example, sediment that appears in ASTM D4740 spot tests, or sediment that is collected during hot filtration tests, may actually be high molecular weight wax crystals that have precipitated out of the fuel. If the true nature of the sediment is not properly identified and addressed, then wax crystallization and agglomeration issues can result in increased sediment formation and system fouling.

Inorganic Contamination

In addition to organic sludge formation, inorganic solids, corrosion by-products and metals can be a major cause of fouling and equipment erosion. Fluid catalytic cracker unit (FCCU) slurry oil often makes for a highly aromatic, low sulfur blend stock for VLSFO. However, high levels of catalysts fines (e.g., ash, silica and alumina) from the reaction bed often carry over into the oil and into a finished fuel unless the contaminants are removed. In addition to the catalyst fines and metals being a primary foulant in fuel logistics systems, as previously discussed, low levels of metals can catalyze degradation mechanisms leading to rapid sludge formation.

Understanding Total Sediment (TSE, TSA, TSP)

The current industry recommendation for determining the stability of marine fuels is to measure the total sediment collected during a hot filtration test. The test methods are:

ISO 10307-1: Total Sediment Existing (TSE)

- ISO-10307-2 Procedure A: Total Sediment Potential (TSP)
- Procedure B: Total Sediment Accelerated (TSA).

Per ISO 8217, the fuel quality requirements for maximum total sediment is a TSA of 0.10% or less, by mass, with the referee method being TSP.[3]

However, in the field it has been found that even if a fuel is below the maximum recommended sediment concentration in one, or all of these tests, it is still possible for a fuel to destabilize over time, form sludge and exhibit poor handling and operability unless properly treated.

Figure 5. Root Cause Analysis via Total Sediment.

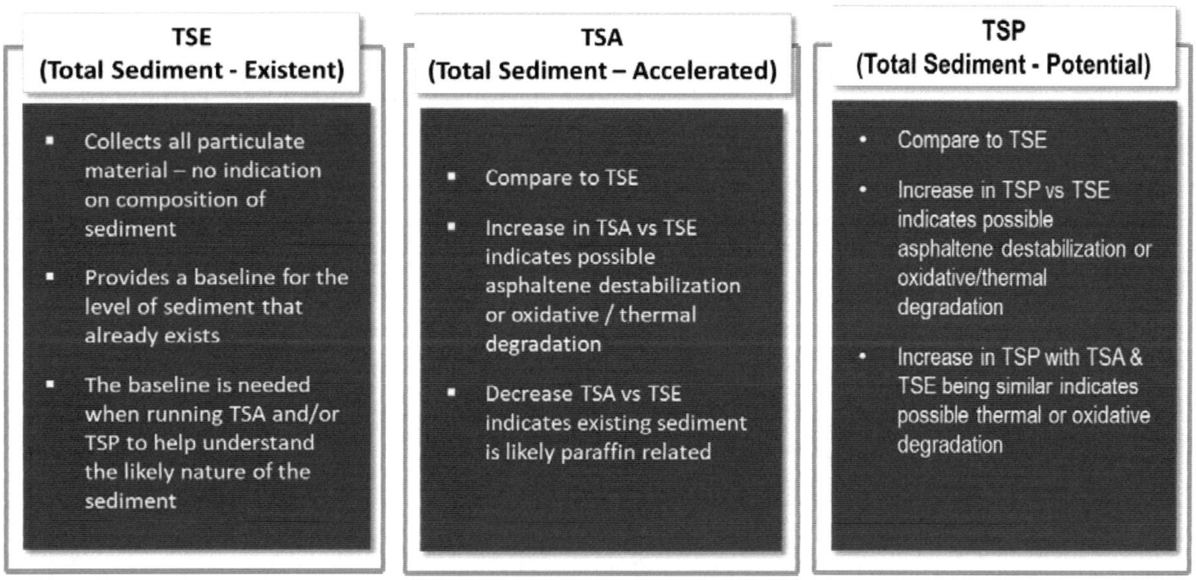

Understanding the nuances of each total sediment test and comparing their results can help with identifying the underlying cause of the sediment formation. Identifying the root cause(s) of the instability is important in designing an effective mitigation strategy.

TSE reflects the existing sediment that is in a bunker fuel or blend stream. In essence, it provides a baseline for the amount of instability that has already occurred. However, this individual measurement does not provide insight or differentiation into whether the nature of the sediment is due to wax crystals, agglomerated asphaltenes, cat fines, gums, etc. Simply put, if the particulate is larger than the pore size of the filter paper, it will be collected and contribute to the total mass.

TSA is an accelerated chemical aging procedure that involves adding 10%-vol of a C18 straight-chained, paraffinic alkane (hexadecane) to the bunker fuel or blend stream and heating the mixture for one hour at 100°C prior to filtration. Since increasing the paraffin content tends to accelerate asphaltene destabilization, the purpose of this approach is to evaluate the inherent, or reserve, stability of the fuel or blend stream.

Therefore, an increase in TSA over TSE would tend to indicate the sediment formation is due to asphaltene destabilization. It also tends to indicate that the sediment is not due to wax since the alkane solvent will help to solubilize paraffins already in the oil. Conversely, a decrease in TSA versus TSE results would tend to indicate that the sediment being collected via TSE is due to wax crystals.

It is also important to note that an increase in TSA over TSE does not rule out thermal or oxidative destabilization being the root cause. If the amount of total sediment continues to increase during TSP, then degradation is a possible root cause.

TSP is also an accelerated aging test. It does not include chemical aging, but it does involve extending the duration of thermal stressing to 24 hours at 100°C before filtering. In this approach, an increase in TSP versus TSE indicates either asphaltene destabilization or thermal/oxidative degradation is occurring as a function of time and temperature.

By completing and comparing the results of all three tests, one can further deduce the most likely fouling mechanisms. For example, if the TSP is higher than the TSE, yet the TSA and TSE are similar, then there is a high probability that the bunker fuel is undergoing thermal and/or oxidative degradation and the resultant sediment is high molecular weight polymers or gums. If the TSP and TSA are similar, and both values are higher than TSE, then there is a higher probability that the mechanism of fouling is asphaltene agglomeration and destabilization.

A Holistic Approach to Fuel Quality Management

The potential underlying and contributing factors that affect bunker fuel quality have become significantly more complex with the implementation of IMO 2020 and the industry's move toward greater sustainability and a lower carbon footprint.

When issues do arise, taking a holistic approach to understanding the application type, the system and operational dynamics, fuel composition and fuel history is critical to identifying the root cause of the problem and to determining whether the cause is proximate or systemic, environmental or compositional. The more thorough a problem and its underlying cause(s) is understood, the easier it is to design a more effective solution that will successfully mitigate the issue and ensure fuel quality.

System Dynamics

In problem solving, it is important to first understand what the immediate issue (i.e. the symptom) is and what key quality or performance criteria is being impacted. Then, identify where within the supply chain the issue is occurring and what type of field application it is.

For example, is the issue occurring while the fuel is in storage, while the fuel is being transferred through the distribution system or while the fuel is being bunkered or consumed onboard a vessel?

Once the type of application is determined, the operational conditions of that system (e.g., system temperatures, storage conditions, etc.) need to be evaluated to determine if they could be contributing factors. Understanding the system and how it is being affected can be helpful in determining if a process change or mechanical control might be an effective preventative measure and, if needed, help to identify what type of additive treatment program might be the most effective in improving quality and ensuring good handling and operability.[4]

Figure 6. Fuel Issues Can Impact the Entire Supply Chain.

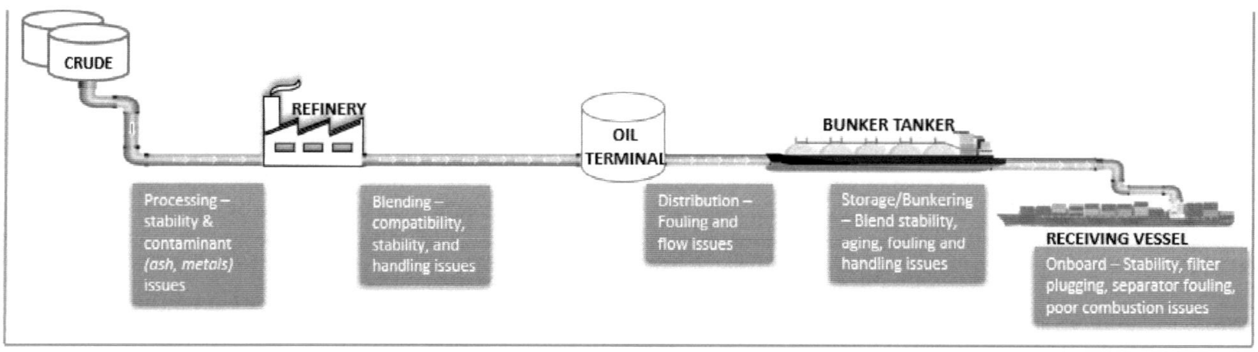

Understanding the Fuel Itself

Is the fuel a new blend? Has it been in storage for a period of time? If stored, under what conditions? Has it been commingled? Determining the history and treatment of a particular fuel stream may not always be possible, but the more that is known about it the better its behavior can be understood.

Understanding the composition of a finished fuel and, in the case of a blend, the individual streams that comprise it, is also important to understanding its behavior and the true nature of any quality-related issue.

Identifying potential instability agents that exist within a blend or component stream and understanding how these trace constituents can interact is important to understanding thermal and oxidative stability.

Saturates, aromatics, resins and asphaltenes (commonly referred to by the acronym 'SARA') are the primary solubility classifications that comprise a bunker fuel. When formulating a fuel blend, understanding and managing the ratio of these parameters is critical to ensuring good compatibility and to minimizing the likelihood of instability.

SARA analysis is a tool that can be very helpful in predicting stability behavior. Fuels that have high levels of aromatics and resins tend to have good inherent stability. Fuels that have high levels of saturates and asphaltenes tend to have instability issues. As the ratio of these components in a fuel change, so too does the behavior of that fuel.

A normal (straight chain) paraffin characterization of a bunker fuel that has an unknown composition can also help in identifying the likely blend streams used to make it, in addition to providing insight into the low temperature flow behavior of the fuel. Distillate streams typically have paraffins with carbon chain lengths from C10 to C28. While intermediates streams, such as gas oils, typically have carbon chain lengths from C20 to C40, and carbon chain lengths greater than C40 are typical of hydrocracker and residual streams.

As an illustration, Figure 7 depicts the paraffin distribution of a VLSFO. From the analysis, it is readily apparent that the fuel is a combination of distillate and heavy blend components. The bi-modal distribution is becoming more common in today's bunker fuels as the variety of streams used to meet the IMO 2020 lower sulfur specification increases.

Figure 7. Paraffin Distribution Analysis of a VLSFO.

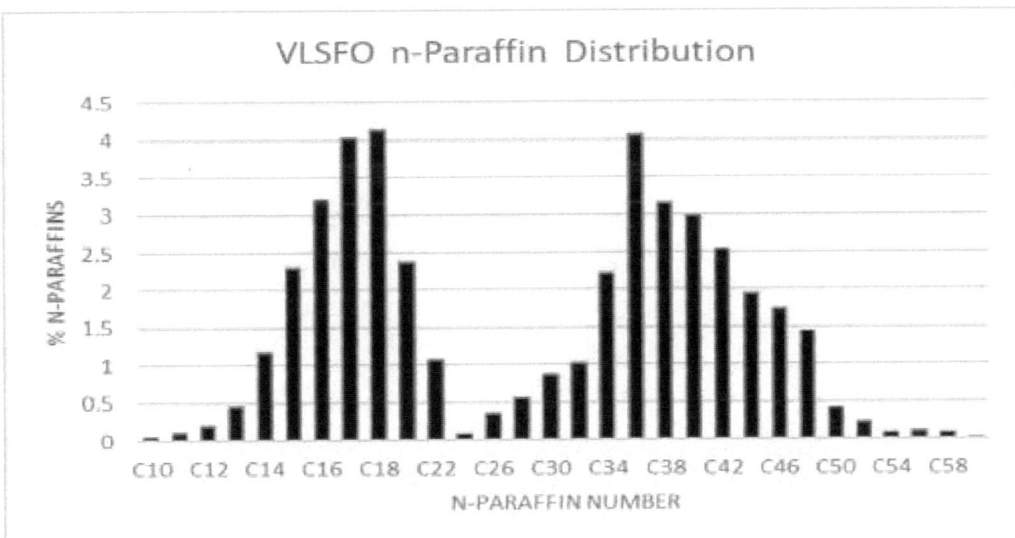

SARA and paraffin analyses can also be combined with stability methods such as ISO 10307 to gain a more complete understanding of the sediment being generated and the fouling mechanism causing it.

An unstable bunker fuel that has a high measured TSP value compared to its TSE; high levels of saturates and asphaltenes and a high percentage of paraffins, particularly in the distillate range, all tend to confirm that asphaltenes are the cause for sedimentation. From this knowledge, there is a higher degree of confidence that implementing an asphaltene mitigation strategy will be the effective solution in preventing or mitigating any associated fouling issues.

In another example, an unstable bunker fuel with a high TSP value but a low TSA value compared to its TSE, coupled with the presence of straight run distillate in the blend, would indicate thermal or oxidative degradation **could** be occurring in the bunker.

Understanding the operational history of the fuel would also help to identify potential fouling mechanisms.

Fuel Stability Testing

The industry recommended ISO 10307 sedimentation tests are helpful tools in evaluating the stability of bunker fuel, but it is important to note that these tests have a high degree of error associated with them and the results only represent a snapshot in time. For reasons stated earlier, it is possible that a bunker will degrade over time. A fuel can initially pass all of the ISO 8217 targets only to degrade during storage or bunkering which then leads to reduced quality, high levels of sludge formation and operational issues. A number of VLSFO producers, suppliers, and consumers have observed incremental increases in total sediment as their VLSFO streams age. This gradual creep in TSA/TSP has led to their fuel going off specification, in as little as a few days in some cases.

From a technical standpoint, It is recommended to evaluate the stability of a bunker fuel in a variety of different ways and to try and mimic actual field conditions as closely as possible when testing. Storing a fuel sample for an extended period of time at an elevated temperature is a good way to help determine if a blend stream or VLSFO is likely to exhibit instability as the fuel ages due to thermal and oxidative degradation. If a ship is going into layup, or a fuel is going to be stored for a period of time, then expanding testing to include an extended stability evaluation is recommended. Proactive treatment should be considered in order to ensure long-term stability and protect against total sediment creep.

Performing alternative stability evaluations, such as determining the reserve stability (RSN) of a fuel, can provide additional insight into the inherent stability of the fuel and they can be useful in demonstrating the efficacy of additive treatment in improving stability and reducing sedimentation. All of these tests performed in combination would provide a more complete picture of a bunker fuel's quality.

From a mitigation or treatment perspective, identifying the foulant is key. Is it asphaltene sludge formation, inorganic material, polymer formation due to thermal degradation or wax crystallization causing the sediment, handling or operational issues?

In summary, compositional information, combined with other fuel analyses and robust performance evaluations, is the basis to gaining a holistic understanding of the true nature and behavior of a fuel, and to being able to identify an effective corrective action or treatment solution should a quality issue arise.

Bottom Line Impact

Fuel quality issues can be costly, and they can impact the entire marine fuel supply chain. As the complexity of bunker fuels increase, so too does the potential for experiencing a problem in the field.

Given the increased challenges, it is increasing important to partner with fuel experts within the industry who are capable of conducting thorough, holistic evaluations of bunker fuel applications in order to identify potential issues that may arise during blending, handling or consumption.

Innospec has over 50 years of experience in the marine industry and has been successful in developing treatment solutions for every segment of the fuel market - from cradle to grave (inception to consumption) across the globe.

Whether it is rapidly responding to fuel issues that occur in the field, mitigating the risk of experiencing fuel quality related issues, or improving the bottom-line economics of on-going applications, Innospec is here to work directly with customers to identify, tailor, and implement effective treatment solutions.

References

1. IPIECA Joint Industry Guidance, The Supply and Use of 0.50% Sulfur Marine Fuel, August 2019
2. BunkerSpot; Deep Dive, Vol 17, No. 5, October/November 2020, pp 55-57
3. CIMAC Guideline 1/2019 Marine fuel handling in connection to stability and compatibility
4. BunkerSpot Economic Benefit of Additives, Vol 18, No. 3, 2021, pp 81-84

IASH 2022, the 17th INTERNATIONAL SYMPOSIUM ON STABILITY, HANDLING AND USE OF LIQUID FUELS
Dresden, Germany
September 11 - 15, 2022

CATALYST FINES REMOVAL FROM SLURRY OIL-A NOVEL APPROACH

Sai Reddy Pinappu*, Sankaran Murugesan

Baker Hughes, 17021 Aldine Westfield, Houston, Texas 77073, USA

ABSTRACT

A new limit on the sulfur content in the fuel oil used on board ships came into force on Jan 1st 2020 (3.5% m/m before to 0.5% m/m current limit). This new limit on Sulfur is expected to improve air quality, preserve environment & protect human health. This made producers to look for new low sulfur fuel oil components, one such stream is, fluid catalytic cracking (FCC) slurry oil.

Slurry oil is attractive to be used for blending down Sulphur content of fuel oils but the concentration of catalyst fines in this stream could exceed 1500 ppm. Per ISO 8217 -specifications of marine fuels, the limit on catalyst fines Aluminum (Al) and Silica (Si) is set at Al+Si <60 ppm. Traditionally, mechanical separation techniques such centrifugation & filtration are used to get the Al, Si concentration under specification but due to the nature of this hydrocarbon , catalyst fines are often coated with organic molecules which hinders the separation of Al & Si.

In this paper, the effect of various physiochemical properties of slurry oil & effect of chemical additives on removal of Al, Si from the slurry oils will be presented. A Principal Component Analysis (PCA) was used to identify the important parameters effecting the separation.

*Corresponding Author: Lead Engineer Sai Reddy Pinappu;
Saireddy.Pinappu@bakerhughes.com; +1505-818-8589

IASH 2022, the 17th INTERNATIONAL SYMPOSIUM ON STABILITY, HANDLING AND USE OF LIQUID FUELS
Dresden, Germany
September 11 - 15, 2022

INVESTIGATION OF WATER DISTRIBUTION AND MICROBIOLOGICAL GROWTH IN MARINE FUEL OIL BLENDED WITH FATTY ACID METHYL ESTER

Leon O'Malley[1], Graham C. Hill[1]*, Gareth J. Williams[1] and Simon Ashton[2]

[1]ECHA Microbiology Ltd., Unit 22&23, Willowbrook Technology Park, Llandogo Road, Cardiff, CF3 0EF, WALES, UK ghill@echamicrobiology.com

[2]Esso Petroleum Ltd., ExxonMobil House, Ermyn Way, Leatherhead, Surrey, KT22 8UX, UK simon.ashton@exxonmobil.com

Blending Fatty Acid Methyl Esters (FAME) with conventional automotive diesel fuel is known to increase the susceptibility to microbiological growth, with the potential for operational and quality issues. FAME is increasingly used in marine distillate fuels, with the 2017 edition of ISO 8217 including three grades allowing FAME up to 7.0 % v/v. More recently, blending FAME in heavier residual marine fuel oils has been trialled. In distillate fuels, microbial growth occurs predominantly in settled water phase. However, for fuels oils, the higher viscosity will result in a tendency for water to remain in suspension and consequently it is postulated that FAME may exacerbate active microbial growth within the bulk fuel oil phase. Using laboratory microcosms, this study investigated microbiological growth in Very Low Sulphur Fuel Oil (VLSFO) and VLSFO blended with FAME at 20% and 50% (v/v) at various fuel:water ratios.

Although water settling was observed at higher water ratios, water distribution characteristics in all fuels were similar at fuel:water ratios reflecting those most realistic of field scenario. Where water did settle, high numbers of viable bacteria were detected in this, for all fuel types. Although the fuel oils were found to retain significant amounts of water in suspension, only limited microbiological contamination was detected in fuel phase for neat VLSFO and VLSFO/FAME blends.

The results of this study suggest that providing marine vessel operators follow good housekeeping practices to minimize water accumulation, blending FAME in fuel oil should not substantially increase susceptibility to microbiological contamination in the fuel.

*Corresponding Author: ghill@echamicrobiology.com

IASH 2022, the 17TH INTERNATIONAL CONFERENCE ON
STABILITY, HANDLING AND USE OF LIQUID FUELS
Dresden, Germany
11 - 15 September 2022

REINFORCEMENT LEARNING FOR THE IDENTIFICATION OF ISOMERS WITH A STRONG SOOTING TENDENCY

Florian Pütz, Clemens Hall, Bastian Rauch and Andreas Huber

[1]*DLR, German Aerospace Center, Institute of Combustion Technology, MAT, 70569 Stuttgart, Germany, florian.puetz@dlr.de*

ABSTRACT

Since the approval of new jet fuel candidates is a very complex and expensive process, the prediction of jet fuel properties based on the fuel composition is an important enabler for the development of sustainable aviation fuels. The fuel composition based on the chemical family and the number of contained carbon atoms can be determined by gas chromatography (GCxGC). However, the GCxGC cannot resolve most of the isomers present in a fuel. The missing information on the isomer composition is one of the reasons for uncertainties in the prediction of fuel properties, which in turn is important for the determination of combustor performance and emissions. For molecules with a short chain length the different isomers can be easily determined, but with increasing C-number, the number of possible structures grows exponentially and the task becomes too computationally expensive.

In this work, Reinforcement Learning (RL) is used to systematically explore the large space of possible chemical compositions of isoalkanes to generate the isoalkanes with the highest Yield Sooting Index (YSI) so that they can be eliminated by fuel manufacturers in the production process. Therefore, two different approaches based on the Deep-Q-Network (DQN) algorithm are compared in this work. It is also investigated how the RL decision maker moves through the state space and generates the isomers during its learning process.

KEYWORDS: Alternative Fuels, Reinforcement Learning, Deep Q-Network, Machine Learning, Yield Sooting Index, Markov Decision Process, Fuel Design

INTRODUCTION

Global warming is one of the great challenges facing humanity [1]. Due to anthropogenic climate change, extreme weather events are increasing around the world and are already having a direct impact on human lives [2–5]. Not only is there an increasing awareness among the general public, but also the political agenda is influenced by anthropogenic climate change. Examples for this are the Paris climate agreement and the European Green Deal. The European Green Deal describes a concept with which the European Union (EU) wants to become climate neutral by the year 2050 [6], while the Paris climate agreement specifies the objective of limiting global warming ideally to 1.5 °C by the end of the century [7]. In order to achieve these ambitious goals, according to Boehm[8] substantial changes must be made in all sectors. In a study published in 2021 Lee et al. estimated the impact of aviation on effective radiative forcing for the time period from 1940 to 2018. The result is that non-CO_2 effects, which are mainly contrails and nitrogen oxides induced, account for about two-thirds of the climate impact of aviation since they change the radiative balance. [9]. In a second study, Brazzola et al. forecasted how aviation will affect global warming by the end of the century, again giving great importance to non-CO_2 effects [10]. Even if the sector becomes fully CO_2 neutral, non-CO_2 effects will still cause a global warming of 0.4 °C by the end of the century if no measures are taken to reduce the effective radiative forcing cause by non-CO_2 effects [10]. This is especially concerning in relation to the 1.5 °C target set in the Paris climate agreement, which applies to the sum of all sectors. Fortunately, with the transition to Sustainable Aviation Fuels (SAF), which the International Civil Aviation Organization (ICAO) has identified as the technology with the highest technical readiness and greatest potential to reduce emissions [11], it is possible to design the fuels in a way that they form fewer contrails [12]. In two flight measurement campaigns conducted by the German Aerospace Center (DLR) and NASA, it was shown that the emitted soot particles can regulate the number of contrail cirrus ice crystals and thus the non-CO_2 effects [12]. Voigt et al. demonstrate that the use of low-aromatic fuels can reduce the emitted soot and ice particles by 50-70 %, which in turn leads to a lower energy deposition in the atmosphere and therefore can reduce the climate impact of aviation [12]. The result shows that SAF should be designed in such a way that less soot is emitted. Within the EU project JETSCREEN (Jet Fuel Screening and Optimization Project), for example, an approximately exponential correlation between hydrogen content and soot emissions was demonstrated [13]. As shown in Figure 1, most measurements closely follow the correlation line showing that the higher the hydrogen content of a fuel, the less it tends to form black carbon [13].

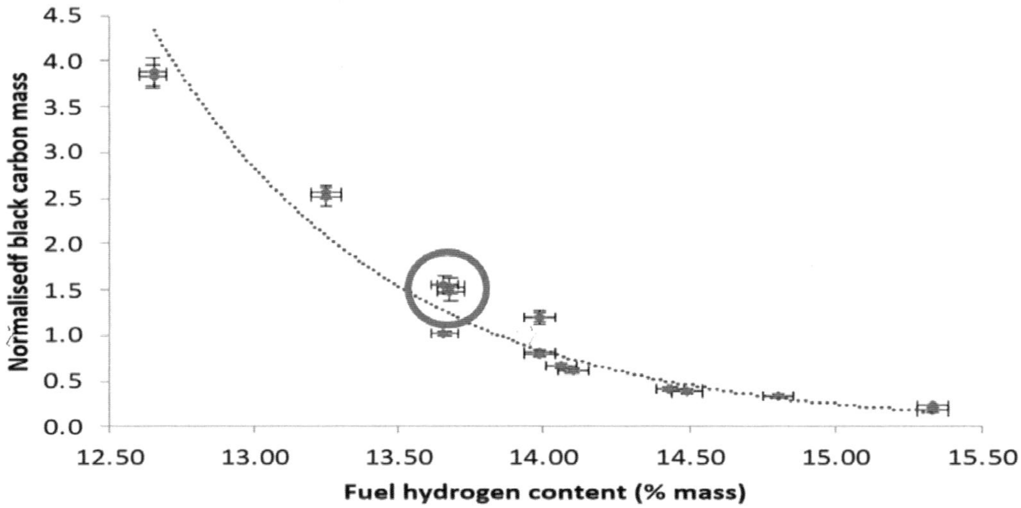

Figure 1: JETSCREEN results Fuel impact on Black Carbon Matter Emissions [13]

However, there are some measurements with a large offset to the correlation line, such as those circled in Figure 1. Therefore, it can be assumed that there are other factors besides the hydrogen content that influence the sooting behavior of a fuel. The key factor determining the fuel properties is, of course, the detailed fuel composition. However, the measured composition contains uncertainties. While the two-dimensional Gas-Chromatography (GCxGC) method can nowadays resolve the fuel composition based on the chemical family and the number of contained carbon atoms, it cannot resolve all the isomers contained in the fuel [14,15]. However, jet fuels are complex mixtures consisting of different hydrocarbon families and, within each family, different isomers. These different isomers can have a significant impact on the fuel properties. This can be observed, for example, in the Yield Sooting Index (YSI), which rates the sooting tendency of a fuel component independent of burner configuration or additive concentration with a significantly lower uncertainty compared to concepts such as smoke point as defined in ASTM D1322 [16]. Based on the National Renewable Energy Laboratory (NREL) estimator, differences in YSI of about 30 % can occur for iso-alkanes with 13 carbon atoms depending on the degree of branching (Figure 2) [17].

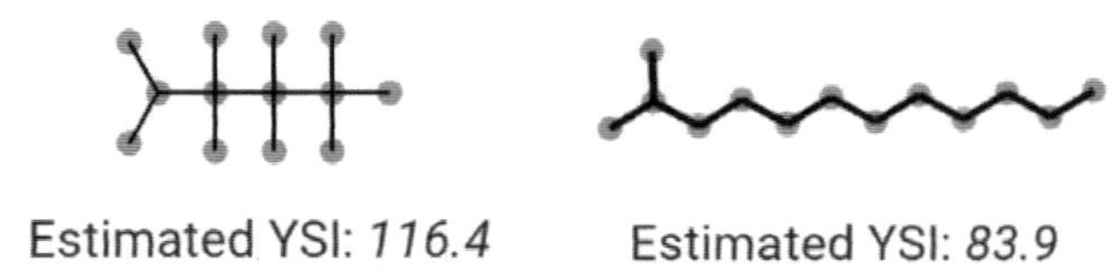

Figure 2: Iso-alkanes with 13 carbon atoms with largest and smallest YSI based on the YSI estimator from the National Renewable Energy Laboratory[17].

The example in Figure 2 shows that different isomers can have an influence on fuel and, explicitly in the example, emission properties. If, for example, the isomers that have the highest impact on the sooting tendency can be identified, it is possible to adjust the fuel production process so that these isomers are not produced. For molecules with a short chain length, the different isomers can be easily determined by hand. However, as the carbon number increases, the number of possible isomers grows exponentially [18]. This quickly leads to a point where the sampling of all isomers becomes computationally too expensive. Therefore, better strategies are needed to determine the isomers that affect fuel properties such as the sooting tendendy. This is especially true if, in addition to the sooting tendency, other performance and safety-relevant properties (such as freezing point, density, heat of combustion ...) must also be considered. If this is the case, the Pareto optimum must be determined for several different requirements.

In recent years, Zhou et al.[19], Popova et al.[20], Putin et al.[21], Guiamares et al.[22] and Olivecrona et al.[23] have used Reinforcement Learning (RL) algorithms in the field of drug discovery to generate molecules with specific properties. One of the reasons these authors have been able to successfully generate these molecules is the ability of RL algorithms to handle a large state space. The large state space in the field of molecular research offers parallels to applications in which the use of RL has enabled major breakthroughs in the past decade. A prominent example is the success of the Google DeepMind research team, which in 2016 managed to beat the current world champion in the board game Go using an RL-based algorithm [24,25]. Since there are $2.08 \cdot 10^{170}$ possible board constellations in the game of Go [26], it was long considered an unsolvable task for a computer, as brute force methods are not suitable for such a large state space [27]. Moreover, RL has proven that it is able to find the balance between different and even contradictory requirements. This ability was demonstrated by Zhou et al.[19] who used multi-objective Reinforcement Learning to generate molecules that exhibit better selectivity but at the same time have a solubility within a certain range.

Considering the great successes achieved with RL in the past decade, in this work RL is used to identify the isomers with the highest sooting tendency so that fuel manufacturers can optimize their fuel production process. The purpose of this study is to evaluate whether RL is suitable for an application in the fuel design process. Therefore, in the following, a task is presented that can also be solved manually so that the performance of the RL decision maker (agent) can be examined. For this study, the molecule space is limited to the iso-alkanes, since these are the dominant hydrocarbon family in the currently approved SAF.

THEORY

In this study, Deep Q-learning with experience replay and a target network is used following the work of Mnih et al.[28]. The RL agent is trained from scratch without pre-training to generate the SMILES[29] keys of iso-alkanes with a high YSI. SMILES is short for Simplified Molecular-Input Line-Entry System and it describes a linear character notation for describing molecules [29]. Like in the work of Guiamares et al.[22] and Olivecrona et al.[23], invalid SMILES keys are penalized.

Furthermore, simple rules are used to prevent the generation of obviously invalid SMILES keys by masking invalid actions so that the agent can learn more efficiently [30].

Yield Sooting Index estimator

The Yield Sooting Index (YSI), which measures how high the sooting tendency of a fuel component is, is predicted with the estimator of the NREL [16,17]. In the first step, the estimator determines the distribution of the different carbon types in a molecule based on its SMILES key. Each of these carbon types is then assigned a sooting tendency from a regression against a uniform database containing YSI measurements for more than 400 individual components. The YSI for the given molecule is then the result of the sum of the YSI for the different carbon types that occur multiplied by their counts. [16]

Markov Decision Process for generating isomers with a high YSI

In order to be able to generate the isomers with the highest YSI with RL, it is necessary to formulate the task as a Markov Decision Process (MDP)[31]. In a MDP, it is required that the current state $s \in S$ contains all the information necessary for an agent to decide on the next action $a \in A(s)$ (Markov Property) [32]. S is the set of possible states and $A(s)$ is the set of possible actions for a specific state s. Mathematically the Markov Property can be expressed as: $\mathbb{P}[s_{t+1}|s_t] = \mathbb{P}[s_{t+1}|s_1, \dots, s_t]$. s_{t+1} denotes the next state while s_t refers to the current state. In this work, the MDP(S, A, P, R) is defined with the following elements:

- The state space S where $s \in S$ is a string for generating SMILES keys. A state provides space for n actions. n is the specified carbon number of the searched isomers. The SMILES keys are generated episodically and two different approaches are investigated for the initial time step $t = 0$:

    1. The initial state is completely empty and n actions must be determined to generate the SMILES key. Accordingly, the number of steps T in the MDP is limited to n.
    2. The initial state contains a random action at a random index. Since the SMILES keys in this work are generated from scratch (episodic task) this action is redefined at the beginning of each episode. In this case, only $n - 1$ actions need to be determined and the episode terminates when $t = n - 1$.

- The action space A, which is a set of characters/character combinations for the construction of a SMILES key. Due to the limitation to iso-alkanes, these four actions are sufficient to build the different isomers:

    1. C
    2. $(C$

3. $C)$
4. (C)

Note that SMILES keys can also be formed with the 3 characters (,), C. By combining the characters as shown above, however, there are advantages in the handling of the states, since each action leads directly to a change of the molecule length. In addition, invalid action sequences such as "))" or "()" are directly excluded. Also, action masking[30] is used to prevent parenthesis from being closed when there are no open parenthesis, or parenthesis from being opened when there are not enough time steps left to close them again. For example, a SMILES key can't end with an open parenthesis and therefore the last action of a state can't be action 2.

- The state transition probability $P_{ss'}^{a} = \mathbb{P}(s_{t+1} = s' | s_t = s, a_t = a)$ describes the probability of landing in state s_{t+1} when the action a_t is executed in the state s_t. In our case, this probability is always 1, which means that when the agent performs an action, the corresponding characters are appended to the state (SMILSE key).

- The reward function R is based on the RDKit[33] python library and the YSI Estimator[16,17] from the NREL. Whenever the agent completes an episode, it is first checked with the RDKit library if the SMILES key describes an iso-alkane. If this is the not the case, the environment returns a small negative reward. If the agent found the SMILES representation of an iso-alkane, the YSI Estimator is used to calculate the corresponding YSI. This estimated YSI is then scaled linearly to a reward between 0 and 1. Isomers with a high YSI receive a reward close to 1 and the isomers with a low YSI will receive a reward close to 0.

  A second variation of the reward function is also tested, where -0.2 is added to the YSI-based reward when the agent generates a SMILES key that it has previously generated. This change is intended to encourage the agent to explore more of its environment.

Reinforcement Learning

Reinforcement learning (RL) is a subarea of machine learning. Unlike other subareas, RL does not necessarily require training datasets. Instead, an agent collects experience on its own and learns similar to a child through trial-and-error while interacting with an environment (Figure 3). The objective of the agent is to maximize a cumulative reward (return). The goal of this study is to solve the MDP described above, which is equivalent to generating the SMILES keys of the iso-alkanes with the highest YSI, since they yield to the highest reward.

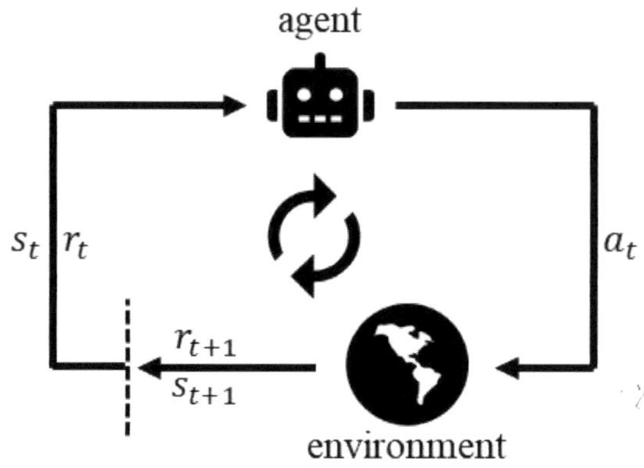

Figure 3: RL model for the interaction between agent and environment.

In order to find these isomers, Q-learning, a value-based temporal difference (TD) learning method, is used. In RL, the objective is to find the optimal policy π^*, which is equivalent to finding the best action for each state. To find an optimal policy, the Q function has to be maximized. The Q-function is equivalent to the expected return if action a is chosen in a state s and then the policy π if followed afterwards. Mathematically, this can be represented as follows:

$$Q^{\pi}(s,a) = \mathbb{E}\left[\sum_{k=0}^{T} \gamma^k r_{t+k+1} \,|s_t = s, a_t = a\right]$$

The discount factor γ, which controls the importance of direct rewards over future rewards, is set to 1 in this study. Since the task is episodic, each episode has a fixed length and the reward for the generated smiles key can only be calculated at the end of the episode, it is not necessary to discount future rewards. The optimal policy π^* then corresponds to:

$$\pi^*\,(\mathrm{s}) = \mathrm{argmax}_a Q^{\pi^*}(s,a)$$

An adaptation of Deep Q-Network (DQN), a deep Q-learning approach by Mnih et al.[28] which estimates the Q-function with a neural network, is used. Therefore, a loss function has to be minimized and since γ is equal to 1, this loss function is:

$$\mathcal{L} = \left((r_{t+1} + max_{a_{t+1}} \hat{Q}(s_{t+1}, a_{t+1}|\hat{\theta}) - Q(s_t, a_t|\theta))\right)^2$$

where $\hat{\theta}$ are the weights of the target network [28].

As an exploration strategy, the simple and widely used ε-greedy algorithm is applied. Thus, the agent selects the greedy action (action with the highest predicted Q-value) with a probability of

$1 - \varepsilon$ and a random action with a probability of ε. The training starts with $\varepsilon = 1$ which is reduced to a minimum of 0.01 with a discount rate of 0.999.

RESULTS AND DISCUSSION

To evaluate whether RL can be used to determine the iso-alkanes that maximize YSI, high YSI iso-alkanes were determined by hand for a carbon number of 13 in this study. The carbon number 13 was chosen because it is approximately the average carbon number of jet fuels. As shown in Figure 4, the maximum of the YSI is 114.6, and there are four isomers for the MDP used with this value. The reason why these iso-alkanes all have the same value is that they all have the same carbon type distribution and the estimator used in this work predicts the YSI based on this distribution. It can also be seen from the figure that, according to the estimator, the highly branched iso-alkanes have a high sooting tendency. In the following, it is investigated how many of these four isomers an RL agent can find independently by trial and error. Furthermore, a visualization of the learning process is used to visualize how the agent moves through the space of possible chemical compositions.

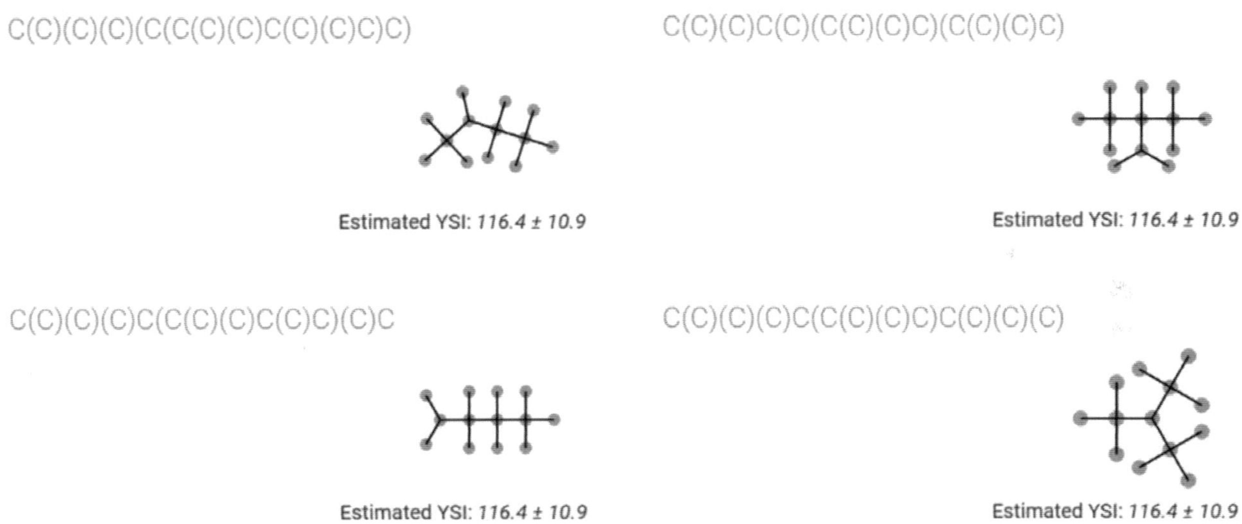

Figure 4: All iso-alkanes with a carbon number of 13, maximizing the YSI.

In this study, two approaches are compared that differ from each other in two aspects, as shown in Figure 5. For the first approach, as described in the section *Markov Decision Process for generating isomers with a high YSI*, the initial state is completely empty. In the second approach, it is also investigated whether the RL agent can benefit from a random action in the initial state. This implies that a random character/character combination is predefined at a random index in the SMILES key. In addition, in the second approach, the agent will receive a reduced reward for generating a SMILES key it has already generated before. In both approaches, the agent trains for 5000 episodes, which is equivalent to the agent generating 5000 SMILES keys. The used model has two hidden layers with 512 neurons each.

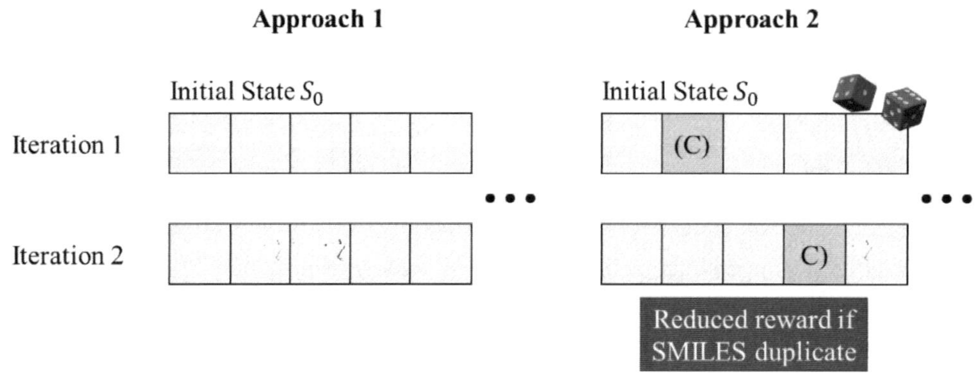

Figure 5: Comparison of the two RL approaches. The second approach differs from the first by a random initial state and reduced rewards for SMILES duplicates.

Approach 1 – empty initial state

In approach 1, the agent was only able to find one of the four isomers. Moreover, as shown in Figure 6, it wastes resources generating the same isomer multiple times with three different SMILES keys. This happened because the SMILES descriptor is not unique. As a consequence, the state space is artificially increased, since the agent makes decisions based on the SMILES keys and not on the isomers. For example, if there are on average 15 different ways to describe an isomer, the state space in the MDP increases by a factor of 15. This implies that the number of SMILES keys found with maximum YSI have a greater significance for the evaluation than the number of isomers found. With approach 1, the agent found a total of 567 valid unique SMILES keys, of which three maximize the YSI.

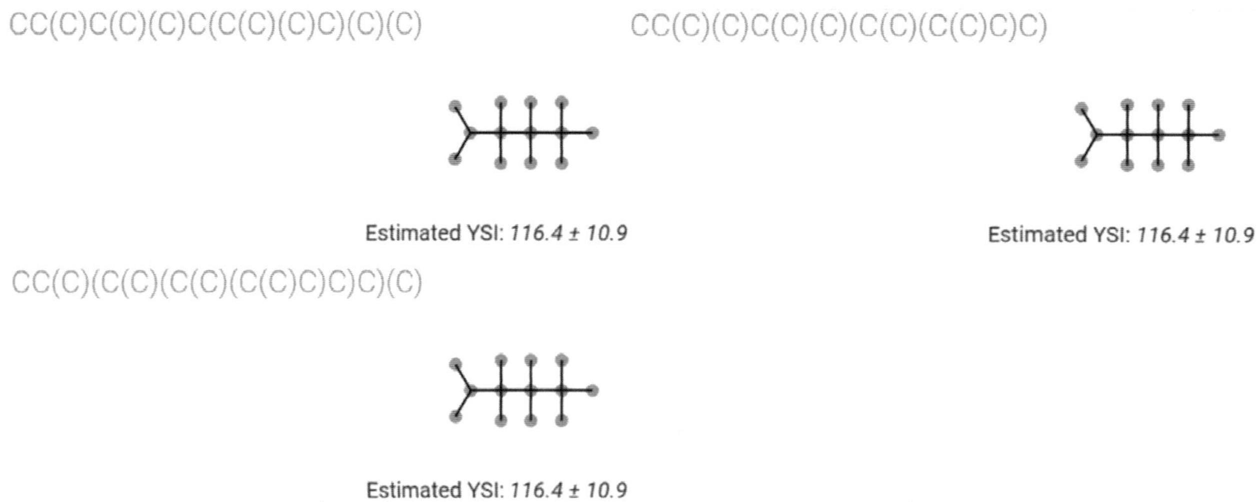

Figure 6: SMILES keys found by the agent for an empty initial state (approach 1).

Approach 2 – random initial state and reduced reward for duplicate SMILES keys

With approach 2, the agent was able to find all four isomers on its own as well as a significantly higher number of smiles keys (Figure 7). A total of 1392 valid SMILES keys were found, 21 of which maximize the YSI.

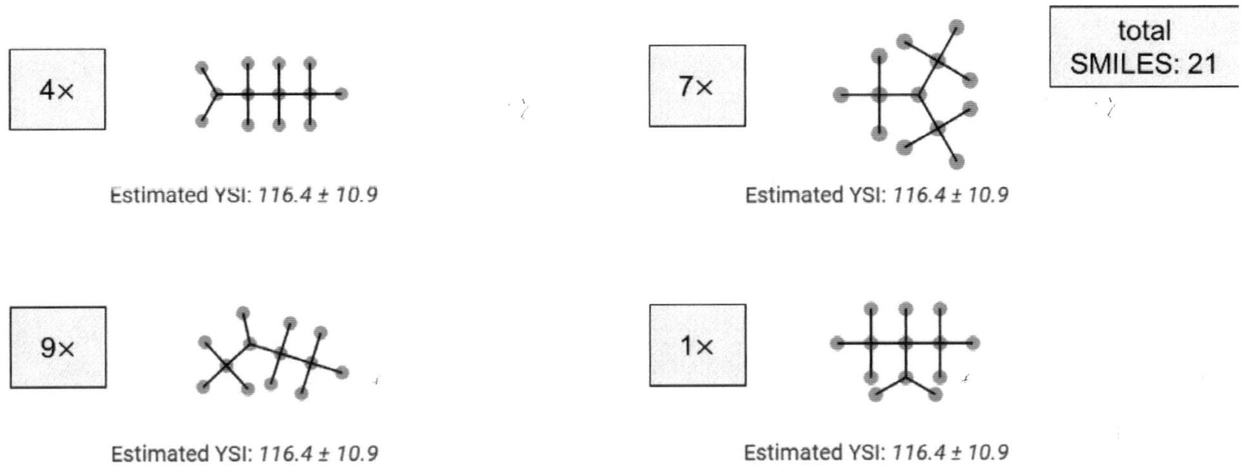

Figure 7: Number of SMILES keys found for the four iso-alkanes with a maximum YSI of 116.4 (approach 2).

Figure 8 shows the visualization of the learning process. It shows that the agent quickly realizes that it can avoid penalties by building isomers with a long chain length. This is because the higher the degree of branching, the more complex the SMILES key and the higher the risk of generating invalid keys (earning a negative reward/ penalty). But as the training proceeds the agents gains more and more knowledge. It starts increasing the branching in order to achieve higher rewards. While the Iso-alkanes get more and more branched, it finds one of the four isomers with the highest YSI after about 1300 epochs. In the following, the agent continued building the highly branched iso-alkanes and managed to find more SMILES keys describing isomers maximizing the YSI.

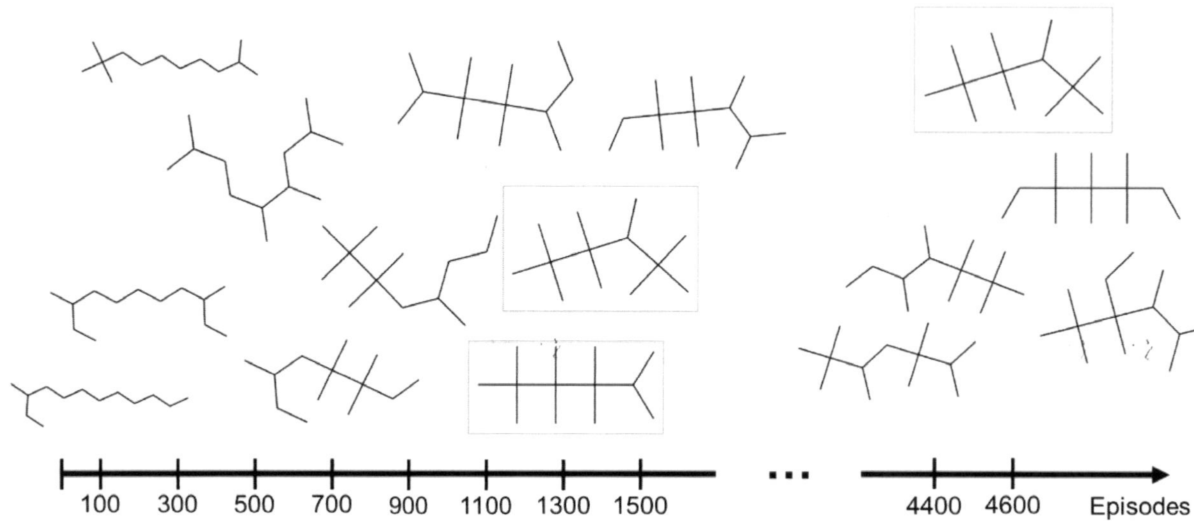

Figure 8: Visualization of the agents learning behavior (approach 2). The isomers shown for the episode number are the ones that the agent generates dominantly at the time of training.

The above analysis shows the results of individual training runs. Since it is well known in RL that the random seed (e.g. randomly initialized network weights) has a big impact on the result [34], another five training runs with different seeds have been performed. The average results are shown in the Table 1.

Table 1: Comparison of approach 1 and approach 2 based on the average results of five training runs (5000 episodes).

	Valid SMILES	Unique SMILES	Episodes until first SMILES with maximum YSI	Number of SMILES with maximum YSI
Approach 1	4533.0	646.6	653.2	7.6
Approach 2	4051.8	1761.6	117.2	15.2

The results show that more valid SMILES keys can be generated with approach 1. However, the number of unique SMILES keys shows that the agent explores significantly more efficiently when approach 2 is applied. Although the agent creates invalid keys more often with approach 2, it was able to generate almost three times as many unique keys. This also leads to it finding SMILES keys with maximum YSI significantly earlier and twice as many.

Since we are not aware of any method to determine the number of different SMILES keys describing the same molecule, it is not possible to quantify how many SMILES keys with maximum YSI an agent could have generated in total. However, when training with approach 2 over 150000 episodes, 117143 valid and 13530 unique SMILES keys can be found, of which 184 maximize the YSI. Assuming that this is the actual number of possible SMILES keys for these four isomers, this implies an artificial increase of the state space by a factor of 46 due to the issue of descriptor ambiguity. Consequently, only one 12 of these 184 SMILES keys could be found on average. This also indicates that the performance of the agent could be improved by further optimization of the hyperparameters as well as by the use of more advanced exploration algorithms. Furthermore, other graphically based unique descriptors such as MOLfile[35] could be useful to generate a unambiguous MDP and thus more interpretable results.

Although the above example shows that RL may be suitable for use in the fuel design process, its use is limited by the condition that models are available that can accurately represent reality. RL is known for its ability to find and exploit errors in its environment. This behavior has been observed, for example, by Baker et al.[40]. In their study of multi-agent RL, agents in a game of hide-and-seek were able to develop behavior that was not intended by the designers of the simulation. The agents exploited a bug in the environment to further maximize their reward beyond the expected level. Thus, to utilize the full potential of RL in fuel design, models that can accurately represent chemical interactions are required.

CONCLUSION

New research in the field of fuel screening shows a great importance of isomers for the modeling of fuel properties of synthetic fuels [41]. Consequently, it is necessary to consider not only the family but individual isomers. However, the number of isomers grows exponentially with an increasing carbon number, which means that sampling all isomers becomes computationally too expensive. A better strategy to explore the space of possible chemical compositions can be self-learning algorithms such as Reinforcement Learning (RL). RL has proven to be very efficient in systematically exploring large state spaces. In recent years, Zhou et al.[19], Popova et al.[20], Putin et al.[21], Guiamares et al.[22] and Olivecrona et al.[23] have taken advantage of this ability in drug discovery, using RL to generate molecules with specific properties. To demonstrate that RL is suitable for application in fuel design, a Markov Decision Process (MDP) based on the SMILES descriptor was developed. The objective was to identify the iso-alkanes with the highest sooting tendency so that they can be avoided in the fuel manufacturing process. For this purpose, an adaptation of the Deep Q-Network (DQN) was successfully applied to identify the isomers with the highest Yield Sooting Index (YSI) for a carbon number of 13. The study exemplifies that the branching of the isomers is a crucial factor for the sooting tendency of the iso-alkanes. In this work two approaches were compared which differ in the initial state and the assignment of rewards. The first approach is based on an empty initial state, while the second uses a predefined random action in the initial state, as well as reduced rewards for duplicate SMILES keys.

It was shown that the SMILES is not the ideal descriptor for the investigated task. Since it is ambiguous, meaning that there are several SMILES keys describing the same isomer, the state space is therefore artificially enlarged, which increases the difficulty for the RL agent. The comparison of the two approaches showed that the RL algorithm benefits from the initial state containing a random predefined action and a reward function that incentivizes more exploration. Regarding the learning process, it was shown that the agent initially focuses on avoiding being punished. This is shown by the fact that the agent mainly builds iso-alkanes with long chain length, since their SMILES keys are less vulnerable to invalidation. As the training proceeds, the agent's knowledge expands and it builds more and more branched isomers as they lead to higher rewards. In this work, the DQN algorithm was used, which is characterized by a simple implementation and a good use for deterministic problems. Therefore, the algorithm is well suited for an initial investigation of whether RL is suitable for a use in fuel design. However, its development dates back to 2015, and the great success achieved with the DQN has led to the further development of this type of algorithm, so that the performance could be increased. Among others, Double DQN (DDQN)[36], prioritized experience replay[37], dueling network architectures[38] and Rainbow[39] are worth mentioning. Rainbow is particularly promising because it combines several optimization approaches that were independently developed for the DQN, which resulted in state-of-the-art performance in the Atari 2600 benchmark [39].

In summary, the present study shows that RL does have potential for application in the fuel design process. However, a fundamental requirement is that property models are available, which predict compound properties as accurately as possible, so that an RL agent can learn by independently generating meaningful training data. The example shown is intended to be an initial proof-of-concept. In order to get closer to the real application, in the next step further molecule families relevant for fuel production like cyclo-alkanes and aromatics, as well as further properties like density, viscosity, freezing point, ... (Multi-Objective RL) have to be considered. Moreover, in the current design, the agent is not able to apply its knowledge to different carbon numbers. For future studies it may be interesting to design the agent in a way that it is able to transfer its knowledge to similar types of problems.

Abbreviations

DLR	German Aerospace Center
DQN	Deep-Q-Network
DDQN	Double DQN
EU	European Union
GCxGC	Two-dimensional Gas-Chromatography
ICAO	International Civil Aviation Organization
JETSCREEN	Jet Fuel Screening and Optimization Project
MDP	Markov Decision Process
NREL	National Renewable Energy Laboratory
RL	Reinforcement Learning
SAF	Sustainable aviation fuel
SMILES	Simplified Molecular Input Line Entry System
TD	Temporal Difference
YSI	Yield Sooting Index

References

1. Houghton, J. Global warming. *Rep. Prog. Phys.* **68,** 1343–1403; 10.1088/0034-4885/68/6/R02 (2005).

2. Hawkins, L. R., Abatzoglou, J. T., Li, S. & Rupp, D. E. Anthropogenic influence on recent severe autumn fire weather in the west coast of the United States. *Geophysical Research Letters* **49,** e2021GL095496 (2022).

3. van der Wiel, K. *et al.* Rapid attribution of the August 2016 flood-inducing extreme precipitation in south Louisiana to climate change. *Hydrol. Earth Syst. Sci.* **21,** 897–921; 10.5194/hess-21-897-2017 (2017).

4. Mitchell, D. *et al.* Attributing human mortality during extreme heat waves to anthropogenic climate change. *Environmental Research Letters* **11,** 74006 (2016).

5. Chiang, F., Mazdiyasni, O. & AghaKouchak, A. Evidence of anthropogenic impacts on global drought frequency, duration, and intensity. *Nature communications* **12,** 1–10 (2021).

6. European Commission. The European Green Deal. COMMUNICATION FROM THE COMMISSION TO THE EUROPEAN PARLIAMENT, THE EUROPEAN COUNCIL, THE COUNCIL, THE EUROPEAN ECONOMIC AND SOCIAL COMMITTEE AND THE COMMITTEE OF THE REGIONS. Available at https://eur-lex.europa.eu/resource.html?uri=cellar:b828d165-1c22-11ea-8c1f-

01aa75ed71a1.0002.02/DOC_1&format=PDF (2019).

7. UNITED NATIONS. PARIS AGREEMENT. Available at https://unfccc.int/sites/default/files/english_paris_agreement.pdf (2015).

8. Boehm, S. *et al.* State of Climate Action 2021: Systems Transformations Required to Limit Global Warming to 1.5°C. *WRIPUB*; 10.46830/wrirpt.21.00048 (2021).

9. D.S. Lee *et al.* The contribution of global aviation to anthropogenic climate forcing for 2000 to 2018. *Atmospheric Environment* **244,** 117834; 10.1016/j.atmosenv.2020.117834 (2021).

10. Brazzola, N., Patt, A. & Wohland, J. Definitions and implications of climate-neutral aviation. *Nat. Clim. Chang.* **12,** 761–767; 10.1038/s41558-022-01404-7 (2022).

11. White, D. Feasibility Study on the Use of Sustainable Aviation Fuels-Kenya (2018).

12. Voigt, C. *et al.* Cleaner burning aviation fuels can reduce contrail cloudiness. *Communications Earth & Environment* **2,** 114; 10.1038/s43247-021-00174-y (2021).

13. Simon Christie *et al.* JETSCREEN D7.3 – Report and analysis of APU emission measurements, 2020.

14. Petr Vozka *et al.* Jet fuel density via GC × GC-FID. *Fuel* **235,** 1052–1060; 10.1016/j.fuel.2018.08.110 (2019).

15. Max Jennerwein, Markus Eschner, Thomas Wilharm, Ramkumar Dhandapani & Tim Nelson. Application of GC×GC-VUV and GC×GC-FID for the analysis of common gasoline samples, middle distillates and crude oil distillation cuts using Zebron® ZB-35HT and ZB-1PLUS GC Columns (2019).

16. Dhrubajyoti D. Das, Peter C. St. John, Charles S. McEnally, Seonah Kim & Lisa D. Pfefferle. Measuring and predicting sooting tendencies of oxygenates, alkanes, alkenes, cycloalkanes, and aromatics on a unified scale. *Combustion and Flame* **190,** 349–364; 10.1016/j.combustflame.2017.12.005 (2018).

17. National Renewable Energy Laboratory. YSI Estimator. Available at https://ysi.ml.nrel.gov/.

18. Fujita, S. Alkanes as stereoisomers. Enumeration by the combination of two dichotomies for three-dimensional trees. *MATCH - Communications in Mathematical and in Computer Chemistry* **57** (2007).

19. Zhou, Z., Kearnes, S., Li, L., Zare, R. N. & Riley, P. Optimization of Molecules via Deep Reinforcement Learning. *Scientific reports* **9,** 10752; 10.1038/s41598-019-47148-x (2019).

20. Popova, M., Isayev, O. & Tropsha, A. Deep reinforcement learning for de novo drug design. *Science advances* **4,** eaap7885; 10.1126/sciadv.aap7885 (2018).

21. Putin, E. *et al.* Reinforced Adversarial Neural Computer for de Novo Molecular Design. *Journal of chemical information and modeling* **58,** 1194–1204; 10.1021/acs.jcim.7b00690 (2018).

22. Guimaraes, G. L., Sanchez-Lengeling, B., Outeiral, C., Farias, P. L. C. & Aspuru-Guzik, A. Objective-Reinforced Generative Adversarial Networks (ORGAN) for Sequence Generation Models, 30.05.2017.

23. Olivecrona, M., Blaschke, T., Engkvist, O. & Chen, H. Molecular de-novo design through deep reinforcement learning. *Journal of cheminformatics* **9,** 48; 10.1186/s13321-017-0235-x (2017).

24. Silver, D. *et al.* Mastering the game of Go with deep neural networks and tree search. *Nature* **529,** 484–489; 10.1038/nature16961 (2016).

25. The Challenge Match. Available at https://www.deepmind.com/research/highlighted-research/alphago/the-challenge-match.

26. Tromp, J. & Farnebäck, G. Combinatorics of Go. In *Computers and games. 5th international conference, CG 2006, Turin, Italy, May 29-31, 2006 ; revised papers,* edited by J. den van Herik, H. J. den van Herik, P. Ciancarini & H. H. L. M. J. Donkers (Springer, Berlin, Heidelberg, 2007), Vol. 4630, pp. 84–99.

27. H.Jaap van den Herik, Jos W.H.M. Uiterwijk & Jack van Rijswijck. Games solved: Now and in the future. *Artificial Intelligence* **134,** 277–311; 10.1016/S0004-3702(01)00152-7 (2002).

28. Mnih, V. *et al.* Human-level control through deep reinforcement learning. *Nature* **518,** 529–533; 10.1038/nature14236 (2015).

29. Weininger, D. SMILES, a chemical language and information system. 1. Introduction to methodology and encoding rules. *Journal of chemical information and computer sciences* **28,** 31–36 (1988).

30. Huang, S. & Ontañón, S. A Closer Look at Invalid Action Masking in Policy Gradient Algorithms. *FLAIRS* **35**; 10.32473/flairs.v35i.130584 (2022).

31. Bellman, R. A Markovian decision process. *Journal of mathematics and mechanics,* 679–684 (1957).

32. Kaelbling, L. P., Littman, M. L. & Moore, A. W. Reinforcement Learning: A Survey. *jair* **4,** 237–285; 10.1613/jair.301 (1996).

33. Landrum, g. RDKit: Open-source cheminformatics. Available at https://www.rdkit.org/ (2021).

34. Henderson, P. *et al.* Deep Reinforcement Learning That Matters. *AAAI* **32**; 10.1609/aaai.v32i1.11694 (2018).

35. Dalby, A. *et al.* Description of several chemical structure file formats used by computer programs developed at Molecular Design Limited. *Journal of chemical information and computer sciences* **32,** 244–255; 10.1021/ci00007a012 (1992).

36. van Hasselt, H., Guez, A. & Silver, D. Deep Reinforcement Learning with Double Q-

Learning. *AAAI* **30**; 10.1609/aaai.v30i1.10295 (2016).

37. Schaul, T., Quan, J., Antonoglou, I. & Silver, D. Prioritized Experience Replay, 18.11.2015.

38. Ziyu Wang *et al.* Dueling Network Architectures for Deep Reinforcement Learning. In *Proceedings of The 33rd International Conference on Machine Learning,* edited by Maria Florina Balcan & Kilian Q. Weinberger (PMLR, Proceedings of Machine Learning Research, 2016), Vol. 48, 1995--2003.

39. Hessel, M. *et al. Rainbow: Combining Improvements in Deep Reinforcement Learning*. Available at https://arxiv.org/pdf/1710.02298 (arXiv, 2017).

40. Baker, B. *et al.* Emergent Tool Use From Multi-Agent Autocurricula, 17.09.2019.

41. John Feldhausen *et al.* Synthetic aromatic kerosene property prediction improvements with isomer specific characterization via GCxGC and vacuum ultraviolet spectroscopy. *Fuel* **326,** 125002; 10.1016/j.fuel.2022.125002 (2022).

IASH 2022, the 17th INTERNATIONAL SYMPOSIUM ON STABILITY, HANDLING AND USE OF LIQUID FUELS
Dresden, Germany
September 11 - 15, 2022

ANALYSIS OF TRACE COMPOUNDS IN MIDDLE DISTILLATE FUELS VIA SOLID PHASE EXTRACTION AND TWO-DIMENSIONAL GAS CHROMATOGRAPHY

Paul Wrzesinski[1]*, Linda Shafer[2], Richard Striebich[2], Zachary West[2]

[1]U.S. Air Force Research Laboratory, 1790 Loop Rd. North, Wright-Patterson AFB, Ohio, 45433-7103 USAForce Research Laboratory
[2]University of Dayton Research Institute, 300 College Park, Dayton, Ohio 45409 USA

ABSTRACT

Analysis of middle distillate fuels, particularly jet and diesel fuel, has benefitted from developments in two-dimensional gas chromatography (GC×GC) for enhanced compositional analysis. While bulk hydrocarbon species analysis has been the primary focus for this technology, more recent, attention has been directed to better develop methods and techniques for determining the trace, heteroatomic components within these fuels. The work discussed here will highlight on-going efforts between DoD labs to develop a standardized approach for the execution of solid phase extraction (SPE) prior to analysis for trace components via GC×GC. The SPE approach both isolates the trace compounds from the bulk fuel matrix and provides a concentration enhancement. Emphasis will be placed on evaluation of different standards, developing an in-depth understanding of extraction efficiencies and observed differences between jet and diesel fuels. The discussion will also highlight the impacts of variations in analytical instrumentation and execution of SPE process.

*Corresponding Author: Senior Research Chemist Paul Wrzesinski; paul.wrzesinski@us.af.mil; 937-255-4867

IASH 2022, the 17th INTERNATIONAL SYMPOSIUM ON STABILITY, HANDLING AND USE OF LIQUID FUELS
Dresden, Germany
September 11 - 15, 2022

IMPACT OF ISOMER SPECIFIC IDENTIFICATION ON FUEL PROPERTY PREDICTIONS

David Bell*, John Feldhausen, Joshua Heyne

University of Dayton, 300 College Park, Dayton, Ohio 45409 USA

ABSTRACT

Petroleum fuels contain thousands of different unique hydrocarbon species. Describing the distribution of these species, frequently done with chromatographic methods, traditionally has culminated with hydrocarbon type analysis, sorting species based on carbon number and group type i.e., n-alkanes, iso-alkanes, aromatics, etc. This analysis can be applied to estimate the properties of the mixture. This level of detail with respect to hydrocarbon characterization has widely been accepted as sufficient. However, within each of these groups, there are hundreds to thousands of isomeric structures, which has a significant influence on the properties of the fuel and the capability of the fuel to perform. This research hopes to highlight the significance of variance in properties within a hydrocarbon group. The impact on properties of the specific species within a hydrocarbon group can be as significant as the hydrocarbon group depending on the property. These observations have led to the development of techniques using two-dimensional gas chromatography with a Vacuum Ultraviolet (VUV) detector. The VUV detector shows potential for isomer specific hydrocarbon identification. Results from this VUV detector demonstrate the improvement of property predictions with this information.

*Corresponding Author: Graduate Research Assistant David Bell; belld5@udayton.edu; 5133162925

IASH 2022, the 17TH INTERNATIONAL CONFERENCE ON STABILITY, HANDLING AND USE OF LIQUID FUELS
Dresden, Germany
11 - 15 September 2022

PROBABILISTIC SEQUENTIAL NEURAL NETWORKS FOR THE MODELLING OF JET FUEL COMPOUND PROPERTIES

Clemens Hall, Bastian Rauch, Uwe Bauder, Manfred Aigner

DLR, German Aerospace Center, Institute of Combustion Technology, MAT, 70569 Stuttgart, Germany, clemens.hall@dlr.de

ABSTRACT

Jet fuels are complex mixtures of hundreds of possible molecular components. The modelling of properties of jet fuels as mixtures often requires the prediction of physical properties of the contained fuel components e.g. derived cetane number or net heat of combustion, if respective measurements are unavailable. The underlying algorithms to predict the molecular components need to correlate the molecular structure of the components with the desired property. Popular modelling methods like Group Contribution (GC) or Quantitative Structure-Property Relationship (QSPR) method only approximate the molecular structure by structural or physio-chemical descriptors, which can lead to insufficient variances in the input representations of e.g. isomers. We present a method that directly correlates the unique line notation of a molecular structure, given by Simplified Molecular-Input Line-Entry System (SMILES), with the desired property, using sequential Long short-term memory (LSTM) Neural Networks. To achieve this, we make use of algorithms developed and used in the field of natural language processing for language translation tasks and transfer the concepts to the regression of molecular properties, considering not only the atoms of the molecules but also their order in the molecular structure. We model properties considered critical for the assessment of jet fuels of up to 1870 molecules and compare the predictive capability of the sequential models with the predictive capability of state-of-the-art QSPR models.

KEYWORDS: Aviation Fuel, Alternative Fuels, Machine Learning, Specification Properties, Modelling, Quantitative Structure-Property Relationship, QSPR, Long short-term memory neural networks, LSTM

INTRODUCTION

The modelling of jet fuel properties is regarded as key enabler in modern concepts of jet fuel screening and development. Building on the findings from extensive research projects like the Jet Fuel Screening and Optimization Project (JETSCREEN) EU project [1] and the National Jet Fuels Combustion Project (NJFCP)[2] Heyne and Rauch developed a screening procedure that utilizes model-based prediction methods for the estimation of critical fuel properties. Combined with the modern analytical method of two-dimensional gas chromatography (GCxGC) this model-based screening allows the assessment of a jet fuel candidate from low volumes below 5 ml. Low volume screening methods are critical especially in the early stage of the jet fuel development, since volumes required for comparable measurement of the critical properties often exceed the capabilities of the production process of the new jet fuel candidates.

The nature of jet fuels as highly complex mixtures composed of hundreds of possible hydrocarbon molecules make the both the analytical measurements and the modelling arduous. State-of-the-art GCxGC methods (using both mass spectroscopy and flame ionization detector) are not able to identify each molecule in present in the fuel [3,4]. There are efforts to further identify the particular molecular species using vacuum ultraviolet spectroscopy e.g. by the University of Dayton [5,6]. However so far, only the exact identification of mono-aromatics components in a high aromatic blending components has been shown successfully [5]. Iso-alkanes and cyclo-alkanes were not identified down to each respective component. Currently detected species in a GCxGC are therefore classified based on their chemical family and the number of contained carbon atoms e.g. iso-alkanes with 10 carbon atoms. The inability of state-of-the-art GCxGC methods to identify isomers induces uncertainty that have to be accounted for in the property modelling, since the value of a physicochemical property can often vary drastically for different isomers. E.g. for 10 carbon atoms a value range of -110.15 °C to 12.6 °C is recorded for the freezing point. To consider potential isomers in the modelling probabilistic modelling methods were developed, which in contrast to traditional deterministic modelling methods do not return one deterministic value but rather a distribution of possible values as predication. The distribution describes the possible range of property values, due the range different property values from the considered isomers. There exist various probabilistic modelling approaches that model the described uncertainties and the resulting property value distribution for the modelled fuel in different ways [7–9]. Yang et al. developed a method, that models the fuel as mixture of fuel components, that approximates the uncertainty by randomly sampling property values of different isomers for a fuel over manly samples using Monte-Carlo sampling. The distribution of possible property values of the fuel is subsequently inferred from the drawn samples under consideration of the respective mass fraction of the fuel components and an adequate mixing rule [9]. The exclusive sampling of measurements however strongly relies on the availability of measurements itself. If no measurements are available the property value of the isomers is not considered, leading to a potentially false distribution of the fuel property value. The use of property models for the individual components can overcome this restriction and with it the need for an extensive measurement database for the sampling. Property models for pure components relate the molecular structure of a component with the corresponding physicochemical property. They have been utilized for long in the modelling of jet fuel property e.g. in the Discrete Component Model by Le Clercq [10,11] for the modeling of the density and the viscosity. There exist various approaches for the outlined correlation of the molecular structure with the property e.g. Group Contribution method or the Quantitative Structure-Property Relationship method (QSPR) [12]. Both methods approximate the molecular structure by quantifying characteristic substructures. This yields a vector, which can be correlated with the physicochemical property value of the components. QSPR is well-established method that was used by Saldana for the development of models for various properties like the density, the viscosity [13], the flash point and cetane number [14] for potential fuel components. The approach however has its flaws, especially for the prediction of different isomers. Depending on the set characteristic substructures for the quantification, the vectors of two different isomers with two different property values are identical, confusing the underlying correlation model and leading to potential deviation. The selection of an adequate set of characteristic substructures is furthermore, difficult and requires expert knowledge in the field or iterative selection over many iterations [13,14]. In this work we present an alternative approach for the modelling of physicochemical properties of fuel components. The approach is heavily inspired by advancements and methods in the field of natural language processing (NLP). NLP is a branch of the field of computer science—and more specifically artificial intelligence, that concerned with giving computers the ability to understand text and spoken words in much the same way human beings can [15]. Using Long Short-Term Memory Neural Networks, we build correlation algorithm that correlate the Simplified Molecular Input Line Entry System representation (SMILES) [16] of a component directly with the physicochemical properties. This method does not rely on a set of characteristic substructures, carefully selected by a field expert and furthermore solves the problem of identical structural descriptions for different isomers. Sequential models the LSTM have already been applied in the field of chemo informatics, e.g. for drug discovery both in classification and regression [17], but not yet for the prediction / regression of physicochemical properties relevant for the modeling of jet fuels. We investigate the potential application of a sequential model for the prediction of properties of up to 1870 molecules for eight properties rated critical by the NJFCP: density, surface tension, kinematic viscosity, flash point, freezing point, cetane number and distillation / boiling point. As reference we compare the predictions with a state-of-the-art QSPR model that uses the utilizes method of quantifying characteristic substructures. We use probabilistic Neural Networks as regression algorithm, which allow the estimation of uncertainty due to conflicting measurements and the dissimilarity of training and testing data as part of the prediction. For the QSPR model we thereby use Monte-Carlo Dropout Neural Networks (MCNN) and for the sequential models, Log Gaussian Loss Neural Networks (LGNN). Both the QSPR and the sequential models will be compared on their predictive capability expressed by three metrics that quantify the both the accuracy of the mean prediction of the models and the validity and precision of the predicted property distribution.

BACKGROUND

Quantitative and sequential structure representation of compounds

The prediction of physicochemical properties of chemical components from their molecular structure is a well-established concept in the field of chemoinformatics. Since computers are not able to correlate human understandable representation of a molecule, like a 2D chemical structure, directly with the physicochemical property of interest, the molecular structure has to be converted in a computer understandable format. Established methods, like the Group Contribution method or the Quantitative Structure-Property Relationship method approximate the molecular structure by quantifying characteristic substructures of the molecule [12]. The difference between the Group Contribution method and the QSPR method is, that for QSPR method one molecule can belong two multiple quantitative substructures, while in the Group Contribution method, the molecule is "decomposed" into different subfractions. In both methods the occurrence of the substructures in the molecule is counted and stored in a corresponding vector. Figure 1 illustrates the general principle for a QSPR method for the calculation of the quantitative structure representation of tert-pentylcyclohexane with five exemplary structural descriptors. The resulting vector representation can be subsequently utilized for the correlation with the desired property. Depending on the method and application the quantitative structure representation can directly be correlated with the physicochemical property or with thermodynamic constants like the critical temperature, acentric factor or activity coefficient, from which the physicochemical property is estimated using an equation of state [12].

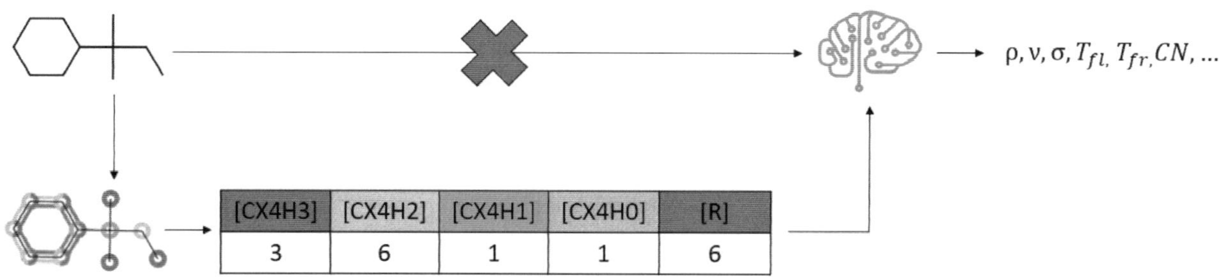

Figure 1: Schematic workflow for the estimation of the mean quantitative structure representation of tert-pentylcyclohexane and correlation with physicochemical properties

As mentioned in the introduction however, in QSPR identical quantitative structural representations especially for isomers. Figure 2 shows an exemplary case for the 3-Methylpentane and isohexane, isomers of the iso-alkanes family containing six carbon atoms. The three considered substructures yield the same quantitative substructure representation for both isomers. For the subsequent correlation two identical inputs with two different outputs poses a problem, since the underlying algorithm has to correlate inconclusive input and output variables. This induces uncertainty and can lead to erroneous predictions. Variance in the input and explicit input and output values for a one component are therefore of upmost importance even before the respective modelling begins. In QSPR modelling it is also possible, to increase the variance by considering more distinct characteristic substructures or by utilizing additional computed features like the Van-der-Waals volume. However, the selection of the quantitative substructures requires expert knowledge and often the iteration over many different sets of substructures. Additional computed features require additional software, additional computation time and still rely on an iterative selection of the best features.

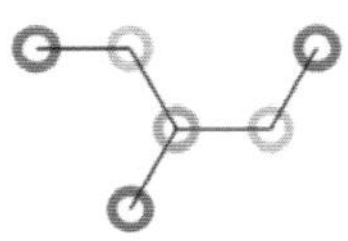

[CX4H3]	[CX4H2]	[CX4H1]
3	6	1

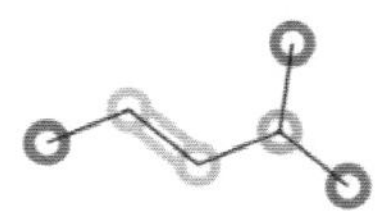

[CX4H3]	[CX4H2]	[CX4H1]
3	6	1

Figure 2: Quantitative substructure representation of 3-Methylpentane (left) and isohexane (right) with three exemplary quantified substructures

The described problem of insufficient variance in the input variables in modelling is not unique to the field of prediction of physicochemical properties of compound. A similar problem was encountered and solved in the field of natural language processing (NLP). NLP strides to give computers the ability to understand text e.g. in the task of questioning and answering. Figure 3 shows a schematic of a typical Q/A task. A model should be trained to answer the question: who eats whom? Based on the training information "Clemens eats the lion". The model has two possibilities to choose from "Clemens" and the "Lion". Since computer cannot directly interpret words, likewise to 2D molecular structures, the training information has to be represented in computer understandable way.

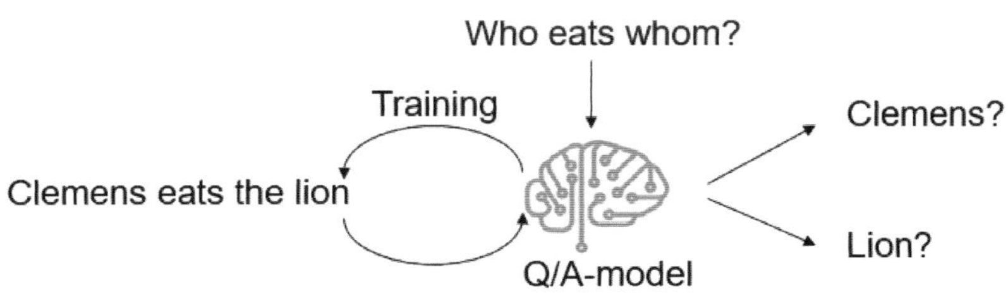

Figure 3: Schematic of a questioning and answering modeling task in Natural Language Processing

Similar to the QSPR method, a selection of characteristic words, in this case "Eat", "Clemens", "Lion" can be quantified, creating a vector that can be utilized for the training of the model. In NLP this approach is called "Bag of words", since only the occurrence of a set of words and not their order in the sentence is considered. The left illustration in Figure 4 shows the Bag of words approach for the given sentence. For the described task this can lead to errors predictions of the Q/A model, since the outlined Bag of words approach yields the same vector for the sentences "Clemens eats the lion" and "The lion eats Clemens". In natural language the order of the words is of critical importance, it holds information that allows the differentiation between the two sentences, using the concept of tokenization rather than quantification. In tokenization the characteristic words: "Eat", "Clemens", "Lion", are assigned to respective numeric tokens. The sentence is now tokenized, by sequentially listing the tokens in a vector with a set length, based on the order of their occurrence. This way the sentence is converted into a computer understandable format while intrinsically maintaining information about the order and providing additional variance than can be utilized by a corresponding model for the differentiation of the two sentences "Clemens eats the lion": [2, 1, 3] and "The lion eats Clemens" [3,2,1].

Quantifying models

Quantification

Clemens eats the lion		
Eat	Clemens	Lion
1	1	1

Sequential models

Tokenization

Clemens eats the lion		
Eat:1	Clemens:2	Lion:3
2	1	3

Figure 4: Comparison of the quantification of words in a sentence ("Bag of words") and tokenization of words in a sentence in Natural Language Processing

The outlined principle can be transferred to the desired task of property modelling using a the Simplified molecular-input line-entry system (SMILES), a string representation unique for each chemical compound [16]. SMILEs contain the same information found in an extended connection table of molecule, but as string representation using only simple vocabulary symbols for atom and bond. The order of the symbols thereby reflects the order of the atoms in the molecular structure. The tokenization of the SMILES keys follows the same procedure as the tokenization of words in a sentence, first all unique characters of the SMILES keys are paired with a numeric token, afterwards the tokens are listed in the same order as the characters in the SMILES key. Two examples tokenization examples for 3-Methylpentane (left) and isohexane are given in Figure 5. Compared to the quantitative structure representation in Figure 2 sequential token representation of the two molecules differ. Both compounds now have distinct input features that allow a correlation with the desired physicochemical properties without inducing uncertainty and potential error. For the correlation of the token vectors with the properties all vectors are padded to the same size, 50 in this work, using an additional number a padding token.

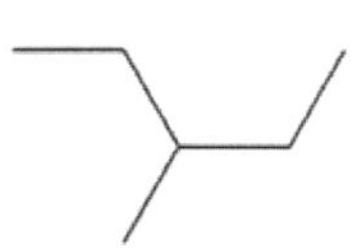

SMILES	CCC(C) CC
Tokens	00010200

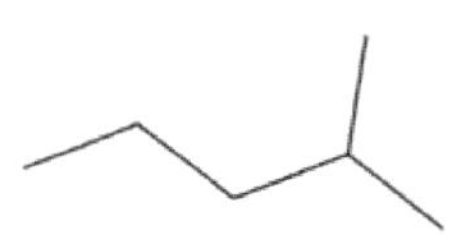

SMILES	CCCC(C) C
Tokens	00001020

Figure 5 Sequential token representation of 3-Methylpentane (left) and isohexane (right) with three exemplary quantified substructures

Probabilistic correlation models

To correlate both the quantitative structure representation and the sequential token representation with the desired physicochemical properties, respective correlation algorithms are needed. We use probabilistic deep Neural Network algorithms, developed by Gal and Ghahramani [18]: the Monte-Carlo Dropout Neural Network (MCNN) for the QSPR model and the Gaussian Loss Neural Network (GLNN) for the sequential algorithm. Both algorithms built up on the Artificial Neural Network (ANN) regression algorithm, an algorithm that mimics the biological neural systems. ANNs are composed of connected layers of individual neurons that propagate an input signal if a threshold of an activation function is exceeded. The connections are weighted to regulate the importance of inputs. The algorithm is trained by adjusting the weights of the individual neurons according to set loss function. MCNN and LGNN differ in their topology and the utilized loss function, which will be explained in more detail in the following corresponding sections. Both the MCNN and the LGNN are probabilistic algorithms, meaning they predict not one deterministic value for the property but a distribution. The predicted distribution is interpreted as a Bayesian approximation that describers the prediction uncertainty due to noise in the data (conflicting measurements) as well as dissimilarity of the training and the test data. The distribution itself is described by the median value y_{pred} and the lower and upper prediction interval y_{PI}^{l} and y_{PI}^{u} for set confidence level, e.g. 95 %. To assess the validity of the predicted uncertainty intervals as prediction intervals (PI), we review the validity and precision of the prediction intervals as part of the predictive capability assessment. The decision for the use of the MCNN for the QSPR modelling and the LGNN for the sequential modelling is based on experience; the algorithms yield the best performance for the respective properties. All algorithms are written in Python using the PyTorch library [19].

QSPR model

In the QSPR modelling the vector with quantitative structure representation of the compound can directly be correlated with the desired property, using the mentioned MCNN algorithm. As characteristic substructures functional group descriptors are utilized, that are largely based on conclusions drawn of work by Saldana [13,14]. The substructures have already been tested in a recent paper of ours, using a related modelling approach [8]. These descriptors quantify relevant functional chemical substructures that characterize the molecules of the considered families and make them usable for the QSPR modelling. The substructures are quantified based on the also based on the SMILES key of the respective molecule, using SMILES arbitrary target specifications (SMARTS). Table 1 list the set of all 49 utilized descriptors. A table with a detailed description for each feature can be found in the Supplementary Material S1. For the quantification of the substructures, the python library rdkit [20] is utilized.

[CX4H3]	[cX3H0](:*)(:*):*	[C]=[C]=[C]
[CX4H2]	[cX3H0]-[cX3]	[!CX1][C]#[C]

[CX4H1]	[cX3H0](:*)(:*)(:*)	[!X1][C]#[C]
[CX4H0]	[cX3H0](:*)(:*)(-[CX4H2R])	[!#1][CH2][!#1]
[CX3H2]	[CX4H2]-[CX4H1]-[CX4H2]	[CX3H1]=[CX3H0]
[CX3H1]	[C][C]([!CX1])([!CX1])[!CX1]	[CX3H0]=[CX3H0]
[CX3H0]	[!C][C]([C])([C])[C]	[CX3H2]=[CX3H1]
[CX2H1]	[C][C]([C])([C])[C]	[CX3H1]=[CX3H1]
[CX2H0]	[C][CR]([!C])([!C])[C]	[CX3H2]=[CX3H0]
[CX4H2R]	[C][CR]([!C])([C])[C]	[c][CX4H3]
[CX4H1R]	[C][CR]([C])([C])[C]	[c][CX4H2]
[CX4H0R]	[cH1]	[c][CX4H1]
[CX3H1R]	[cH0]	[CX2H1]#[CX2H0]
[CX3H0R]	[C]=[C]([!CX1])[!CX1]	[CX2H0]#[CX2H0]
[cX3H1](:*):*	[C]=[C]([C])[!C]	[R]
[cX3H0](:*)(:*)*	[C]=[C]([C])[C]	

Table 1: Utilized SMARTS codes of characteristic molecular substructure quantified for QSPR model

The MCNN correlation algorithm, is an extension of the regular ANN. The MCNN uses the popular regularization technique dropout technique of randomly deactivating neurons of the Neural Network, originally developed to prevent overfitting of the model, not only during the training but also during the prediction. The prediction of a fuel property value is thereby repeated multiple times, each time deactivating random neuros. This produces varying property values, that resemble a probability distribution. A schematic illustration of prediction process of the MCNN is given in Figure 6. The predicted distribution therefore reflects the certainty of the prediction, with a narrow distribution indicating a certain prediction and a wider predicted distribution indicating an uncertain one [18].

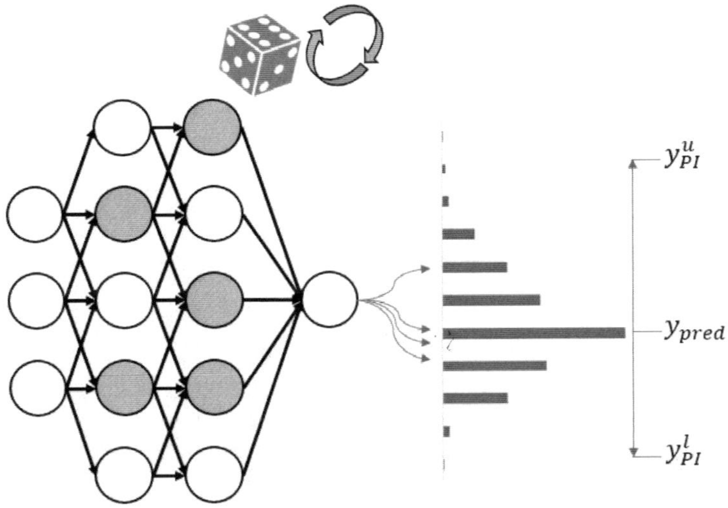

Figure 6: Schematic representation of Monte-Carlo Dropout Neural Network. Network neurons are deactivated randomly (gray) to generate a distribution of prediction values.

Sequential model

By default, ANN cannot process sequential information as input, since their intrinsic structure does not know about the concept of order. All input features are processed at once, without the neuros in a layer knowing about the information processed by the other neurons in the layer. To process sequential information an additional form of artificial neuron is introduced, the Long Short-Term Memory neuron (LSTM). These neurons are prepended in extra layers in front of the MCNN. Characteristic of LSTM neurons are their interconnections; a LSTM neuron is connected to the previous (unidirectional) or both (bidirectional) neighboring neurons, which allows the exchange of information with neurons in the same layer and therefore an understanding of the concept of order. A schematic representation of a sequential model with a prepended LSTM layer is given in Figure 7.

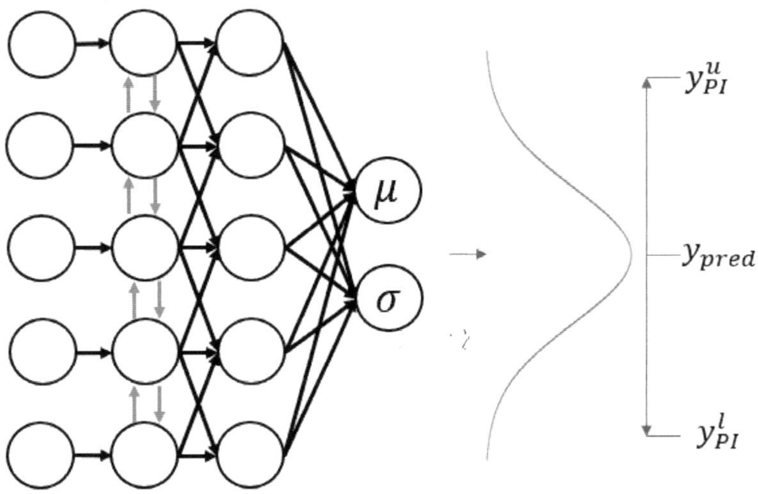

Figure 7: Schematic representation of a Log Gaussian Loss Neural Network with prepended layer of Long Short-Term Memory neurons

LSTM neurons are a special kind of interconnected neurons. As the name implies they are built to consider both long- and short-term information, meaning information from both directly neighbored neurons and distant ones [21]. This is achieved by a special design of the internal cell structure. A schematic illustration of the LSTM structure is given in Figure 8. A unidirectional LSTM neuron has three inputs and three outputs. The three inputs and outputs receive and propagate information form the direct feature input x_t, from the previous neighboring cell h_{t-1} with index $t-1$, resembling the short-term information, and the long long-term information c_{t-1}. During the prediction the LSTM neuron processes information from all the three inputs and decides which information to propagate along the three outputs h_t to the next layer, h_t to the appending LSTM neuron and c_t along the long-term information connection. For this the LSTM cell has different functionality gates conduct the operations. To process both the input form the feature x_t and the short-term information from the previous neuron h_{t-1}, the LSTM has a so called forget gate, which decides if information from the x_t or h_{t-1} is directly relevant for the long-term, and should be considered for the computation of the new long-term output signal c_t. In the following operations in the input gate, the precursor values for the calculation of the long-term cell state c_t and the calculation of the formation short-term cell state h_t are calculated. c_t and h_t are subsequently computed in the long-term memory gate and the output gate [21].

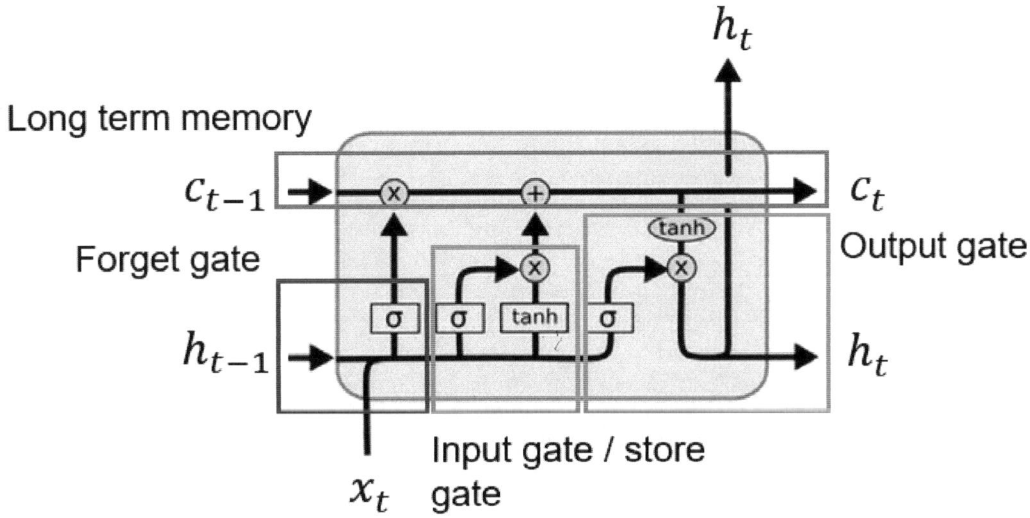

Figure 8: Schematic structure and functionalities of an LSTM neuron, modified and adapted from [21]

The information processed by the LSTM layers is processed further using and appended ANN to finally compute the desired physicochemical property. Training weights in the individual gates of the LSTM neurons are adjusted during the training process with the weights of the appended ANN. As mentioned in the introduction of this section, we utilize a Log Gaussian Loss Neural Network for the correlation with the sequential model. In contrast to the MCNN, the LGNN does not utilize the technique of dropout for the estimation of uncertainty but directly estimates the parameter of a gaussian distribution the mean value μ and the standard deviation σ by two output neurons, as indicates in Figure 7. σ is thereby estimated as $\log(\sigma)$ for easier convergence. For the training of the LGNN an special loss function is utilized, that uses both the estimated μ and σ value in the name giving logarithmic Gaussian loss function, given in Equation 1. The loss is thereby calculated over all datapoints n.

$$Loss = -\frac{n}{2} ln(2\pi) - \frac{n}{2} ln(\sigma^2) - \frac{1}{2\sigma^2} ln \sum_{i}^{n} (y_i - \hat{y}_i)^2 \qquad \textit{Equation 1}$$

Likewise to the MCNN algorithm, the predicted variables μ and σ allow the description of a distribution that reflects the certainty of the prediction, with a narrow distribution indicating a certain prediction and a wider predicted distribution indicating an uncertain one.

Hyperparameter Optimization

Neural networks are highly dependent on their hyperparameters, parameters which are not adjusted during the training process, e.g. the topology of the network with the number of layers as well as learning parameters. These hyperparameters have to be adjusted before the actual training of the networks for the property prediction. To find an optimal configuration for each modelling task, we used a Bayesian optimization with a Gaussian Process. The procedure is laid out more detailed in our recent publication [8]. A complete list of the categorical values considered for the hyperparameter optimization is given in Supplementary Material Table S2 and Table S3.

Model validation and Predictive capability metrics

To validate both the QSPR and the sequential models and compare their predictive capabilities a validation concept is utilized that we introduced in a recent paper [7]. Since both are probabilistic Machine Learning models, not only the accuracy of the mean prediction but also aspects of the predicted distribution have to be considered and quantified. Our concept therefore uses metrics to quantify the accuracy of the mean value, as well as the validity and precision of the distribution. The validity thereby describes the average percentage of measurements enclosed by the PI of the distribution and the precision measures the size of the distribution. Both models are compared on these three aspects to review their predictive capability and the potential benefits of the sequential modelling approach.

The metrics are calculated from the predictions of the testing set of the cross-validation (CV), a validation method frequently used for the assessment of Machine Learning models. The CV is explained in more detail in the following.

As metric of the accuracy we calculate the mean absolute error (MAE) of the median of the predicted distribution $y_{pred,i}$ and the measurement value $y_{test,i}$, see Equation 2. As a metric for the validity of the predicted distribution, described by the PI, we calculate the Prediction Interval Coverage Probability (PICP), see Equation 2. The PICP calculates the average probability, that a measured value lies inside the predicted upper and lower PI ${y_{PI}}^l$ and ${y_{PI}}^u$. In Equation 3 c_i is a boolean value; it is 1 if $y_{pred,i}$ lies inside the interval and therefore ${y_{PI}}^l < y_{test} < {y_{PI}}^u$ and 0 otherwise. If the PICP and the set confidence level of PI of the prediction are comparable, predictions do on average lie inside the PI and the PI can be considered valid. If this is true for training and testing, the PI is considered reliable. E.g. for a set confidence of 95 %, 95 % of the measurements have to be enclosed by the PI in order to consider them as valid. As metric for the precision of the distribution we calculate the Normalized Mean Prediction Interval Width (NMPIW), see Equation 4. The NMPIW calculates the mean width of the PI relative to a reference Δ_{ref}, see Equation 4. Note, that different references are possible, e.g. the reproducibility of a property measurement or a set value range for the decision of a model adequacy as demonstrated in our recent paper [7]. For the scope of this work, where the goal is the comparison of the two modelling methods, the Δ_{ref} is set to 1. NMPIW is therefore calculated in the same unit of the property and states the average width of the set PI.

$$MAE = \frac{1}{n_{Test}} \sum_{i=1}^{n_{Test}} y_{pred,i} - y_{test,i}$$ *Equation 2*

$$PICP = \frac{1}{n_{Test}} \sum_{i=1}^{n_{Test}} c_i * 100\ \%$$ *Equation 3*

$$\text{NMPIW} = \frac{1}{n_{Test}} \sum_{i=1}^{n_{Test}} \frac{y_{PI,i}{}^{u} - y_{PI,i}{}^{l}}{\Delta_{ref}} * 100\ \%$$ *Equation 4*

Cross-validation

In the field Machine Learning the concept of cross-validation (CV) is established and widely used for the validation of a model on a set amount of original data. The CV process simulates the procedure of training and testing several times, using different training and test datasets derived from the same original data set. The different training and test datasets are created by randomly shuffling the data and splitting the them into set fractions for training and testing, e.g. 85 % for training 15 % for testing. The process is repeated over several folds. In each fold the model is trained on training data and subsequently validated on the testing data. Figure 9 schematically illustrates the splitting of the data for a 4-fold CV.

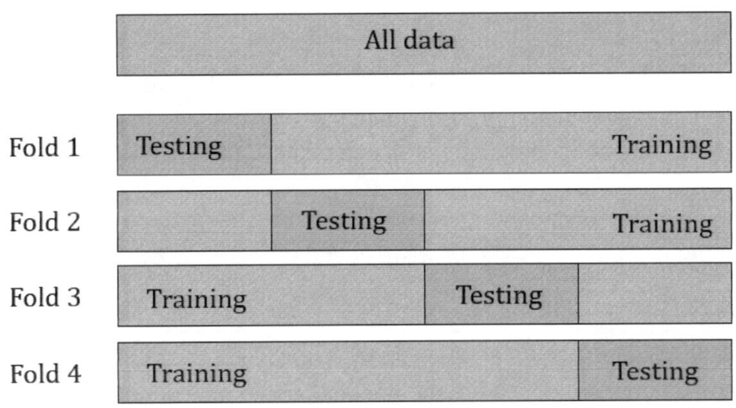

Figure 9: Schematic illustration of the data split in a cross-validation with four randomly shuffled folds with sets for training and testing

By comparing the predictions from several folds, the predictive capability of the model can be estimated independent of the training data composition. The predictive capability metrics are subsequently calculated based of the predictions of the all test set. For the actual application of a model, the model would be subsequently trained on all of the original dataset.

Data

For the training of models from the two modelling methods, we use a dataset that holds close to 60 000 measurements of up to 1870 pure compounds from eight relevant hydrocarbon families for all eight considered properties. The pure compounds belong to the hydrocarbon families of n-alkanes, iso-alkanes, mono-cyclo-alkanes, bi-cyclo-alkanes, tri-cyclo-alkanes, mono-aromatics, cyclo-aromatics and bi-cyclo-aromatics and are classified based on their molecular formula and structural characteristics of their chemical family. Details about the classification and are given in Supplementary Material Table S1.

The data was gathered from multiple different commercial and public databases: NIST Standard Reference Database 103a [22], DIPPIR 801 database [23], databases Pubchem [24] and Chemspider [25]. Data from Pubchem and Chemspider are especially used for the properties flash point and freezing point. Data for the cetane number is taken from publications of the National Renewable Energy Laboratory [26]. Table 2 summerizes both the number of the compounds and the corersponding number of measurements.

Property	# Compounds	# Measurements
Density	1865	27347
Surface tension	294	4455
Kinematic viscosity	1795	20046
Net heat of combustion	1503	1875
Flash point	268	640
Freezing point	402	1053
Cetane number	159	211
Distillation / boiling point	1825	4333

Table 2: Number of unique compounds and measurements for each of the eight physicochemical properties

For temperature depended properties e.g. density, measurements in a range of -40 to 140 °C at a pressure of 1 atm are used, since this is the range most relevant for the screening of jet fuel properties after the protocol of Heyne and Rauch [2]. Flash point, freezing point, net heat of combustion, distillation line and cetane number are measured under laboratory standard conditions.

Since the data originates from different databases, the datapoints were preprocessed to remove outliers that significantly deviated from the average measurement values. Two methods for the outlier detection and removal are utilized, one for the detection of significant outliers from the whole dataset of a property and one for the for detection of outliers for measurements of each individual compound. For the detection of outliers from the whole dataset of a property, measurements with a deviation of more than 3.5 standard deviations are removed. Datapoints with a score greater than 3.5 are removed from the dataset. Outliers of each individual compound on the other hand are removed using the modified Z-score by Iglewicz and Hoaglin, an outlier detection method for small sample sizes as recommended by NIST [27]. For the temperature-dependent properties e.g. the density, the outlier detection is carried out for measurements within a temperature window of ±1 °C both for the whole dataset of a property and the measurements of an individual compound.

RESULTS AND DISCUSSION

The prediction results from the testing of the cross-validation are shown in form of unity plots, where the mean predictions are plotted against the measurements. The estimated prediction intervals (PI) for a 95 % confidence level are indicated as error bars. If prediction and measurement are in perfect accordance, the marker lies on the unity line indicated in black. For a prediction to be valid, the PI has to cross the indicated unity line. Respective unity plots are shown for the QSPR model on the left and the sequential SMILES property relationship model (SSPR) on the right The unity plots are provided separately for every property in the Figure 10 to Figure 17. Table 3 summarizes the outlined predictive capability metrics for both models. The prediction results of from the training in the cross-validation is given in the Supplementary Material in Figure S8.

Generally, the predictions / measurement pairs of both models follow the unity lines for all properties. This illustrates, that both models are able to fundamentally model all properties based on the provided input, the molecular structure representations as quantitative structure representation for the QSPR models and sequential token representation for the SSPR models. The comparison between the unity plots of the QSPR models on the left and the SSPR models on the right shows that the markers of the SSPR model lie on average closer to the unity line. Furthermore, the scattering of the markers around the SSPR comparably smaller the and number and deviation of the outliers is reduced. This is especially visible for the density, the kinematic viscosity, freezing point and the boiling point in the respective Figure 10, Figure 11, Figure 15, Figure 17. For the other properties the increase in accuracies is less visually striking. The size computed PI, illustrated by the error bars of the individual markers is similar or slightly smaller, indicating, that the precision of the SSPR model is similar or slightly higher compared to QSPR model. The computed predictive capability allows the assessment and comparison of the prediction results based on the computed predictive capability metrics, which quantify the accuracy of the mean prediction and the validity and precision of the PI. The observations of the unity plots are reflected in the corresponding metrics in Table 3. For the accuracy metric, the MAE, average errors are lower and the accuracy therefore higher for the SSPR model compared to the QSPR model. For the density, kinematic viscosity, freezing point and boiling point the accuracy of the SSPR model is twice as high as for the QSPR model. For other properties the accuracy of the SSPR model is comparable or only slightly higher. The validity of the SSPR models expressed by the PICP shows higher values especially for the density, flash point, freezing point and boiling point. For the density, kinematic viscosity and boiling point only the PI of the SSPR model are valid, while the PICP of the QSPR model lies below the desired limit of 95 %. The PI of the surface tension are valid for both models whereas for the freezing point and especially the cetane number the calculated PI lie below the set confidence limit. The precision of the predicted PI measured by the NMPIW is computed to be similar or slightly higher for the SSPR model, as already observed in the unity plots. Higher accuracy and validity are illustrated in Table 3 by a bold font.

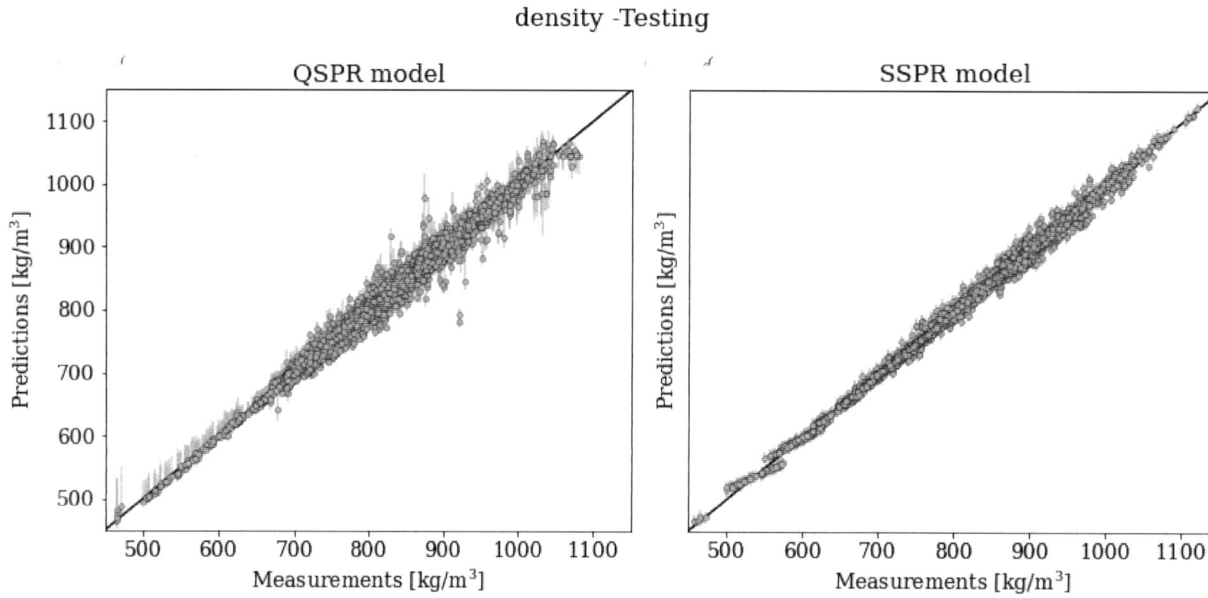

Figure 10: Unity plots with prediction results for the density from the testing of the cross-validation for the QSPR model (left) and the SSPR model (right)

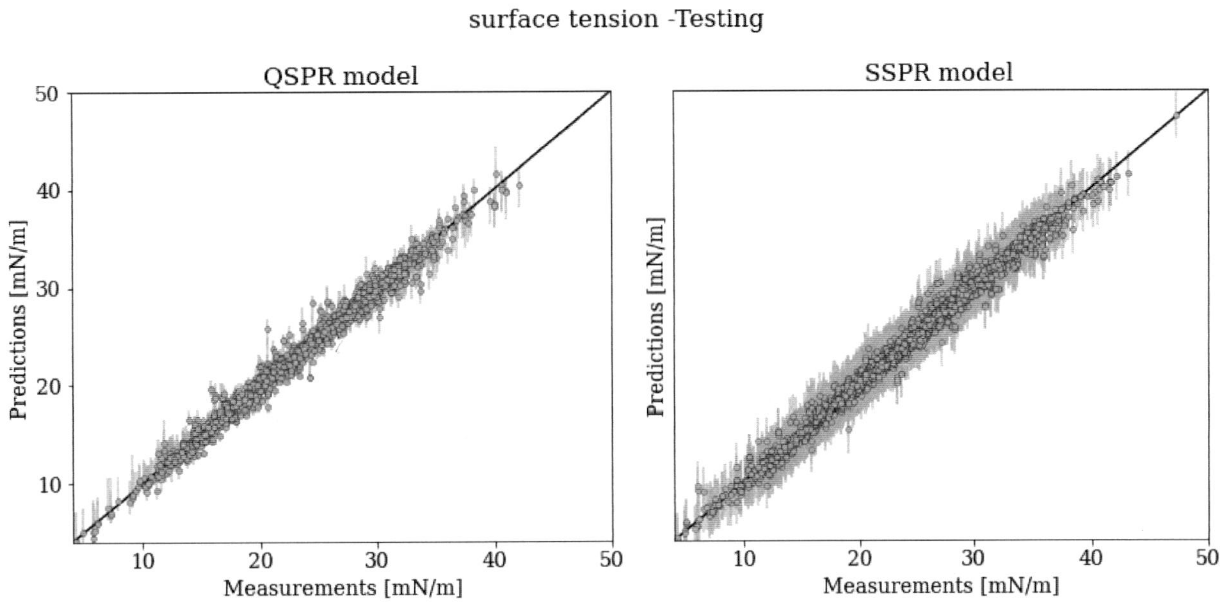

Figure 11: Unity plots with prediction results for the surface tension from the testing of the cross-validation for the QSPR model (left) and the SSPR model (right)

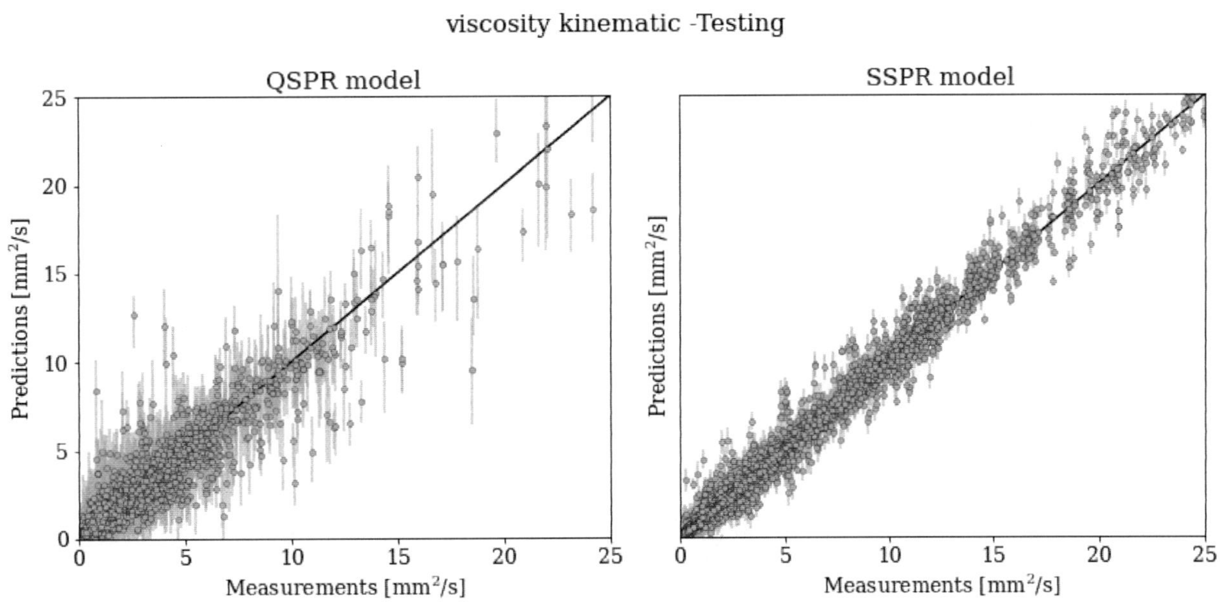

Figure 12: Unity plots with prediction results for the kinematic viscosity from the testing of the cross-validation for the QSPR model (left) and the SSPR model (right)

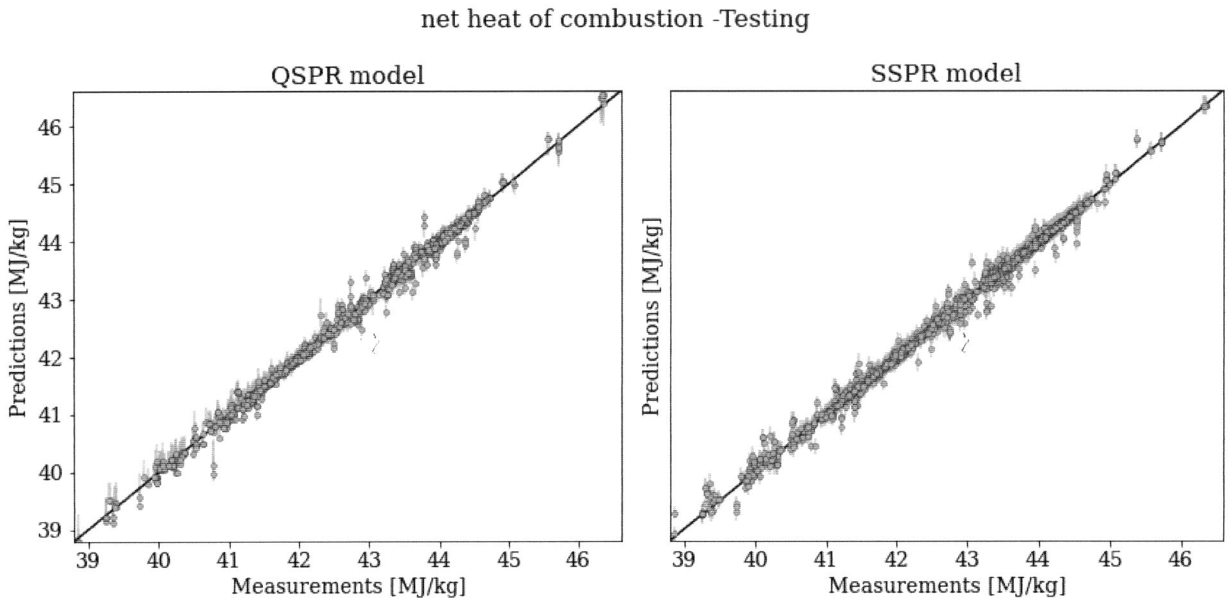

Figure 13: Unity plots with prediction results for the net heat of combustion from the testing of the cross-validation for the QSPR model (left) and the SSPR model (right)

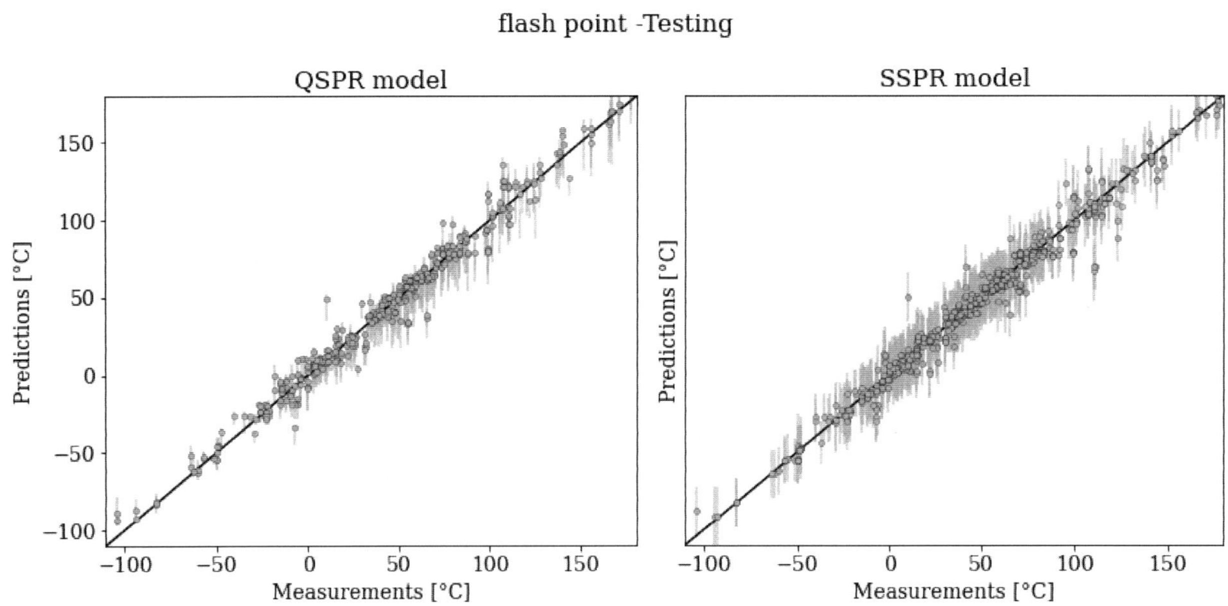

Figure 14: Unity plots with prediction results for the flash point from the testing of the cross-validation for the QSPR model (left) and the SSPR model (right)

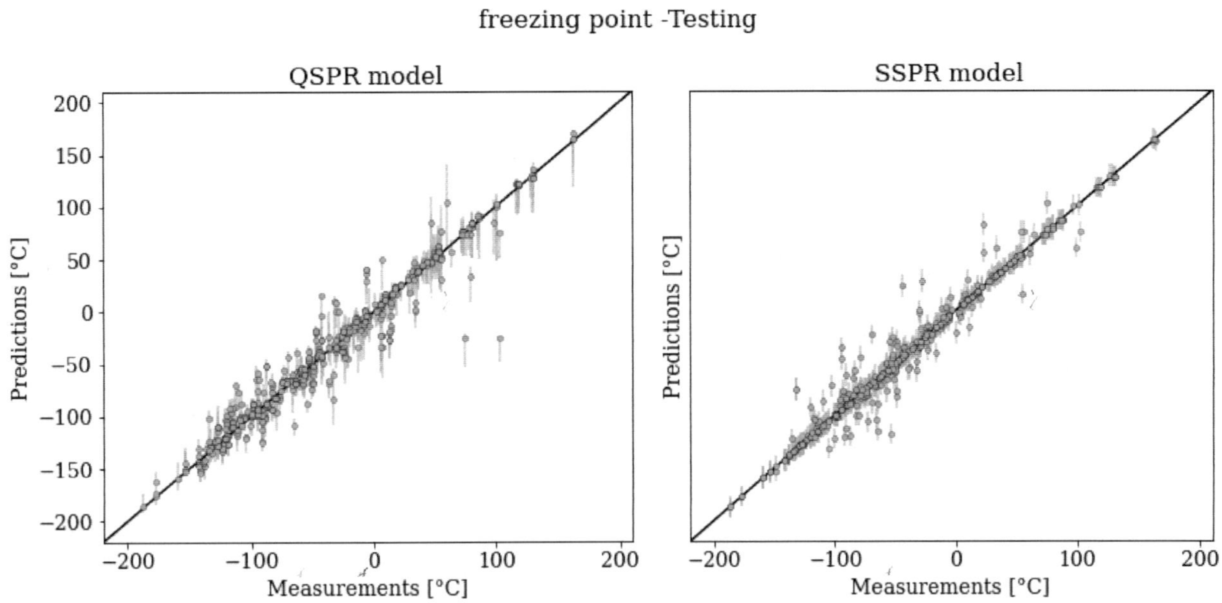

Figure 15: Unity plots with prediction results for the freezing point from the testing of the cross-validation for the QSPR model (left) and the SSPR model (right)

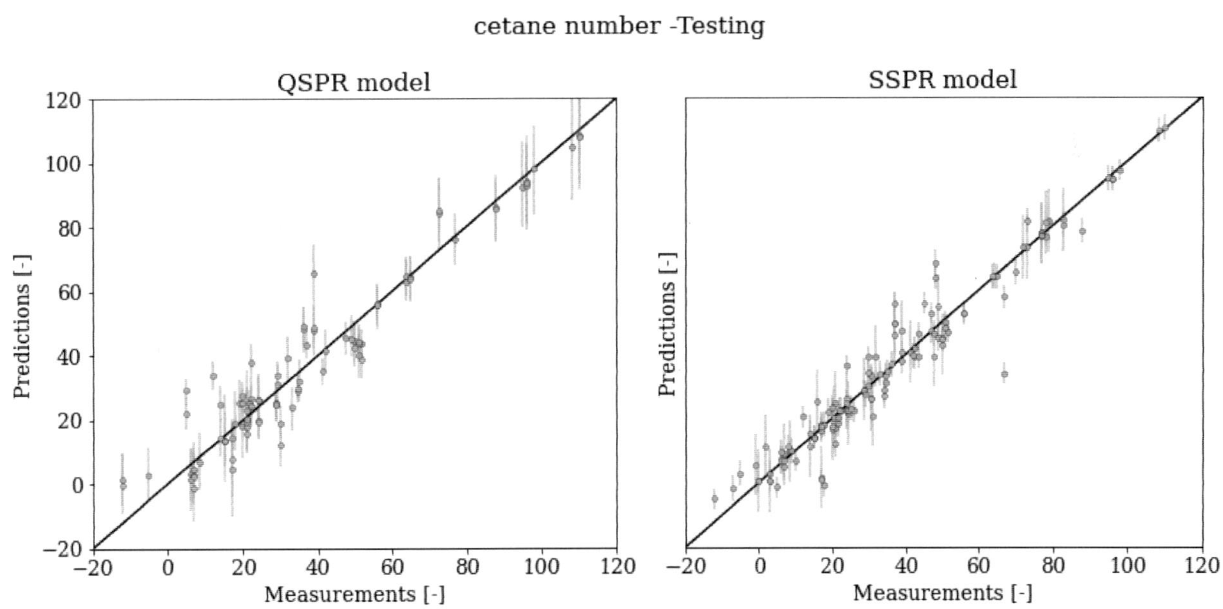

Figure 16: Unity plots with prediction results for the cetane number from the testing of the cross-validation for the QSPR model (left) and the SSPR model (right)

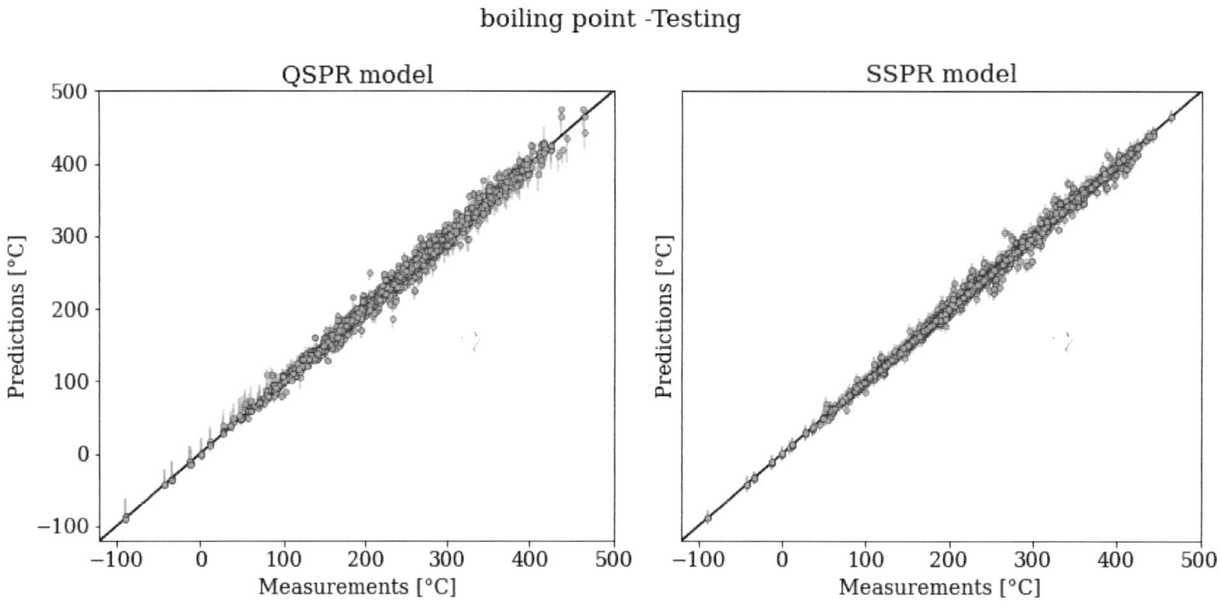

Figure 17: Unity plots with prediction results for the boiling point from the testing of the cross-validation for the QSPR model (left) and the SSPR model (right)

	QSPR			SSPR		
	MAE	**PICP [%]**	**NMPIW**	**MAE**	**PICP [%]**	**NMPIW**
Density	**5.9**	**77.41**	**19.43**	**2.53**	**96.23**	**20.24**
Surface tension	**0.53**	74.31	1.58	**0.4**	**99.08**	4.84
Kinematic viscosity	**0.41**	92.35	1.96	**0.13**	**95.51**	1.12
Net heat of combustion	**0.06**	86.06	0.24	**0.05**	**91.93**	0.29
Flash point	**5.5**	69.73	17.09	**4.77**	**94.55**	32.53
Freezing point	**9.59**	71.49	21.69	**3.89**	**88.47**	17.89
Cetane number	**5.78**	63.33	15.33	**4.16**	**74.62**	10.41
Distillation / boiling point	**4.37**	75.38	14.33	**2.09**	**95.42**	19.07

Table 3: Predictive capability metrics of the QSPR and SSPR model for all considered properties

The results show, that increases in the predictive capability with the SSPR model is predominantly achieved for properties with higher amounts of available training data and significant differences in the property values of the isomers. This is the case for the density, kinematic viscosity, freezing point and boiling point. Based on this observation it can be inferred, that for those properties the utilized QSPR descriptors do not provide sufficient variance in the structure representation especially for the isomers, which are present in higher number in the respective datasets, compared to the flash point, surface tension and the cetane number. The sequential token representation on the other hand provides additional variance for the isomers with a direct representation of the atom order in the molecular structure. For the net heat of combustion, also a property with a relatively large number of measurements and unique compounds, the accuracy only slightly increases, because the property values of the different isomers differ only slightly compared e.g. to the freezing point or kinematic viscosity. The predictive capabilities of both models based on all three metrics are therefore comparable.

For the flash point, surface tension and cetane number not the variance in the structure representation is the limiting factor, but the overall number and variance of the available measurements, which is significantly smaller compared to the other properties. This is illustrated by the comparable predictive capability and only slightly increased accuracies of the SSPR model. The modelling of these properties therefore generally requires a greater number and variety in of measurements and not and alternative modelling method.

CONCLUSION

Modern screening concepts allow the assessment of new jet fuel candidate on low fuel volume, using model-based prediction methods together with compositional measurements. The requirements for the models are high accuracy and reliability, to correctly estimate the chances of jet fuel candidate based on the predicted properties. If a fuel is modelled as a mixture of multiple hundreds of possible thousands of different fuel components, the individual properties often have to be predicted, since only a fraction of the components have recorded measurements for the required properties. In this work we presented a new alternative method for the prediction of physicochemical properties, that is heavily inspired on the methods created and utilized in the field of natural language processing. The presented approach tokenizes the Simplified molecular-input line-entry system (SMILES), a string representation of a molecule, into a sequential token representation and uses a sequential model to correlate the token vector with the desired physicochemical property. The method of tokenizing the SMILES allows the direct consideration of the order of the atoms in the molecular structure for the correlation. The method stands in contrast to established methods like Group Contribution and Quantitative Structure-Property Relationship (QSPR) methods, that indirectly approximate the molecular structure by a set of characteristic substructure descriptors. To compare the potential of the sequential modelling method we build models for the prediction of eight critical fuel properties and compared their predictive capability with models of the established QSPR modelling technique. Both models were trained on the same database of almost 60 000 measurements from 1870 unique pure compounds of eight hydrocarbon families relevant for the modelling of jet fuels. As underlying correlation algorithms the Monte-Carlo Neural Network (MCNN) was used for the QSPR modelling and Log Gaussian Loss Neural Networks (LGNN) for the modelling with the sequential models. Both are probabilistic Machine Learning algorithms, which estimate not only one deterministic value as part of their prediction but rather a distribution that reflect uncertainties due to multiple conflicting property measurements or the dissimilarity of training and validation data. For the sequential model additional layers Long Short-Term Memory neurons were implemented, that allow the processing of the sequential information. The predictive capability of the models is quantified and compared based on the accuracy of the mean prediction and the validity and precision of the predicted distribution.

The comparison of the sequential models yield overall higher accuracies and validities of the predictions compared to the ones the QSPR reference model, for all eight properties. For the density, kinematic viscosity, freezing point and the boiling point significantly higher accuracies of up to 50% and validities up to 20 % were observed and explained with the higher variance of the sequential token representation compared the quantitative structure approximation of the QSPR modelling. Especially for isomers the sequential token representation provides extra variance the allows for distinct differentiation by the algorithm and precise correlation with the desired property. For the flash point, surface tension and cetane number the predictive capability especially the accuracy increased only slightly. Compared to the aforementioned properties the available datasets for training and validation of those properties are significantly smaller, with a smaller fraction of isomers. For these properties the comparably smaller number of datapoints is more constraining than the variance in the structure representation of the pure compounds. Their modelling therefore requires an increase in number and variety of the overall property measurements to increase the predictive capability of the models beyond their current state.

In sum the presented modelling proved to be a promising alternative for the modelling of physicochemical properties of fuel components. Especially for properties with large number of datapoints that might suffer from insufficient variance in the structure representation of isomers this method can produce models with increased predictive capability. The next step is the utilization of the presented method for the actual task of the modelling of jet fuel components to see effects of the SSPR model as an underlying correlation model for the estimation of jet fuel properties directly. Furthermore, a comparison with alternative modelling approaches like graph-based methods, which also increase the variance in the structure representation of the fuel components, would be of great interest.

Abbreviations

ANN	Artificial Neural Network
GCxGC	Two-dimensional Gas-Chromatography
GLNN	Gaussian Loss Neural Network
MAE	Mean absolute error
MCNN	Monte-Carlo Dropout Neural Network
LSTM	Long Short-Term Memory
NMPIW	Normalized mean prediction interval width
PI	Prediction interval
PICP	Prediction interval coverage probability
QSPR	Quantitative Structure-Property Relationship
SMILES	Simplified Molecular Input Line Entry System
SMARTS	SMILES arbitrary target specifications
SSPR	Sequential SMILES Property Relationship

REFERENCES

(1) *Convention · JETSCREEN-h2020/FuelDatabase Wiki · GitHub*. https://github.com/JETSCREEN-h2020/FuelDatabase/wiki/Convention (accessed 2022-07-25).

(2) Heyne, J.; Rauch, B.; Le Clercq, P.; Colket, M. Sustainable aviation fuel prescreening tools and procedures. *Fuel* **2021**, *290*, 120004. DOI: 10.1016/j.fuel.2020.120004.

(3) Vozka, P.; Modereger, B. A.; Park, A. C.; Zhang, W. T. J.; Trice, R. W.; Kenttämaa, H. I.; Kilaz, G. Jet fuel density via GC × GC-FID. *Fuel* **2019**, *235*, 1052–1060. DOI: 10.1016/J.FUEL.2018.08.110.

(4) Tim Nelson. Application of GC×GC-VUV and GC×GC-FID for the analysis of common gasoline samples, middle distillates and crude oil distillation cuts using Zebron® ZB-35HT and ZB-1PLUS GC Columns **2019**.

(5) Feldhausen, J.; Bell, D. C.; Yang, Z.; Faulhaber, C.; Boehm, R.; Heyne, J. Synthetic aromatic kerosene property prediction improvements with isomer specific characterization via GCxGC and vacuum ultraviolet spectroscopy. *Fuel* **2022**, *326*, 125002. DOI: 10.1016/j.fuel.2022.125002.

(6) Heyne, J.; Bell, D.; Feldhausen, J.; Yang, Z.; Boehm, R. Towards fuel composition and properties from Two-dimensional gas chromatography with flame ionization and vacuum ultraviolet spectroscopy. *Fuel* **2022**, *312*, 122709. DOI: 10.1016/j.fuel.2021.122709.

(7) Hall, C.; Rauch, B.; Bauder, U.; Le Clercq, P.; Aigner, M. Predictive Capability Assessment of Probabilistic Machine Learning Models for Density Prediction of Conventional and Synthetic Jet Fuels. *Energy Fuels* **2021**, *35* (3), 2520–2530. DOI: 10.1021/acs.energyfuels.0c03779.

(8) Hall, C.; Creton, B.; Rauch, B.; Bauder, U.; Aigner, M. Probabilistic Mean Quantitative Structure–Property Relationship Modeling of Jet Fuel Properties. *Energy Fuels* **2022**, *36* (1), 463–479. DOI: 10.1021/acs.energyfuels.1c03334.

(9) Yang, Z.; Kosir, S.; Stachler, R.; Shafer, L.; Anderson, C.; Heyne, J. S. A GC × GC Tier α combustor operability prescreening method for sustainable aviation fuel candidates. *Fuel* **2021**, *292*, 120345. DOI: 10.1016/j.fuel.2021.120345.

(10) Hall, C.; Rauch, B.; Bauder, U.; Le Clercq, P.; Aigner, M. Application of Machine Learning algorithms for the prediction of fuel properties in comparison with state of the art physical models. *Proceedings of the 16th International Conference on Stability, Handling and Use of Liquid Fuels 2019* **2019**.

(11) Harstad, K.; Le Clercq, P.; Bellan, J. A Robust Statistical Model for the Evaporation of Multicomponent-Fuel Drops. In *41st Aerospace Sciences Meeting and Exhibit*; American Institute of Aeronautics and Astronautics: Reston, Virigina, 2003. DOI: 10.2514/6.2003-1321.

(12) Nieto-Draghi, C.; Fayet, G.; Creton, B.; Rozanska, X.; Rotureau, P.; Hemptinne, J.-C. de; Ungerer, P.; Rousseau, B.; Adamo, C. A General Guidebook for the Theoretical Prediction of Physicochemical Properties of Chemicals for Regulatory Purposes. *Chemical reviews* **2015**, *115* (24), 13093–13164. DOI: 10.1021/acs.chemrev.5b00215. Published Online: Dec. 1, 2015.

(13) Saldana, D. A.; Starck, L.; Mougin, P.; Rousseau, B.; Ferrando, N.; Creton, B. Prediction of Density and Viscosity of Biofuel Compounds Using Machine Learning Methods. *Energy Fuels* **2012**, *26* (4), 2416–2426. DOI: 10.1021/ef3001339.

(14) Saldana, D. A.; Starck, L.; Mougin, P.; Rousseau, B.; Pidol, L.; Jeuland, N.; Creton, B. Flash Point and Cetane Number Predictions for Fuel Compounds Using Quantitative Structure Property Relationship (QSPR) Methods. *Energy Fuels* **2011**, *25* (9), 3900–3908. DOI: 10.1021/ef200795j.

(15) Natural Language Processing (NLP), Jul 2, 2020. https://www.ibm.com/cloud/learn/natural-language-processing (accessed 2022-08-01).

(16) *Daylight Theory: SMILES.* https://www.daylight.com/ (accessed 2022-03-04).

(17) Goh, G.; Hodas, N.; Siegel, C.; Vishnu, A. SMILES2vec Predicting Chemical Properties from Text Representation **2018**.

(18) Gal, Y.; Ghahramani, Z. *Dropout as a Bayesian Approximation: Representing Model*

Uncertainty in Deep Learning. http://arxiv.org/pdf/1506.02142v6.

(19) *PyTorch*. https://pytorch.org/ (accessed 2022-03-09).

(20) *RDKit*. https://www.rdkit.org/ (accessed 2022-03-04).

(21) Travis Kessler. *GitHub - ecrl/padelpy: A Python wrapper for PaDEL-Descriptor software*. https://github.com/ecrl/padelpy (accessed 2022-04-20).

(22) NIST. *NIST Standard Reference Database 103a*. https://www.nist.gov/mml/acmd/trc/thermodata-engine/srd-nist-tde-103a (accessed 2022-01-17).

(23) *DIPPR 801 Database*. https://www.aiche.org/dippr/events-products/801-database (accessed 2022-01-18).

(24) PubChem. *PubChem*. https://pubchem.ncbi.nlm.nih.gov/ (accessed 2022-03-04).

(25) *ChemSpider*. https://www.chemspider.com/ (accessed 2022-03-04).

(26) J. Yanowitz; M.A. Ratcliff, R.L. McCormick, and J.D. Taylor; M.J. Murphy. Compendium of Experimental Cetane Numbers **2017**.

(27) *1.3.5.17. Detection of Outliers*. https://www.itl.nist.gov/div898/handbook/eda/section3/eda35h.htm (accessed 2022-07-26).

COMPARISON OF PROBABILISTIC JET FUEL PROPERTY MODELS FOR THE FUEL SCREENING AND DESIGN

Clemens Hall, Bastian Rauch, Uwe Bauder, Manfred Aigner

DLR, German Aerospace Center, Institute of Combustion Technology, MAT, 70569 Stuttgart, Germany, clemens.hall@dlr.de

Table S1: Description of utilized structural molecular features

Feature	**Description**
[CX4H3]	Aliphatic C with 1 further total connection with 3 further hydrogen
[CX4H2]	Aliphatic C with 2 further total connections with 2 further hydrogen
[CX4H1]	Aliphatic C with 3 further total connections with 1 further hydrogen
[CX4H0]	Aliphatic C with 4 further total connections with 0 further hydrogen
[CX4H1]	Aliphatic C with 1 further total connection with 2 further hydrogen
[CX3H1]	Aliphatic C with 2 further total connections with 1 further hydrogen
[CX3H0]	Aliphatic C with 3 further total connections with 0 further hydrogen
[CX2H1]	Aliphatic C with 1 further total connection with 1 further hydrogen
[CX2H0]	Aliphatic C with 2 further total connections with 0 further hydrogen
[CX4H2R]	Aliphatic C with 2 further total connections 2 further hydrogen, in a ring
[CX4H1R]	Aliphatic C with 3 further total connections 1 further hydrogen, in a ring
[CX4H0R]	Aliphatic C with 4 further total connections 0 further hydrogen, in a ring
[CX3H1R]	Aliphatic C with 2 further total connections 1 further hydrogen, in a ring

[CX3H0R]	Aliphatic C with 3 further total connections 0 further hydrogen, in a ring
[cX3H1](:*):*	Aromatic C with 0 further total connections with 1 further hydrogen
[cX3H0](:*)(:*)*	Aromatic C with 0 further total connections with 0 further hydrogen
[cX3H0](:*)(:*):*	Aromatic C with 0 further total connections with 0 further hydrogen
[cX3H0]-[cX3]	Aromatic C with 2 further total connections with 0 further hydrogen
[cX3H0](:*)(:*)(:*)	Aromatic C with 0 further total connections with 0 further hydrogen
[cX3H0](:*)(:*)(-[CX4H2R])	Aromatic C with 0 further total connections with 0 further hydrogen with connection to an aliphatic C with 1 further total connection with 2 further hydrogen, in a ring
[CX4H2]-[CX4H1]-[CX4H2]	Bound of aliphatic C with 1 further total connection with 1 further hydrogen
[C][C]([!CX1])([!CX1])[!CX1]	Bound of aliphatic C with one further aliphatic C and 3 non C atoms with 0 further total connections
[!C][C]([C])([C])[C]	Bound of aliphatic C with two 3 further aliphatic C and one non C atom
[C][C]([C])([C])[C]	Bound of aliphatic C with 4 further aliphatic C
[C][CR]([!C])([!C])[C]	Bound of aliphatic C in a ring with two further aliphatic C and two non C atoms
[C][CR]([!C])([C])[C]	Bound of aliphatic C in a ring with three further aliphatic C and one non C atoms
[C][CR]([C])([C])[C]	Aliphatic C in a ring with 4 further aliphatic C
[cH1]	Aromatic C with 1 further connection
[cH0]	Aromatic C with 0 further connections
[C]=[C]([!CX1])[!CX1]	Bound of 2 aliphatic C with branched double bound and 2 non C atoms with 0 further connections
[C]=[C]([C])[!C]	Bound of 3 aliphatic C and one non C atom, with branched double bound between first and second C

[C]=[C]([C])[C]	Bound of 4 aliphatic C, with branched double bound between second and third C
[C]=[C]=[C]	3 aliphatic C with double bounds
[!CX1][C]#[C]	Bound of aliphatic C with triple bound to aliphatic C and C with 0 further total connections
[!X1][C]#[C]	Bound of aliphatic C with triple bound to aliphatic C and one atom with 0 further connections
[!#1][CH2][!#1]	Bound of aliphatic C with two further hydrogen and two atoms that are not hydrogen atoms
[CX3H1]=[CX3H0]	Bound of aliphatic C with 1 further total connection with 1 further hydrogen and aliphatic C with 2 further total connections with 0 further hydrogen
[CX3H0]=[CX3H0]	Bound of two aliphatic C with 2 further total connections with 0 further hydrogen
[CX3H2]=[CX3H1]	Bound of aliphatic C with 0 further total connections with 2 further hydrogen with aliphatic C with 1 further total connection with 1 further hydrogen
[CX3H1]=[CX3H1]	Bound of two aliphatic C with 1 further total connection with 1 hydrogen
[CX3H2]=[CX3H0]	Bound of aliphatic C with 0 further total connections with 2 further hydrogen with aliphatic C with 2 further total connections with 0 further hydrogen
[c][CX4H3]	Bound of aromatic C with aliphatic C with 0 further total connections with 3 further hydrogen
[c][CX4H2]	Bound of aromatic C with aliphatic C with 1 further total connection with 2 further hydrogen
[c][CX4H1]	Bound of aromatic C with aliphatic C with 2 further total connections with 1 further hydrogen
[CX2H1]#[CX2H0]	Triple bound of aliphatic C with 0 further total connections with 1 further hydrogen and aliphatic C with 1 further total connection with 0 further hydrogen
[CX2H0]#[CX2H0]	Aliphatic C with triple bond and 1 further total connection 0 further hydrogen
[R]	Number of atoms in ring

Tables S2: Utilized parameters for hyperparameter optimization of Monte-Carlo Neural Networks

Parameter	*Optimization range*
(Neurons per layer)	(num_{feat}), (num_{feat}, num_{feat}), $(num_{feat}, num_{feat}, num_{feat})$, $(2*num_{feat})$, $(2*num_{feat}, 2*num_{feat})$, $(2*num_{feat}, 2*num_{feat}, 2*num_{feat})$,
Droprate	*0.001, 0.01, 0.05, 0.1, 0.2, 0.3, 0.4, 0.5*
Learning rate	*1e-4, 1e-3, 1e-2*
Learning weight decay	*0, 1e-8, 1e-6, 1e-4, 1e-2*
Number of training epochs	*500, 1000, 1500, 2000*
Activation function	*ReLU, ELU, Sigmoid*
X feature scaler	*None, MinMaxScaler, RobustScaler*
y feature scaler	*None, MinMaxScaler, RobustScaler*

Tables S2: Utilized parameters for hyperparameter optimization of Log Gaussian Loss Neural Network

Parameter	*Optimization range*
(Neurons per layer)	(num_{feat}), (num_{feat}, num_{feat}), $(num_{feat}, num_{feat}, num_{feat})$, $(2*num_{feat})$, $(2*num_{feat}, 2*num_{feat})$, $(2*num_{feat}, 2*num_{feat}, 2*num_{feat})$,
(Number of Long Short Term Memory layers)	*1, 2*
Droprate	*1e-6, 1e-5*
Learning rate	*1e-5, 1e-4, 1e-3, 1e-2*
Learning weight decay	*0, 1e-8, 1e-6, 1e-4, 1e-2*
Number of training epochs	*1000, 2000, 3000, 5000*
Activation function	*ReLU, ELU, Sigmoid*
X feature scaler	*None, MinMaxScaler, RobustScaler*
y feature scaler	*None, MinMaxScaler, RobustScaler*

Utilized Scalers:

$$MinMaxScaler = \frac{x_i - \overline{x_i}}{x_{i,max} - x_{i,min}}$$

$$RobustScaler = \frac{x_i - Q_{50}(x_i)}{Q_{75}(x_i) - Q_{25}(x_i)}$$

Figure S4: Cross-validation training results of the density for the QSPR and SSPR model, plots and metrics

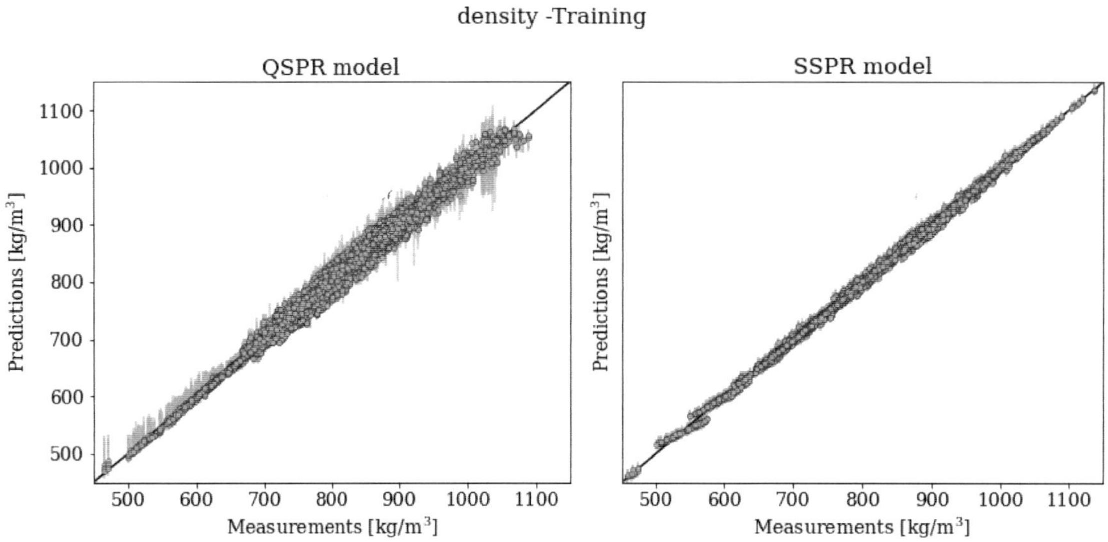

density

	MAE	PICP	NMPIW	MAOE
QSPR model	5.51	78.54	19.33	5.14
SSPR model	1.78	98.61	20.24	2.4

Figure S5: Cross-validation training results of the surface tension for the QSPR and SSPR model, plots and metrics

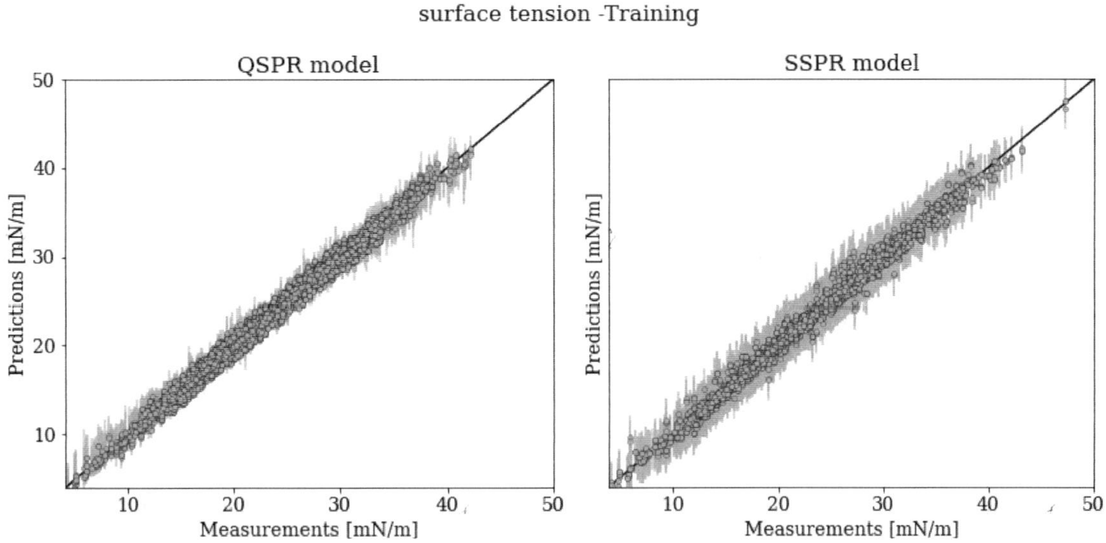

surface tension

	MAE	PICP	NMPIW	MAOE
QSPR model	0.43	78.13	1.57	0.4
SSPR model	0.36	99.49	4.84	0.43

Figure S6: Cross-validation training results of the kinematic viscosity for the QSPR and SSPR model, plots and metrics

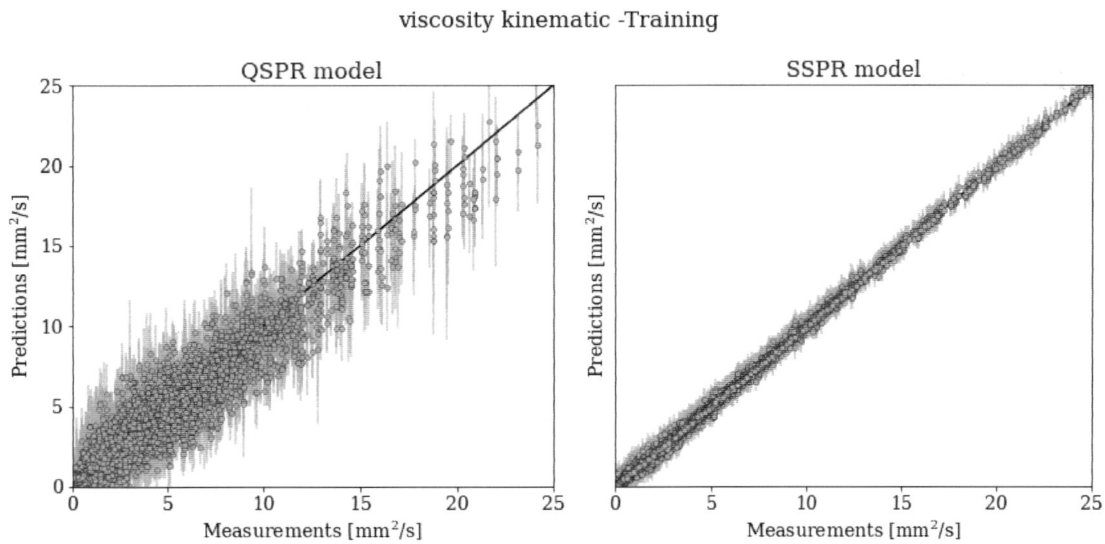

viscosity kinematic

	MAE	PICP	NMPIW	MAOE
QSPR model	0.37	93.94	1.95	0.54
SSPR model	0.07	99.64	1.12	0.14

Figure S7: Cross-validation training results of the net heat of combustion for the QSPR and SSPR model, plots and metrics

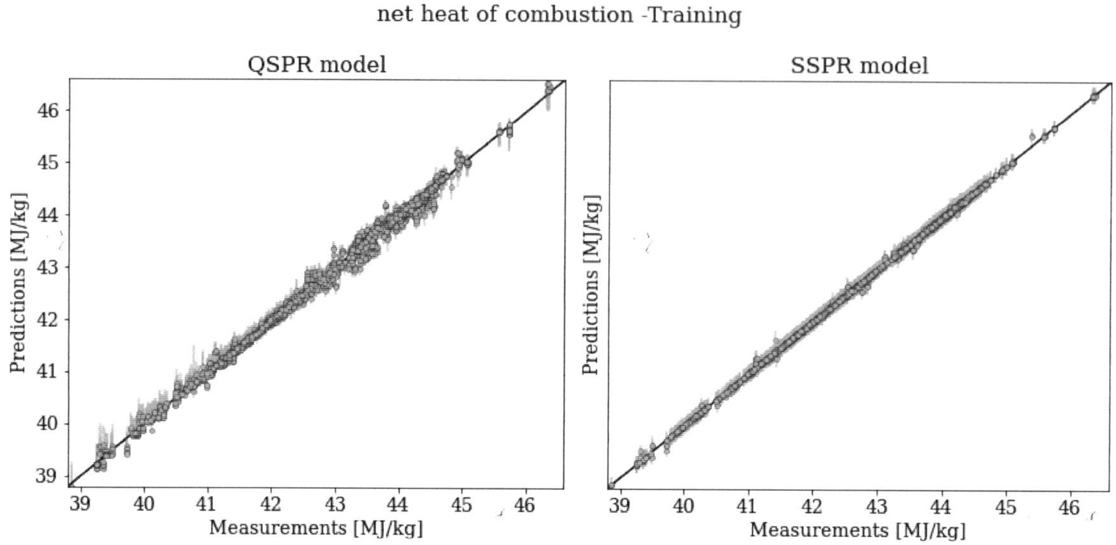

net heat of combustion

	MAE	PICP	NMPIW	MAOE
QSPR model	0.05	89.53	0.24	0.07
SSPR model	0.01	99.52	0.29	0.05

Figure S8: Cross-validation training results of the flash point for the QSPR and SSPR model, plots and metrics

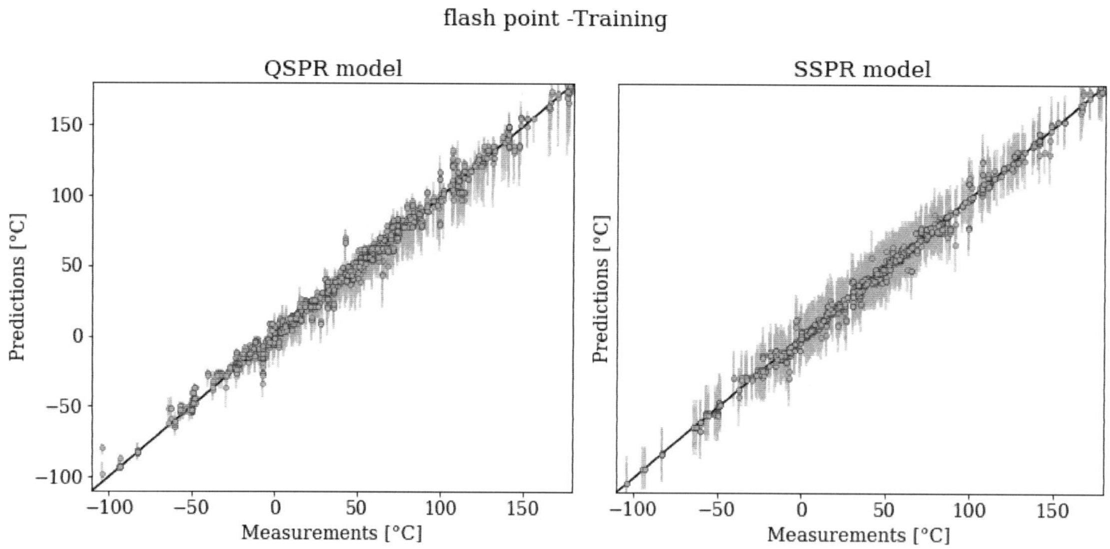

flash point

	MAE	PICP	NMPIW	MAOE
QSPR model	4.04	78.37	16.98	4.03
SSPR model	2.43	98.99	32.51	3.27

Figure S9: Cross-validation training results of the freezing point for the QSPR and SSPR model, plots and metrics

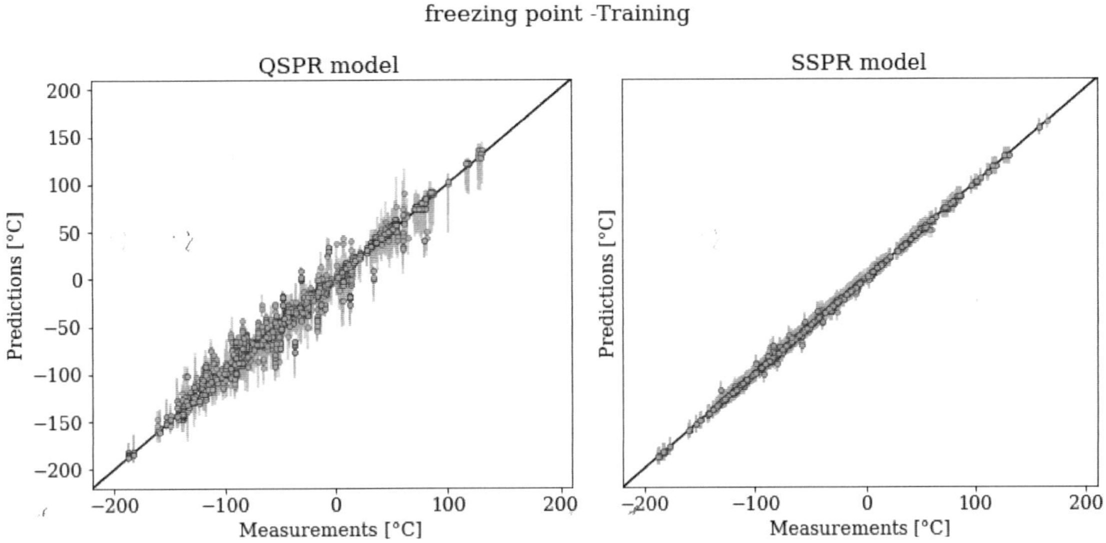

freezing point

	MAE	PICP	NMPIW	MAOE
QSPR model	7.17	77.79	22.12	10.86
SSPR model	0.62	98.85	17.87	2.05

Figure S10: Cross-validation training results of the cetane number for the QSPR and SSPR model, plots and metrics

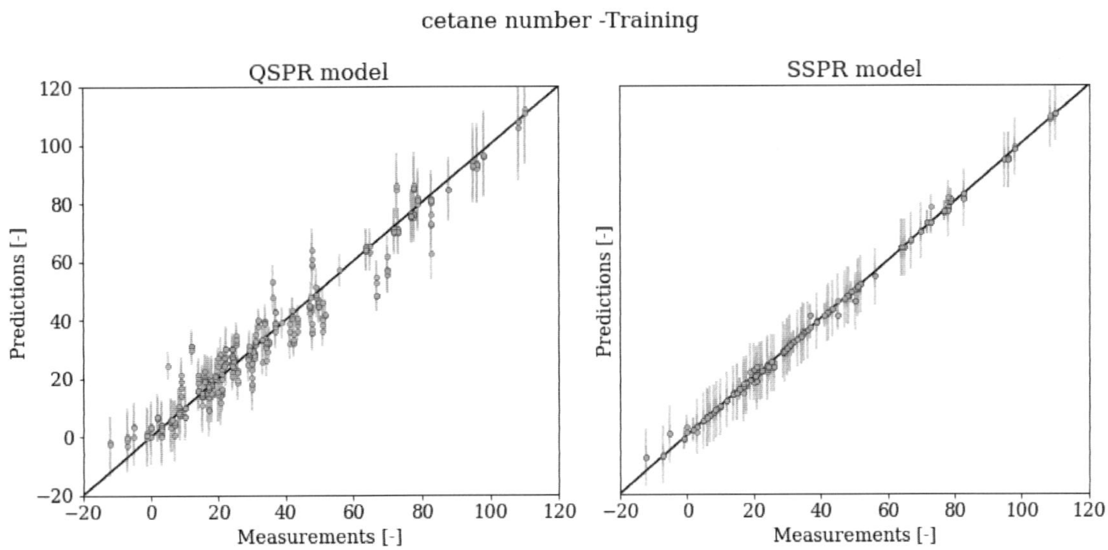

cetane number

	MAE	PICP	NMPIW	MAOE
QSPR model	4.42	74.12	14.23	4.3
SSPR model	0.61	96.45	10.47	0.62

Figure S11: Cross-validation training results of the boiling point for the QSPR and SSPR model, plots and metrics

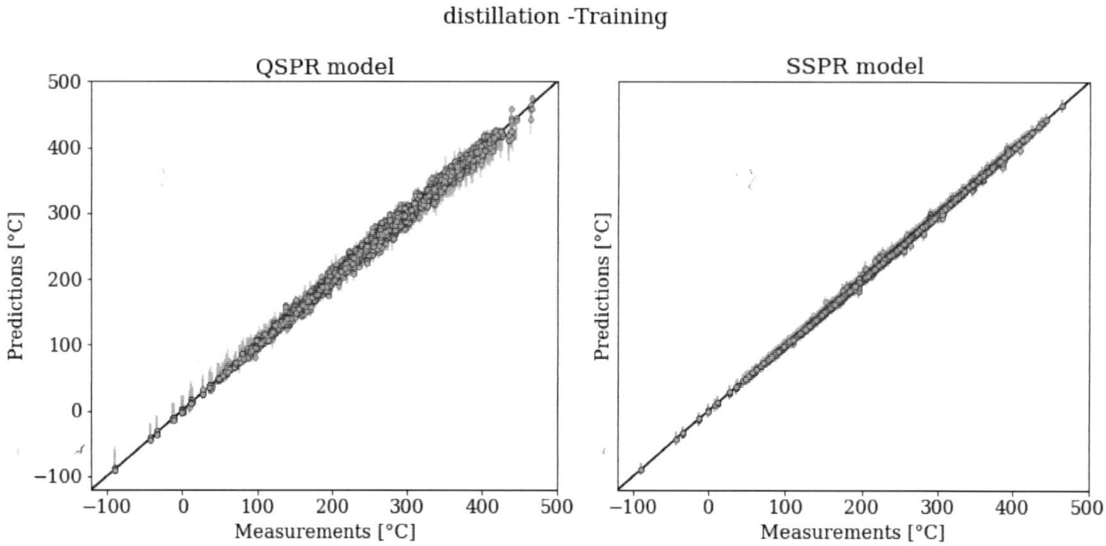

boiling point

	MAE	PICP	NMPIW	MAOE
QSPR model	4.37	75.38	14.33	4.36
SSPR model	2.09	95.42	19.07	6.04

IASH 2022, the 17TH INTERNATIONAL CONFERENCE ON STABILITY, HANDLING AND USE OF LIQUID FUELS
Dresden, Germany
11 - 15 September 2022

ANOMALY DETECTION VIA CHROMATOGRAPHY AND ONE-CLASS SUPPORT VECTOR MACHINES

Jeffrey A. Cramer[1], Robert E. Morris[2], and Thomas N. Loegel[1].

[1]*U.S. Naval Research Laboratory, Navy Technology Center for Safety and Survivability, Chemistry Division, 4555 Overlook Ave. SW, Washington, D.C. 20375 USA.*
jeffrey.cramer@nrl.navy.mil
[2]*Nova Research Inc., Alexandria, VA 22308. RMorris@novaresearchonline.com*

ABSTRACT

Failure and anomaly investigations are a regular occurrence in both routine fuel handling operations and non-routine fuels research. When fuel problems arise in storage or during use, a failure analysis in generally conducted with the goal of ascertaining why the fuel is underperforming so that a remediation strategy can be formulated. While underlying fuel problems can arise from gross contamination issues, subtle compositional differences can also lead to unwelcome synergistic interactions, especially during long-term fuel storage or during the course of downstream fuel commingling. Although near-infrared (NIR) spectroscopy has shown itself to be capable of sensitive anomaly detection, the broadness and convolution of the technique's combination and overtone bands complicate precise *ab initio* diagnostics. More informative anomaly detection procedures might be obtained via recently developed analytical methods based on gas chromatography (GC). However, quickly and accurately assessing the large and compositionally rich data sets provided by GC-based fuel analysis methods remains an ongoing challenge, especially if automated anomaly detection and diagnosis is the goal. A case study will be presented wherein a slightly anomalous diesel fuel, which evaded detection via less robust GC-based data assessment strategies, was automatically detected and characterized by modeling two-dimensional GC with flame ionization detection (GCxGC-FID) data with one-class support vector machines (SVM). This outcome demonstrates the practical potential of combining compositionally detailed fuel data sets with machine learning algorithms.

KEYWORDS

Anomaly detection, outlier detection, machine learning, support vector machines, gas chromatography, flame ionization detection, GCxGC-FID, near-infrared spectroscopy, NIR, mass spectrometry, GC-MS

INTRODUCTION

While fuel problems can arise from gross contamination issues, subtle compositional differences can also lead to unwelcome synergistic interactions, especially during long-term fuel storage or during the course of downstream fuel commingling. Such subtle compositional differences might lead to circumstances in which, for instance, fuels that can pass initial stability testing will become unstable during routine operations.[1,2] The ability to detect subtle compositional differences, then, would facilitate fuel failure and anomaly investigations in a general sense, and would also provide the capability to detect potential fuel issues before they become apparent in an operational capacity.

Near-infrared (NIR) spectroscopy has been shown to be effective at modeling fuel performance generally[3] and characterizing compositional differences in fuels specifically,[4] including at least some types of trace compositional differences.[5] NIR spectroscopy's capabilities in the context of detecting trace compositional differences stem from the fact that NIR spectra consist of broad combination and overtone bands, as opposed to more compositionally specific bands as might be found in, say, mid-infrared spectroscopy. A change in composition of any kind, even a subtle one, is likely to result in a relatively unsubtle change in NIR spectroscopic data shape. Even if the specific cause of the data shape change is unknown, data analysis and modeling methods that rely upon finding clear linear patterns in data sets, such as principal component analysis (PCA),[6] can be used to detect the presence of some type of NIR data anomaly and, thus, some type of compositional anomaly.

Unfortunately, the very broadness and convolution of NIR spectroscopy's combination and overtone bands, which facilitate anomaly detection more generally, complicate precise *ab initio* diagnostics. This is a non-trivial consideration, given that any given fuel failure investigation is likely to require a full accounting of the fuel failure's cause in order to prevent the same type of fuel failure from happening again. What would be more desirable is an anomaly detection method that combines the sensitivity of NIR spectroscopy with the information content of recently developed analytical methods based on gas chromatography (GC). However, quickly and accurately assessing the large and compositionally rich data sets provided by GC-based fuel analysis methods remains an ongoing challenge, especially if automated anomaly detection and diagnosis is the goal.

Presented here is a case study in which a slightly anomalous diesel fuel, which was first perceived in NRL's internal fuel archives via NIR spectroscopy, evaded detection via less robust GC-based data assessment strategies. Upon further analytical development, this diesel fuel was automatically detected and characterized by modeling two-dimensional GC with flame ionization detection (GCxGC-FID) data with one-class support vector machines (SVMs). This outcome demonstrates the practical potential of combining compositionally detailed fuel data sets with machine learning algorithms.

EXPERIMENTAL

Initial manual assessments of sample F0100. NRL maintains an ever-expanding physical archive of refinery fuel samples from worldwide DLA Energy procurements. These samples represent both diesel (primarily F-76) and jet (primarily JP-5) fuel populations. The following case study focuses upon a sample designated internally as F0100, which was clearly indicated to be an F-76 diesel fuel sample in the logistics data maintained alongside the physical fuel forensics samples themselves.

As a matter of course, NIR spectroscopic data are collected for all fuel forensics samples via the Navy Fuel Property Monitor (NFPM).[3] During semi-routine NIR data analysis and modeling operations utilizing PCA, it was discovered that F0100 tended to either cluster with jet fuel samples, as opposed to diesel fuel samples, or did not correspond well to either existing jet or diesel fuel clusters, depending upon which principal components (PCs) were being assessed. This unexpected clustering behavior can be seen in Figures 1 and 2.

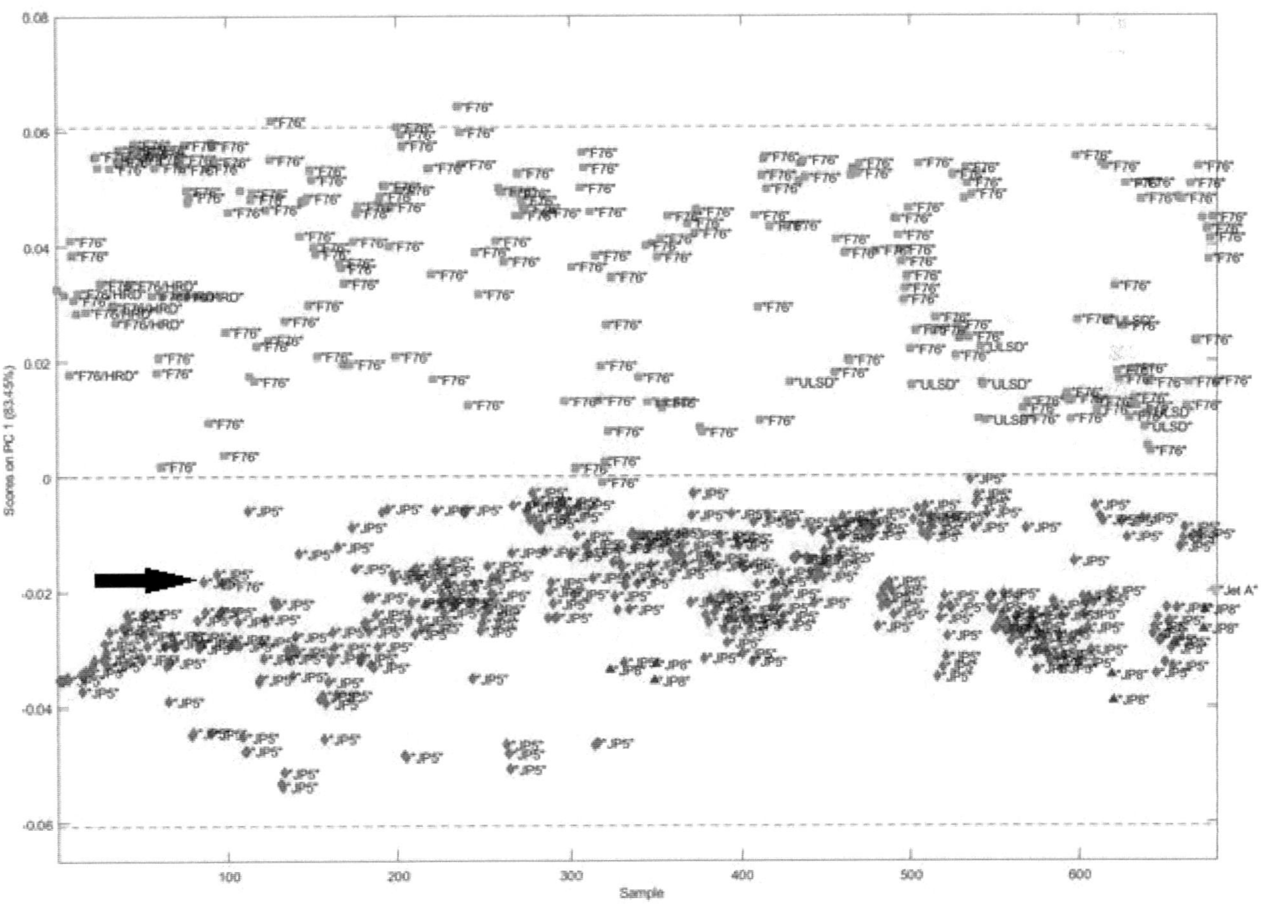

Figure 1. PC1 (scores) vs. sample number, obtained from combined (jet and diesel) NIR data population of fuel forensics samples. F0100 (an F-76 diesel fuel) clusters much more closely to

the bulk of jet fuels than would be expected under normal circumstances.

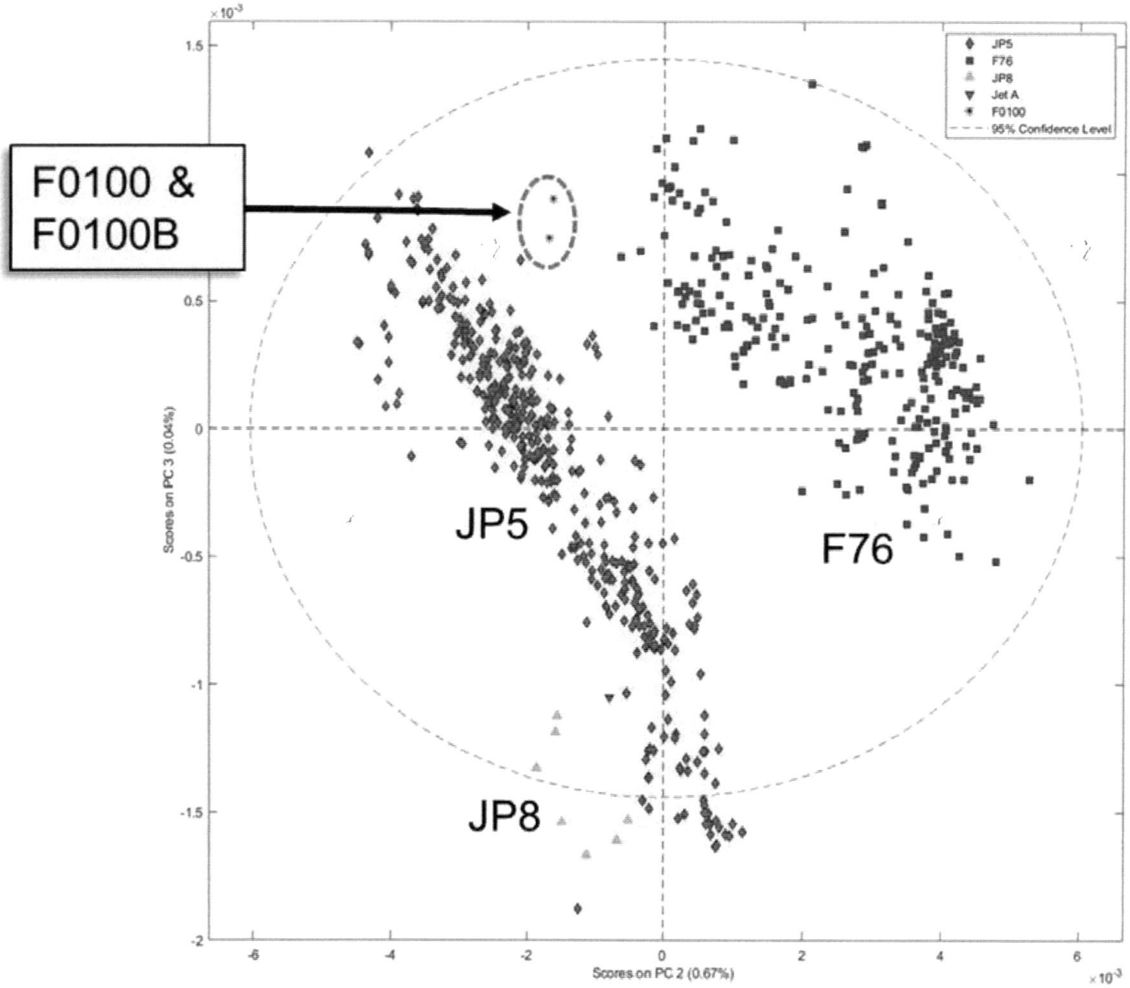

Figure 2. PC3 vs. PC2 (scores), obtained from combined (jet and diesel) NIR data population of fuel forensics samples. F0100 (an F-76 diesel fuel), and a more recent replicate collected to determine if the original NIR data were collected improperly, both cluster apart from the bulk of the jet and diesel populations.

Upon noting this behavior, an *ad hoc* manual investigation was undertaken in which the GCxGC-FID-based, high precision (Tier III) compositional profile data[7] for this sample, as maintained in the Navy Fuel Analysis and Characterization Tool (NFACT),[8] was plotted alongside the average compositional data for the entire diesel fuel population that was being utilized at the time. The results of this operation can be found in Figure 3. These results would seem to indicate that F0100 is a somewhat unusual diesel fuel, insofar as it possesses higher-than-expected amounts of the following compound classes, as determined via manual assessment: C14 dicycloparaffins, >C15 tricycloparaffins, >C13 alkylbenzenes, >C16 cycloaromatics, and >C14 diaromatics.

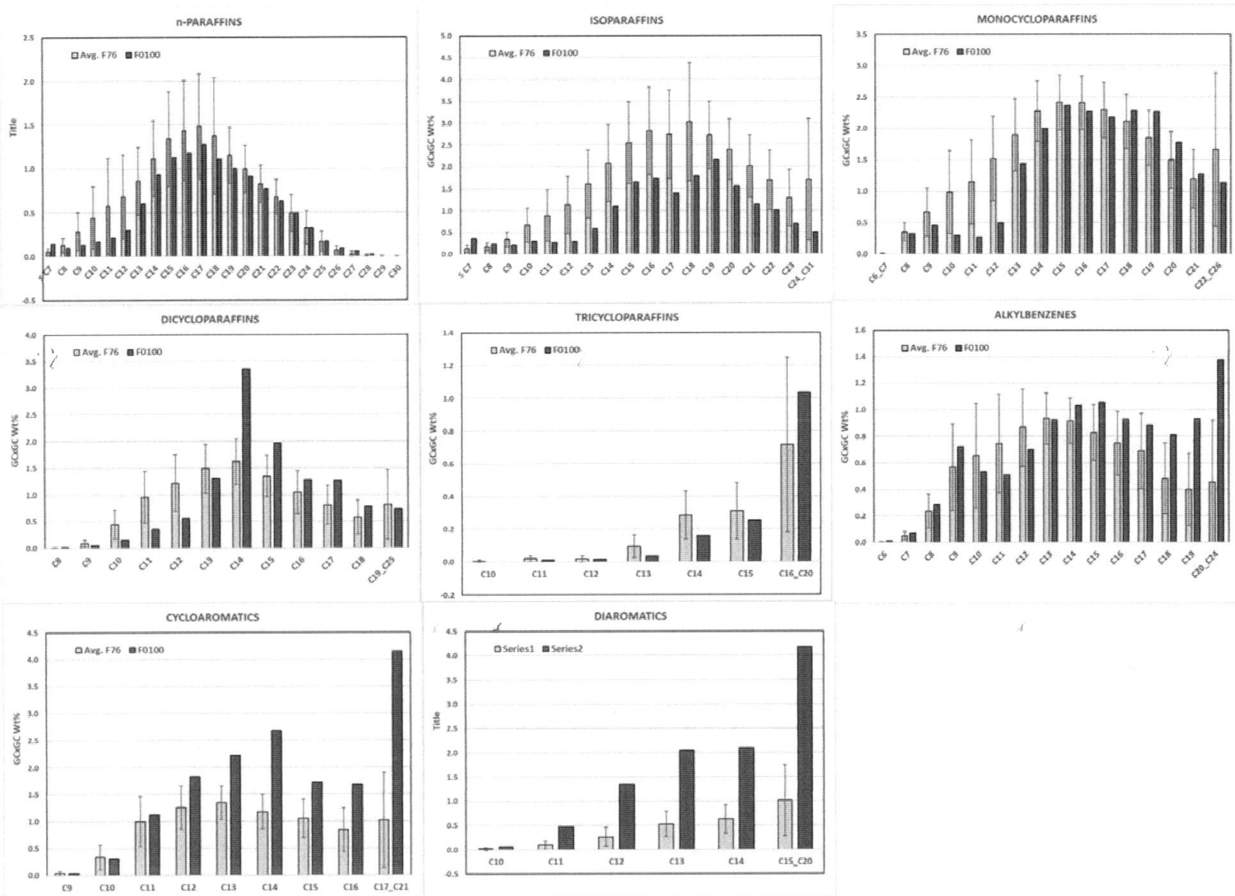

Figure 3. High precision (Tier III) NFACT compositional profile data for an average F-76 sample (light), plotted alongside this same data for F0100 (dark). Error bars: ± 1 st. dev.

This operation began to raise questions regarding how such an outlier might be directly detected, in an automated fashion, via the same GC-based methods that were utilized to produce the NFACT compositional profile data in the first place. Two types of GC data were thus assessed in that context:

- Gas chromatography coupled with mass spectrometry (GC-MS) data are collected as a matter of course from the fuel forensics samples. These data sets are maintained in a database separate from the NFACT database, developed with the Fuel Composition and Screening Tool (FCAST) software developed in-house.[9]
- NFACT compositional profile data are based on two-dimensional gas chromatography coupled with flame ionization detection (GCxGC-FID). Again, these data sets are collected as a matter of course from the fuel forensics samples. However, because the parent GCxGC-FID data sets are so large, assessments were made based on high precision "Tier III" post-profiling formats (i.e. how the data sets exist in the NFACT software) only. Regardless, due to the use of two columns, these GCxGC-FID data sets would be expected to convey richer compositional information that the GC-MS data sets.

GC-MS data. The FCAST software itself was utilized to analyze sample F0100 as if it were an unknown sample. This analysis was carried out in two parts. First, as per existing and native FCAST software capabilities, a distillation curve was calculated from the "unknown" sample and compared with calculated distillation curves from non-anomalous jet and diesel fuel forensics samples obtained during roughly the same time frame. The results of this operation can be seen in Figure 4. Interestingly, despite the compositional oddities noted previously in Figure 3, the calculated distillation curve for F0100 lines up very well with what one would expect from a standard diesel fuel. In other words, F0100 would not appear to be anomalous as a consequence of this operation.

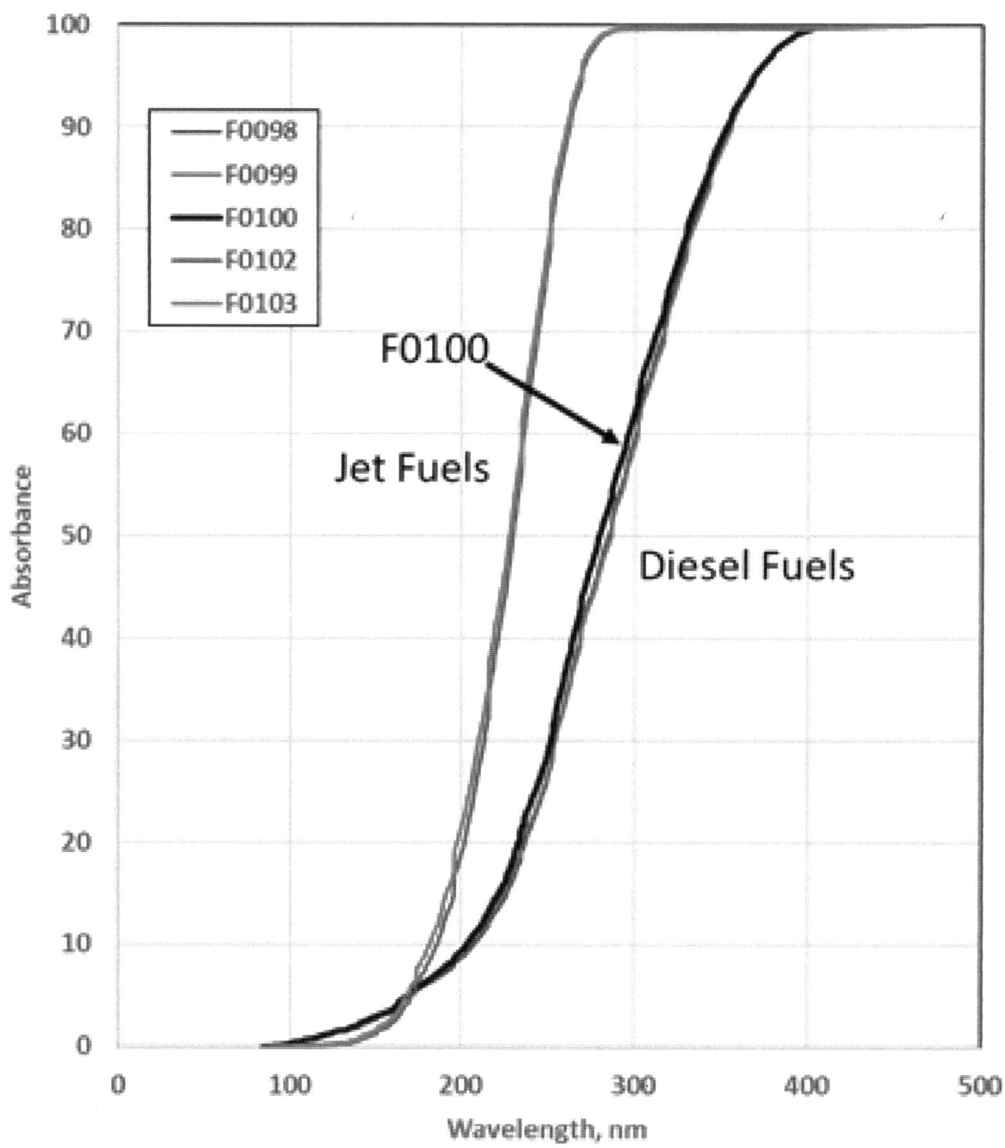

Figure 4. FCAST calculated distillation curve obtained for F0100 and compared alongside non-anomalous jet and diesel calculated distillation curves as obtained for fuel forensics samples obtained during roughly the same time frame.

Next, the total ion chromatographs (TICs) obtained from F0100 and fuels forensics samples collected during roughly the same time frame were subjected to PCA. This operation resulted in a dendrogram representation that separated the samples primarily into jet and diesel sample populations, as can be seen in Figure 5. In this representation, F0100 clustered with the non-anomalous F-76 samples, thus, again, counter-indicating the F0100 compositional anomalies indicated previously.

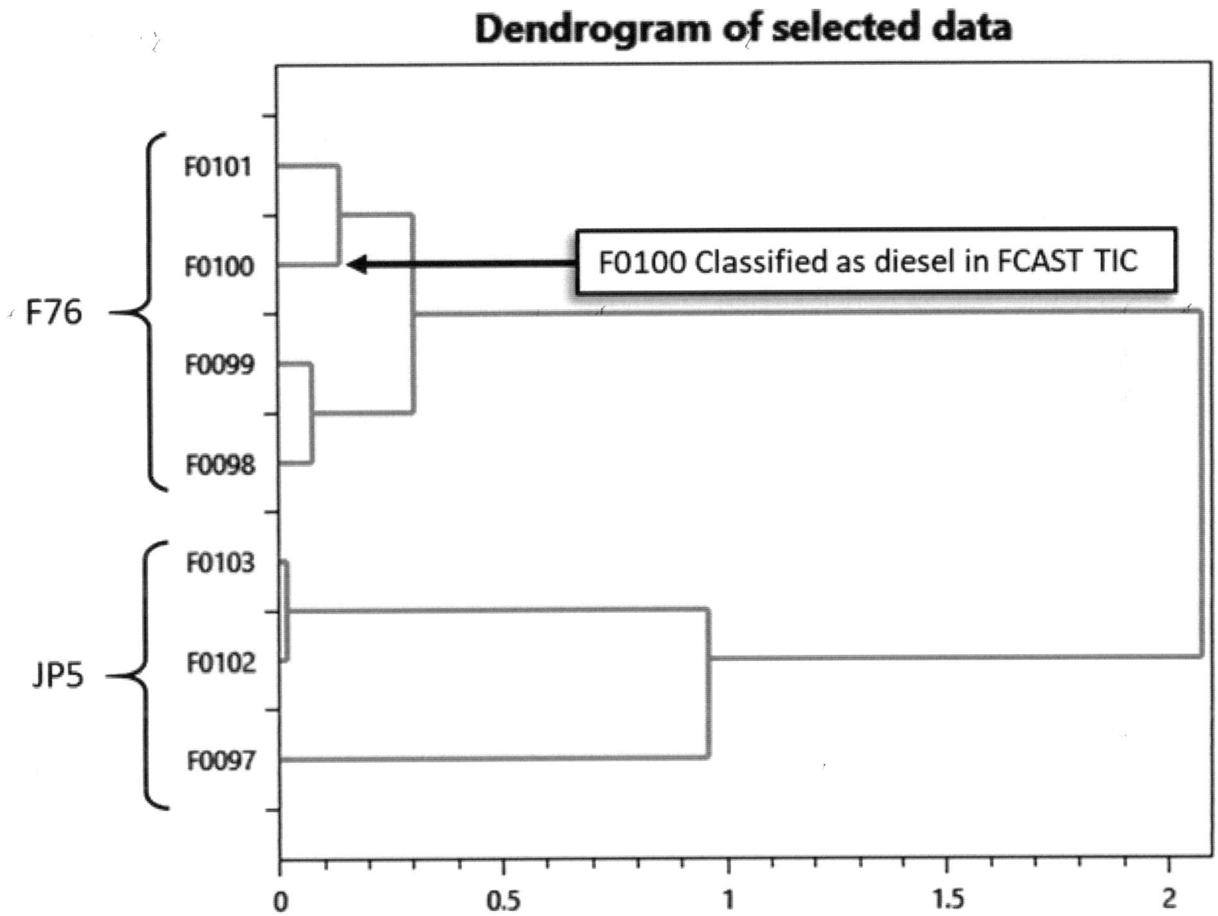

Figure 5. FCAST TIC PCA dendrogram results obtained for F0100 and non-anomalous jet and diesel fuel forensics samples obtained during roughly the same time frame.

GCxGC-FID data. There are three "tiers" of compositional information provided by the NFACT software. Tier I describes the chemical class (e.g., aromatics); Tier II contains chemical subclasses (e.g., alkylbenzenes, cycloaromatics); and Tier III contains chemical sizes (e.g., C5-benzenes). In the present work, the highest-precision Tier III information is utilized. This tier sorts the compositional information contained within the raw GCxGC-FID data sets into over one hundred bins, separated by both compound class and carbon number, to provide a comprehensive snapshot of a fuel's composition. This binned Tier III information was subjected to PCA in order to determine if a relatively straightforward operation, which could presumably be performed automatically in future iterations of the NFACT software itself if as much were deemed to be necessary, could reveal the presence of F0100-like compositional anomalies. PCA was performed on both a combined population of fuel forensics JP-5 and F-76 fuels and an F-76-only fuel forensics population. The results of these operations can be found in Figures 6 and 7. These two PC2 vs. PC1 plots both classify F0100 as an F-76 fuel, though it might be noted that this sample appears at the edge of the diesel cluster when performing PCA on only diesel fuels. This might be indicative of GCxGC-FID data being at least marginally more sensitive to the anomalous properties of F0100. Nonetheless, nothing found in either PCA plot clearly indicates the existence of compositional anomalies in fuel F0100.

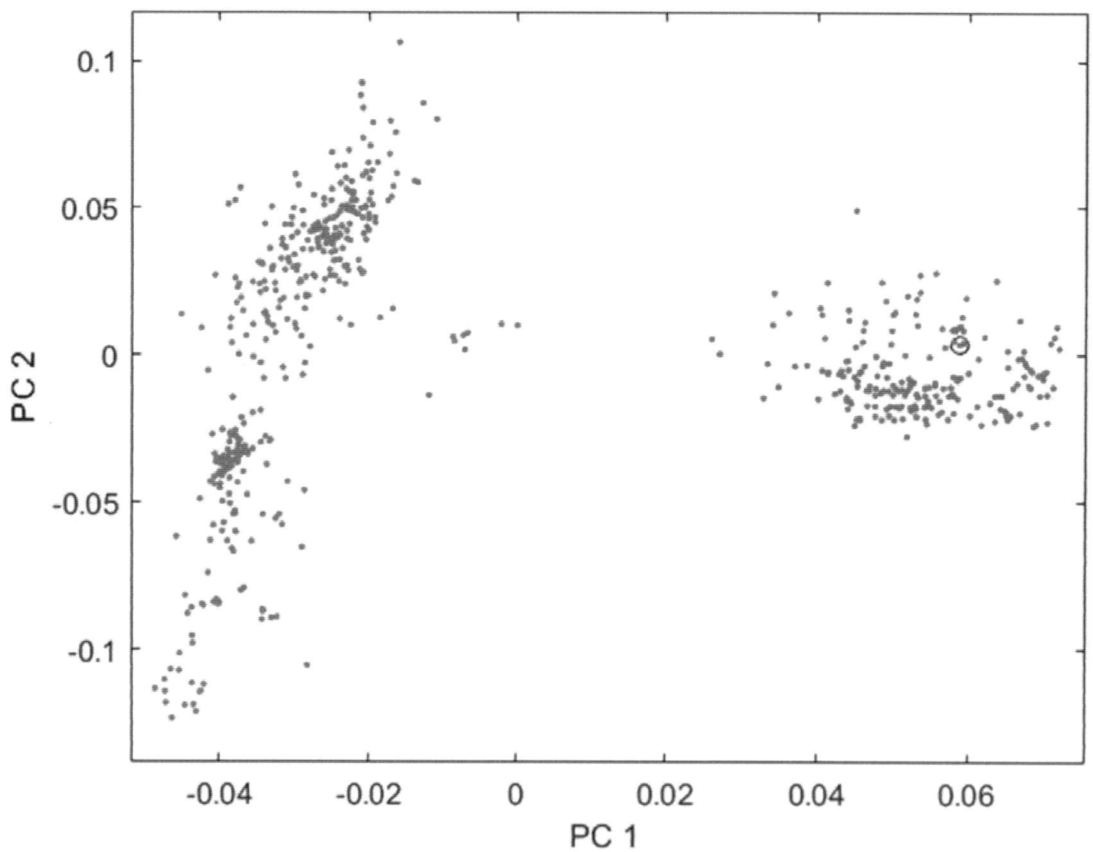

Figure 6. Tier III compositional data PCA results obtained for F0100 and non-anomalous jet and diesel fuel forensics samples. JP-5 samples are plotted in red, F-76 samples are plotted in magenta, and F0100 is circled.

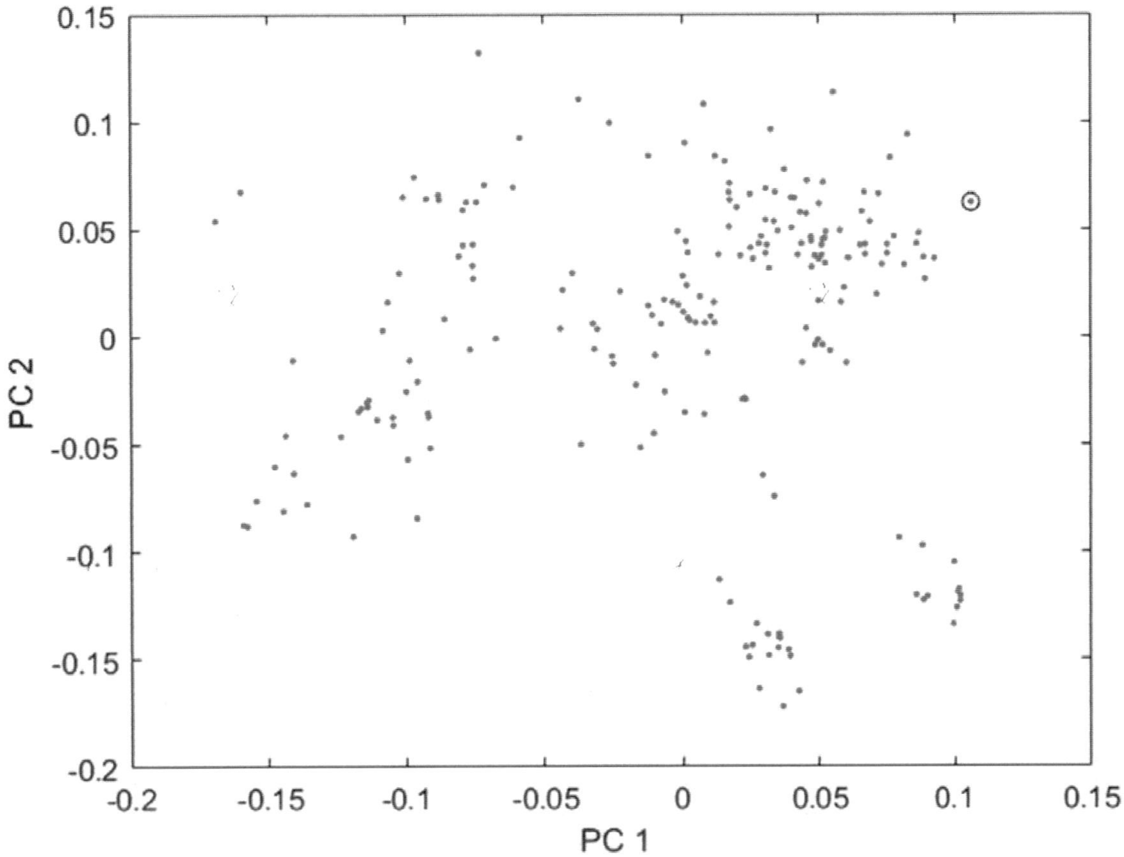

Figure 7. Tier III compositional data PCA results obtained for F0100 and non-anomalous diesel fuel forensics samples. F0100 is circled.

Fit-for-purpose (FFP) fuel modeling. In considering how best to detect subtle compositional anomalies in fuel samples, one might consider previous in-house efforts to perform fuel quality assessments in the context of FFP modeling.[9,10] While this paradigm was originally developed to address the challenges associated with adopting alternative fuel stocks into fuel supplies not explicitly designed to accommodate them, the overarching idea of screening a fuel for suitability for further testing remains relevant in the present context. One specific concept developed in-house was the defining of a multidimensional FFP compositional space into which unknown fuel samples could be mathematically projected. This compositional space would be defined by a comprehensive population of fuels known to be "fit" (non-anomalous and passing standard specification testing, ASTM and otherwise) for a given application. If an unknown fuel sample fell within this space, then it would be known to be compositionally similar enough to known fit fuels to warrant further testing.

Theory behind the use of one-class SVMs. If the goal is to develop such a compositional space, one of the primary challenges is to clearly define the borders separating fit fuels from unfit fuels.

As one might imagine, this is a non-trivial challenge in the field of fuel analysis because a fuel might be unfit for any number of underlying compositional reasons, even within relatively restricted fuel populations. Essentially, if one were to attempt to separate two classes of fuels, fit and unfit, the challenge quickly becomes intractable because any explicitly defined class of unfit fuels would be, at least for the purposes of robust fuel assessments, infinite in size.

A more productive approach would instead utilize one-class modeling, based only on explicitly modeling fit fuels, thus consigning any fuel that does not correlate well to explicitly modeled fitness criteria (in this case, based on compositional information derived from GCxGC-FID data) into an intentionally under-defined class of unfitness. Such data modeling cannot be adequately pursued using linear modeling techniques such as PCA because the search for underlying linear factors within a data set is not well-suited to characterizing a single class while leaving entire classes of samples intentionally under-defined, as might be implied by the PCA results found in Figure 7. However, nonlinear modeling capabilities provided by machine learning, and specifically SVMs,[11] can be leveraged to model a single targeted class in such a fashion. A one-class SVM model can "score" fuels based on how closely they resemble the bulk of any given training data set of fit fuels, and fuels possessing low scores can be flagged for potential unfitness.

Anomaly detection via one-class SVMs. The data sets to be utilized in the present work were extracted from the NFACT database on 09JUN2021. The present work was performed using Matlab version 9.7 R2019b (The Mathworks, Inc.) with functionality proved via the Statistics and Machine Learning Toolbox version 11.6 (The Mathworks, Inc.). The function fitcsvm was specifically utilized to produce one-class SVM models, and this function was assigned the following parameters:

- Solver: Sequential Minimal Optimization (SMO)
- Kernel function: Gaussian
- Kernel scale: automatically determined within fitcsvm
- Initial alpha coefficient estimates: 10
- Delta gradient tolerance: 0
- KTT tolerance: 0
- Box Constraint: 1
- Iteration limit: 10^5
- Outlier fraction: 0

It should be noted here that the SMO solver involves randomness, so results from 100 modeling runs are utilized in the present work to ensure robust and reliable modeling results.

A program was written to start with a parent data set of compositionally binned and profiled GCxGC-FID data (including data from any samples to be explicitly investigated, such as F0100), produce 100 models based on that parent data set using fitcsvm, and then re-introduce all of the parent data back into these models to determine if any samples' resultant scores fall outside of a 95% confidence interval range as defined by the scores produced by the entire parent data set. A fuel sample is only reported as being anomalous if all 100 models respectively indicate that the fuel's score is outside of the defined score range.

As indicated previously, one of the advantages of using GCxGC-FID for anomaly detection is that it functions in a diagnostic capacity, allowing one to pinpoint which portions of compositional data are yielding the compositionally-relevant anomalies. To full leverage this advantage, multiple parent data subset permutations were utilized in the present work. Firstly, both a combined jet and diesel training data set and a diesel-only training data set were utilized. Each of these data sets were divided into eight compositional categories (monoaromatics, cycloaromatics, di- and tri-aromatics, n-paraffins, isoparaffins and light hydrocarbons [LHC], monocycloparaffins, dicycloparaffins, tricycloparaffins).

The data sets corresponding to these categories were then further sub-divided into halves and thirds based on the carbon number ranges being assessed. For example, with respect to the monoaromatics compositional category, which spans binned carbon numbers C6 to C20+ (the latter value being included in a single bin) over 15 variables, six separate anomaly detection trials are performed: one for all bins (C6 to C20+), one for the first half of these bins (C6 to C13), one for the second half of these bins (C13 to C20+), one for the first third of these bins (C6 to C10), one for the second third of these bins (C11 to C15), and one for the last third of these bins (C16 to C20+).

In summary: for both the combined jet and diesel training data set and a diesel-only training data set (2), six carbon number range-specific operations (6) are performed per compositional category across eight total compositional categories (8), plus one non-compositional "control" operation utilizing all of the available data without composition-based divisions or carbon number-based sub-divisions (1). Thus, these data divisions and sub-division yield a total of:

$$2 * ((6 * 8) + 1) = 98 \qquad \text{Eq. 1}$$

separate anomaly detection operations, each of which requiring 100 individual model constructions.

If anomalies are detected for any given compositional category, the carbon number ranges within which anomalous activities were detected were combined in two different ways to produce regions of interest. First, the maximum and minimum carbon numbers indicated for any given compositional category were simply utilized as region of interest boundaries, though this often led to overly broad region definitions. Second, the carbon numbers associated with anomalous data-subdivisions for any given compositional category were assessed for any unambiguous carbon number overlap to, on balance, produce narrower regions of interest for each compositional category. If such an overlap was not to be found, the total carbon number range is simply reported as the overlap-based region of interest in order to err on the side of caution.

Diesel F0100 Anomalous Compositional Information, as identified via one-class SVMs:		
data subsets	***F0100 anomalous in diesel training data?***	***F0100 anomalous in diesel/jet training data?***
all bins, control (109 bins, C6 to C30)	**yes**	**yes**
monoaromatics (15 bins, C6 to C20+)	no	no
- first half of bins (C6 to C13)	no	no
- second half of bins (C13 to C20+)	no	no
- first third of bins (C6 to C10)	no	no
- second third of bins (C11 to C15)	no	no
- third third of bins (C16 to C20+)	no	no
cycloaromatics (9 bins, C9 to C17+)	**yes**	**yes**
- first half of bins (C9 to C13)	no	no
- second half of bins (C13 to C17+)	**yes**	**yes**
- first third of bins (C9 to C11)	no	no
- second third of bins (C12 to C14)	**yes**	**yes**
- third third of bins (C15 to C17+)	**yes**	no
di- & tri-aromatics (7 bins, C10 to C15+ di-, C14+ tri-)	**yes**	**yes**
- first half of bins (C10 to C13)	**yes**	**yes**
- second half of bins (C13 to C15+)	**yes**	no
- first third of bins (C10 to C12)	**yes**	**yes**
- second third of bins (C12 to C14)	**yes**	**yes**
- third third of bins (C14 to C15+)	**yes**	no
n-paraffins (24 bins, C7 and lower to C30)	no	no
- first half of bins (C7 to C18)	no	no
- second half of bins (C19 to C30)	no	no
- first third of bins (C7 to C14)	no	no
- second third of bins (C15 to C22)	no	no
- third third of bins (C23 to C30)	no	no
isoparaffins and LHC (19 bins, C7 & lower to C24+ iso., 1 LHC category)	no	no
- first half of bins (C6 to C15)	no	no
- second half of bins (C15 to C24+)	no	no
- first third of bins (C6 to C12)	no	no
- second third of bins (C12 to C18)	no	no
- third third of bins (C18 to C24+)	no	no
monocycloparaffins (16 bins, C7 to C22+)	no	no
- first half of bins (C7 to C14)	no	no
- second half of bins (C15 to C22+)	no	no
- first third of bins (C7 to C12)	no	no
- second third of bins (C12 to C17)	no	no
- third third of bins (C17 to C22+)	no	no
dicycloparaffins (12 bins, C8 to C19+)	**yes**	**yes**
- first half of bins (C8 to C13)	no	no
- second half of bins (C14 to C19+)	**yes**	**yes**
- first third of bins (C8 to C11)	no	no
- second third of bins (C12 to C15)	**yes**	**yes**
- third third of bins (C16 to C19+)	no	no
tricycloparaffins (7 bins, C10 to C16+)	no	no
- first half of bins (C10 to C13)	no	no
- second half of bins (C13 to C16+)	no	no
- first third of bins (C10 to C12)	no	no
- second third of bins (C12 to C14)	no	no
- third third of bins (C14 to C16+)	no	no

Table 1. Results of SVM-based assessments of whether or not F-76 diesel fuel F0100 is anomalous in the context of various subsets of Tier III compositional data.

F0100 anomalous activity, summary of regions of interest:	diesel training data		diesel/jet training data	
	per min/max carbon #s of sub-divisions	per sub-division overlap	per min/max carbon #s of sub-divisions	per sub-division overlap
cycloaromatics	C12-C17+	C9-C17+	C12-C17+	C13-C14
diaromatics / triaromatics	C10-C15+	C10-C15+	C10-C14	C12
dicycloparaffins	C12-C19+	C14-C15	C12-C19+	C14-C15

Table 2. Summary of the regions of interest indicated by the results shown in Table 1.

RESULTS AND DISCUSSION

The diagnostic results indicating whether or not diesel sample F0100 was determined to be anomalous within any given data permutation, as well as the overall regions of interest derived from these individual results, can be found in Tables 1 and 2, respectively. These results and regions of interest can be compared against Figure 3 to show that the SVM-based GCxGC-FID data assessment methodology successfully identified the anomalous nature of F0100's cycloaromatic contents, diaromatic and triaromatic contents, and dicycloparaffin contents.

F0100's tricycloparaffin and alkylbenzene contents were not similarly identified as being anomalous, as was determined during the *ad hoc* manual analysis of the sample. However, this does not negate the practical utility of SVM-based modeling as an automated diagnostic data screening tool, as the sample was identified as being anomalous in a fashion consistent with the majority of the manual analysis. Nonetheless, future development work will likely be required to address this disparity in order to ensure that automated anomaly detection proceeds in a fashion consistent with what an experienced human practitioner could be relied upon to perform.

Regardless of whether or not the aforementioned disparity is truly indicative of model shortcomings, one specific course of action that might be utilized to address potential SVM-based model shortcomings during any future development work is the implementation of safeguards against model overfitting. As fuels are very complex, it is critical for models to be trained in such a manner as to avoid inappropriately extrapolating mathematical and statistical relationships from meaningless data features. In-house linear modeling implementations have long utilized the statistical F-test[12] to mitigate the possibility of model overfitting in an automated fashion. Unfortunately, a similar statistical test is not appropriate in the context of nonlinear machine learning. However, because machine learning is no less susceptible to data overfitting than other methodologies,[13] the technique of regularization, which imposes a penalty upon model complexity to promote a more generally applicable performance level,[14] was developed and can be implemented into machine learning processes.

CONCLUSIONS

This work demonstrates the power of statistical analysis, coupled with detailed compositional data to systematically detect and identify subtle compositional artifacts that can lead to poor fuel performance. Although NIR spectroscopy has shown itself to be capable of sensitive anomaly detection, the broadness and convolution of the technique's combination and overtone bands complicate precise *ab initio* diagnostics. More informative anomaly detection procedures might be obtained via recently developed analytical methods based on GC. A promising case study has been presented wherein a slightly anomalous F-76 diesel fuel, which evaded detection via less robust GC-based data assessment strategies, was automatically detected and characterized by modeling compositionally profiled GCxGC-FID data with one-class SVMs. While this characterization fell short of fully replicating the original manual anomaly assessment, this outcome demonstrates the practical potential of combining compositionally detailed fuel data sets with machine learning algorithms, and further investigations are considered to be warranted.

REFERENCES

[1] Morris, R.E.; Westbrook, S.R.; Baniszewski, D.; Serino, P. NRL-DLA Fuel Forensics and F-76 Investigation. Proceedings of IASH 2015, the 14th International Symposium on Stability, Handling and Use of Liquid Fuels, 2015.

[2] Carico, A.; Mattingly, D. Report of WESTPAC F-76 High Particulate Quality Investigation, November 2014. NF&LCFT Report No. 441/15-005, February 2015.

[3] Morris, R.E.; Hammond, M.H.; Cramer, J.A.; Johnson, K.J.; Giordano, B.C.; Kramer, K.E.; Rose-Pehrsson, S.L. Rapid Fuel Quality Surveillance through Chemometric Modeling of Near-Infrared Spectra. Energy and Fuels, 2009, 23, 1610-1618.

[4] Cramer, J.A.; Hammond, M.H.; Myers, K.M.; Leska, I.A.; Morris, R.E. Expanded Framework for the Prediction of Alternative Fuel Content and Alternative Fuel Blend Performance Properties Using Near-Infrared Spectroscopic Data. Energy and Fuels, 2015, 29, 7026–7035.

[5] Pilar Dorado, M.; Pinzi, S.; de Haro, A.; Font, R.; Garcia-Olmo, J. Visible and NIR Spectroscopy to assess biodiesel quality: Determination of alcohol and glycerol traces. Fuel, 2011, 90 (6), 2321-2325.

[6] Bro, R.; Smilde, A.K. Principal Component Analysis. Analytical Methods, 2014, 6, 2812-2831.

[7] Metz, A.E.; Loegel, T.N.; Wrzesinski, P.J.; Baniszewski, D.J.; Esposito, M.C.; Shafer, L.M.; Truong, D.M.; Striebich, R.C.; Simms, G.R.; Peretich, M.E.; McDaniel, A.M.; Kamin, R.A.. GCXGC Hydrocarbon Method Alignment Across Multiple DOD Facilities. Proceedings of the 16th International Symposium on Stability, Handling, and Use of Liquid Fuels, 2019.

[8] McDaniel, A.; Fischer, S.; Metz, A.; Cramer, J.; Gilman, R.; Antaki, K.; Bychkowski, M.; Morris, R. Application of Analytics Technology for Fuels Data: IASH 2019. Proceedings of the 16th International Symposium on Stability, Handling, and Use of Liquid Fuels, 2019.

[9] Hammond, M.H.; Cramer, J.A.; Loegel, T.N.; Johnson, K.J.; Morris, R.E.; Myers, K.M. Navy Fuel Composition and Screening Tool (FCAST) v3.0. NRL Memorandum Reports NRL/MR/6181--19-9860.

[10] Cramer, J.A.; Hammond, M.H.; Myers, K.M.; Loegel, T.N.; Morris, R.E. Novel Data Abstraction Strategy Utilizing Gas Chromatography–Mass Spectrometry Data for Fuel Property Modeling. Energy and Fuels, 2014, 28, 1781–1791.

[11] Noble, W. What is a support vector machine? Nat. Biotechnol., 2006, 24, 1565–1567.

[12] Kramer, K.E.; Morris, R.E.; Rose-Pehrsson, S.L.; Cramer, J.; Johnson, K.J. Statistical Significance Testing as a Guide to Partial Least-Squares (PLS) Modeling of Nonideal Data Sets for Fuel Property Predictions. Energy and Fuels, 2008, 22, 523-534.

[13] Handelman, G.S.; Kok, H.K.; Chandra, R.V.; Razavi, A.H.; Huang, S.; Brooks, M.; Lee, M.J.; Asadi, H. Peering Into the Black Box of Artificial Intelligence: Evaluation Metrics of Machine Learning Methods. American Journal of Roentgenology, 2019, 212 (1), 38-43.

[14] Tian, Y.; Zhang, Y. A comprehensive survey on regularization strategies in machine learning. Information Fusion, 2022, 80, 146-166.

Issues Relevant to Aviation Fuel Quality Assurance Throughout the Supply Chain and End-Use

Panel Session Discussion

Moderator:

Dr Alisdair Clark
(BP)

Members

Dr Joanna Bauldreay
(Bauldreay Jet Fuel Consulting Ltd)

Mr Patrick Bosmans
(CEPS – Central Europe Pipeline System)

Dr Moshe Rabaev
(Israeli Air Force)

Issues Relevant to Aviation Fuel Quality Assurance Throughout the Supply Chain and End-Use

Introduction to Panel Session

Wednesday 14 September

Dr Joanna Bauldreay

Prevent or Control Product Quality (PQ) Problems from Production to Usage, including:

- Problems
- Challenges
- Solutions

Production
- Crudes
- Process
- Additives
- MoC

Initial Storage

Transportation
- Rail, Shipping, Trucks
- Pipelines

Terminals and Depots

Airport Operations

Aircraft

Some sources of problems and challenges

Housekeeping
- Water, dirt and microbiological control measures

Contamination
- Dirt, water, etc.
- Other (mainly fuel) products in system

Degradation
- Maybe follows from above, but not necessarily
- Two on-spec fuels can mix and produce off-spec fuel

Sampling & Testing
- Is the sample valid and in an appropriate container
- Is the test result reliable and believable?

Does customer actually recognise odd results and challenge them?

Industry Challenges – Technical and More

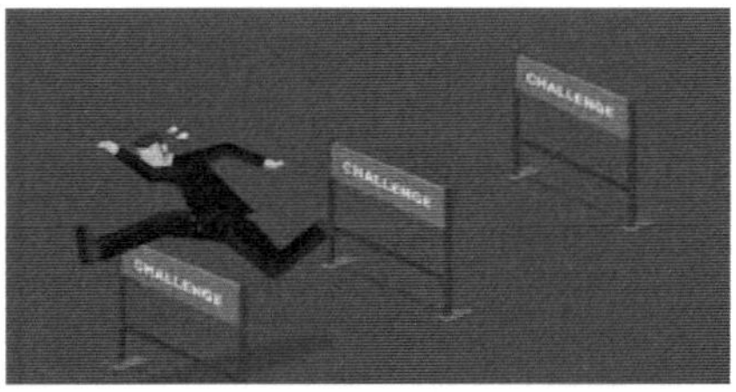

- **Players involved in the PQ chain**
  - Fewer International Oil Companies involved with whole PQ chain
  - More new players involved with small parts, with less access to "the full story"
  - "New fuel" producers
- **Loss of corporate and relevant knowledge**
  - Loss of technical experts, in IOCs and other sectors in the PQ chain
  - Loss of or difficulty in accessing "paper" records
  - Destruction of electronic records, including emails, to save storage costs ($x/GB) and/or avoid litigation risks
  - More people who don't know what they don't know!
- **New technologies, in particular SAFs, increasing complexity of supply systems and changing where risks may be increasing**

Where to go for guidance?

- **For good control measures and targets, ideally to PREVENT problems**
  -May not help recognising and dealing with issues in the first place
  - EI Filtration standards
  - IATA Guidance materials (especially for microbiological matters)
  - EI/JIG 1530
  - JIG 1 and 2
  - Specifications and Methods
- **Process and equipment suppliers**

- **Problem-focussed materials very limited (mainly find expertise is kept "in-house")**
  - Maxwell Smith, "Aviation Fuels", 1970 – Excellent case studies but out-of-print and 50 years behind on technology and practices
- **A new book is in production that aims to address the needs of a 2020s aviation fuel community, with current chapters indicated overleaf**

Topics and Locations to Consider

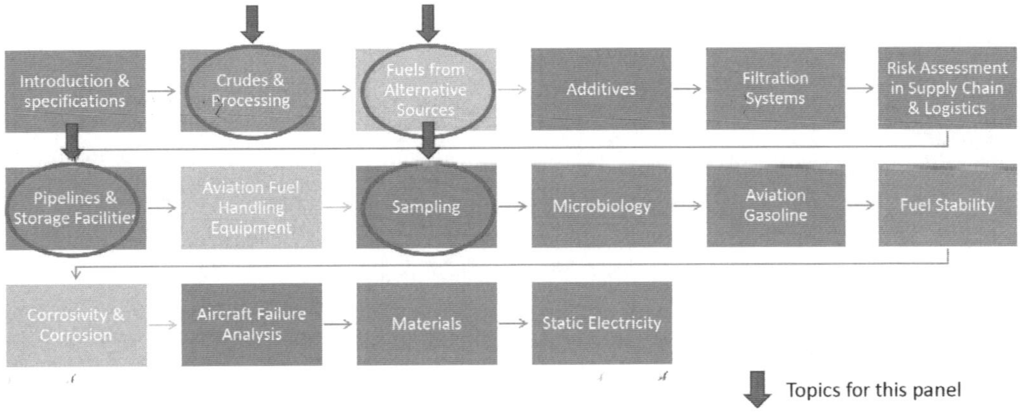

Panel Discussion

- **Moderator and Panel members have broad-ranging expertise related to PQ**
  - **Alisdair Clark** Fuel Specifications, Manufacture and Quality Control
  - **Joanna Bauldreay** R&D + TS support Production, Distribution and Marketing
  - **Patrick Bosmans** CEPS NATO Pipeline System, Supplying many Western European Airports and Military Bases
  - **Moshe Rabaev** Maintenance, Research, Test Methods and User
- <u>Specific Topics for Discussion</u>:

Crudes & Processing | Fuels from Alternative Sources | Pipelines & Storage Facilities | Sampling

IASH 2022, the 17 TH INTERNATIONAL CONFERENCE ON STABILITY, HANDLING AND USE OF LIQUID FUELS
Dresden, Germany
11 - 15 September 2022

PREDICTING THERMAL STABILITY BEHAVIOR WITH QUANTUM CHEMISTRY

Charlie Adams,*,† Marco Conte,‡ Ehsan Alborzi,*,† Anthony JHM Meijer,‡ Kevin Hughes,† and Mohamed Pourkashanien*,†

†*Department of Mechanical Engineering, The University of Sheffield, Sheffield, S3 7RD, UK*
‡*Department of Chemistry, The University of Sheffield, Sheffield, S3 7RD, UK*

E-mail: cadams2@sheffield.ac.uk; e.alborzi@sheffield.ac.uk; mohamed.pourkashanian@sheffield.ac.uk

ABSTRACT

To understand the formation of insolubles in fuels from first principles, a series of DFT calculations were run to calculated the barriers of the autoxidation and coupling reactions for several common fuel aromatics/heteroatoms. The compounds chosen were: phenol, toluene, naphthalene, pyrrole, quinoline, and indole. A homolytic aromatic substition mechanism was indentified which when integrated into a simplified psuedo-detailed mechanism, could predict deposition behavior of the compounds tested in a simple *n*-dodecane surrogate. The homolytic aromatic substition mechanism identified starts with the formation of an AHA· intermediate, formed from the attack of a A· to an AH species. It was found that ROOH could re-aromatize the AHA· intermediate forming a deposit dimer. Sensitivity analysis of our psuedo-detailed mechanism indicated that R· and RO· abstraction steps had a large influence on the final mass of deposit. However, the A· $^{+}$AH coupling step also had the largest influence on the rate of deposit formation. An aromatic/heteratom model containing phenol and toluene was also built, which showed that phenol suppressed deposition from toluene, and reduced peaked in deposit mass at a 25:75 phenol:toluene ratio.

1. INTRODUCTION

The formation of insolubles in the liquid-phase of jet fuels is mainly driven by the formation of oligomers from fuel components.[1, 2] Oligomers successively grow starting from dimers, trimers, tetramers and so on. Each oligomer growth step is predominantly characterized by the formation of C–C and C–O bonds. Previous research has emphasized the termination of antioxidant and/or fuel components ($A\cdot + A \rightarrow A_2$) as a key source of fuel insolubles, and by that extension C–C/C–O bond formation.[3, 4, 5] This has led Heneghan and Zabarnick to highlight an inverse correlation between the ease of oxidation and the formation of deposits. In other words, a fuel with a lower oxidation rate arises from the higher concentration of chain-breaking A· radicals, and therefore have a higher concentration of A· to undergo termination steps producing deposit.[3] However, in some instances, this relation does not always hold. Notably, some nitrogen compounds like pyrroles and indoles tend to show low/no effects on autoxidation rates yet are severe deposit promoters.[6, 7, 8] To explain these exceptions, Heneghan and Zabarnick propose that some fuels produce termination A–A products with a higher solubility.[3] Nevertheless, there may be other chemical factors that need to be explored.

The formation of A–A via the termination of two free-radicals A· in solution is part of an oxidative coupling process. However, it is possible to form A–A without a termination step under oxidative conditions. Focusing on transition metal free processes, homolytic aromatic substition (HAS) has received considerable attention as a method for cross-coupling aromatic compounds under free-radical conditions.[9, 10, 11, 12] Additionally, HAS mechanisms have been used as a way to understand the rate of radical reactions.[13, 14] The general mechanism of HAS is shown in Figure 1.1. First, an attack of an aryl radical to an arene compound generates a -intermediate. The -intermediate is then re-aromatized via elimination of a leaving group forming a dimer. In this sense, HAS reactions are analogous to electrophilic aromatic substitution (EAS) reactions, except that the -intermediate is not charged. EAS reactions have been proposed as a mechanism in the formation of jet fuel deposits.[15] However, in aprotic non-polar jet fuel, the stabilization of charged EAS -intermediates is precluded.[16] By contrast, HAS reactions have been shown to readily occur in non-polar aprotic solvents between arenes under molecular oxygen.[17] In general HAS reactions have presented a challenge to organic chemists due to poor selectivity, leading to intractable mixtures.[11] Nevertheless, in the context of jet fuel fouling, fuel deposits and gums are characterized by highly disordered coupled products composed of mainly aromatic groups.[18, 19] Overall, the HAS reactions between aromatic fuel species should be investigated as possible route to deposits. Particularly as HAS offers a route to form C–C bonds without terminating the free-radical chain mechanism, and can instead be considered a propagation step.

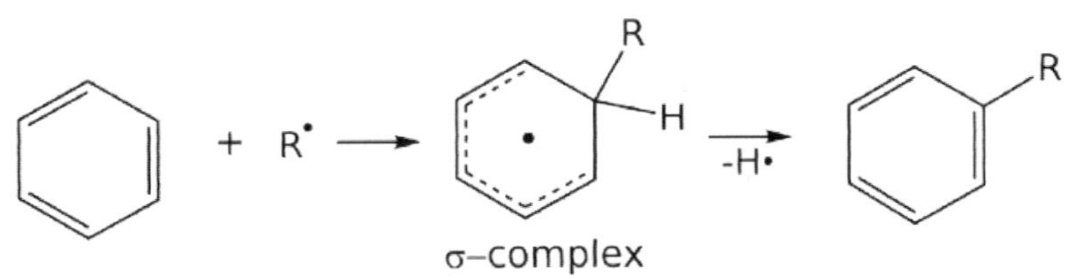

Figure 1.1: Generalized HAS mechanism, where an A· radical attacks a generic aromatic species. This attack subsequently forms a sigma-complex, which is in a doublet state. The sigma complex is re-aromatized by the loss of a hydrogen *via* an unspecified mechanism.

Several reactions for the formation of insolubles/deposits are represented in pseudo-detailed mechanisms in the public literature.[5, 20] However, at present, they are implicit. As a result, they do not represent specific chemical transformations, but are composed of 'pseudo' species with parameters fitted to experiments. In the future, a wider range of fuel chemistries and blends will require predictive mechanisms with greater sensitivity to the starting components, without relying too heavily on fitting parameters. However, due to complexity and range of insoluble structures, a compromise needs to be met between the range of reactions and products represented.

In recent years, density functional theory (DFT) has become a popular tool to build pseudo-detailed mechanisms jet fuel deposition from 'first-principles'.[21, 22] DFT allows direct calculations of thermochemical and kinetic data, and evaluation of competing chemical reaction pathways. By contrast, previous pseudo-detailed mechanisms have been produced by producing activation energies from fitted experimental data.[5, 20] However, this leads to mechanisms which are only suitable for specific types of fuel.

It is the aim of this paper to: 1) explore the possibility of HAS as a route to insolubles formation and 2) attempt to predict insoluble formation tendencies using DFT methods. First, several two-component fuels containing bulk and heteroatoms will be stressed to produce an insoluble mass. Then, the energetic pathways from fuel heteroatoms and bulk species to dimers are calculated and compared to insolubles generated by the surrogate fuels. As a means of understanding the deposition process further, the solubility of oligomers will be considered too.

2 **METHODS AND MATERIALS**

2.1 Surrogate Fuels for this Study

Six surrogate fuels were built with a range of compounds designed to represent the different heteroatom and aromatic compounds found in fuels. Out of the nitrogen class of compounds pyrrole (Sigma Aldrich, >98% purity), quinoline (Oakwood chemicals, >98% purity), and indole (Sigma Aldrich, >99% purity) were selected.

Pyrrole and indole are two 5-membered nitrogen heteratom compounds known to promote insoluble formation, with pyrrole being a particularly problematic insoluble promoter.[8, 23] Quinoline has been shown to also promote insolubles, but to a lesser degree than pyrrole and indole.[8, 7] Next, phenol (ACROS Organics, >99%) was chosen to represent the phenolic class of compounds, again shown to promote insolubles in a variety of real and surrogate fuels.[8, 24, 25, 26] Finally, two aromatic components, naphthalene (Fluorochem, >99%) and toluene (SLS, 99.5%) were chosen for their presence in the mono-aromatic and di-aromatic class in fuels.[27] Each of the above six components were added as 0.1moll^{-1} to *n*-dodecane (ACROS Organics, >99%).

2.2 Method of Thermal Stressing and Deposit Measurement

To produce the insoluble masses, 5 ml of fuel was added to a 50 ml borosilicate round bottom pressurized flask. The flask was heated to 140°C for 24 h under 1 bar. After heating, the flask was allowed to cool and insolubles were then filtered through a 0.1 μm glass fibre filter to give the total weight of insolubles in the bulk. The flask was then washed with trisolvent and then washed into a flask. The flask was then dried in a vacuum oven to remove any liquid residue, weighed, and then compared with the weight of the clean flask- giving the total weight of adherent insolubles left in the flask.[28] The sum of the insolubles weights on the filter and in the flask gave a mass total insolubles per surrogate. The deposit experiments were repeated at least 3 times for each surrogate to improve the accuracy of the total insolubles measurement.

2.3 Hansen Solubility Parameters and Computational Details

All calculations were performed in Gaussian09 (E.01) using the B3LYP functional.[29][30] Grimme's DFT-D3(BJ) dispersion correction was applied to all the calculations to account for long-range effects.[31] A PCM solvation model, with n-dodecane as the chosen solvent, was selected to replicate the hydrocarbon bulk.[32] The basis set chosen was cc-pVTZ on an ultrafine grid. This basis set adds polarization functions, allowing orbital hybridization to be taken into account.[33] Transition states were optimized using the QST1/3 method depending on the reaction studied. All transitions states were verified by the presence of one imaginary frequency corresponding to the saddle point. Additionally, intrinsic reaction coordinate (IRC) calculations were performed to verify the transition state corresponded to the expected reactants and products. Unrestricted (broken symmetry) calculations were performed on open-shell systems, where the HOMO and LUMO were mixed (guess=mix op- tion). Entropy values were corrected using the GoodVibes script, which employs a quasi-harmonic correction corrected at 298K.[34, 35, 36]

In Hansen solubility theory, three Hansen solubility parameters (HSPs) are assigned to each molecule: D for dispersion, P for polarity, and H for hydrogen bonding. As a consequence, each solvent exists in a 3-dimensional space of HSPs. The HSP distance between two molecules in the 3d HSP space is given by:

$$Ra^2 = 4(D_1 - D_2)^2 + (P_1 - P_2)^2 + (H_1 - H_2)^2 (1.1)$$

Where R_a is the Hansen distance.

The HSP parameters for the test solvents were either obtained from the HSPIP dataset or calculated using the software. The HSPIP software requires a series of test solvents/solutes, tested for their solubility in the *n*-dodecane. A score 0 is assigned to insoluble solvents and a score of 1 is assigned to soluble solvents.

In each case, 0.1 moll ^{1} of each test compound was added to 5ml of *n*-dodecane. Resulting from the series of tests is a 'sphere' in 3d HSP space, whose dimensions are determined by the solubility of the sphere. The radius of sphere R_0 is then used to asses the solubility of a proposed solvent, where R_a of the test molecule is used in the following equation:

$$RED = R_a/R_0. \quad (10.2)$$

Where RED stands for Relative Energy Difference. A RED > 1 implies the molecule is likely insoluble in your main solvent (*n*-dodecane in this case), whereas a RED < 1 implies the molecule is soluble. Our calculated sphere was composed of 13 lab grade compounds known for their presence in fuel and gave a fit of 1 according to HSPIP software. The results for these tests are presented in Table 2.1 . Here it is important to acknowledge that the solubility of the deposit precursors in pure *n*-dodecane this study will likely be different to a conventional fuel due to the presence of aromatic compounds and minor components. Nevertheless, these parameters are useful to understand insoluble formation our surrogate fuel, and to improve the HSP method for future studies.

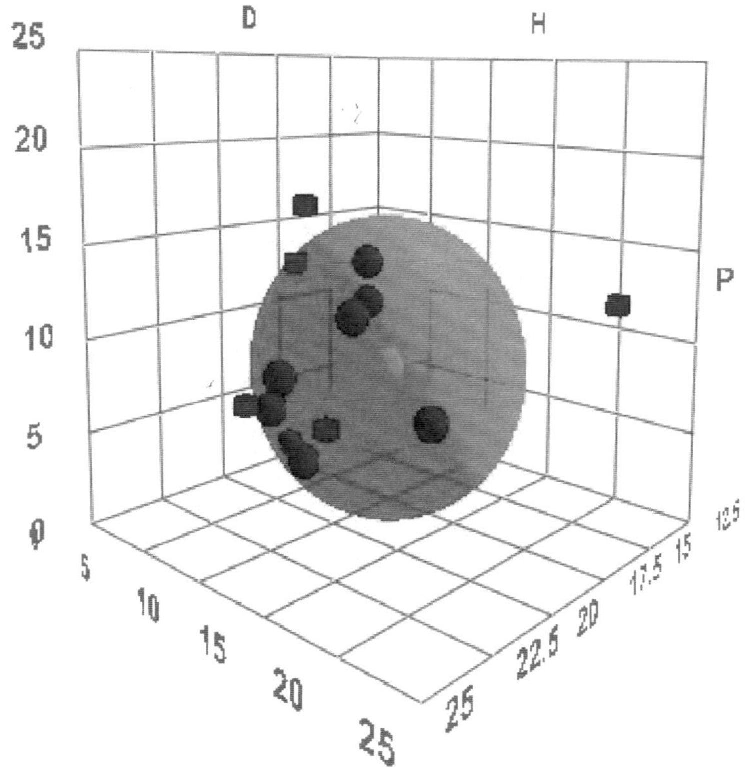

Figure 2.1: HSPIP Sphere

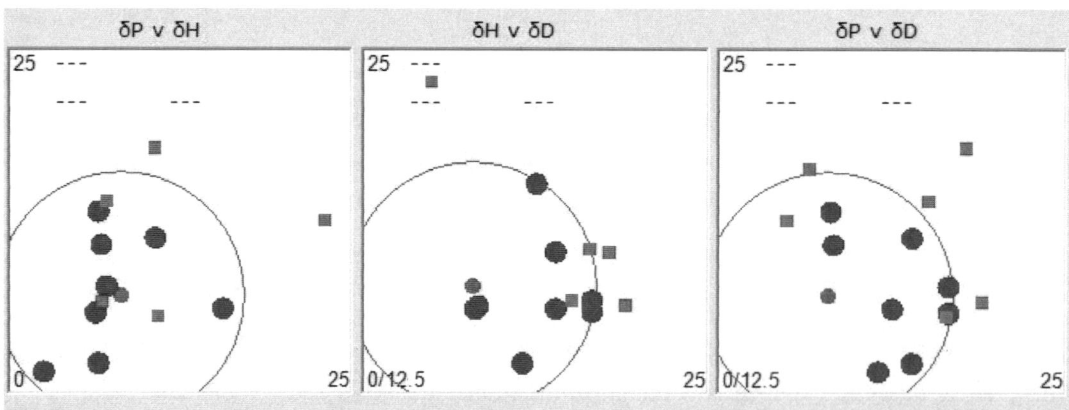

Figure 2.2: HSPIP Sphere parameter fits

Table 2.1: Hansen Solubility test solvents and their Associated HSPs. The sphere gave a fit of 1.000

Solvent	Database/Calculated HSP	D	P	H	Score	RED
Indigo	Calculated	21.10	17.4	10.0	0	1.639
Carbazole	Database	21.7	6.4	6.2	0	1.243
p-Benzoquinone	Database	19.8	13.7	6.5	0	1.121
Water	Database	15.5	16.0	42.3	0	4.042
2-Naphthol	Database	20.4	5.4	10.2	0	1.000
Methanol	Database	14.7	12.3	22.3	0	1.800
Di-*n*-Butyl Sulfoxide	Database	16.4	10.5	6.1	1	0.443
Naphthalene	Database	19.2	2.0	5.9	1	0.884
Pyrrole	Database	19.2	11.0	10.0	1	0.863
Dipropyl Sulfone	Database	16.3	12.9	5.9	1	0.705
Toluene	Database	18.0	1.4	2.0	1	0.961
Indole	Database	20.5	7.5	6.5	1	0.973
Phenol	Database	18.5	5.9	14.9	1	0.979
Quinoline	Database	20.5	5.6	5.7	1	0.995

2.4 Pseudo-Detailed Mechanism in Fuels

In order to compare the total insolubles to the number of dimers predicted by DFT, several new reaction steps were proposed which will be elucidated in the results and discussion. To capture the autoxidation reactions in the bulk, the basic autoxidation scheme (BAS) was used which gave good agreement with oxygen and hydroperoxide depletion with experiment. The BAS scheme was optimized for a range of C10-C14 hydrocarbons, whereby the thermochemical and kinetic parameters were obtained using *n*-dodecane as the model fuel. Further details of the BAS scheme can be found in reference[22]. To construct the mechanism, the Eyring equation was used, with A being formed from the entropy barrier and E_a formed from the enthalpy barrier. All the forward and reverse barriers were calculated from a stable pre-reaction and post-reaction complex.

For each surrogate fuel, different mechanisms were constructed to study oxygen depletion and deposit formation. For oxygen depletion, the level of oxygen was fixed at 1.8mmoll ^{1}.[37] For the deposition mechanism, oxygen was removed as a limiting reagent (kept constant at 1.8mmoll ^{1}) to reflect the continuous supply of oxygen in our deposit forming rig. The mechanism was integrated in MATLAB using the ode45 solver. The mechanism gave a molar concentration of a dimer, which was then related to a mass via the molecular weight of the proposed dimer.

3 RESULTS AND DISCUSSION

3.1 Homolytic Aromatic Substitution Mechanism

The key bottleneck in any HAS reaction is the re-aromatization and liberation of hydrogen from the -intermediate. The loss of H· in this step is not well understood. Nevertheless, one paper exploring HAS reactions between aryl iodides and arenes using oxygen as an oxidant, proposed a re-aromatization step involving oxygen this is shown as reaction 1) in Figure 3.1 Because hydroperoxides (ROOH) and oxygenated species like alcohols (ROH) form under oxidative conditions in fuels, these were also considered as possible reagents to remove hydrogen from the -intermediate. ROOH reacting with the -intermediate is proposed to form RO· and H_2O, is shown as reaction 2) in Figure 3.1. ROH reacting with the -intermediate is proposed to form R· and H_2O, is shown as reaction 3) in Figure 3.1.

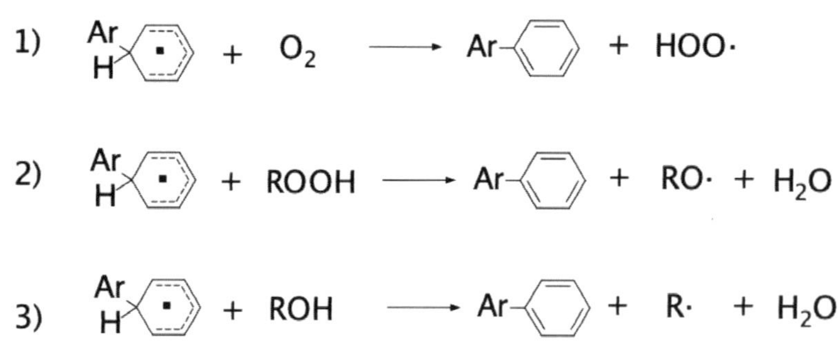

Figure 3.1: HAS reactions considered for fuel coupling reactions. Each step shows the -intermediate formed from an A· + AH reaction, and the species chosen to re-aromatize the intermediate.

Considering the mechanisms presented in Figure 3.1, we explored the possibility of these reactions in phenolic coupling. Phenols have widely been recognized as being detrimental to fuel thermal stability.[8] HAS mechanisms have previously been considered as a possible pathway the oxidative coupling of phenols.[14]

The calculated HAS pathways for phenol are presented in Figure 1.2. First, the formation of the -intermediate is endergonic and has a free-energy barrier of $‡G$=+25.64kcalmol ^{1} leading to an intermediate PhOPhO·. The para- position of the phenol was chosen as the site of oxidative coupling of another phenoxy radical given that this is the generally the more favorable site.[38] The first mechanism in Figure 1.1, involving the re-aromatization with oxygen, could not be identified for phenol. A key challenge with the oxygen transition state is the choice of spin multiplicity. Given that oxygen is in the triplet state, and the system -intermediate is in the doublet state, an open-shell doublet or a quartet surface can be chosen.

The other two mechanisms presented in Figure 1.1 involve ROH and ROOH. The ROOH pathway in Figure 1.2 shows a lower barrier to re-aromatization of the intermediate compared to the ROH by 30.22 kcalmol ^{1}. Additionally, the ROOH is thermodynamically favored, namely due to the enhanced stability of the RO· radical compared to R· The IRCs for both these pathways are presented in Figures S.1 and S.2 in the SI. The ROOH and ROH re-aromatization transition states are characterized by a rotation of the terminal OH moiety towards an available hydrogen at the para-coupling site.

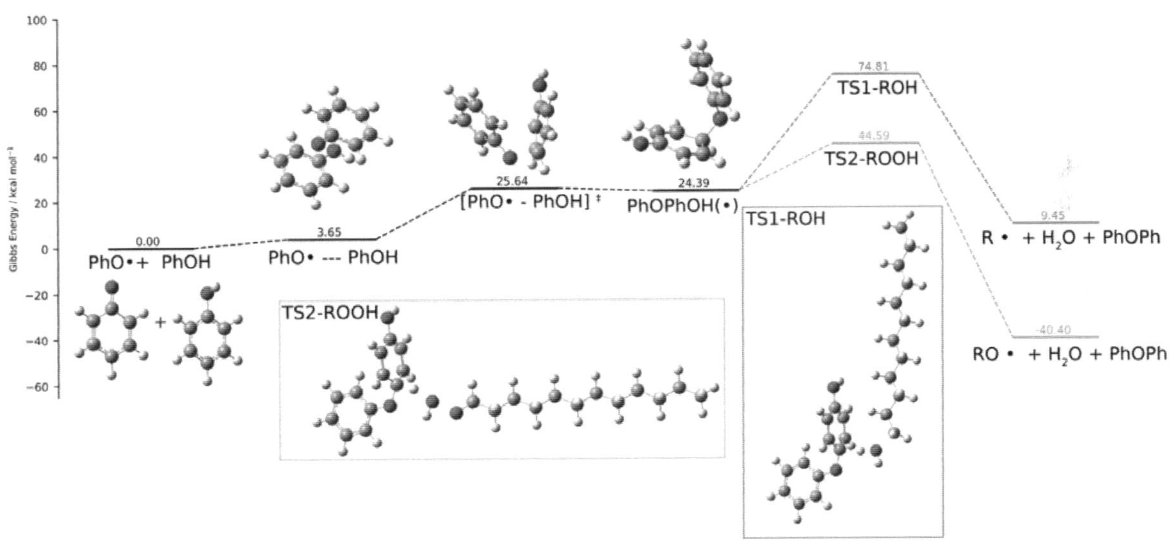

Figure 1.2: Comparison of different HAS pathways at the B3LYP-D3//cc-pVTZ level of theory using *n*-dodecane (PCM) as a solvent. The first step of the process in black shows the A· + AH ! AHA· reaction of a phenol and a phenoxy radical, leading to the -intermediate. The subsequent levels in green shows the re-aromatization step with ROH and the level in orange shows the re-aromatization step with an ROOH species.

3.2 Comparison of Antioxidant Oligomer Solubility and Pathway to Deposit

To justify the assumption that the formation of insolubles results from the coupling of AH species, we calculated Hansen solubility parameters for oligomers for our chosen heterotoms of increasing size. Previous research has hypothesized that aromatic compounds in fuel rapidly become insoluble as they grow in size.[43] Figure 3.3 shows the change in RED score as the oligomer chain grows, where relative energy difference (RED) > 1 indicates an oligomer that has become insoluble in *n*-dodecane. For real fuels, these results will vary, particularly as the aromatic content will increase the 'likeness' of the solvent to other extended aromatic structures.[39] Nevertheless, because our model fuels in this study contain *n*-dodecane as the base solvent, the Hansen plots here are a useful tool to explore deposition tendency in our model fuels. Oligomer structures were chosen based on the favored coupling site for each heteratom based on literature data and our calculations.[38, 40, 41, 42] The phenol chain growth is shown as an example. What is clear is that in general as the oligomer grows in size, the solubility in *n*-dodecane decreases. However, different oligomers reach the insolubility threshold in fewer units, where a single unit is the monomer, 2 units is a dimer and so on. For example, comparing toluene and naphthalene, naphthalene reaches the insolubility threshold after 2 units have coupled. This is consistent with previous observations that di-aromatics form deposit more rapidly because they 'require fewer consecutive reaction steps to produce high-molecular-weight'.[43]

Interestingly, heteroatom size does not influence the solubility of the resultant oligomers. Instead, the lower solubility is related to the dD parameter in the HAS framework, representing Van Der Waals forces between the solvent and heteroatom. As the oligomers grow, the difference in polarity and hydrogen bonding begins to decrease (indicated by the decreasing dP and dH parameters), but the difference in dispersion forces dD increases.

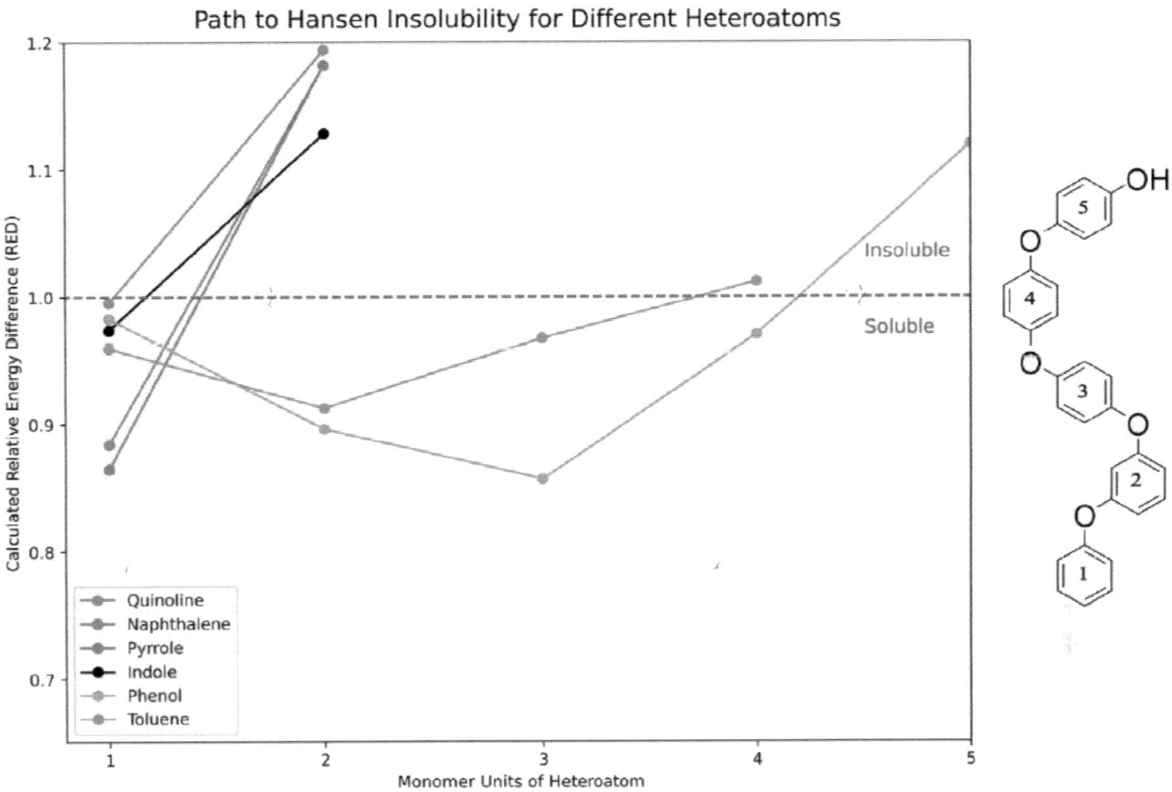

Figure 2.3: Effect of oligomer unit size on solubility calculated using the HSPiP software. It is clear some heteroatoms reach insolble threshold in fewer units than others. Phenol and toluene is particularly intersting, becoming initially more soluble in *n*-dodecane before reaching the threshold. For phenol this is due to the growth of oligomer leading to a lower proportion of the molecule containing the *H*-bonding –OH group, but as the molecule grows the Van Der Waals dD difference grows eventually leading to an insoluble.

3.3 Predicting Deposition formation using DFT

3.3.1 H-abstraction and Coupling Mechanisms for Each Heteratom

Based on the solubility modelling presented in Figure 3.3, it was clear that the coupling of fuel heteratoms would lead to insoluble formation in our *n*-dodecane surrogate. Following the proposed HAS pathway, we calculated the H-abstraction and subsequent barriers to form dimers via the HAS pathway. The barriers for these reactions are presented in detail in Table S.1 of the appendix. The termination reaction was barrierless for all the species, and the pre-exponential factor was kept constant for each of the species at 3×10^9.[5]

The H-abstraction site and the coupling site for each species was selected based on literature and our own testing, and is summarized in Figure 3.4. For phenol, H-abstraction occurs at the O-site, with coupling between the subsequent phenoxy radical and the para-carbon site of a phenol.[38] For naphthalene, the C3 site was favored over the C2 position in terms of both abstraction and coupling barrier heights. For quinoline, the C2 site is the most favored site for coupling and H-abstraction, and the barriers for abstraction were lower at the C2 site compared to the C3 site.[44] For toluene, the para position was selected for coupling and H-abstraction based on our own testing and literature data.[45]

For indole and pyrrole, the favored coupling and H-abstraction site were different. For indole, the N1 site was found to be the most favorable site for H-abstraction, but the C3 site is the most favored for coupling. Nevertheless, C–N linkages are detected very rarely and are found to be thermodynamically prohibited for indole oligomers.[46, 47] As a consequence, we compared the overall pathway to dimers at the C3 position via H-abstraction at the N1 and C3 positions. The results for the indole dimer formation calculations are found in Figure S.4 of the SI. Overall, the C3 H-abstraction pathway shows the highest barriers. Therefore, the N1 H-abstraction pathway for indole was chosen. Similarly, for pyrrole, H-abstraction at the N1 position is favored, yet polypyrrole is formed of C2-C2 linkages.[40] Comparing both pyrrole coupling C2 coupling pathways with H-abstraction at the N1 or C2 position, presented in Figure S.3, both pathways have similar barrier heights. Nevertheless, the initial H-abstraction reaction at the N1 position is more favorable, and leads to an intermediate I1b *G* 24.49kcalmol ^{1} lower in energy that the C2 pathway. As a consequence, following the N1 pathway, pyrrole would have antioxidant properties, which is reflected in the petroOxy measurements (Figure 3.6b). Therefore the N1 abstraction pathway for pyrrole was chosen.

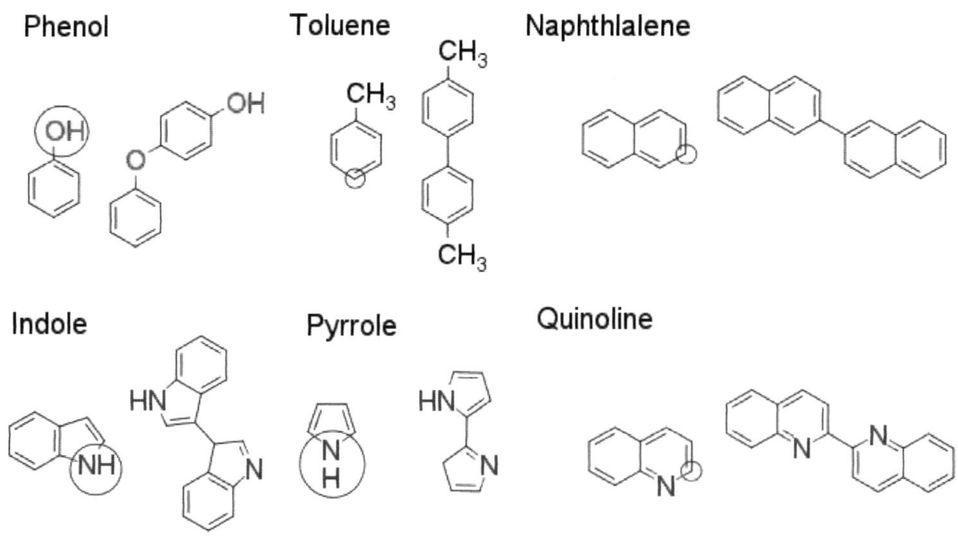

Figure 3.4: H-abstraction Sites and Dimers Selected for the Pseudo-Detailed Mechanisms. The selected H-abstraction site is circled, and resultant dimer is shown for each species.

Based on the above proposed coupling and H-abstraction sites, we computed the barriers for AH species reacting with *n*-dodecane R·,RO·, and ROO· radicals and subsequently forming dimers via a HAS process. The Gibbs potential energy surfaces for our chosen heteroatoms undergoing H-abstraction to the formation of dimers are presented in Figure 3.3 for RO·. The values for the reactions with the other radicals are presented in Table S.1 of the SI.

The first section (reaction A) of Figure 3.5, showing the abstraction of hydrogen from our selected AH species. Here, we can see there is a clear difference in the reactivity and thermochemistry. To begin with, phenol shows the lowest barriers for hydrogen abstraction out of all the species tested. Phenol has long been known as a powerful antioxidant in fuels.[5] By contrast, toluene exhibits the highest barrier and the overall Gibbs energy abstraction reaction is endergonic at +28.32kcalmol 1. The overall order of the Gibbs energy barrier to hydrogen abstraction (section A) is, in descending order: toluene > naphthalene > quinoline > pyrrole > indole > phenol.

The resultant radicals formed indicate that phenol, pyrrole, and indole all exhibit antioxidant properties, by undergoing an exergonic H-abstraction reactions. By contrast, naphthalene, toluene, and quinoline undergo endergonic H-abstraction reactions showing no antioxidant qualities. At this point in the reaction process, two A· radicals can terminate to form a dimer. Nevertheless, this termination process is a rare occurrence due to the low-concentration of A· species in fuel (for example, for pyrrole the peak concentration in our mechanism was 5.17E-13 moll 1).

The HAS reaction between an A· and AH first proceeds with an attack of the radical. The radical attack is shown as pathway B) in Figure 3.5. For this step, the antioxidant species formation is endergonic and proceeds with high Gibbs energy barriers for phenol and indole of ‡*G* 25.63kcalmol 1 and ‡*G*=24.36kcalmol 1 respectively. By contrast, pyrrole proceeds with a lower barrier of ‡*G*=19.20kcalmol 1. On the other hand, the non-antioxidant species show lower barriers for pathway B). In fact, naphthalene and toluene show the highest barriers for the H-abstraction step, but lowest barriers out of all the species for step B) at ‡*G*=12.87kcalmol 1 and ‡*G*=13.50kcalmol 1 respectively. Additionally, the formation of the intermediate is the most exergonic for naphthalene and toluene. The relationship between the antioxidant tendency and the ease of coupling is likely related to the aromatic structure. Species with more antioxidant qualities will also have higher levels of aromaticity to stabilize radicals, but also will have higher barriers to aromaticity-breaking radical attack.

The final step in the dimer forming process is the re-aromatization step *via* loss of hydrogen from the intermediate. Interestingly, the barriers for this step are similar for our aromatics tested, in the range of 23.87 17.47kcalmol 1. This step is clearly less related to the structure of the AH compound. Nevertheless, the concentration of ROOH available to complete this step, is related to the antioxidant qualities of the AH species in fuels. For example, for phenols, due to their antioxidant properties, the concentration of ROOH is expected to be low.[24] As a consequence, termination reactions likely provide a higher source of C–C/C–

O coupling. To explore how these interrelated pathways yield deposits for different fuel components, we created different pseudo-detailed mechanisms to compare each fuel.

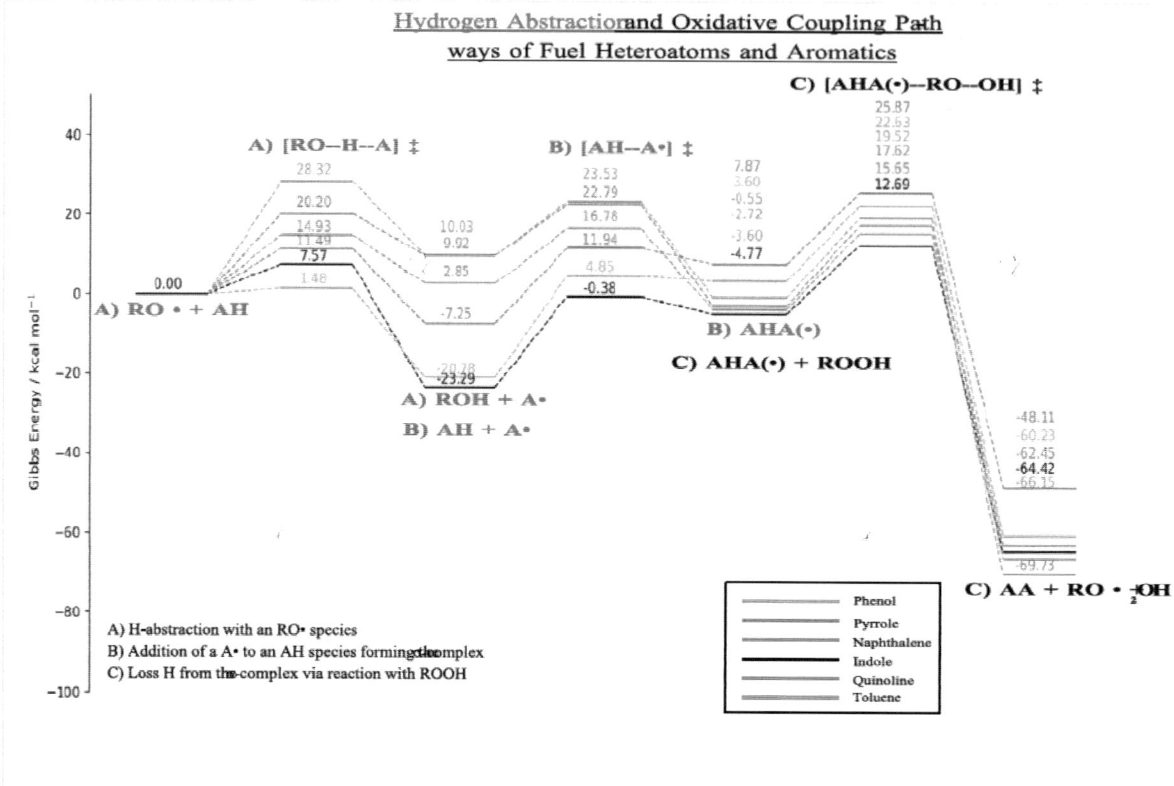

Figure 3.5: Growth to a deposit dimer for our chosen heteroatoms, calculated at the B3LYP-D3/cc-pVTZ *n*-dodecane PCM level of theory. The barriers were calculated from stable pre-reaction complexes. The R species was *n*-dodecane. The antioxidant species (phenol, indole and pyrrole) show exergonic H-abstraction in step A.

3.3.2 Comparison of Experimental Results and Psuedo-Detailed Models

Antioxidant Behavior

First, the antioxidant behavior of our surrogate fuels was explored in relation to our models. The O_2 depletion predicted by our model alongside the petroOxy depletion curves are presented in Figure 3.6. Although the petroOxy device cannot measure the O_2 depletion in the bulk, the device can be used to assess the antioxidant behavior of different fuels.[76] Our model successfully predicts that pyrrole, indole, and phenol are all antioxidant species, as shown by the increased induction period in thepetroOxy curves. In addition, the petroOxy device indicates that indole and phenols are stronger antioxidants than pyrrole, which is also reflected in the O_2 depletion curves in our mechanism. However, the difference between indole and phenol in thepetroOxy curves is not reflected in the mechanism. Nevertheless, this difference between the surrogates is better reflected in the AH + RO· abstraction barrier in Figure 7.5. Exploring the GCMS results for indole in our previous study, in Figure 10.22, no *n*-dodecane autoxidation products are observed after 8h. By contrast, pyrrole and quinoline both show lesser antioxidant qualities and lead to *n*-dodecane autoxidation products in Figure 5.7. Similarly, for our model, no oxygen is consumed leading to no autoxidation products.

Deposition Behavior

The amount of insoluble dimers predicted by our model compared with the total insolubles measured is presented in Figure 3.7. The model predicts the insoluble formation behavior between our different surrogates well, with pyrrole producing the largest amounts of insolubles. Interestingly, phenol and indole are shown to produce the lowest amount of deposit in both our experiments and model. By contrast, in real fuels, indoles, and particularly phenol concentration are shown to correlate well with final insoluble mass.[8] A key difference between our simplified surrogates and real fuels are the synergistic effects between fuel classes, which are not captured in our two-component surrogates. In particular, sulfur is shown to interact strongly with 5-membered nitrogen heterocycles, and phenols with other nitrogen compounds.[7, 48] For example, in a simplified surrogate, previous work has shown trends between deposition propensity of 5- and 6-membered heterocycles break down, where quinoline-like compounds have shown to produce more deposit than indole compounds.[7] Nevertheless, the use of surrogates is still important to understand the chemical interactions within species classes.

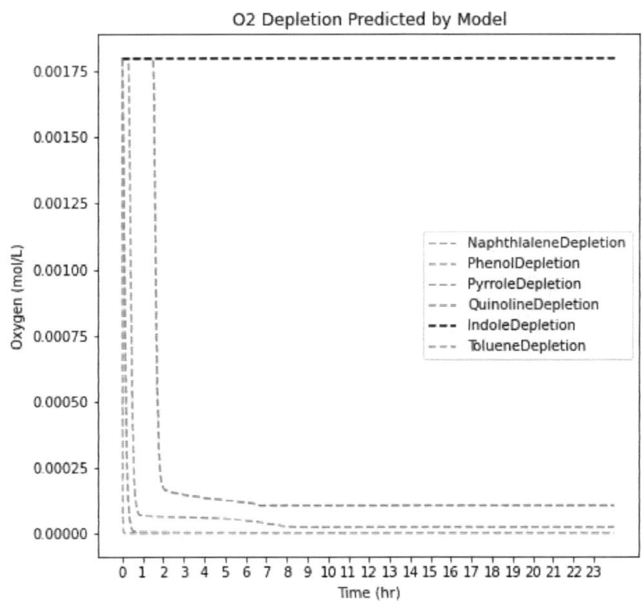

(a) O_2 Depletion Predicted by our model

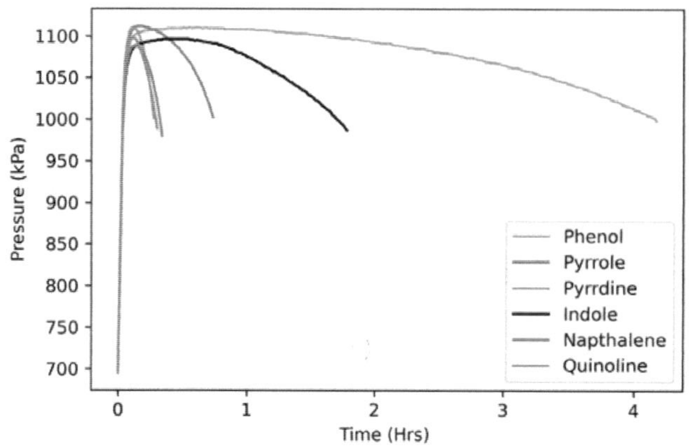

(b) Petroxy depletion behavior

Figure 3.6: O_2 depletion behavior of our surrogate- comparison between model and experiment. The model shows the change in O_2 concentration over time. The petroOxy depletion curves show the change in headspace pressure, and although they are not a direct measure of O_2 depletion, they can differentiate between antioxidant qualities of different species.

Interestingly, in our experiment and models, toluene and naphthalene produced more insolubles than phenol and indole. This is interesting as fuel thermal degradation is often viewed as solely driven by heteroatoms.[49] Exploring the barriers for toluene and naphthalene in detail in Table S.1, we see that lower barriers to the AH + A·→ AHA· reaction are found with the aromatic species. Sensitivity analysis (Figure 3.9) of our models suggests this coupling step is strongly related to the final insoluble mass. However, in a 2-component fuel, combining an antioxidant heteroatom and aromatic, the formation of aromatic insolubles will be significantly reduced due to the size of the H-abstraction barrier in comparison to the other heteroatoms used in this study (Figure 3.5). Instead, the heteroatom antioxidant class will form the majority of the A· radicals going on to form deposit. To strengthen this hypothesis, using our calculated values, a fuel model was built containing both phenol and toluene.

The effect of heteroatoms suppressing deposition from non-antioxidant species is demonstrated in Figure 3.8, where insolubles are generated from pseudo-detail model fuels containing different ratios of toluene and phenol is tested. The model was built using the same BAS scheme base described in the section 2.4, with the addition of both toluene and phenol pathways. In this simplified model, toluene-derived insolubles are suppressed upon introduction of phenol. This effect can help explain why fuel deposits/insolubles often have an elemental composition of heteroatoms higher than conventional fuels also containing aromatic hydrocarbons.[50] Another interesting effect of blending is the peak in phenol insolubles at 25:75 phenol:toluene ratios above a 100% phenol mix, which implies that above a certain concentration, phenols will prevent the formation of insolubles. The peak concentration of hydroperoxide steadily drops at different % phenol concentrations, which is related to the extent the autoxidation chain is suppressed. The peak of phenol deposition could be related to the ROOH concentration, allowing more -intermediates to be re-aromatized. This shows our model can replicate the effect of changing concentrations of antioxidant on total insolubles. Finally, our results here lend credence to a HAS pathway leading to deposit. When the HAS pathway is removed from the mechanism in Figure 3.7b, no discernable trend is found between the formation of dimers and insolubles measured. In fact, the HAS pathway appears particularly important for species with low/no antioxidant tendency, but with a tendency to still form insolubles.

Sensitivity analysis of our bespoke mechanisms in presented in Figure 3.9. The reaction numbers cited here correspond to those given in Table S.1. For our mechanisms, the reaction 29, A· + AH → AHA·, has the largest influence on the level of deposit with the exception of phenol. Increasing the rate of reaction 29 leads to more AHA· species which can readily form deposit. Reaction 32, A· + ROOH → AH + ROO·, removing hydroperoxides from the system also has an influence on the level of deposit. As the rate of reaction 32 increases, hydroperoxides are removed from the system leading to fewer AHA· +ROOH re-aromatization reactions leading to deposit dimers. The change of ROOH and insoluble concentration presented in Figure 3.10 demonstrates that as insolubles are produced ROOH is depleted. For all the species tested, the RO· (reaction 26) and R· (reaction 27) hydrogen abstraction steps have a large influence on the level of deposit, with faster rates leading to more A· species able to undergo coupling reactions. By contrast, the rate of the ROO· abstraction barrier (reaction 16) has a negligible influence on deposition for most species because this reaction in general has high barriers already.

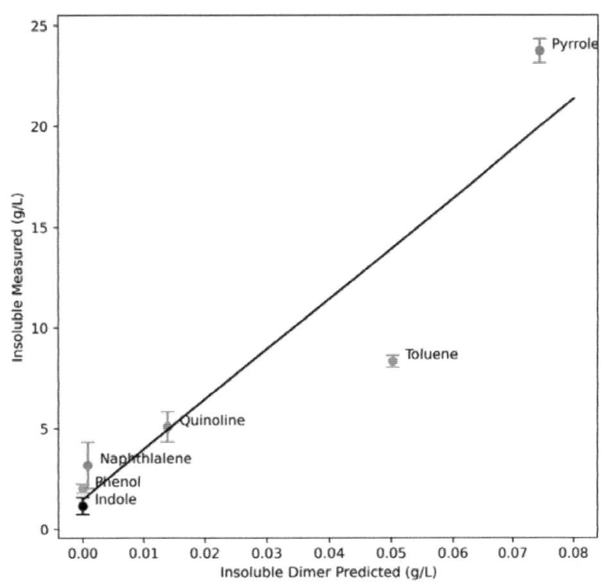

(a) Correlation between mass of dimer predicted by our model and insolubles measured.

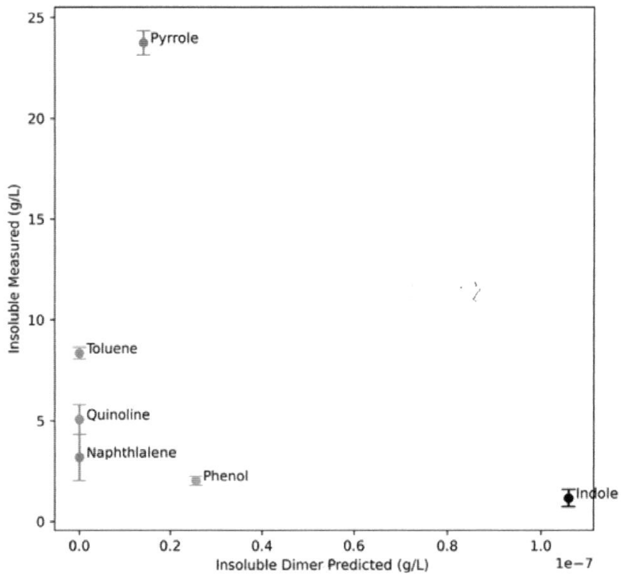

(b) Correlation between mass of dimer predicted by our model and insolubles measured when the HAS pathway is excluded.

Figure 3.7: Insoluble formation behavior- comparisons between model and experiment. Temperature of the experiment and psuedo-detailed *n*-dodecane chemical mechanism containing the A and E_a values shown in Table S.1. The temperature was set at 431 K both the model and the experiment.

Finally, for phenol, reaction 26, AH + R→ A· + RH also influences the formation of deposit more than the other species, and is more important than the coupling step (reaction 29). This is likely because the termination pathway dominates the formation of deposits here, thus leading to a direct correlation between the concentration of A· and the amount of deposit dimer.

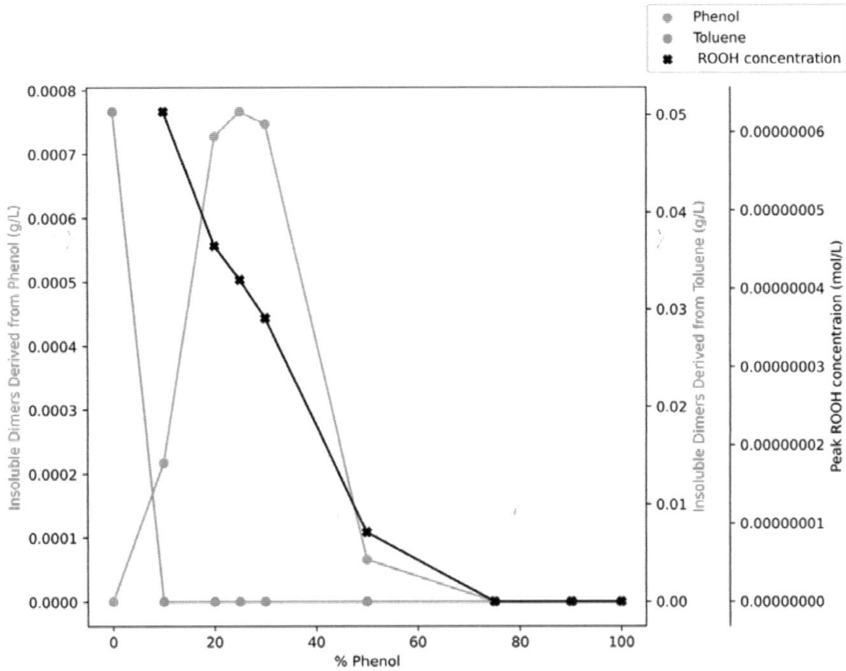

Figure 3.8: Effect of phenol % in a phenol toluene aromatic total of 0.1molL[1] on the insolubles derived from phenol and toluene.

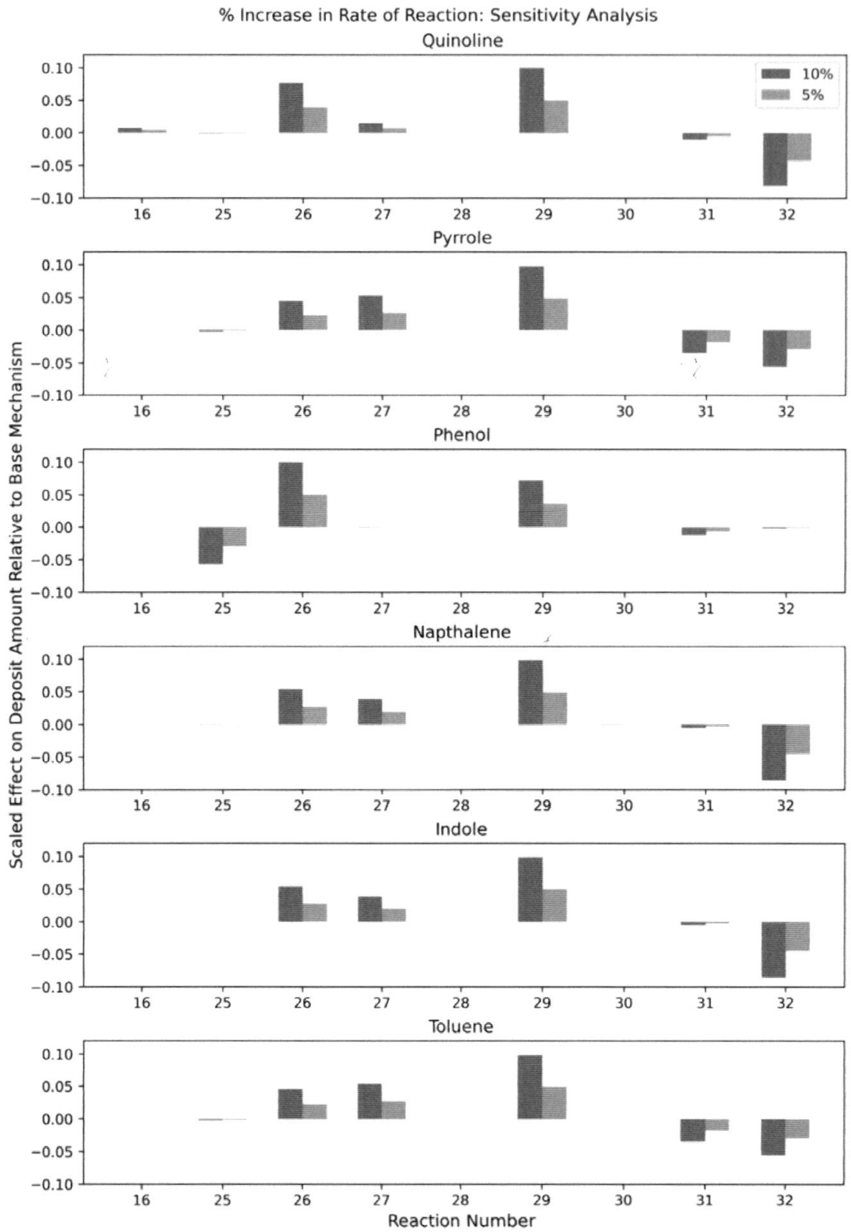

Figure 3.9: Sensitivity analysis of the bespoke insoluble pseudo-detailed mechanisms. For the sensitivity analysis, the rate of reaction is increased by a factor of 10% (blue) and 5% (orange) and the resultant increase in insolubles is shown relative to the base pseudo-detailed mechanism. The reaction number corresponds to Table S.1 in the SI

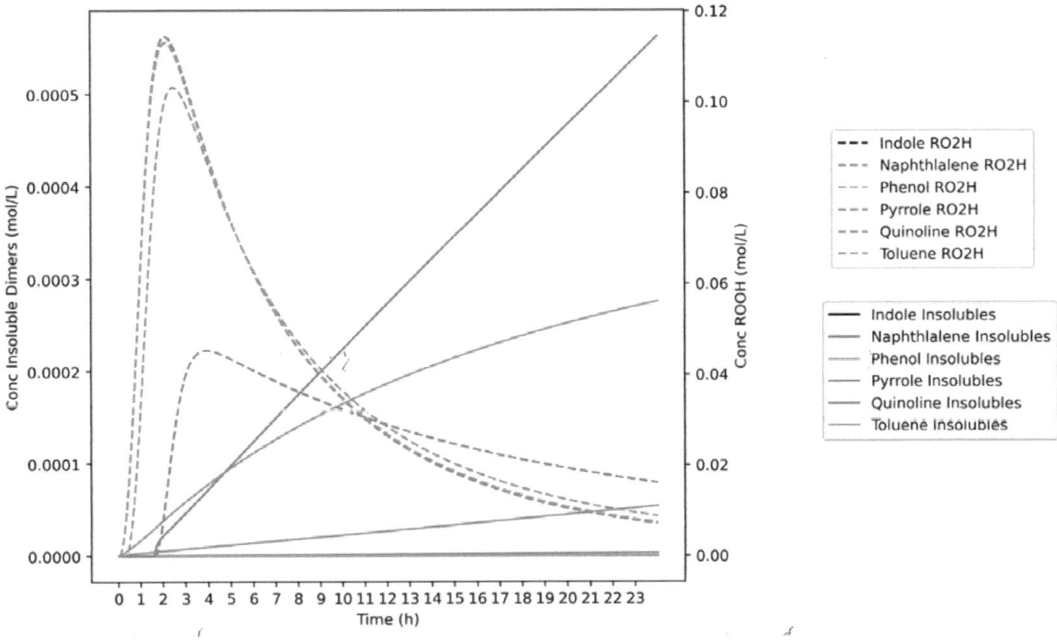

Figure 3.10: Change in the concentration of insolubles and ROOH over the course of the 24h pseudo-detailed models. The thick lines show the change in insoluble dimer concentration, whereas the dotted lines show the change in ROOH concentration. The results clearly show that for each species, as ROOH is depleted, the level insolubles begin to increase.

3.4 Implication for Fuels

The formation of fuel insolubles is influenced by both the H-abstraction and coupling propensity of the heteroatoms/aromatics present in fuel. By simplifying the insoluble formation process into the formation of dimers, we were able to show clear differences between the insoluble formation and antioxidant tendency between surrogate fuels. Nevertheless, the formation of dimers represents a small subsection of the insoluble/deposit structure. To expand the mechanism to reach an insolubles formation model which can provide a quantitatively accurate description, additional pathways need to be considered. This is a challenge, particularly as there are likely thousands of possible side reactions between and within species classes.[2] Even in our surrogate fuels, side reactions are likely to occur between AH and RH species, where the oxidation of *n*-dodecane is not significantly repressed.

To address this challenge, several approaches using techniques benchmarked in this study can be used. The first approach involves scaling the deposit forming dimer steps to experimental deposit measurements. This would lead to a semi-empirical mechanism, which moves away from the first-principles approach using DFT. Nevertheless, the DFT thermochemical and kinetic parameters will allow the gross differences between species class and structure to be explored. This approach relies on the assumption that the formation of deposit dimers relates to a general property of 'coupling propensity'. Some preliminary calculations imply this assumption is generally correct, where a propanal doublet attacking various nitrogen compounds give barrier sizes (presented in Figure S.5 in the SI) in the same order of magnitude as the dimerization barriers for step B) in Figure 3.5.

A second, more computationally intensive approach, would be a set of high throughput calculations on additional deposit forming pathways. This approach would use HAS and termination reactions to calculate the pathways to various oligomers. Hansen solubility parameters will guide a 'cut-off' point for energetic pathways to oligomers to form insolubles. Nevertheless, insoluble species are still likely to react to form larger molecular weight species, which adds an additional challenge to this approach.

4.5 **CONCLUSIONS**

The formation of insolubles in fuel is driven by coupling reactions between aromatics. In this work we have shown the importance of HAS in explaining the formation of these insolubles. DFT calculations revealed that the hydroperoxides play a key role in facilitating the HAS pathway. As a result of these findings, we showed that kinetic values calculated for a simplified two-component (dodecane and aromatic) surrogate can reveal differences in antioxidant and deposition tendency of different aromatics compared to the experiment. In addition, when the HAS pathway is excluded, the predicted deposition trends break down. Exploring the behavior of the mechanism further, we show that the addition of antioxidant to an antioxidant+aromatic model leads the suppression of insolubles composed of the aromatic. Instead, the antioxidant+aromatic model leads to insolubles only composed of the antioxidant. Finally, sensitivity analysis reveals that the formation of hydroperoxides (AH + ROO→ A· +ROOH) and the homolytic coupling (AH+A→AHA·) have a strong influence on the final deposit mass.

BIBLIOGRAPHY

1- T. Jia, L. Pan, S. Gong, J. Xie, X. Wang, Y. Fang, J. J. Zou, and X. Zhang, "Mechanistic insights into the thermal deposition of highly thermal-stable jet fuel," *Fuel*, vol. 276, no. May, p. 118100, 2020.

2- M. Commodo, I. Fabris, C. P. T. Groth, and Ö. L. Gülder, "Analysis of aviation fuel thermal oxidative stability by electrospray ionization mass spectrometry (ESI-MS)," *Energy and Fuels*, vol. 25, no. 5, pp. 2142–2150, 2011.

3- S. P. Heneghan and S. Zabarnick, "Oxidation of jet fuels and the formation of deposit," *Fuel*, vol. 73, pp. 35–43, jan 1994.

4- F. R. Mayo and B. Y. Lan, "Gum and deposit formation from jet turbine and diesel fuels at 130.degree.C," *Industrial Engineering Chemistry Product Research and Development*, vol. 25, pp. 333–348, jun 1986.

5- N. J. Kuprowicz, S. Zabarnick, Z. J. West, and J. S. Ervin, "Use of measured species class concentrations with chemical kinetic modeling for the prediction of autoxidation and deposition of jet fuels," *Energy and Fuels*, vol. 21, no. 2, pp. 530–544, 2007.

6- R. H. Clark and L. Smith, "Further studies of the effects of polar compounds on the thermal stability of jet fuel," in *3rd International Conference on Stability And Handling of Liquid Fuels*, no. 2, 1988.

7- S. Zabarnick, Z. J. West, L. M. Shafer, S. S. Mueller, R. C. Striebich, and P. J. Wrzesinski, "Studies of the role of heteroatomic species in jet fuel thermal stability: model fuel mixtures and real fuels," *Energy and Fuels*, vol. 33, no. 9, pp. 8557–8565, 2019.

8- L. M. Balster, S. Zabarnick, R. C. Striebich, L. M. Shafer, and Z. J. West, "Analysis of polar species in jet fuel and determination of their role in autoxidative deposit formation," *Energy and Fuels*, vol. 20, no. 6, pp. 2564–2571, 2006.

9- M. Grzybowski, K. Skonieczny, H. Butenschön, and D. T. Gryko, "Comparison of oxidative aromatic coupling and the scholl reaction," *Angewandte Chemie - International Edition*, vol. 52, no. 38, pp. 9900–9930, 2013.

10- P. D. McDonald and G. A. Hamilton, *Mechanisms of Phenolic Oxidative Coupling Reactions*, vol. 5. Academic Press, INC., 1973.

11- W. R. Bowman and J. M. Storey, "Synthesis using aromatic homolytic substitution—recent advances," *Chemical Society Reviews*, vol. 36, no. 11, pp. 1803–1822, 2007.

12- C.-L. Sun and Z.-J. Shi, "Transition-metal-free coupling reactions," *Chemical Reviews*, vol. 114, no. 18, pp. 9219–9280, 2014. PMID: 25184859.

13- M. Monge-Palacios, E. Grajales-González, G. Kukkadapu, and S. M. Sarathy, "Kinetics of the benzyl + HO2and benzoxyl + OH barrierless association reactions: fate of the benzyl hydroperoxide adduct under combustion and atmospheric conditions," *Physical Chemistry Chemical Physics*, vol. 22, no. 16, pp. 9029–9039, 2020.

14- F. Minisci, "Recent aspects of homolytic aromatic substitutions.," *Topics in current chemistry*, vol. 62, pp. 1–48, 1976.

15- B. Beaver, L. Gao, C. Burgess-Clifford, and M. Sobkowiak, "On the mechanisms of formation of thermal oxidative deposits in jet fuels. Are unified mechanisms possible for both storage and thermal oxidative deposit formation for middle distillate fuels?," *Energy and Fuels*, vol. 19, no. 4, pp. 1574–1579, 2005.

16- T. Stuyver, D. Danovich, F. De Proft, and S. Shaik, "Electrophilic Aromatic Substitution Reactions: Mechanistic Landscape, Electrostatic and ElectricField Control of Reaction Rates, and Mechanistic Crossovers," *Journal of the American Chemical Society*, vol. 141, no. 24, pp. 9719–9730, 2019.

17- D. P. Curran and A. I. Keller, "Radical additions of aryl iodides to arenes are facilitated by oxidative rearomatization with dioxygen," *Journal of the American Chemical Society*, vol. 128, no. 42, pp. 13706–13707, 2006.

18- R. Venkataraman and S. Eser, "Characterisation of solid deposits from the thermal-oxidative degradation of jet fuel," *International Journal of Oil, Gas and Coal Technology*, vol. 1, no. 1-2, pp. 126–137, 2008.

19- O. Altin and S. Eser, “Carbon deposition from thermal stressing of petroleum fuels,” *ACS National Meeting Book of Abstracts*, vol. 228, no. 1, 2004.

20- Z. Liu, S. Tang, Z. Li, Z. Qin, S. Yuan, L. Wang, L. Wang, X. Zhang, and G. Liu, “An improved kinetic model for deposition by thermal oxidation of aviation hydrocarbon fuels,” *Fuel*, vol. 258, no. June, p. 116139, 2019.

21- C. M. Parks, A. J. Meijer, S. G. Blakey, E. Alborzi, and M. Pourkashanian, “Computational studies on the reactions of thiols, sulfides and disulfides with hydroperoxides. Relevance for jet fuel autoxidation,” *Fuel*, vol. 316, no. January, p. 123326, 2022.

22- E. Alborzi, M. R. Dwyer, C. M. , A. Sheikhansari, D. C. Mielczarek, M. Zanganeh, A. J. Meijer, S. G. Blakey, and M. Pourkashanian, “Construction of a reduced chemical kinetic mechanism for autoxidation of n-paraffinic solvent – A model for aviation fuel,” *Fuel*, vol. 294, no. January, p. 120170, 2021.

23- A. C. Antoine, “EFFECT OF SOME NITROGEN COMPOUNDS ON THERMAL STABILITY OF JET A.,” *NASA Technical Memorandum*, jun 1982.

24- L. Jones and N. C. Li, “Ageing of SRC II middle distillate from Illinois No. 6 coal,” *Fuel*, vol. 62, no. 10, pp. 1156–1160, 1983.

25- R. N. Hazlett and A. J. Power, “Phenolic compounds in Bass Strait distillate fuels: their effect on deposit formation,” *Fuel*, vol. 68, no. 9, pp. 1112–1117, 1989.

26- T. Jia, M. Zhao, L. Pan, C. Deng, J.-j. Zou, and X. Zhang, “Effect of phenolic antioxidants on the thermal oxidation stability of high-energy-density fuel,” *Chemical Engineering Science*, p. 117056, 2021.

27- John M. Andrésen, James J. Strohm, , Lu Sun, and C. Song*, “Relationship between the Formation of Aromatic Compounds and Solid Deposition during Thermal Degradation of Jet Fuels in the Pyrolytic Regime,” 2001.

28- S. Jain and M. P. Sharma, “Review of different test methods for the evaluation of stability of biodiesel,” *Renewable and Sustainable Energy Reviews*, vol. 14, no. 7, pp. 1937–1947, 2010.

29- M. J. Frisch, G. W. Trucks, H. B. Schlegel, G. E. Scuseria, M. A. Robb, J. R. Cheeseman, G. Scalmani, V. Barone, B. Mennucci, G. A. Petersson, H. Nakatsuji, M. Caricato, X. Li, H. P. Hratchian, A. F. Izmaylov, J. Bloino, G. Zheng, J. L. Sonnenberg, M. Hada, M. Ehara, K. Toyota, R. Fukuda, J. Hasegawa, M. Ishida, T. Nakajima, Y. Honda, O. Kitao, H. Nakai, T. Vreven, J. A. Montgomery, Jr., J. E. Peralta, F. Ogliaro, M. Bearpark, J. J. Heyd, E. Brothers, K. N. Kudin, V. N. Staroverov, R. Kobayashi, J. Normand, K. Raghavachari, A. Rendell, J. C. Burant, S. S. Iyengar, J. Tomasi, M. Cossi, N. Rega, J. M. Millam, M. Klene, J. E. Knox, J. B. Cross, V. Bakken, C. Adamo, J. Jaramillo, R. Gomperts, R. E. Stratmann, O. Yazyev, A. J. Austin, R. Cammi, C. Pomelli, J. W. Ochterski, R. L. Martin, K. Morokuma, V. G. Zakrzewski, G. A. Voth, P. Salvador, J. J. Dannenberg, S. Dapprich, A. D. Daniels, Ö. Farkas, J. B. Foresman, J. V. Ortiz, J. Cioslowski, and D. J. Fox, “Gaussian 09 Revision E.01,” 2009. Gaussian Inc. Wallingford CT 2009.

30- A. D. Becke, "Density-functional thermochemistry. III. The role of exact exchange," *The Journal of Chemical Physics*, vol. 98, no. 7, pp. 5648–5652, 1993.

31- S. Grimme, S. Ehrlich, and L. Goerigk, "Effect of the damping function in dispersion corrected density functional theory," *Journal of computational chemistry*, vol. 32, no. 7, pp. 1456–1465, 2011.

32- S. Miertuš, E. Scrocco, and J. Tomasi, "Electrostatic interaction of a solute with a continuum. a direct utilizaion of ab initio molecular potentials for the prevision of solvent effects," *Chemical Physics*, vol. 55, no. 1, pp. 117–129, 1981.

33- T. H. Dunning Jr, "Gaussian basis sets for use in correlated molecular calculations. i. the atoms boron through neon and hydrogen," *The Journal of chemical physics*, vol. 90, no. 2, pp. 1007–1023, 1989.

34- S. Grimme, "Supramolecular binding thermodynamics by dispersion-corrected density functional theory," *Chemistry–A European Journal*, vol. 18, no. 32, pp. 9955–9964, 2012.

35- R. F. Ribeiro, A. V. Marenich, C. J. Cramer, and D. G. Truhlar, "Use of solution-phase vibrational frequencies in continuum models for the free energy of solvation," *The Journal of Physical Chemistry B*, vol. 115, no. 49, pp. 14556– 14562, 2011.

36- R. Paton, "Goodvibes," 2016.

37- S. Zabarnick, "Chemical kinetic modeling of jet fuel autoxidation and antioxidant chemistry," *Industrial Engineering Chemistry Research*, vol. 32, pp. 1012–1017, jun 1993.

38- S. Kobayashi and H. Higashimura, "Oxidative polymerization of phenols revisited," *Progress in Polymer Science (Oxford)*, vol. 28, no. 6, pp. 1015–1048, 2003.

39- S. Baena-Zambrana, S. L. Repetto, C. P. Lawson, and J. K. Lam, "Behaviour of water in jet fuel - A literature review," *Progress in Aerospace Sciences*, vol. 60, pp. 35–44, 2013.

40- J. K. Howard, K. J. Rihak, A. C. Bissember, and J. A. Smith, "The Oxidation of Pyrrole," *Chemistry - An Asian Journal*, vol. 11, no. 2, pp. 155–167, 2016.

41- Z. Zhang, X. Yue, Y. Duan, and Z. Rao, "A study on the mechanism of oxidized quinoline removal from acid solutions based on persulfate-iron systems," *RSC Advances*, vol. 10, no. 21, pp. 12504–12510, 2020.

42- J. D. Mosley, A. M. Ricks, P. V. Schleyer, J. I. Wu, and M. A. Duncan, "IR spectroscopy of α- And -protonated pyrrole via argon complex photodissociation," *Journal of Physical Chemistry A*, vol. 116, no. 39, pp. 9689–9695, 2012.

43- M. J. DeWitt, Z. West, S. Zabarnick, L. Shafer, R. Striebich, A. Higgins, and T. Edwards, "Effect of Aromatics on the Thermal-Oxidative Stability of Synthetic Paraffinic Kerosene," *Energy & Fuels*, vol. 28, pp. 3696–3703, jun 2014.

44- S. Zabarnick, "Chemical kinetic modeling of jet fuel autoxidation and antioxidant chemistry," *Industrial Engineering Chemistry Research*, vol. 32, pp. 1012–1017, jun 1993.

45- . Wang, A. Violi, D. H. Kim, and J. A. Mullholland, "Formation of naphthalene, indene, and benzene from cyclopentadiene pyrolysis: A DFT study," *Journal of Physical Chemistry A*, vol. 110, no. 14, pp. 4719–4725, 2006.

46- J. Vercammen, M. Bocus, S. Neale, A. Bugaev, P. Tomkins, J. Hajek, S. Van Minnebruggen, A. Soldatov, A. Krajnc, G. Mali, V. Van Speybroeck, and D. E. De Vos, "Shape-selective C–H activation of aromatics to biarylic compounds using molecular palladium in zeolites," *Nature Catalysis*, vol. 3, no. 12, pp. 1002–1009, 2020.

47- P. J. Steel, "Aromatic biheterocycles: Syntheses, structures, and properties," *Advances in heterocyclic chemistry*, vol. 67, pp. 2–118, 1997.

48- W. E. Harrison, H. C. Mongia, S. P. Heneghan, and D. R. Ballal, "Advanced jet fuels - JP-4 through JP-8 and beyond," *Proceedings of the ASME Turbo Expo*, vol. 3, 1995.

49- R. N. Hazlett and A. J. Power, "Phenolic compounds in Bass Strait distillate fuels: their effect on deposit formation," *Fuel*, vol. 68, no. 9, pp. 1112–1117, 1989.

50- R. N. Hazlett, J. A. Schreifels, W. M. Stalick, R. E. Morris, and G. W. Mushrush, "Distillate fuel insolubles: formation conditions and characterization," *Energy and Fuels*, vol. 5, no. 2, pp. 269–273, 1991.

SUPPORTING INFORMATION

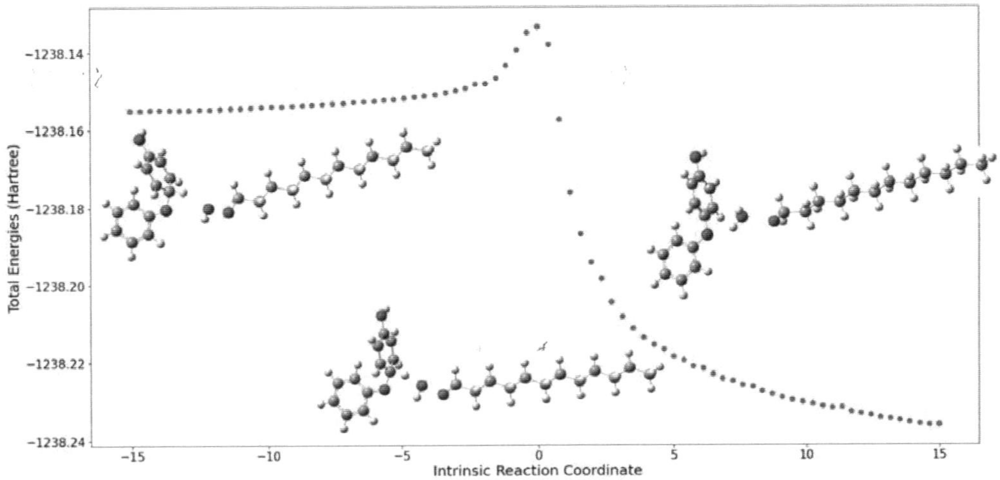

Figure S.1: IRC calculation showing the transition state of the ROOH rearomatization of the phenol -intermediate. In this case, R refers to a dodecane moiety. The transition state yields a phenol dimer, an RO· and H_2O.

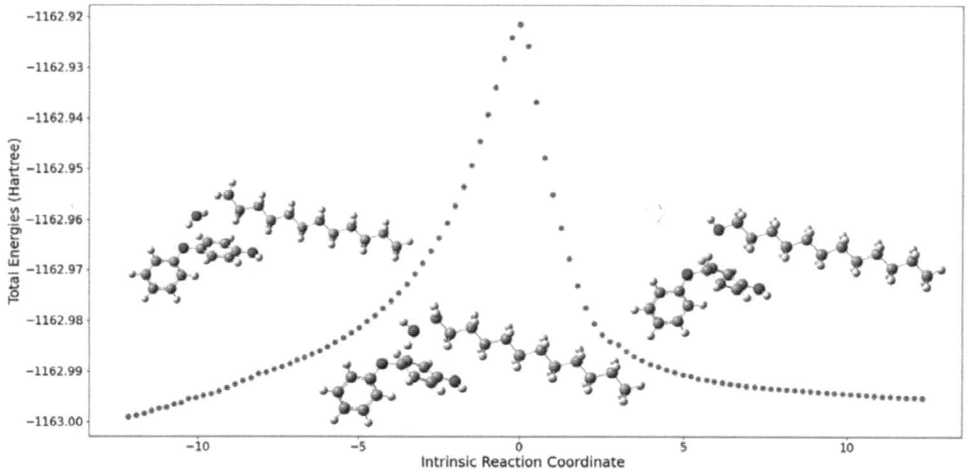

Figure S.2: IRC calculation showing the transition state of the ROH rearomatization of the phenol -intermediate. In this case, R refers to a dodecane moeity. The transition state yields a phenol dimer, an R· and H_2O.

Table S.1: All kinetic data calculated at the B3lYP-D3//cc-pVTZ level of theory for the reaction mechanisms added to the BAS scheme. For bimolecular reactions, the units of the pre-exponential factor A are lmol ^{1} s ^{1} for bimolecular reactions and s ^{1} for unimolecular reactions. The units of E_a are kcalmol ^{1}.

		Indole		Pyrrole		Toluene		Phenol		Quinoline		Naphthalene	
Reaction Number	Reaction	A	E_a	A	E_a	A	E_a	A	E_a	A	E_a	A	E_a
16	AH + ROO· !	2.57E+12	10.87	1.56E+13	18.23	7.41E+12	29.09	2.67E+12	8.94	1.3581E+13	25.13	7.80E+12	23.39
25	A· + ROOH	9.80E+08	20.62	2.67E+12	20.86	7.14E+12	7.08	1.72E+12	21.10	8.6083E+12	12.34	7.64E+12	8.50
26	A· + RH ! AH + R· AH + R· ! A· + RH	1.76E+12	10.95	2.09E+12	16.53	3.94E+12	19.29	8.61E+12	8.24	4.5206E+12	17.62	5.50E+12	20.98
27	AH+ RO· ! A· + ROH	6.69E+12	2.97	2.22E+12	5.68	3.59E+12	11.14	3.78E+12	0.29	9.7775E+12	10.33	1.02E+13	14.76
28	A· + A· ! Insoluble	3.00E+09	0.00	3.00E+09	0.00	3.00E+09	0.00	3.00E+09	0.00	3.00E+09	0.00	.00E+09	0.00
29	AH + A· ! AHA·	2.42E+12	16.67	2.7758E+08	6.29	1.02E+13	5.31	1.82E+12	20.05	4.57E+12	8.49	5.30E+12	6.62
30	AHA· + RO_2H !	5.84E+08	5.50	9.28E+08	0.077	3.75E+08	7.55	4.30E+08	7.86	2.30E+08	7.45	4.78E+08	11.65

	Insoluble + RO· + H_2O												
31	A· +ROH ! AH+ RO·	7.04E+12	12.60	3.38E+12	12.97	6.81E+12	2.62	2.36E+12	15.94	8.60E+12	5.79	6.85E+12	4.19
32	A· + ROOH ! AH + ROO·	5.09E+12	5.97	6.24E+12	7.09	2.30E+12	0.00	4.99E+12	8.49	8.60E+12	1.35	7.70E+12	0.48

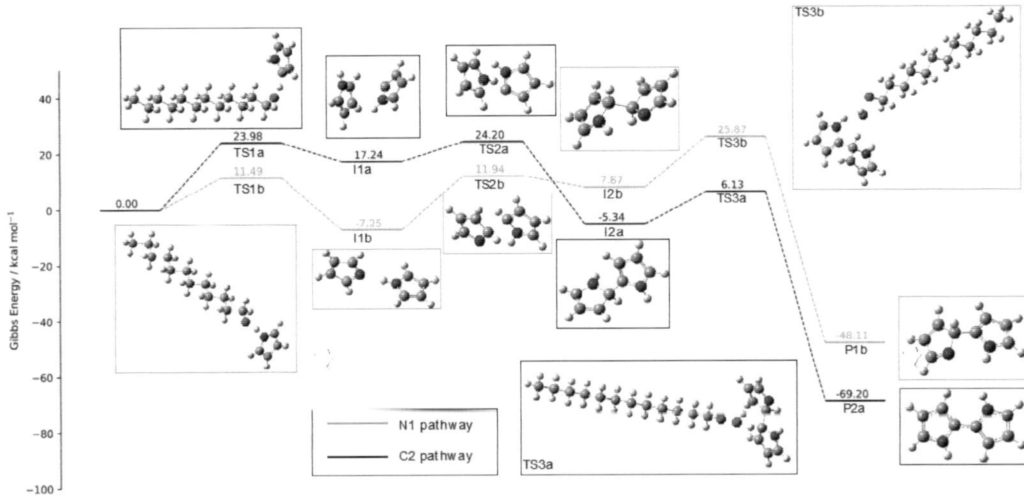

Figure S.3: Potential energy surface for the pyrrole coupling at the C2 position, comparing H-abstraction at the N1 or C2 position

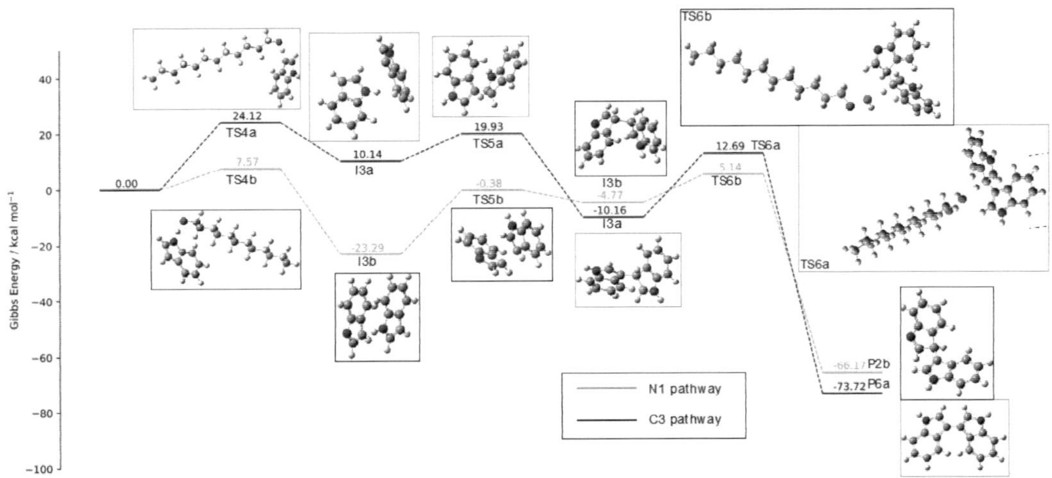

Figure S.4: Potential energy surface for the indole coupling at the C3 position, comparing H-at the N1 or C3 position

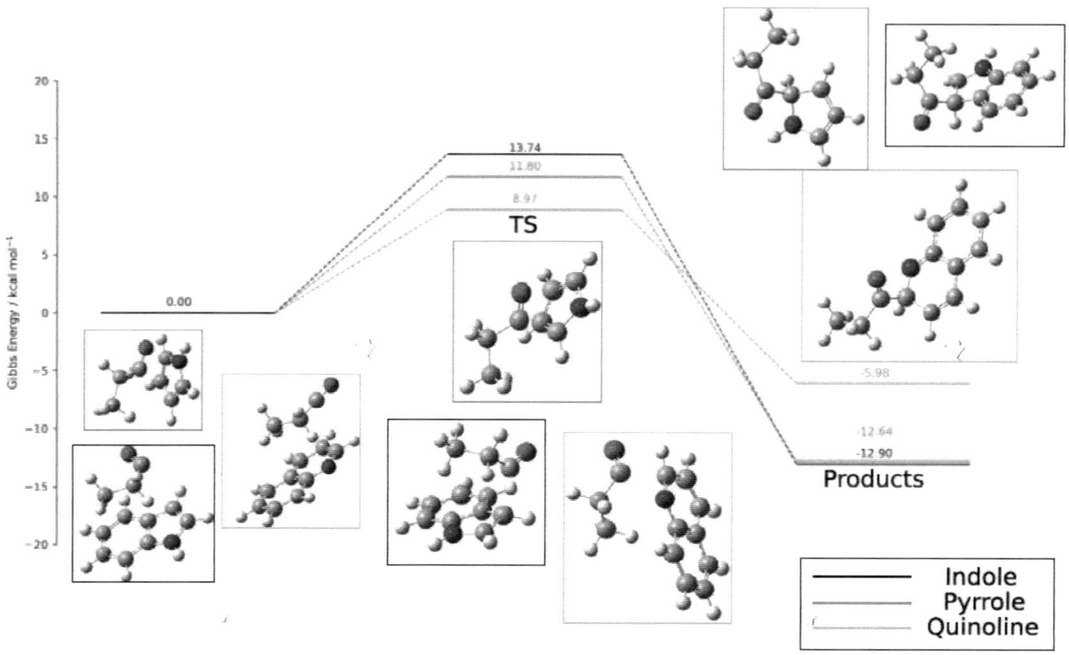

Figure S.5: Propanal doublet HAS attack step pathway calculated at the B3lYPD3//cc-pVTZ level. The barriers heights are in the order from largest to smallest; indole > pyrrole > quinoline, which is also reflected in the A· +AH barrier heights.

IASH 2022, the 17th INTERNATIONAL SYMPOSIUM ON STABILITY, HANDLING AND USE OF LIQUID FUELS
Dresden, Germany
September 11 - 15, 2022

COMPARISON OF SELECTIVITY/REACTIVITY OF ZEOLITE COATED MONOLITHS ON IMPROVEMENT OF AVIATION FUEL THERMAL STABILITY

Ehsan Alborzi[1]*, Abby Samson[1], Simon Blakey[2], Mohamed Pourkashanian[1]

[1]University of Sheffield, Sheffield S10 2TN, United Kingdom
[2]University of Birmingham, Edgbaston, Birmingham B15 2TT United Kingdom

ABSTRACT

Strong selective adsorption of a number of heteroatomic species, as well as partial removal of dissolved O2 was observed upon treatment of aAviationa Jet A-1 fuel sample treatment with zeolite 3.7Å (also known as chabazite). exhibited a strong selective adsorption of a number of heteroatomic species, as well as partial removal of dissolved O2 . This is due to the crystalline structure of zeolites with which exhibit a network of cavities and pores, which collectively offerings a preferential adsorption for the molecules smaller than the size of pore diameters. Another important criterion that determines the nature of adsorption of species by zeolites is the ratio of Si/Al. Experimental evidence supported by quantum chemistry calculations suggest that the interaction of dissolved O2 with chabazite is of a physisorption nature, as opposed to the underlying process of oxygenated and other polar species removal by chabazite which is dominantly chemisorption.

Chabazite monoliths, coated with three different Si/Al ratios were used in a lab-scale packed bed reactor, to compare the effect of selectivity and reactivity of the sorbets on improvement of the thermal oxidative stability of the Jet A-1 fuel. Three types of chabazite coated monolith with different Si/Al ratio w ere used in a lab-scale packed bed reactor to compare the effect of selectivity and reactivity of the sorbents on the improvement of thermal oxidative of a Jet A-1 fuel sample. The effect of bed length on improvement of fuel thermal oxidative stability was also studied. The thermal stability assessment of the treated fuels was carried out using a "High Reynolds Number Thermal Stability (HiReTS)". assessments showed thatThe treated fuel samples showed displayed a significant improvement of in reduced surface deposition propensity. via assessment with a "High Reynolds Number Thermal Stability (HiReTS)" Furthermore, the effect of bed length on improvement of fuel thermal oxidative stability was studied. Quantification of post-treatment samples demonstrated a substantial adsorption of dissolved Fe molecules. Quantum chemistry calculations were carried out using the ORCA 5 program package to elucidate the nature of the interactions of polar species with the inner and outer surfaces of chabazite. The polar species used in the ab initio quantum chemistry model calculations include Fe naphtanate, aniline, di butyl di sulphide, Butylated hydroxytoluene (BHT), hexanol, hexanone and hexanoic acid.

*Corresponding Author: Dr Ehsan Alborzi; e.alborzi@sheffield.ac.uk; ++44(0)7863711711

IASH 2022, the 17th INTERNATIONAL SYMPOSIUM ON STABILITY, HANDLING AND USE OF LIQUID FUELS
Dresden, Germany
September 11 - 15, 2022

HIGHLIGHTING THE "STRUCTURE-REACTIVITY" RELATIONSHIP FOR THE AUTOXIDATION REACTION OF HYDROCARBONS

Soraya Aminane[1]*, Mickaël Sicard[1], Lorette Sicard[2], Frédéric Ser[1]

[1]DMPE, ONERA Université Paris Saclay, Palaiseau, F-91123, France
[2]Laboratoire ITODYS, 15 Rue Jean Antoine de Baïf, 75013 Paris, France

ABSTRACT

One concern of the aviation industry is the jet fuel thermal stability. When not guaranteed, it leads to the formation of insoluble deposits in the fuel system and injectors which causing clogging and failures. The aim of this study was to investigate the mechanisms and kinetics reaction for hydrocarbon molecules composing jet fuel. Considering the complexity of interactions and the contribution of minor species on deposit formation, five model hydrocarbons were selected (n-dodecane and its isomers blend, n-butylcyclohexane, 1,2,4-trimethylbenzene and 1-methylnaphthalene) and subjected individually to oxidation using the PetroOXY device, focusing on the beginning of autoxidation ($\Delta P/P_{max}$ = 2 to 10 %). An experimental protocol was developed for the first time to identify and quantify precisely the oxidation products formed in the gas and liquid phases, using analytical (micro-GC, GC-MS, FTIR, SEC, HPLC) and chemical (PV, TAN, water content) techniques. This diversity of techniques made it possible to monitor the consumption of reactants, namely oxygen, which is always higher than that of the parent hydrocarbons suggesting competitive reactions. Although the oxygen consumption is globally similar for the all alkanes, it is variable for the aromatics. This illustrates their reactivity confirmed by the induction periods (IP). The characterization results showed a link between the initial hydrocarbon's structure and the nature and number of carbon atoms in the oxidized products. In addition, the tests performed at different temperatures (140,150 and 160°C) enabled the calculation of the overall and hydroperoxide dissociation kinetics constants reaction. Arrhenius parameters values obtained were higher for alkanes (Ea>120 kJ.mol-1, A~10+12 s-1) than for aromatics (Ea<110 kJ.mol-1 and A<10+09 s-1), confirming the "structure-reactivity" relationship of hydrocarbons.

*Corresponding Author: Highlighting the "structure-reactivity" relationship for the autoxidation reaction of hydrocarbons Soraya Aminane; soraya.aminane@onera.fr; +33180386005

IASH 2022, the 17th INTERNATIONAL SYMPOSIUM ON STABILITY, HANDLING AND USE OF LIQUID FUELS
Dresden, Germany
September 11 - 15, 2022

WHY DEPOSITS LOOK ABNORMAL

David Abdallah*

ExxonMobil Research & Engineering, 1545 US Highway 22 East, Annandale, NJ 08801

ABSTRACT

The JFTOT specification defines abnormal deposits as a deposit with a color that is neither peacock nor like those of the Color Standard. A deposits appearance consists of a blend of physical and chemical characteristics of the deposit. We will look at both contribution in an attempt to better understand the visual rating mysteries of many abnormal deposits.

*Corresponding Author: Fuels researcher David Abdallah; david.j.abdallah@exxonmobil.com; 9085074535

IASH 2022, the 17th INTERNATIONAL SYMPOSIUM ON STABILITY, HANDLING AND USE OF LIQUID FUELS
Dresden, Germany
September 11 - 15, 2022

FUNDAMENTAL STUDY OF JET FUEL OXIDATIVE DEPOSIT FORMATION

Krege M Christison*, Gregory B. Boursalian

Chevron, 100 Chevron Way, Richmond, California 94801 USA

ABSTRACT

When analyzed by Quartz Crystal Microbalance (QCM) under oxidative conditions, there is no deposit formation in hydrocracked jet fuel component due to the severe hydrotreating used in its production, which generally eliminates reactive, polar, and heteroatom-containing molecules. Using a hydrocracked jet fuel component as a baseline, compounds from classes that have been identified as significantly involved in oxidative deposit formation were spiked into jet fuel component and the resultant mixture was analyzed by QCM.[1,2] Several analogues from each compound class were subjected to this analysis to determine structure/activity relationships in precursors to oxidative deposit formation.

For the spiked hydrocracked jet fuel components that formed appreciable deposits in the QCM, the deposits were collected for further analysis. Pyrolysis GC/MS is a technique that allows for the analysis of these deposits through thermolytic cleavage of chemical bonds, which breaks the deposits down into their fundamental components. These data offer insights into the mechanism of deposit formation. The spiked fuels were also analyzed by comprehensive GCxGC and LC/MS before and after QCM to monitor the amount of the precursor compounds and to detect any soluble reaction products.

References:

1. Christison, K. M.; Lorenz, R. M.; Xue, L.; Sparkman, O. D., Exploring the Molecular Origin of Jet Fuel Thermal Oxidative Instability through Statistical Analysis of Mass Spectral Data. Energy & Fuels 2019.
2. Zabarnick, S.; West, Z. J.; Shafer, L. M.; Mueller, S. S.; Striebich, R. C.; Wrzesinski, P. J., Studies of the Role of Heteroatomic Species in Jet Fuel Thermal Stability: Model Fuel Mixtures and Real Fuels. Energy & Fuels 2019.

*Corresponding Author: Lead Research Chemist Krege Christison;
kregechristison@chevron.com; +1 (510) 242-2061

IASH 2022, the 17th INTERNATIONAL SYMPOSIUM ON STABILITY, HANDLING AND USE OF LIQUID FUELS
Dresden, Germany
September 11 - 15, 2022

AN INTERACTIVE, INTERDISCIPLINARY AND COLLABORATIVE DIGITAL PLATFORM FOR THE ASSESSMENT AND OPTIMIZATION OF JET FUELS

Uwe Bauder*, Clemens Hall, Florian Pütz, Benedict Enderle

German Aerospace Center, DLR, Pfaffenwaldring 38-40, Stuttgart 70569 Germany

ABSTRACT

Fuel-aircraft interaction is a complex and interdisciplinary problem, which becomes clearly visible for example in the scope of ASTM D4054 approval process. Optimization or even changes within such a complex system are difficult, since changes of the fuel composition to improve a fuel performance property might in contrary lead to decrease of other performance properties or even to safety critical properties. Also, the current specification limits have been empirically evolved and might have led to an over-determined or non-optimal set, limiting the potential of SAF.

The poster will provide an overview about the SimFuel platform, a digital tool that was designed to help solving complex problems like the one above. SimFuel connects distributed models and combines them with datasets and knowledge to gain additional insight. The platform consists of four main components: databases (20’000 fuels, 5300 single components), models, a distributed model environment and the human-in-the-loop concept. With its web-based dashboards, the platform can support expert groups (human-in-the-loop) to collaboratively, interactively and systematically use the data, models and workflows of the platform in their decision process. Major use cases on assessing different machine learning models for fuel property prediction, and applying SimFuel for fuel prescreening and optimization are presented in detail.

*Corresponding Author: Dr.-Ing. Uwe Bauder; uwe.bauder@dlr.de; +497116862422

IASH 2022, the 17th INTERNATIONAL SYMPOSIUM ON STABILITY, HANDLING AND USE OF LIQUID FUELS
Dresden, Germany
September 11 - 15, 2022

NEW TOOLS FOR UNDERSTANDING FUEL AUTOXIDATION AND DEPOSITION

Steven Zabarnick[1]*, Zachary J. West[1], Paul J. Wrzesinski[2]

[1]University of Dayton Research Institute, 300 College Park, Dayton, Ohio 45409 USA
[2]U.S. Air Force Research Laboratory, 1790 Loop Rd. North, Wright-Patterson AFB, Ohio,45433-7103 USA

ABSTRACT

Three new computational tools will be described which assist in developing a better understanding of fuel oxidation and deposition measurements. These tools are (1) a Windows dynamically-linked library (DLL) which simulates jet fuel autoxidation kinetics as a function of time and temperature which can be called from any host application, such as Excel, Matlab, or Python; (2) a Partial-Least Squares (PLS) method for using fuel composition and fuel deposition measurements to determine species classes most responsible for causing deposits; and (3) a non-linear data fitting method for determining Generalized Logistics Function (GLF) parameters for Quartz Crystal Microbalance (QCM) oxidation and deposition profiles.

The autoxidation DLL brings the ability to predict fuel oxidation vs. time and temperature without the need for advanced chemical kinetic software. The PLS method enables determination of important fuel species classes from large fuel composition data sets. The GLF fitting method provides the ability to obtain functional form fits to QCM oxidation and deposition, and also to determine global oxidation and deposition rates from these profiles. While these tools were developed for jet fuels, their use is applicable to a wide variety of fuel types.

*Corresponding Author: Division Head, Fuels & Combustion Steven Zabarnick; Steven.Zabarnick@udri.udayton.edu; 1-937-266-7231

IASH 2022, the 17th INTERNATIONAL SYMPOSIUM ON STABILITY, HANDLING AND USE OF LIQUID FUELS
Dresden, Germany
September 11 - 15, 2022

IMPACT OF CLAY TREATMENT ON THE THERMAL STABILITY CHARACTERISTICS OF AVIATION TURBINE FUELS

Zachary West*, Richard Striebich, Samuel Tanner, Susan Mueller

University of Dayton Research Institute, 300 College Park, Dayton, Ohio 45409 USA

ABSTRACT

The thermal stability characteristics of aviation turbine fuels remain a topic of intense interest. Fuels that fail thermal stability requirements after entering the fungible fuel distribution system can be especially problematic to deal with. Disposition strategies range in cost, expedience, and fuel handling assets available, e.g., diluting the batch via blending operations, returning the batch to a refinery, and dosing the fuel with additives are possible options. Another option that has been proposed for field remediation of poor thermal stability fuels is clay treatment. Clay treatment vessels are present at various fuel handling locations, and are typically used to remove surfactants from aviation turbine fuels prior to filter-coalescer treatment. Clay treatment is also known to remove polar organic compounds that have been implicated as strong, negative factors influencing thermal stability of aviation turbine fuels. Recent projects, funded by DLA-Energy, have focused on the effects of clay treatment with respect to the thermal stability of aviation turbine fuels. An emphasis is placed on understanding the resulting chemical compositions as well as the overall impact/efficiency of clay treatment as a remediation strategy. Clay treatment was found to be a plausible remediation strategy for some fuels, under specific conditions; however, numerous caveats and questions remain for field practitioners who wish to employ this strategy. This presentation will outline our research findings to-date and provide recommendations for future efforts.

*Corresponding Author: Principal Research Engineer Zachary West; zwest1@udayton.edu; (937) 255-4062

IASH 2022, the 17th INTERNATIONAL SYMPOSIUM ON STABILITY, HANDLING AND USE OF LIQUID FUELS
Dresden, Germany
September 11 - 15, 2022

AVIATION FUELS PROBLEMS AND SOLUTION- AIRCRAFT FAILURE ANALYSIS INDUCED BY THE FUEL OR FUEL SYSTEM

Albert Schmidt*, Dr. Moshe Rabaev, Yalfal Sium, Haim Juda

ABSTRACT

NTSB's CAROL database contains information about aviation investigations since 1983. CAROL database hold 22,384 records of aviation safety occurrences from 2008 to 2020, from which 1265 are classified as fuel related. Fuel related occurrence aviation safety occurrences is defined as an occurrence in which "One or more powerplants experienced reduced or no power output due to fuel exhaustion, fuel starvation/mismanagement, fuel contamination/wrong fuel, or carburetor and/or induction icing". In this chapter we review the fuel system major regulatory airworthiness requirements, survey dozens of real fuel related aviation safety accidents and incidents, analyze each safety occurrence, demonstrate a systemized analysis of complex scenarios involving fuel and fuel systems and describe investigation practices and techniques.

*Corresponding Author: Mr. Yalfal Sium; yalfalsi@gmail.com; +972509349610

IASH 2022, the 17th INTERNATIONAL SYMPOSIUM ON STABILITY, HANDLING AND USE OF LIQUID FUELS
Dresden, Germany
September 11 - 15, 2022

AN INTERACTIVE, INTERDISCIPLINARY AND COLLABORATIVE DIGITAL PLATFORM FOR THE ASSESSMENT AND OPTIMIZATION OF JET FUELS

Uwe Bauder*, Clemens Hall, Florian Pütz, Benedict Enderle

German Aerospace Center, DLR, Pfaffenwaldring 38-40, Stuttgart 70569 Germany

ABSTRACT

Fuel-aircraft interaction is a complex and interdisciplinary problem, which becomes clearly visible for example in the scope of ASTM D4054 approval process. Optimization or even changes within such a complex system are difficult, since changes of the fuel composition to improve a fuel performance property might in contrary lead to decrease of other performance properties or even to safety critical properties. Also, the current specification limits have been empirically evolved and might have led to an over-determined or non-optimal set, limiting the potential of SAF.

The poster will provide an overview about the SimFuel platform, a digital tool that was designed to help solving complex problems like the one above. SimFuel connects distributed models and combines them with datasets and knowledge to gain additional insight. The platform consists of four main components: databases (20'000 fuels, 5300 single components), models, a distributed model environment and the human-in-the-loop concept. With its web-based dashboards, the platform can support expert groups (human-in-the-loop) to collaboratively, interactively and systematically use the data, models and workflows of the platform in their decision process. Major use cases on assessing different machine learning models for fuel property prediction, and applying SimFuel for fuel prescreening and optimization are presented in detail.

*Corresponding Author: Dr.-Ing. Uwe Bauder; uwe.bauder@dlr.de; +497116862422

IASH 2022, the 17th INTERNATIONAL SYMPOSIUM ON STABILITY, HANDLING AND USE OF LIQUID FUELS
Dresden, Germany
September 11 - 15, 2022

THE USE OF ULSD AND RENEWABLE DIESEL BLENDS AS AN EMERGENCY FUEL FOR AVIATION

David Evans*, Tony Marasigan, Nathan Matheson, Paul M. Rawson

Defence Science & Technology Group, 506 Lorimer Street, Fishermans Bend, 3207 Australia

ABSTRACT

Military helicopters can utilise diesel fuels such as F-76 as an emergency fuel albeit with maintenance penalties and altitude restrictions. The initial certification of F-76 as an emergency fuel occurred when the diesel fuel was typically straight run and high sulfur. Since then, the fuel industry has undergone significant changes from a refining perspective such as the degree of hydroprocessing to reduce the sulfur level, inclusion of FAME in commercial diesel and most recently, the introduction of renewable diesel. This body of work investigates the performance of ultralow sulfur diesel (ULSD) and renewable diesel blends as an emergency fuel for aviation use. Initial testing highlights the improved thermal stability characteristics of ULSD and their renewable diesel blends compared to low and high sulfur diesel as well as traditional jet fuels.

*Corresponding Author: DR David Evans; David.Evans@dst.defence.gov.au; 0438653371

IASH 2022, the 17th INTERNATIONAL SYMPOSIUM ON STABILITY, HANDLING AND USE OF LIQUID FUELS
Dresden, Germany
September 11 - 15, 2022

AFRL RESEARCH ACTIVITIES ON FUEL EFFECTS ON COMBUSTION PERFORMANCE AND EMISSIONS

Edwin Corporan[1]*, Scott Stouffer[2], Tyler Hendershott[2], Veronica Williams[1]

[1]U.S. Air Force Research Laboratory, 1790 Loop Rd. North, Wright-Patterson AFB, Ohio 45433-7103 USA
[2]University of Dayton Research Institute, 300 College Park, Dayton, Ohio 45409 USA

ABSTRACT

Interest in higher concentrations of sustainable aviation fuels (SAF) in jet fuel blends to reduce carbon footprint and environmental impact of aviation activities has increased the need to understand the impacts of fuel chemical and physical properties on gas turbine combustor operability (e.g., ignition, lean blowout (LBO)) and emissions. SAF chemistries and physical properties may vary significantly relative to conventional fuels and thus, merit investigation to ensure adequate combustion performance and safety of flight. A summary of major findings of research activities at the Air Force Research Laboratory's Fuels and Energy Branch on fuel impacts on combustion operability on a single-nozzle swirl-stabilized combustor (referee combustor) and a T63 engine, is presented. Many of these activities supported the recently completed National Jet Fuel Combustion Program (NJFCP). Fuels studied include: SAFs, jet fuel blends, surrogates and single component fuels, which encompass a wide range of fuel chemical and physical properties. In addition, results of a study investigating the correlation of subcritical fuel decomposition products with LBO performance, and preliminary results on the impact of high temperature fuel injection (up to 500°F) on combustion performance and emissions of the referee combustor are discussed.

*Corresponding Author: Principal Mechanical Engineer Edwin Corporan; edwin.corporan@us.af.mil; 1-937-255-2008

IASH 2022, the 17th INTERNATIONAL SYMPOSIUM ON STABILITY, HANDLING AND USE OF LIQUID FUELS
Dresden, Germany
September 11 - 15, 2022

FUEL QUALITY IMPACT ON VINTAGE AIRCRAFT

Paul Rawson[1]*, Nathan Matheson[1], Glenn Pinnuck[2], David Evans[1]

[1]Defence Science & Technology Group, 506 Lorimer Street, Fishermans Bend, 3207 Australia
[2]Defence Flight Safety Bureau, Canberra, Australia

ABSTRACT

Heritage aircraft often have propulsion and fuel systems that were designed to run on JP-4 fuel which has different performance specifications to those used by current military aircraft. Whilst they remain kerosene fractions the impact on contamination and thermal stability can have dramatic consequences for these old engine types. A Vampire aircraft operated by 100 SQN of the Royal Australian Air Force suffered an uncontained engine failure inflight in 2022. Examination of the engine and the fuel from the aircraft indicated fuel thermal instability was a possible contributor to the failure. Fuel delivered to the aircraft was of very high thermal stability as measure by JFTOT and QCM stability techniques whilst that recovered from the aircraft was of very poor stability characteristics. Analysis found High concentrations of plasticizer's and materials potentially extracted from the fuel system which may have caused significant deterioration of the fuels stability leading to disrupted fuel spray patterns and damage to the combustor cans and nozzle guide vanes.

*Corresponding Author: DR David Evans; David.Evans@dst.defence.gov.au; +61438653371

IASH 2022, the 17th INTERNATIONAL SYMPOSIUM ON STABILITY, HANDLING AND USE OF LIQUID FUELS
Dresden, Germany
September 11 - 15, 2022

PREDICTION OF WATER SOLUBILITY IN PETROLEUM AND SYNTHETIC JET FUELS WITH COSMO-RS

Jinxia Fu*, Scott Q. Turn

University of Hawaii at Manoa, 1680 Eastwest Road, Post 109, Honolulu, Hawaii 96826 USA

ABSTRACT

Comprehensive GC×GC chromatography provides novel characterization of complex liquid fuels. Standard GC×GC methods have been developed and included in ASTM D4054 for certification of new aviation turbine fuels and fuel additives, generating copious compositional data. The present work utilizes the conductor-like screening model for realistic solvation (COSMO-RS) method to calculate water solubility in petroleum and sustainable aviation fuels (PAFs and SAFs) over a temperature range of -40 to 50 °C using GC×GC composition data. The calculation results are in good agreement with experimental values at temperatures below 20°C, with RMSE = 17.1 and 17.7 ppm for PAFs and SAFs, respectively. Results calculated for these complex fuel chemistries predicted experimental values more closely than calculations performed for simpler PAF and SAF surrogates. COSMO-RS was also employed to calculate water solubility in PAF and SAF blends (RMSE = 18.3 ppm). Results demonstrate that the COSMO-RS method has potential as a tool in new SAF certification.

*Corresponding Author: Assistant Researcher Jinxia Fu; jinxiafu@hawaii.edu; 8089565944

IASH 2022, the 17th INTERNATIONAL SYMPOSIUM ON STABILITY, HANDLING AND USE OF LIQUID FUELS
Dresden, Germany
September 11 - 15, 2022

STATIC DISSIPATIVE ADDITIVE PERFORMANCE AND STORAGE STABILITY IN BIODIESEL BLENDED DIESEL

Sankaran Murugesan*, Asha Patel, Matthew Barnes, Don Wolfe

Baker Hughes, 17021 Aldine Westfield, Houston, Texas 77073, USA

ABSTRACT

Static Dissipative Additives (SDA) are used to facilitate the safe dissipation of electrostatic charges through liquid hydrocarbon by reducing the accumulation of hazardous surface charges. Recently, there has been growing interest in the production and use of biofuels and renewable fuels as an alternative to conventional distillate fuels. It is expected that any transition from convention fuels to lower carbon intensity fuels will transition slowly in the marketplace. As such, there will be a period of time where the various fuels will be comingled. Addition of biofuels to distillate fuels can cause both synergistic and antagonistic effects.

Industry has examined the impact of adding biofuels to conventional fuels for some common performance specifications such as lubricity. However, it would appear that such a study investigating the role of biodiesel on conductivity performance has not been explored. This presentation gives a comprehensive study of conductivity performance of biodiesel blended with Ultra Low Sulfur Diesel (ULSD) treated with SDA. Furthermore, this provides a detailed discussion on SDA dose variation with temperature and the synergistic effect caused by the presence of biodiesel.

*Corresponding Author: Senior Scientist Sankaran Murugesan; sankaran.murugesan@bakerhughes.com; 2813864422

IASH 2022, the 17th INTERNATIONAL SYMPOSIUM ON STABILITY, HANDLING AND USE OF LIQUID FUELS
Dresden, Germany
September 11 - 15, 2022

DEVELOPMENT OF A SHIPBOARD TEST KIT FOR DISSOLVED COPPER IN JET FUEL PROTOTYPE FIELD TEST KIT

Kristina M. Myers[1]*, Thomas N. Loegel[2], Robert E. Morris[1]

[1]Nova Research, Inc., 1900 Elkin Street, Suite 230, Alexandria, VA 22308
[2]Naval Research Laboratory, Code 6181, 4555 overlook Ave. S.W. Washington, DC 20375

ABSTRACT

The U.S. Navy is currently investigating shipboard inline copper sequestration technologies to remediate the deleterious impact of dissolved copper on ship propulsion and aircraft engines. In the event that such technologies are deployed, a convenient means to monitor copper content would inform ship crews as to when to change out adsorbent media. A spectrophotometric procedure was previously developed at NRL to measure low PPB levels of copper in jet fuel, which utilized the underivitized form of bathocuproine (BCP), with ascorbic acid as a reducing agent. Since BCP only responds to Cu(I), this eliminates the need for post-analysis data treatment to resolve other interfering metals. In addition, uncomplexed BCP is colorless, eliminating the need to add BCP to the reference sample or remove copper from the reference fuel sample of interest. By utilizing standard additions of a copper standard, the BCP procedure will determine total copper content without the need for a copper-free reference, even in the presence of strong chelating agents, e.g., metal deactivating additives. Minor improvements to the previous method were made to improve the chemistry and extend of the method to measure copper contents of diesel and MGO fuels.

*Corresponding Author: Staff Scientist III Kristina Myers; kristina.myers.ctr@nrl.navy.mil; 1-202-404-3317

IASH 2022, the 17th INTERNATIONAL SYMPOSIUM ON STABILITY, HANDLING AND USE OF LIQUID FUELS
Dresden, Germany
September 11 - 15, 2022

HANDLING, MAINTENANCE AND SAFETY USE OF AVIATION FUEL-INFORMATIVE BULLETINS

Maor Saban*, Dr. Moshe Rabaev, Yalfal Siyum, Liel Mazar

Israeli Air Force, Depot 22, Materials Division, Fuel and Chemistry Department, POB 02538

ABSTRACT

Optimal care and maintenance of aviation fuels and infrastructure is of crucial importance for ensuring the safety of passengers, personal and the environment. In order to Increase the awareness of fuel technicians, engineers and decision makers in our organization for the safety issues regarding the maintenance and handling of fuel in the field, we have produced numerus Informative bulletins written in simple and convenient language, which makes it possible to convey important and essential knowledge in easy, digestible manner for every reader. Topics covered in the bulletins include information about "Apple Jelly" (see appendix 1), microbial contamination (see appendix 2), fuel filters, static electricity, aviation fuel additives, laboratory tests, critical fuel properties, fuel contaminations and real fuel related aircrafts safety events - emphasizing how a minor human mistake can have major safety implications. The bulletins have greatly raised the awareness of our organization personal while handling aviation fuel, and therefore we find it essential to share them with the wide global community.

*Corresponding Author: Doctor Moshe Rabaev; rabaevm@gmail.com; +972547807472

IASH 2022, the 17th INTERNATIONAL SYMPOSIUM ON STABILITY, HANDLING AND USE OF LIQUID FUELS
Dresden, Germany
September 11 - 15, 2022

INVESTIGATIONS INTO AUTOMATIC GENERATION OF A MECHANISM FOR AUTOXIDATION OF N-DECANE USING REACTION MECHANISM GENERATOR (RMG)

Seyed Yoosof Sadat*, Dr. Ehsan Alborzi, Dr. Kevin Hughes

University of Sheffield, Sheffield S10 2TN, United Kingdom

ABSTRACT

The detailed mechanisms explaining the autoxidation of jet fuel available in the public domain are normally constructed based on one single block surrogate fuel. With the introduction of alternative aviation fuels to the industry, an understanding of how different building blocks of hydrocarbons would behave in the system is crucial. Introducing extra blocks of surrogate fuel such as cycloparaffins to the model would add an extensive amount of computational work, which could be significantly expensive in terms of time, and the manual handling of it would be almost impossible. In order to avoid the expensive computational power required and add complexity to the detailed mechanism, an open-source computer code written in Python called the Reaction Mechanism Generator (RMG) was selected for auto-generation of a detailed mechanism. RMG uses a basic understanding of how chemical species behave on a molecular level to construct a detailed reaction mechanism and its kinetic parameters. The software has been widely used in the gaseous phase, however understanding of it remains limited in the liquid phase. In this study, RMG has been deployed to autogenerate a chemical kinetic mechanism for autoxidation of n-decane and it has been further merged with Decalin as a cyclo-paraffinic block. Following that a numerical technique has been used to reduce the mechanism by grouping and lumping method. The reduction method is able to overcome the complexity while the robustness of the mechanism is preserved to make it ready for further applications. Following the successful validation of the mechanism with experimental data, further hydrocarbon blocks could be integrated to it.

*Corresponding Author: Mr Seyed Yoosof Sadat; ys1394@gmail.com; 07432859450

IASH 2022, the 17th INTERNATIONAL SYMPOSIUM ON STABILITY, HANDLING AND USE OF LIQUID FUELS
Dresden, Germany
September 11 - 15, 2022

FOSSIL AND ALTERNATIVE FUEL DEPOSIT FORMATIONS UNDER REPRESENTATIVE OPERATING CONDITIONS

Mickaël Sicard*, Bruno Raepsaet, Soraya Aminane, Yanis Melliti

ONERA, Chemin de la Huniere, Palaiseau 91123, France

ABSTRACT

In order to simulate the shutdown phase of an aircraft engine and the degradation of the fuel film in a hot environment (engine radiation), a specific test bench has been designed at ONERA to study the coking behavior of fossil and alternative jet fuels. Dynamic conditions, that is to say with a continuous flow of fuel, are followed by "stagnant" phases. A thin film of fuel remains inside the pipe and is maintained under air at around 200 °C.

The relatively slow kinetics of deposits formation requires to accumulate a significant number of tests and thus stagnant phases. At the same time, it also helps to observe the cumulative effect of deposit growth. At the end of a long series of tests (30 flow and 30 stagnation phases for each fuel), the formation of solid deposits on the flat part of the diaphragm injector was observed. The aim of this study was to characterize and quantify the deposits formed by various techniques (DOM, SEM, EDS, FTIR, AFM…) to observe their growth by cumulative effect and their evolution with time.

As expected, the most important quantity of deposits was obtained with the Jet A-1 but deposits were also observed for alternative fuels. The deposit starts with a few “islets” caught in a gel-like substance. During accumulations, these individual islets coalesce to form a continuous layer that reaches several microns thick. The quantitative evolution of deposit is an important challenge because it provides data mandatory to calibrate the models for numerical simulation.

*Corresponding Author: Dr. Mickael SICARD; mickael.sicard@onera.fr; 33180386052

IASH 2022, the 17TH INTERNATIONAL CONFERENCE ON STABILITY, HANDLING AND USE OF LIQUID FUELS
Dresden, Germany
11 – 15 September 2022

EXPERIMENTAL STUDY OF THE IMPACT OF BIOALCOHOL STRUCTURE ON THE OXIDATION STABILITY OF A SURROGATE JET-FUEL

Ryma Benrabah[1], Zaki El Sayah[1], Minh Duy Le[1], Pierre-Alexandre Glaude[1], Baptiste Sirjean[1]

[1]*Université de Lorraine, CNRS, LRGP, F-54000 Nancy, FR.*

ABSTRACT

The addition of biofuel in jet fuels can be seen as a promising answer to reduce the carbon footprint of the aviation transport sector. In this work, we investigated the impact of the addition of alcohol on the oxidation stability of a jet fuel surrogate. Different chemical structures of bioalcohols were considered and blended with *n*-decane used as the jet-fuel surrogate. The thermal oxidation stability of the mixtures was measured in a standard PetroOxy apparatus, which gives an induction period (IP). In addition to this global indicator of the fuel thermal stability, we quantified the total hydroperoxide concentration at the end of the test by iodometric titration. The influence of the length of the carbon chain in alcohols (*n*-butanol to *n*-decanol) and the isomeric structure of butanol on the IPs of pure *n*-decane have been quantified. We show that the addition of a low percentage by volume of alcohols in the fuel increases its oxidation stability. This effect is non-linear, and the IP value increases with the volume percentage of alcohol in the fuel up to 15%. Beyond this value, a rapid decrease in the stability of the mixture is observed. The influence of the length of the carbon chain is shown to be negligible. In contrast, the structure of butanol isomers strongly impacts the stabilization efficiency of the alcohol in the fuel. Conclusions on the fundamental chemical mechanism are drawn from these experiments and highlight the central role of aldehyde formation in alcohol oxidation kinetics.

KEYWORDS: autoxidation, bioalcohols, jet fuel

INTRODUCTION

New environmental regulations for the reduction of the carbon footprint of the transportation sector encourage the use of biofuels in this sector and especially in the aviation field since aircraft propulsion systems are known to be almost impossible to electrify. [1,2] However, blending oxygenated biofuels with a jet fuel is difficult since in aircrafts, the fuel is also used as a thermal sink and is submitted to high temperatures. Therefore, thermal and oxidation stability of jet fuels are crucial parameters.[3–5] The present paper focuses on the oxidation stability of a conventional fuel surrogate, *n*-decane, which has been selected as a surrogate to represent a real jet fuels. It is well established that *n*-decane is among the major components of petroleum-derived fuels (jet fuels) and it has been shown that this hydrocarbon suitably represents the reactivity of jet fuels.[6-9] The first part of this study is dedicated to the oxidation of *n*-decane and to the impact on its stability of the addition of different types of oxygenated biofuels, at different proportions (0.2 – 20 %). The measurement of the induction period (IP) as a function of additive percentage will quantify the oxidation stability of these blends. IP is defined as the time necessary to observe a 10 % pressure decrease from the maximum pressure recorded in the PetroOxy during the auto-oxidation of a fuel.[3] Several molecules of the next-generation biofuel additives (bioalcohol) have been studied at a temperature of 413 K and oxygen pressure of 700 kPa. Several parameters in the bioalcohol chemical structures such as the size of the chain (C_4, C_6, C_8, C_{10}) in linear alcohols, position of the –OH group (butan-1-ol, butan-2-ol), and level of branching in the hydrocarbon chain (*n*-butanol, iso-butanol, tert-butanol) were studied.

EXPERIMENTAL SECTION

Oxidation stability of fuel blends was tested in a PetroOxy from Anton Paar, one of the references of the Rapid Small Scale Oxidation Test of the ASTM 7545 methods. This method involves the introduction of a small amount, 5ml, of liquid fuel in a gold-plated cell, which is pressurized with pure O_2 under 7 bar at room temperature. The fuel blends inside the cell are then heated to the test temperature (413 K in this case) and the evolution of the pressure is recorded as a function of time at a constant temperature. The induction period (IP), which corresponds to the time required to reach a 10 % pressure decrease (O_2 consumption in the liquid fuel) is measured and characterizes the oxidation stability of the fuel. Figure 1 represents the IP result obtained for a pure *n*-decane sample tested at 413 K.

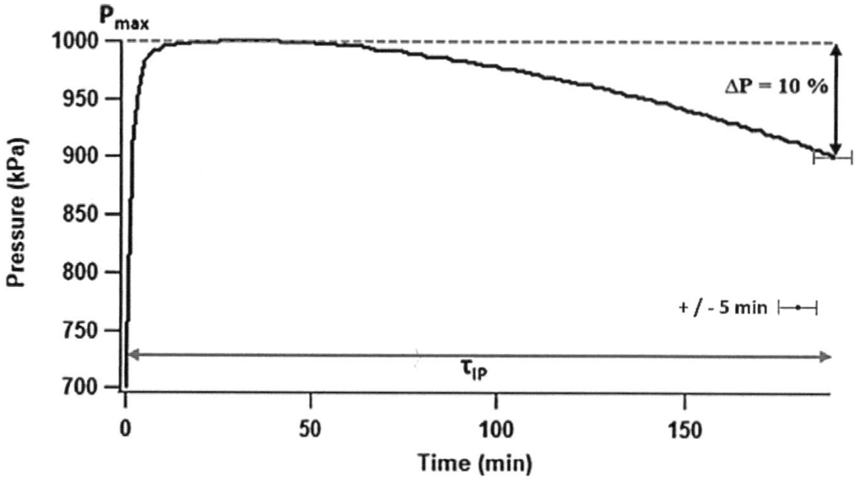

Figure 1: Typical induction period measurement with the PetroOxy device for n-decane at 413 K and 700 kPa of molecular oxygen

The PetroOxy measurements present a high reproducibility. The absolute average deviation of results, measured on fuel samples at the same temperature and pressure, can be expressed as a function of the relative standard uncertainty of IP determined which is +/- 2.5 % IP in this case. For each test, the volume of the sample is the the same and equal to 5 mL. The joint (viton) of the cell was changed between each test.

Reactants

n-Decane (≥ 99 %), 1-butanol (99.8 %), 1-decanol (≥ 99 %), tert-butanol (≥ 99.5 %), 2-butanol (99.5 %) and isobutanol (99.5 %) were all anhydrous and purchased from Sigma Aldrich. 1-hexanol (99 %) cames from Alfa Aesar and 1-octanol (99 %) was provided by Fisher Scientific.

RESULTS AND DISCUSSION

Influence of the length of alcohol's carbonated chains

The impact of the various alcohols was studied on the auto-oxidation of *n*-decane. Pure alcohols were studied first in the PetroOxy to compare their behavior and results are given in Figure 2.

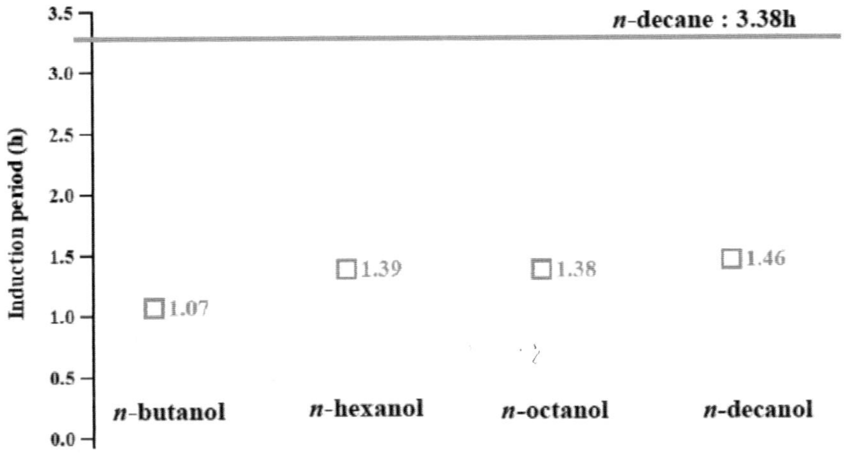

Figure 2: Influence of length of carbonated chain of primary alcohols, 140 °C | 7 bar

The length of the carbon chain of the alcohols is shown to lightly increase the induction period but the IP of each alcohol is much shorter than that of the neat surrogate fuel. As a first guess, one can predict than the blending of *n*-decane with the alcohols should then decrease its oxidation stability.

The evolution of the measured IPs of all these linear alcohols (*n*-butanol, 1-hexanol, 1-octanol, 1-decanol) blended with *n*-decane (5% alcohol / 95% *n*-decane), as a function of their size is presented in Figure 3

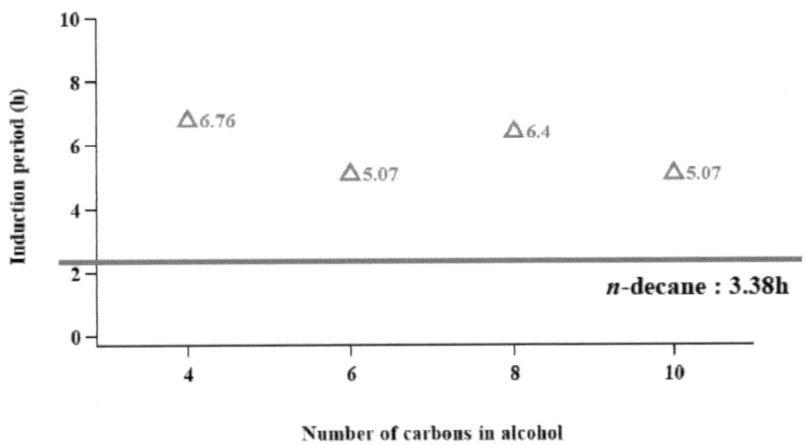

Figure 3: IP for 5% volume of alcohol added to n-decane 140 °C | 7 bar

The addition of alcohols is shown to strongly increase the stability of the jet fuel surrogate, for all alcohols. However, the evolution of the IP with the length of the carbonated chain show that this parameter has a little impact on the oxidation stability. The results observed for all the linear alcohols studied in this work highlight the importance and efficiency of these alcohols as additives with *n*-decane to improve the oxidation stability of the fuel surrogate. In the following section, we explore the effect of the proportion of linear alcohol on the IP of *n*-decane, using *n*-decanol.

Influence of the quantity of n-decanol in the blend

IP variation of blended *n*-decane was measured with different proportions of *n*-decanol. Results given in Figure 4 show important variations in the IP of *n*-decane / *n*-decanol blends with the proportion of *n*-decanol. The upper plot in Figure 4 show %vol. of *n*-decanol between 0 and 20%, while the lower plot zooms in on a %vol. range from 0 to 1%. The IP increases from 3.38 h for pure *n*-decane to 8.07 h when 10 % *n*-decanol is added.

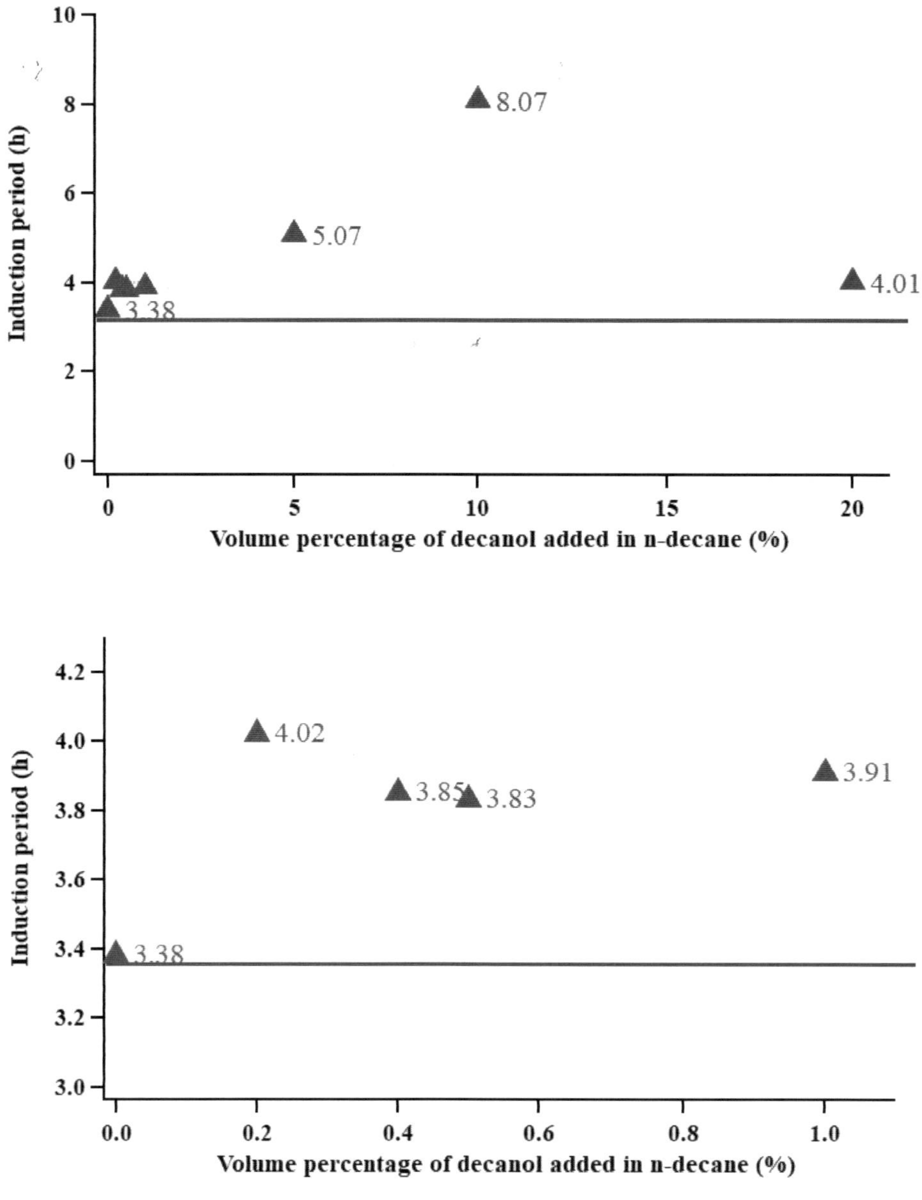

Figure 4: Effect of the addition of n-decanol on induction period of n-decane, 140 °C | 7 bar. Lower plot is for %vol. < 1

We show that the addition of a low percentage by volume of alcohols in the fuel increases its oxidation stability. This effect is non-linear and the IP value increases with the volume percentage of alcohol in the fuel up to 10%. Beyond this value, a decrease in the stability of the mixture is observed. This result is shown with other alcohols which have been blended. Indeed, above 10% added, the induction period and so the oxidation stability of the fuel decreases.

Stability of butanol's isomers

The previous paragraph showed that all alcohols have the same behavior regarding the oxidation stability, either pure or added to the *n*-decane fuel, independently of their sizes. Another structural parameter that can be explored is the influence of the branching and the position of the OH group in alcohols. Butanol isomers represent an adequate test set to probe these effects. The induction periods of pure butanol isomers are represented in Figure 5.

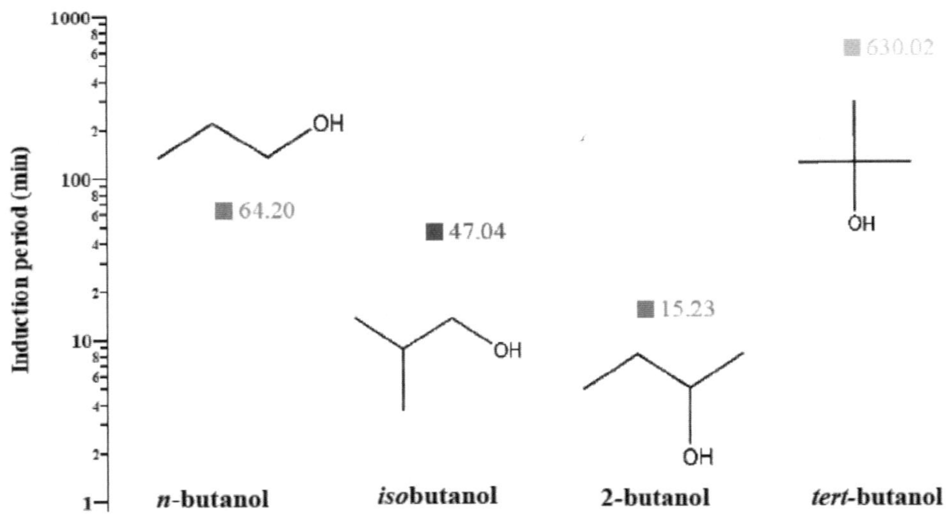

Figure 5: Induction periods of the isomers of butanol, 140 °C | 7 bar

The structure of butanol isomers strongly impacts the induction period the alcohol. *Tert*-butanol is shown to be the most stable butanol isomer, while 2-butanol is the least stable. The addition of butanol isomers in *n*-decane was also studied and the results are shown in Figure 6

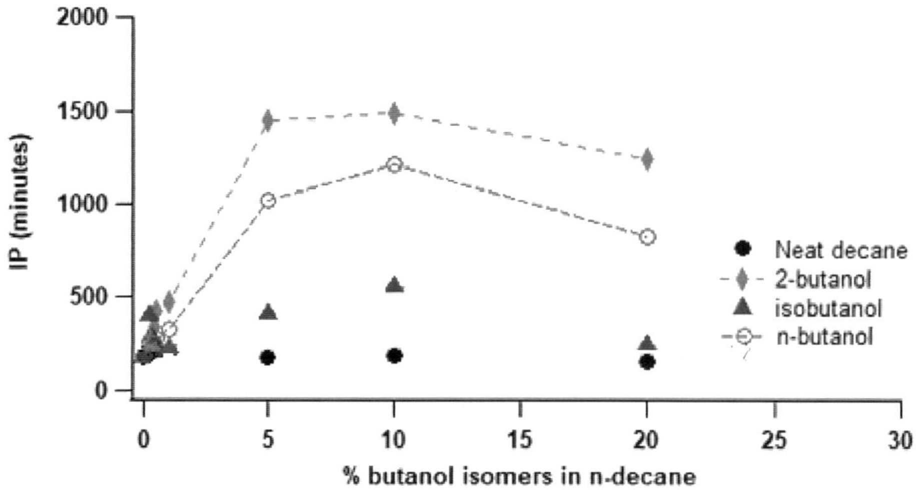

Figure 6: Induction periods of blends of n-decane and butanols as a function of the %vol. added in n-decane, 140 °C | 7 bar.

The addition of these butanol isomers to the fuel surrogate leads to a strong increase in oxidation stability, for all proportions added, except for *tert*-butanol. The largest impact is observed at 10% of butanol blended which is similar to 1-decanol results.

The increase in oxidation stability (higher IPs) non-linearly depends on the proportions of butanols blended with the fuel surrogate. The position of the alcohol group and the level of substitution of the alcohol (*n*-butanol isomers) are shown to affect the oxidation stability of the fuel surrogate, in a non-linear way.

CONCLUSIONS

The oxidation stability of a fuel surrogate, *n*-decane, doped with different alcohols has been investigated. The induction period of *n*-decane at 413 K and an initial pressure of pure oxygen of 700 kPa is 3.38 h. The IP of all pure alcohols (*n*-butanol to *n*-decanol, *i*- and 2-butanol) are shown to be lower than the IP of n-decane, by at least a factor of 2. Pure alcohols are therefore less stable than *n*-decane regarding oxidation stability. However, when added to *n*-decane, alcohol fuels are demonstrated to strongly enhance its oxidation stability with a maximum enhancement of stability at 10% volume percentage. Oxidation stability of the biofuel/fuel blends is impacted to different extents by the variations in the alcohol chemical structures. In linear alcohols (C_4, C_6, C_8, C_{10}), an increase of the chain length seems to have a little effect on the induction period. The position of the alcohol group and the level of substitution of the alcohol showed a greater impact in the oxidation stability of the fuel surrogate.

ACKNOWLEDGMENTS

The project leading to this application has received funding from the European Research Council (ERC) under the European Union's Horizon 2020 research and innovation program (grant agreement No 101003318). This work was also financially supported by the BioACe project, grant ANR-18-CE05-002 of the French National Research Agency, and by the Grand-Est region.

REFERENCES

(1) Westbrook, C. K. Biofuels Combustion. *Annu. Rev. Phys. Chem.* **2013**, *64* (1), 201–219. https://doi.org/10.1146/annurev-physchem-040412-110009.

(2) Bergthorson, J. M.; Thomson, M. J. A Review of the Combustion and Emissions Properties of Advanced Transportation Biofuels and Their Impact on Existing and Future Engines. *Renew. Sustain. Energy Rev.* **2015**, *42*, 1393–1417. https://doi.org/10.1016/j.rser.2014.10.034.

(3) Aminane, S.; Sicard, M.; Melliti, Y.; Ser, F.; Sicard, L. Experimental Study of the Kinetics of Degradation of N-Dodecane under Thermo-Oxidative Stress at Low Temperature and Mechanism Inferred. *Fuel* **2022**, *307*, 121669. https://doi.org/10.1016/j.fuel.2021.121669.

(4) Sicard, M.; Boulicault, J.; Coulon, K.; Thomasset, C.; Ancelle, J.; Raepsaet, B.; Ser, F. Oxidation Stability of Jet Fuel Model Molecules Evaluated by Rapid Small Scale Oxidation Tests. Oxidation Stability of Jet Fuel Model Molecules Evaluated by Rapid Small Scale Oxidation Tests.. The 13th International Conference on Stability, Handling and Use of Liquid Fuels 2013 IASH 2013, Oct 2013, RHODES, Greece. Ï¿¿hal-01057478ï¿¿.

(5) West, Z. J. Studies of Jet Fuel Autoxidation Chemistry: Catalytic Hydroperoxide Decomposition & High Heat Flux Effects. *Univ. Dayt.* **2011**, No. December.

(6) Le Dortz R, Strozzi C, Sotton J, Bellenoue M. Evaluation of the surrogates capacity to reproduce the laminar burning velocities and the sensitivity to stretching of a commercial kerosene under constant volume combustion conditions. Fuel 2021;287:119426.

(7) Naik CV, Puduppakkam KV, Modak A, Meeks E, Wang YL, Feng Q, et al. Detailed chemical kinetic mechanism for surrogates of alternative jet fuels. Combustion and Flame 2011;158(3):434-45.

(8) Mzé-Ahmed A, Hadj-Ali K, Diévart P, Dagaut P. Kinetics of Oxidation of a Synthetic Jet Fuel in a Jet-Stirred Reactor: Experimental and Modeling Study. Energy & Fuels 2010;24(9):4904-11.

(9) El-Sayah Z, Glaude P-A, Fournet R, Sirjean B. Comparison of the Effects of Different Biofuels on the Oxidation Stability of a Hydrocarbon Fuel. SAE Technical Paper 2020; 2020-01-2101.

IASH 2022, the 17th INTERNATIONAL SYMPOSIUM ON
STABILITY, HANDLING AND USE OF LIQUID FUELS
Dresden, Germany
September 11 - 15, 2022

HYDROCARBON MULTIDIMENSIONAL GAS CHROMATOGRAPH TEMPLATE OPTIMIZATION STUDY

John Feldhausen*, David Bell, Joshua Heyne

University of Dayton Research Institute, 300 College Park, Dayton, Ohio 45409 USA

ABSTRACT

Multidimensional gas chromatography is one of the most popular analytical chemistry separation techniques and is valuable in characterizing the hydrocarbon composition of jet fuels. Important operability and safety properties can be predicted from the results of a chromatograph provided the stencil making the various hydrocarbon classifications was properly calibrated. For those interested in building a hydrocarbon template from the beginning, in such a way that it reproduces the exact results of calibrated template, can be challenging. Existing chromatography literature attempts to instruct template development, but inconsistencies between various instruments is still likely, and can be difficult to source and quantify. This work proposes a solution to the struggle of bridging various separate gas chromatography instruments with an optimization approach that receives an initial guess of group coordinates from one template and minimizes the difference in the results against a separate specified baseline template. In this way, a proven template can be leveraged to guide the construction other templates. With this technique, compositional results from separate groups will be much more reproducible and the differences between groups can be used to determine relative uncertainties.

*Corresponding Author: Graduate Research Assistant John Feldhausen;
feldhausenj1@udayton.edu; 9205745899

IASH 2022, the 17th INTERNATIONAL SYMPOSIUM ON STABILITY, HANDLING AND USE OF LIQUID FUELS
Dresden, Germany
September 11 - 15, 2022

PREDICTING THE INSOLUBLE FORMATION TENDENCIES OF FUEL HETEROATOMS AND AROMATICS WITH COMPUTATIONAL TECHNIQUES

Charlie Adams*, Ehsan Alborzi, Kevin Hughes, Anthony JHM Meijer

University of Sheffield, Mechanical Engineering, Sheffield S10 2TN, United Kingdom

ABSTRACT

In recent years, quantum chemistry (QC) methods have started to be used as a tool to build predictive thermal oxidative degradation models for jet fuels from 'first-principles'. Nevertheless, QC work has solely focused on the first stage of the degradation process involving substrate activation and autoxidation. The second stage, agglomeration, and subsequent insoluble formation has not yet been studied with QC techniques. The complexity of the agglomeration process, where C-C/C-O bonds are formed leading to successively larger macromolecules, has meant it has been difficult to isolate a specific set of reactions to predict deposition tendency. To overcome this complexity, we proposed a chain-growth free radical reaction forming a deposit dimer from an array of fuel aromatics and heteroatomics. Furthermore, it was assumed that the deposit consists of heteroatom monomer units, allowing moles of deposit to be calculated. The calculated energetic barriers for the chain-growth reaction were then compared with total insolubles and extraction induced precipitate (EIP). The two component fuels contained n-dodecane and the selected aromatic/heteroatom of interest. The fuels were stressed in a glass flask at 140 °C with 1 bar molecular oxygen headspace for a duration of 24 h. It was found that the largest calculated transition state barrier provided a reasonable correlation with the molar amounts of deposit generated. However, a much stronger correlation was found between the EIP moles and the rate determining step.

*Corresponding Author: Mr Charlie Adams; cadams2@sheffield.ac.uk; +44 7741467159

IASH 2022, the 17th INTERNATIONAL SYMPOSIUM ON STABILITY, HANDLING AND USE OF LIQUID FUELS
Dresden, Germany
September 11 - 15, 2022

THE EFFECT OF SPACING FUEL BLEEDING REGIMES ON AIRCRAFT TANKS MICROBIAL CONTAMINATION LEVELS

Liel Mazar*, Alexi Bikbulatov, Yalfal Sium, Dr. Moshe Rabaev, Haim Yehuda

ABSTRACT

Microbial contamination in aircraft fuel tanks is a known problem, resulting in major time-consuming maintenance procedures and, in severe cases, can lead to safety accidents and incidents. There are several ways of controlling microbial contamination levels in which the primal one being regular fuel tank bleeding. Our Organization performs daily fuel bleedings on most of its aircraft models to mitigate microbial contamination problems. With a high availability, large fleets, daily fuel bleeding is an expensive procedure from work-load perspective, thus the incentive to decrease the frequency of bleeding regime.

Our research investigates experimental procedure for determining the effectiveness of daily bleeding on the occurrence and severity of microbial contamination as opposed to weekly bleeding. The experimental procedure investigated can be used as rigorous scientific tool for determining a lower fuel bleeding frequency, without compromise microbial contamination control.

Three types of aircrafts were divided into two populations each, those performing daily and weekly bleeding regimes, respectively. The amount of colony forming units was measured in the two populations in the beginning of the experiments and after about one year. Then, two-part model hypothesis test as proposed by Lachenbruch (1976) was applied on the data, including post-hoc power calculations.

A statistically significant difference was found in one type of aircraft, while samples from the two other types were determined as not having enough power to draw significant conclusions, in those cases a proper sample size was determined using numerical simulations. A short discussion of engineering significance of the conclusions is also provided.

*Corresponding Author: Liel Mazar, Liel.mazar@gmail.com, +972526609003

IASH 2022, the 17th INTERNATIONAL SYMPOSIUM ON STABILITY, HANDLING AND USE OF LIQUID FUELS
Dresden, Germany
September 11 - 15, 2022

AVIATION FUELS PROBLEMS AND SOLUTION - CHAPTER 14 - RESEARCH TEST METHODS FOR CONTAMINANTS IN AVIATION FUELS

Konstantin Tartakovsky*, Dr. Regina Sinelnikov, Dr. Moshe Rabaev

ABSTRACT

Fuel contamination is a catch-all term for anything that causes a fuel test to fail quality assurance testing. Aircraft fuel contaminations are dangerous and may lead to aviation safety accidents and incidents. Fortunately, most contaminations can be detected via simple and quick tests. Contaminants that cannot be detected by standard tests, which usually pose an immediate hazard, will be discussed in this chapter. Inability to detect such contaminants can cause irreversible damage to the fuel system and engines. Understanding the nature and source of the contamination can prevent its reoccurrence, help pinpoint a problem within the fuel system before it becomes critical, and treat the contamination. We will demonstrate that to accurately and effectively identify the pollution and its source, a combination of three working groups is required: a fuel laboratory, a fuel systems engineer, and an analytical research laboratory.

*Corresponding Author: Mr. Konstantin Tartakovsky; kosta002@gmail.com; +14317788272

IASH 2022, the 17th INTERNATIONAL CONFERENCE ON STABILITY, HANDLING AND USE OF LIQUID FUELS
Dresden, Germany
11-15 September 2022

GAINING A DEEPER UNDERSTANDING OF FUEL CHEMICAL COMPOSITION IN THE CONTEXT OF POLAR COMPOUND EXTRACTION PROCESSES

Grant S. Ochoa[1], Matthew C. Billingsley[2], Robert E. Synovec[1]

[1]Department of Chemistry, Box 351700, University of Washington, Seattle, WA 98195, USA
[2]Air Force Research Laboratory/RQRC, Edwards AFB, CA 93524, USA

INTRODUCTION

Kerosene-based rocket propellants must meet strict compositional criteria to ensure consistent and optimal performance. Small concentrations of polar compounds can lead to buildup of carbon deposits leading to poor engine performance. Current contaminant testing is time intensive and may lack detailed information of contaminant identities. An effective approach to analyze the composition of fuels is comprehensive two-dimensional gas chromatography with time-of-flight mass spectrometry detection (GC×GC-TOFMS), allowing for more confident compound class identification. Polar contaminants however are often in low concentration compared to the bulk fuel and are difficult to discern. Use of solid phase extraction (SPE) with fuels allows for the removal and/or isolation of targeted contaminant compounds from the fuel, in effect creating a new sample class, i.e., the pass fuel, which can be directly compared to the neat fuel sample. Fisher ratio (F-ratio) analysis, a powerful supervised non-targeted discovery technique, is well suited for comparing these classes to locate all compounds which were affected by the SPE. Through this a sample's polar compound profile may be determined without the need to clean up and analyze the extract fraction. Furthermore, this neat and pass sample scheme may be utilized to compare multiple fuels at once to determine all of their differences, in contaminants and bulk components. This opens the path for quick and efficient fuel testing. For visualization, an algorithm synthesizing F-ratio information into one chromatogram is utilized.

PROCEDURE

All rocket fuel samples, RP-1 and RF-2, were processed with SPE to create a pass fuel sample class that has reduced polars concentration compared to the neat fuels. Silica phase SPE cartridges of nominal ~1 mL volume were rinsed with 5 mL methanol to clean the cartridge. The cartridge was then rinsed with 15 mL of hexanes followed by loading the cartridge with 8 mL of sample at a flow rate of 1-3 mL/min. The first mL of sample passed through was collected and kept as the “pass” sample (Figure 1). The cartridge was then rinsed with 15 mL of hexanes and the pre-concentrated polar compounds were eluted with 0.5 mL of methanol. The GC×GC system (Pegasus 4D BT, LECO Corp, St. Joseph, MI, USA) was configured with a reverse column configuration with a mid-polarity Rxi-17Sil MS ^{1}D column (26.4 m × 250 μm i.d. × 0.25 μm film thickness) to a non-polar Rxi-1 MS ^{2}D column (1.9 m × 180 μm i.d. × 0.18 μm film thickness). The oven temperature was initially held at 40°C for 1.5 min, then increased to 200°C at 5°C/min, and then finally held at 200°C for 1 min. The ^{2}D column oven and modulator block utilized the same temperature program with offsets of +30°C and +12°C respectively. The modulation period was 3s, with a hot and cold pulse time of 0.75 s each. All data were exported and processed with MATLAB 2019A (The MathWorks, Inc., Natick, MA, USA).

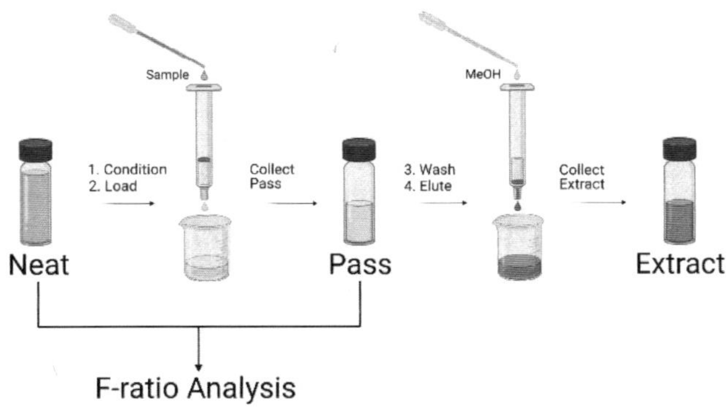

Figure 1. Graphical diagram showing the workflow for the SPE procedure for removing polar compounds from the fuel samples.

RESULTS AND DISCUSSION

GC×GC-TOFMS chromatograms were collected for each part of the extraction process for an RP-1 fuel that was determined to have poor thermal performance. The reverse column configuration was used to achieve compound class separation of paraffins, iso paraffins, aromatics, and polar compounds, with 2D retention time decreasing for each class, respectively. There is visually little different between the total ion current (TIC) chromatograms of the neat and pass fuel samples, however the extract chromatogram demonstrates that polar compounds were extracted and thus their absence should be detectable in the original fuel (Figure 2).

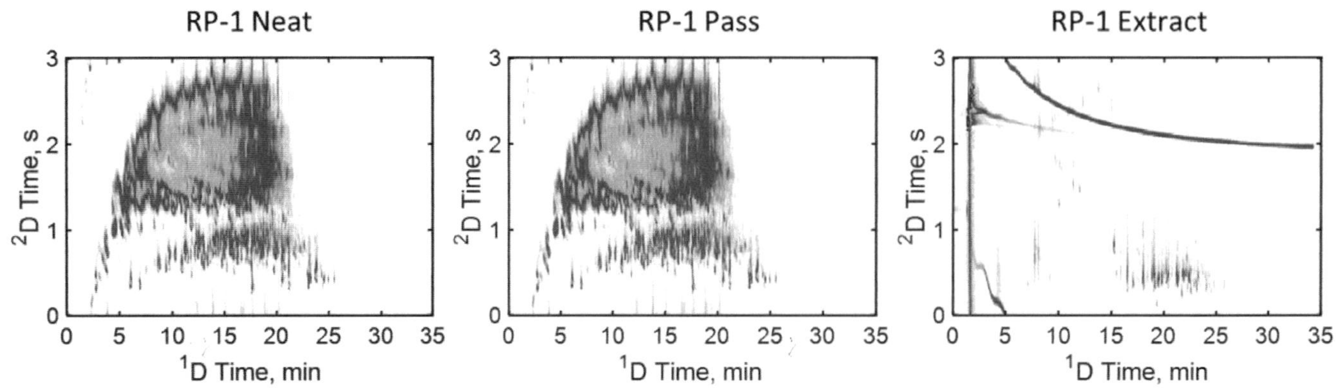

Figure 2. GC×GC-TOFMS TIC chromatograms for the neat and pass sample classes and SPE of the RP-1 fuel.

Fisher ratio analysis was applied to a 6 versus 6 set of the neat and pass samples of the RP-1 fuel, discovering hits that were significantly different between the two classes. Because the SPE procedure was applied these hits are likely due to reduction or absence of compounds in the neat fuel as they were extracted to create the pass fuel. Figure 3 shows the chromatographic location of the top 39 hits from the F-ratio analysis. These hits are predominantly located beneath the aromatics, indicating they are more polar than aromatic compounds. Comparing this region in the neat to the extract chromatogram shows remarkable match between hits and extracted compounds (Figure 4). The polar compounds found were identified as oxygenate compounds, with the majority phenolic in character. In order to better visualize the F-ratio results a "stitch" chromatogram is constructed using an algorithm adapted from Cain et al, allowing the raw neat data to be visualized without coeluting interferences (Figure 5).

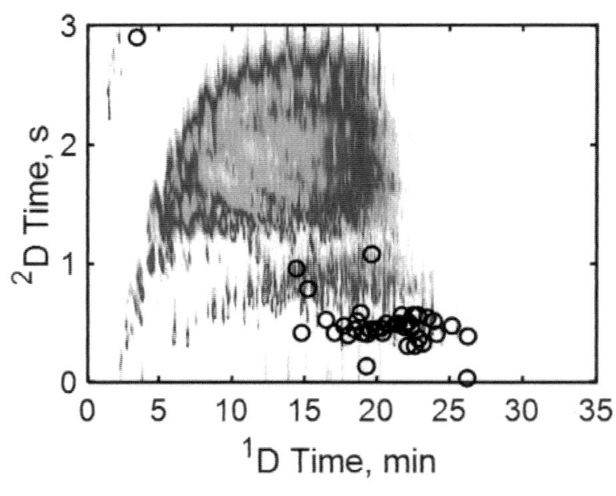

Figure 3. GC×GC-TOFMS TIC chromatogram of RP-1 fuel with locations of top 39 hits from F-ratio analysis overlaid.

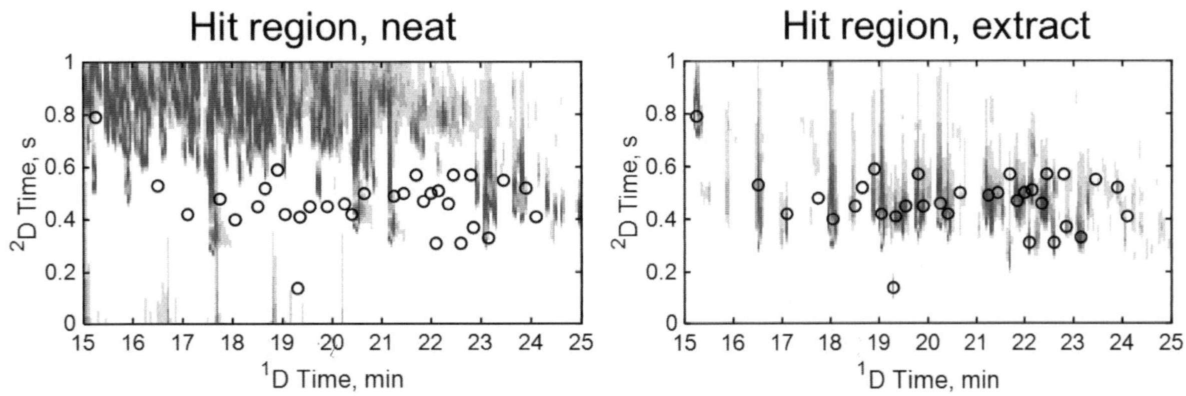

Figure 4. Region of the chromatogram where the majority of hit markers are present for the neat sample class and the extract sample. Overlaid F-ratio hit markers demonstrate close agreement with peaks present in the extract sample chromatogram.

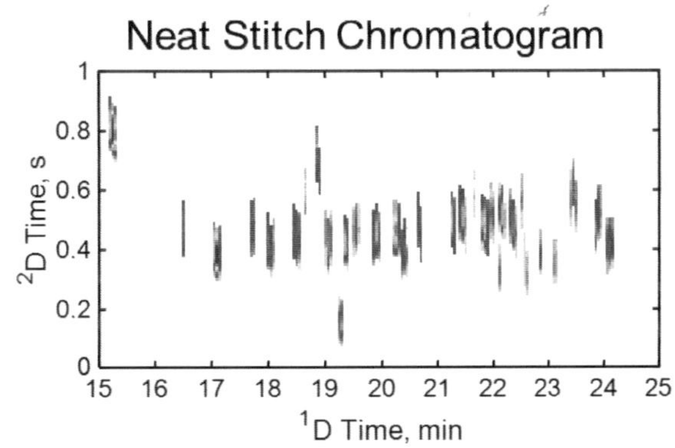

Figure 5. Stitch chromatogram constructed using data pulled from the top F-ratio *m/z* for each hit resulting from the F-ratio analysis allows for eased visualization of class differentiating chromatographic peak features, and aligns closely to the extract chromatogram.

Fuels deviating from specification can be compared against reference (clean) fuels via F-ratio analysis to discover the chemical features responsible for the errant behavior (Figure 6). Combining the SPE approach to create a pass fuel allows for a comprehensive comparative analysis to be performed. The fuels may be compared to each other through both sample class types, yielding a hitlist for the neat sample versus neat sample comparison and a hitlist for the pass sample versus pass sample comparison. The pass versus pass hitlist will be the same as the neat versus neat except for the lack of hits related to the minor polar compounds. Thus, by comparing these two hitlists we separate the bulk hydrocarbon hits from minor polar contaminant hits (Figure 7). In this example two rocket fuel formulations, RF-A and RF-C, are compared where RF-C was found through the neat versus pass approach above to have low polar content.

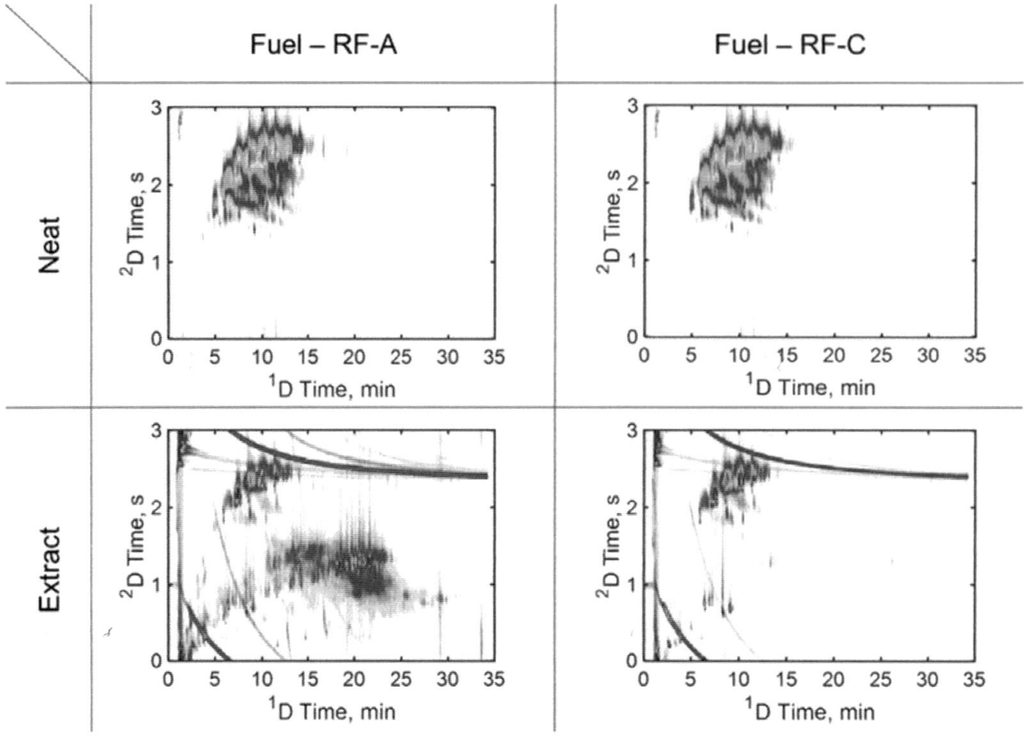

Figure 6. GC×GC-TOFMS TIC chromatograms of the neat sample and silica SPE extracts of two RF-2 fuel formulations. Fuel RF-A has a relatively high polar content whereas fuel RF-C has minimal polars, making it a good candidate to be a reference fuel to test RF-A against it.

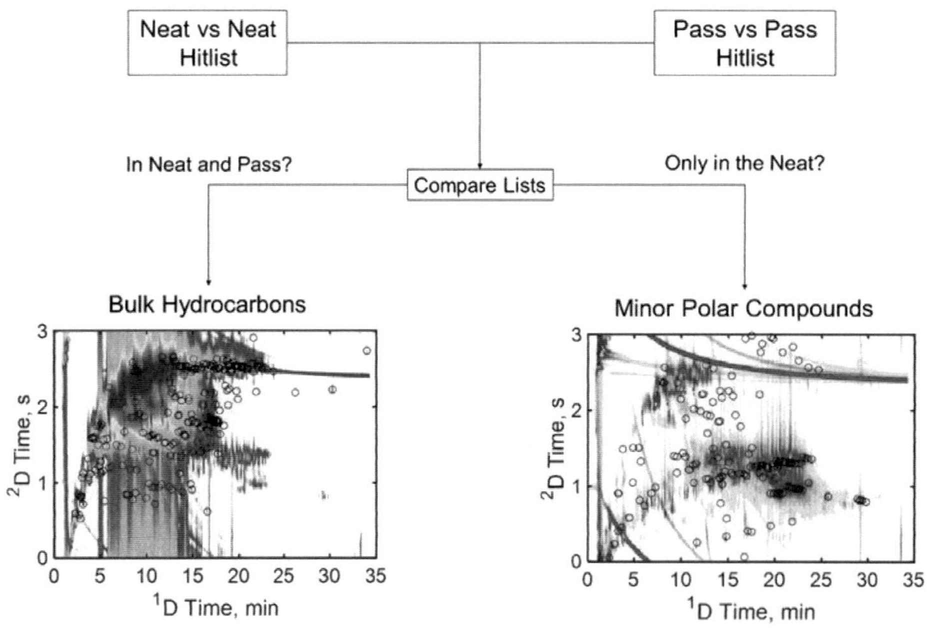

Figure 7. Flow chart depicting how the hitlists are compared to separate the bulk hydrocarbon hits and the minor polar compound hits. These hits are plotted over the GC×GC-TOFMS TIC chromatogram and the extract of fuel RF-A respectively.

Inter-fuel analysis allows for a comprehensive set of differences between fuels to be obtained. Utilizing the stitch chromatogram approach alongside the concentration ratio chromatograms may be obtained which show the retention position of all hits as well as their relative concentrations. A stitch chromatogram is obtained for each fuel for each hitlist, and divided by each other, yielding the concentration ratio stitch chromatograms to the left. Oftentimes the fuels under comparison have very few visual differences and performing this visualization step allows for the differences to be interpreted chromatographically, leading to a deeper understanding of the fuel pairs' distinctive characteristics.

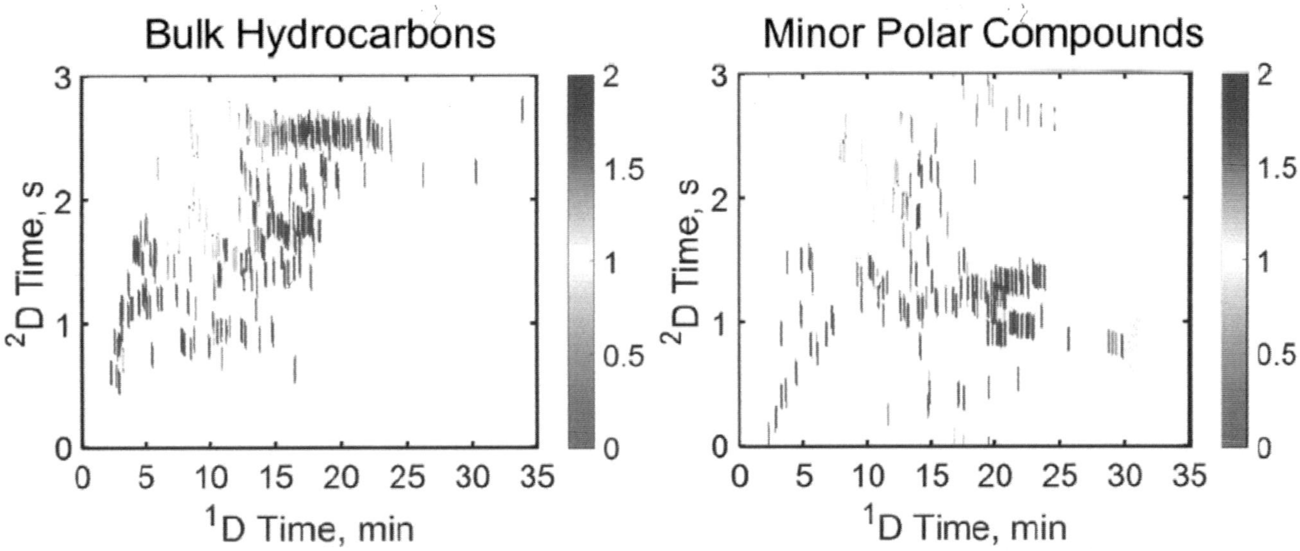

Figure 8. Concentration ratio stitch chromatograms of the bulk hydrocarbon hits (left) and minor polar compound hits (right) from the inter-fuel analysis of fuel RF-A and fuel RF-C. Regions colored blue have higher concentration in RF-A while regions colored red have higher concentration in RF-C.

CONCLUSIONS

A procedure for targeted polar compound discovery and comprehensive inter-fuel comparison via F-ratio analysis of fuels treated with SPE was presented for use with GC×GC-TOFMS data. This process may be utilized to help to identify low concentration polar contaminants or any targeted compound group in a single fuel by altering the SPE phase polarity and chemistry. Exploring these compounds will help rapidly determine compounds which have a detrimental effect on a fuel's thermal performance. Furthermore, the procedure was applied to inter-fuel analysis, allowing for fuels to be compared while separating low level contaminant differences from bulk component differences. This application provides a quick and efficient method for testing fuels against references to check for compliance. The concentration ratio chromatograms offer an easier interpretation of concentration differences between fuel determined by the inter-fuel method. In future work we plan to apply the extraction method to a large set of rocket fuels to model and study the effect of polar compounds on fuel thermal performance metrics.

REFERENCES

1. Ochoa, G.S; Billingsley M.C.; Synovec, R.E.; *Anal. Bioanal. Chem.* Accepted **2022**
2. Striebich, R. C.; Contreras, J.; Balster, L. M.; West, Z.; Shafer, L. M.; Zabarnick, S. *Energy and Fuels* **2009**, *23* (11), 5474–5482.
   Cain, C. N.; Schöneich, S.; Synovec, R. E. *Anal. Chem.* **2020**, *92* (16), 11365–11373.
3. Berrier, K. L.; Freye, C. E.; Billingsley, M. C.; Synovec, R. E. *Energy and Fuels* **2020**, *34* (4), 4084–4094.
4. Marney, L. C.; Christopher Siegler, W.; Parsons, B. A.; Hoggard, J. C.; Wright, B. W.; Synovec, R. E. *Talanta* **2013**, *115*, 887–895.

IASH 2022, the 17th INTERNATIONAL SYMPOSIUM ON STABILITY, HANDLING AND USE OF LIQUID FUELS
Dresden, Germany
September 11 - 15, 2022

INVESTIGATION OF SUSCEPTIBILITY OF SYNTHETIC AVIATION FUELS TO MICROBIOLOGICAL GROWTH

Graham Hill*, Gareth Williams

ECHA Microbiology Ltd., Unit 22&23, Willowbrook Technology Park, Llandogo Road, Cardiff, CF3 0EF, WALES, UK

ABSTRACT

Microbial growth in aircraft fuel tanks and aviation fuel distribution systems can cause problems of filter clogging, malfunction of fuel quantity indicator systems and airframe corrosion and can have a dramatic impact on fuel quality and operating safety. Microorganisms can utilise nutrients in any free water present in the tanks, but the hydrocarbon of the fuel provides the principal nutrient for proliferation. Any changes in chemical composition of fuels can thus influence the rate and extent of microbial growth which occurs. Relatively little research has been published on the susceptibility of synthetic and sustainable aviation fuels to microbial growth.

This paper describes a laboratory investigation of the relative susceptibilities to microbial growth of four synthetic jet fuel blends; a coal to liquid (CTL) fully synthetic jet fuel, a gas to liquid (GTL) Fischer Tropsch synthetic paraffinic kerosene and the GTL with blends of 50% naphthenic cut and 20 % hexanol. Susceptibility to microbial growth was compared to a conventional MEROX treated Jet A1.

The study utilised laboratory microcosms simulating tanks containing fuel and very small amounts of pure water which were assessed visually and by various microbiological methods over a 2 month period. All fuels, except the GTL blend with 20% hexanol, were found to be susceptible to some degree of microbial growth. The most susceptible fuel was MEROX treated Jet A1, and GTL with 50% naphthenic cut was the susceptible of the synthetic fuels.

*Corresponding Author: Mr. Graham Hill; ghill@echamicrobiology.com; +44 2920 365930

AUTHOR INDEX